Population Genetics & Fishery Management

Nils Ryman and Fred Utter · Editors

THE BLACKBURN PRESS

Ryman, Nils, and Utter, Fred, eds. 1987. *Population Genetics & Fishery Management*.
Washington Sea Grant Program, Seattle, WA. (Reprinted 2009 by The Blackburn Press,
Caldwell, NJ). 420 pp.

Reprint of 1987 Edition by University of Washington, Washington Sea Grant Program

Population Genetics and Fishery Management

ISBN-10: 1-932846-22-0
ISBN-13: 978-1-932846-22-5

Library of Congress Control Number: 2009936705

THE BLACKBURN PRESS
P. O. Box 287
Caldwell, New Jersey 07006 U.S.A.
973-228-7077
www.BlackburnPress.com

Contents

Preface to the 2009 Printing

Since the first printing of *Population Genetics and Fishery Management* in 1987, large technological, analytical and conceptual changes have occurred in the topic area of the book, and it is now long out of print. Yet, despite the emergence of capabilities to reveal and analyze up to genomic levels of variation both rapidly and inexpensively, the book has not drifted into obsolescence. Rather, a demand for this book has persisted, as reflected by its translation into Russian in 1991, and the high current price of the occasional copy that becomes available on the used book market.

Reasons among colleagues who have encouraged this reprinting include: (1) by explaining basic population genetics in a fisheries context, the book continues to serve as an excellent starting point for approaching complex recent developments, and (2) the book as a whole provides a projection from the 1980s to anticipate future problems, understandings and guidelines in fish conservation and management, many of which remain valid and others which have changed dramatically.

We reaffirm the acknowledgements in the original printings, particularly noting the encouragement and assistance of Professors Douglas Chapman and Ole Mathisen, both now deceased, and we emphasize the necessary and detailed editorial assistance of Patricia Peyton and Alma Johnson of the Washington Sea Grant office. We are grateful to Blackburn Press and, particularly, Maryanne Kenny, who agreed that our book has a future beyond its initial printing and who has assisted us throughout the process of implementing the present printing.

Nils Ryman
Fred Utter
September 2009

Foreword

Scientific fisheries management is a relatively new branch of the biological sciences. Because of the rapid changes in the field, few textbooks on fisheries management have been written and those that have been, rapidly become outdated. As an alternative the Washington Sea Grant Program instigated a lecture series in 1972 to address various aspects of fisheries management. Each year since then experts have been invited to discuss selected topics and their lectures have been published as books and bulletins. In this way it has been possible to contribute to scientific fisheries management theory and practice in many fields.

One scientific field that clearly has importance for fisheries management is genetics. This is so for at least two major reasons. While stocks of fish have been impacted by man since the dawn of history, recent technology and fishing fleet development have provided fisheries with the capability of substantial reduction, even decimation of large stocks of demersal or pelagic fishes. It is essential to management to know the genetic implications of such reductions.

In another field there are new dimensions in the complexities of management, namely in salmon management. These have come about with the development of a highly efficient hatchery technology, enhancement programs and ocean ranching of salmon. Thus the natural selection and adaptation of local salmon races, which had been in progress since the last glacial period, suddenly were affected by the overlay of transferred stocks and the introduction of new genetic material.

Against this background, we decided to invite a geneticist to address the question of population genetics in relation to fisheries management. The choice was Dr. Nils Ryman from Stockholm University. After he had concluded his lecture series, he suggested that, instead of the customary publication of lecture notes, the scope should be expanded to include the whole field of population genetics and fishery management. Thus, with Dr. Fred Utter, of the U.S. National Marine Fisheries Service, as co-editor, contributions from 24 authors from four continents were sought. The result is the present volume which we believe will take its place as a most useful contribution in man's perennial endeavor to conquer the Earth and its waters.

We consider it an honor that this volume would be the final one in the lecture series.

D. G. Chapman
School of Fisheries
University of Washington

Ole A. Mathisen
School of Fisheries and Science
University of Alaska—Juneau

Preface

Although the fields of population genetics and fisheries management have coexisted for almost a century, there has not been substantial interaction between these disciplines. Even though the means for adequate genetic characterization of populations ultimately were provided by the techniques developed in molecular genetics, the two fields have largely remained distinct from one another.

Regardless of the reasons for this separation, some of which are discussed in Chapter 1, it has become apparent that fisheries may diminish if the tools and theory of population genetics are not properly applied to the management of these resources. Responsible stewardship of these resources for future generations demands optimal use of existing knowledge.

This book arose from a series of lectures presented by one of us (N.R.) at the University of Washington School of Fisheries during the fall of 1982 under the sponsorship of the Washington Sea Grant Program. The theme of those lectures was population genetics in relation to fisheries management, and upon request for a book covering that subject we decided to edit a multi-authored volume of invited papers rather than tamper with too many topics beyond our own immediate fields of experience. It was our intention that the book fill at least part of the void between techniques and concepts of population genetics and those of fisheries management.

The book's extended gestation period stems partly from our attempts to give cohesiveness to a multi-authored volume. Some overlaps have been unavoidable, but others have been encouraged in order to provide as completely as possible a picture of central topics such as inbreeding and the biological interpretation of various measures of genetic divergence.

Chapter 1 sets the stage for the book by outlining the need for applying the principles of population genetics to the management of natural fish populations and stocks which spend at least part of their life cycle in the wild. We have excluded that part of fisheries management which concerns raising fish under controlled conditions throughout their entire life cycle (pen rearing for direct consumption), because the genetic problems related to that form of aquaculture are similar to those associated with the husbandry of any animal.

Identification of biologically meaningful management units (stocks) is the classical and most frequently encountered problem in fisheries management. A considerable part of the volume therefore deals with topics relevant for the delineation and interpretation of inter- and intraspecific genetic variability patterns.

Chapter 2 focuses on the central role played by electrophoretic procedures in providing population genetics data applicable to fisheries management. The conceptual basis for the ubiquitous concern of inbreeding is presented in Chapter 3, and the statistical procedures for delineation of the genetic variability patterns in subdivided populations are reviewed in Chapter 4. A case study of the genetic structure of an intensely managed fish species over most of its natural range is presented in Chapter 5. Chapter 6 outlines population genetic principles to be applied to the management of hatchery stocks. Chapter 7 reviews methods for detection of natural hybridization in fishes and describes how the principles of population genetics can be applied to quantitative investigations of hybridization and introgression.

The next two chapters relate to the estimation of the amount of genetic divergence among populations and to the interpretation and application of such estimates. Chapter 8 provides a review of statistical methods for estimating genetic distance and of the problems associated with the reconstruction of phylogenetic trees using genetic distances. In Chapter 9 these concepts are applied and discussed in relation to the biology and life history divergence between a number of congeneric fish species of the Atlantic and Pacific Oceans. Chapter 10 examines procedures for using genetic characteristics of conspecific populations to estimate their contribution in a mixed harvest.

The fisheries management applications of genetic markers and techniques other than those provided by electrophoresis are considered in Chapters 11, 12, and 13. A description of technical and theoretical aspects of restriction enzyme analysis of mitochondrial DNA (Chapter 11) is followed by a presentation of inter- and intraspecific variability patterns observed in salmonids (Chapter 12). Chapter 13 provides a review of methods for induction of polyploidy and gynogenesis, and describes the most likely management roles of chromosome markers and manipulations in fishes.

The book ends with two chapters related to the understanding, use, and conservation of genetic resources. The unique local adaptation of natural populations is discussed in Chapter 14 and is suggested as a basis for modifying current management activities. Chapter 15 considers the effects of fishing and hatchery practices on the gene pools of exploited species, pointing out problems inherent in current harvesting policies and recommending appropriate courses of remedial action.

The contributing authors are principally geneticists rather than fisheries managers; the book therefore views fisheries management from a geneticist's perspective. We deliberately chose this approach because we believe that fisheries management is hindered by a general absence of genetic considerations resulting from the lack of interaction between the two disciplines. As discussed in Chapter 1, we consider appropriate training in genetics for those who are and will become responsible for the management of fisheries resources an imperative to ensure the persistence of these resources.

We have directed this volume to a multidisciplinary audience, and we hope it will serve as a text appropriate for upper division or graduate university courses in subjects related to the title. Some readers may consider parts of the material too technical for their purposes; nevertheless, we hope that most of the text will be helpful to fisheries managers as well as geneticists. Single chapters are also intended as reference sources for individuals having varied backgrounds and interest in genetics or management. We hope that this volume will stimulate further books and articles that will encourage the application of the principles of population genetics to fisheries management.

Nils Ryman
Fred Utter
April 1985

Contributors

Paul Aebersold, Northwest & Alaska Fisheries Center, National Marine Fisheries Service, Seattle, Wash., U.S.A.

Standish K. Allen, Jr., School of Fisheries, University of Washington, Seattle, Wash., U.S.A.

Fred W. Allendorf, Department of Zoology, University of Montana, Missoula, Mont., U.S.A.

Yuri P. Altukhov, Institute of General Genetics, USSR Academy of Sciences, Moscow, U.S.S.R.

William J. Berg, Department of Animal Science, University of California, Davis, Calif., U.S.A.

Donald E. Campton, Department of Fisheries and Aquaculture, University of Florida, Gainesville, Fla., U.S.A.

Ranajit Chakraborty, Center for Demographic and Population Genetics, University of Texas, Graduate School of Biomedical Science, Houston, Tex., U.S.A.

Stephen D. Ferris, BioRad, Richmond, Calif., U.S.A.

Graham A. E. Gall, Department of Animal Science, University of California, Davis, Calif., U.S.A

W. Stewart Grant, Department of Microbiology, University of Cape Town, Rondebosch, South Africa

Ulf Gyllensten, Department of Genetics, University of Stockholm, Stockholm, Sweden

Olof Leimar, Department of Genetics, University of Stockholm, Stockholm, Sweden

George B. Milner, Northwest & Alaska Fisheries Center, National Marine Fisheries Service, Seattle, Wash., U.S.A.

Masatoshi Nei, Center for Demographic and Population Genetics, The University of Texas Health Science Center at Houston, Tex., U.S.A.

Keith Nelson, Bodega Marine Laboratory, Bodega Bay, Calif., U.S.A.

Jerome J. Pella, Northwest & Alaska Fisheries Center, National Marine Fisheries Service, Auke Bay Laboratory, Auke Bay, Alaska, U.S.A.

Nils Ryman, Department of Genetics, University of Stockholm, Stockholm, Sweden

Elena A. Salmenkova, Institute of General Genetics, USSR Academy of Sciences, Moscow, U.S.S.R.

Michael Soulé, School of Natural Resources, University of Michigan, Ann Arbor, Mich., U.S.A.

Gunnar Ståhl, Department of Genetics, University of Stockholm, Stockholm, Sweden

Gary H. Thorgaard, Program in Genetics and Cell Biology and Department of Zoology, Washington State University, Pullman, Wash., U.S.A.

Fred M. Utter, Northwest & Alaska Fisheries Center, National Marine Fisheries Service, Seattle, Wash., U.S.A.

Allan C. Wilson, Department of Biochemistry, University of California, Berkeley, Calif., U.S.A.

Gary Winans, Northwest & Alaska Fisheries Center, National Marine Fisheries Service, Seattle, Wash., U.S.A.

Acknowledgments

The completion of this book was dependent on the cooperation of many individuals and organizations. The editors gratefully commend each of the contributing authors for responding as participants and for their efforts in producing valuable and pertinent contributions.

The Washington Sea Grant Program and the University of Washington provided the basic sponsorship for the production of the book. The Swedish Natural Science Research Council and the Northwest & Alaska Fisheries Center have given the necessary support for carrying out the editors' responsibilities.

Professor Douglas Chapman of the School of Fisheries, University of Washington, and Professor Ole Mathisen, now of the School of Fisheries and Science, University of Alaska, provided valuable interest and encouragement, particularly during the earlier phases of the project. Personnel of the Washington Sea Grant Communications Office have consistently given necessary assistance in all phases of the production of the volume.

Peer review is essential to the production of scientifically oriented literature. The editors are greatly indebted to all their colleagues who provided solicited reviews of the different chapters.

Acknowledgments relating to specific chapters include:

Chapter 2. F. M. Utter, P. Aebersold and G. Winans
The authors thank Carol Hastings for her preparation of the figures.

Chapter 4. R. Chakraborty and O. Leimar
The authors are indebted to D. E. Campton, P. Smouse, and G. Ståhl for helpful comments on the manuscript. This study was supported by the U.S. National Institutes of Health, grant GM 20293 (RC), and the Swedish Natural Science Research Council, grant B-BU 3746-112(OL).

Chapter 5. G. Ståhl
The author acknowledges K. Hindar, C. Hjort, U. Lagercrantz, R. Leary, O. Leimar, S. Lindell, E. Loudenslager, N. Ryman, and F. Utter for valuable comments on the manuscript; and G. Ahlbäck, R. Andersen, T. Heggberget, Å. Hjalmarsson, A. Johnels, N. Johansson, Ö. Karlström, J. Mork, W. Persson, G. Rådström, J. Sahlin, R. Saunders, and J. Taggart for their aid with the collection of samples. The investigation was supported by grants from the Swedish

National Environment Protection Board, grant 5313131-4, and the Swedish Natural Science Research Council, grant B-BU 3746-107).

Chapter 6. F. W. Allendorf and N. Ryman

FWA was supported by the U.S. National Science Foundation, grant BSR-8300039, while the manuscript was written; NR thanks the Swedish Natural Science Research Council for their support. The authors thank G. Gall, K. Hindar, R. Leary, and O. Leimar for their comments on the manuscript; they also thank P. Ihssen for his comments on the manuscript and for pointing out the potential importance of relaxed selection under hatchery conditions.

Chapter 7. D. E. Campton

The author thanks J. C. Avise, T. E. Dowling, D. P. Philipp, and the editors of *Evolution, Copeia, Transactions of the American Fisheries Society,* and *Canadian Journal of Fisheries and Aquatic Sciences* for their permission to reproduce Figs. 7.1, 7.2, and 7.4–7.6 from their respective journals. The author also thanks W. J. Berg, J. Felsenstein, and R. F. Leary for their many helpful comments and suggestions on an earlier draft of this chapter. The author was supported by a U.S. National Institutes of Health predoctoral traineeship, grant GM07467, and by the Department of Animal Science, University of California, Davis, during the preparation of this manuscript.

Chapter 8. M. Nei

The author acknowledges N. Takahata and C. Stephens for their comments on the manuscript. This work was supported by research grants from the U.S. National Institutes of Health and the U.S. National Science Foundation.

Chapter 9. W. S. Grant

This paper greatly benefited from comments by E. Oliviera, J. Graves, and D. Hopkins. Support for this work was provided by grants from the Benguela Ecology Programme of the South African Council for Science and Industry and from the Sea Fisheries Research Institute.

Chapter 10. J. Pella and G. B. Milner

The authors appreciate the efforts and interest of editors N. Ryman and F. Utter, and owe special thanks to A. J. Gharrett and R. B. Millar for their constructive comments. J. Felsentein is thanked for suggesting use of the EM algorithm during preliminary investigations.

Chapter 11. S. D. Ferris and W. J. Berg

The authors thank J. Avise and U. Gyllensten for suggestions on an earlier draft. Support during preparation of this manuscript was in part by a grant from the U.S. National Science Foundation, DEB 81-12412; a National Institutes of Health National Service Award, 5T32-GM 07467; and a traineeship from the California Sea Grant College Program, Project E/G-2, 790520-22575.

Chapter 12. U. Gyllensten and C. A. Wilson

The authors thank F. W. Allendorf, A. Ferguson, J. Hutchins, U. Lagercrantz, R. F. Leary, B. Ragnasson, and J. Taggart for help in collecting fish and D. E.

Campton, W. J. Berg, O. Leimar, S. R. Palumbi, E. M. Prager, N. Ryman, and G. F. Shields for comments on the manuscript. This study was supported by grants from the Sweden-America Foundation, the Wallenberg Foundation (to UG), the U.S. National Science Foundation, grant DEB 81-12412, the U.S. National Institutes of Health, grant GM-21509 (to ACW), and by grants from the Swedish Natural Science Research Council, grant B-BU 3756-112, to N. Ryman, Department of Genetics, University of Stockholm, Sweden.

Chapter 13. G. H. Thorgaard and S. K. Allen

The authors thank R. Phillips for generously providing the photographs of banded trout chromosomes. GHT was supported by the U.S. National Science Foundation, grant PCM 8108787, Washington Sea Grant College Program, project R/A-31, and U.S. Department of Agriculture, grant 82-CRSR-2-1058, while this manuscript was being written.

Chapter 14. Yu. Altukhov and E. A. Salmenkova

The authors take this opportunity to express thanks to the editors of this book for the invitation to participate in it. J. Clayton, N. Ryman, and F. Utter are thanked for useful comments on this chapter.

Chapter 15. K. Nelson and M. Soulé

The authors wish to thank the editors of this book and the following people for helpful comments upon all or part of the chapter: R. J. Behnke, R. J. H. Beverton, D. G. Buth, P. T. Handford, D. Hedgecock, P. A. Larkin, J. Mork, W. E. Ricker, and S. Stearns.

1.

Genetics And Fishery Management
Past, Present, and Future

Fred W. Allendorf, Nils Ryman, and Fred M. Utter

Of all the animals and plants in the sea, fishes are the most important source of human food. They are the major source of protein for many people and constitute the main part of the diet in many cultures. Most of the fish used for human consumption is obtained through exploitation of wild populations.

In this book the principles of population genetics are applied to fishery management. Fishery management has been defined as the application of scientific knowledge to the problems of providing the optimum yield of commercial fisheries products or angling pleasure (Everhart and Youngs 1981). We have excluded from consideration that segment of aquaculture concerning the raising of fishes under controlled conditions throughout their life cycle. The problems relating to these conditions are very different from those of managing wild populations, and the genetic problems of aquaculture are similar to those associated with the husbandry of any animal. The management of wild populations comprising commercial or sport fisheries, on the other hand, presents genetic problems that are unique to fisheries management.

The management of a fishery requires an understanding of the biological principles underlying the resource. Fishery management has largely been concerned with the immediate resource of interest, that is, the abundance and size of fish available for harvesting. Because of this short-term focus, the biological perspective of fishery management has been dominated by ecology and population dynamics. Little attention has been directed toward an understanding of the genetics of these populations.

The extent of this domination is apparent in almost any text in fishery management. For example, the most recent edition of a widely used textbook, *Principles of Fishery Science* (Everhart and Youngs 1981), has chapters on population management, age and growth, estimating population size, mortality, and recruitment and yield, but mentions genetics only in regard to selective breeding of hatchery fish. Another recent text, *Fisheries Management* (Lackey and Nielsen 1980), does not consider genetics at all. The absence of genetic considerations in fishery management is also reflected in the educational system. For example, the University of Washington School of Fisheries does not require its fisheries majors (except for those students who specialize in fish culture) to take a course in genetics.

Such a narrow perspective may be economically advantageous in the short run, but is doomed to fail the test of time. The abundance and characteristics of a population cannot be ensured for the future simply by selecting an appropriate balance between harvest and recruitment to maximize yield. Differential survival and reproduction of fish with different genotypes will change the genetic composition of the harvested population. A successful management program, therefore, cannot ignore the genetic effects of management decisions.

In 1976 the United States passed the Fishery Conservation and Management Act, by which it took control of nearly all fisheries out to 200 miles off its coast and extended its authority over salmon to wherever they migrated at sea. This law requires that the fishery resources in the area of control be managed. It defines management as those measures that will "produce the optimum yield, which is prescribed on the basis of maximum *sustainable* yield from such fishery" (McHugh 1980, emphasis ours). This Act typifies the objectives of fisheries management to maximize the harvest and to conserve the resource.

Thus, in the long view fishery management is the conservation of existing resources to ensure a sustainable yield. Fisheries managers have often failed to achieve this goal. A large number of productive commercial fisheries have collapsed in the present century (Clark 1976). According to a recent report by the United Nations, "It is obvious that much less has been done on the conservation of fish genetic resources than other categories of living organisms" (1980). In this chapter we show that actions intended to ensure or maximize the future yield of a fishery must consider genetic factors. We also review the history of genetic management of fisheries and introduce the major areas on which we believe that efforts should be focused to ensure the conservation of fishery resources.

HISTORICAL PERSPECTIVE OF GENETICS AND FISHERIES MANAGEMENT

The preceding section may give the impression that genetics has been totally ignored in fisheries management. On the contrary, we think that the potential role of genetic factors in affecting intraspecific phenotypic variability has been more strongly recognized for fish than for the majority of other wild or semi-domesticated vertebrates that are heavily managed. An obvious example is the fish "stock" concept demonstrating the general recognition of genetic differentiation coupled with phenotypic divergence among conspecific subpopulations of fish (e.g., Heincke 1898, Schmidt 1917, Alm 1959, Ricker 1972; see also Berst and Simon 1981 for reviews).

Nevertheless, the persons responsible for the actual management have often been unaware of or failed to recognize the pertinence of the information provided by fishery scientists working in the field of genetics. A recent example is Larkin's (1981) suggestion to reverse the current goal of management of Pacific salmon based on accumulating knowledge of genetic structure of stocks because of the anticipated difficulties involved in managing a complex of genetically distinct populations.

In a recent survey of U.S. managers of state and federal wildlife resources, Schonewald-Cox and Bayless (1983) found that only 15% of the principal concerns relating to the genetics of animals referred to fish. This percentage is strikingly small considering that the genetics of fish populations appears to have been manipulated and affected by human activities to a much larger extent than any other group of vertebrates. For instance, it is hard to find parallels among other vertebrates to the massive, continuous, and frequently uncontrolled spread of genetic material over large areas that became possible through the development of hatchery procedures that have been practiced in the past century (e.g., Johansson 1981, Ros 1981, California Gene Resources Program 1982, Philipp et al. 1983). The disproportionately small effort directed towards genetic concerns in fishery biology and management is surprising and needs explanation.

We believe there are a number of reasons for the delayed application of basic genetic principles to fisheries management:

- The systematics of fisheries management has traditionally been dominated by taxonomists, a group that is generally not concerned with differences between individuals.
- The genotype-phenotype relationships in fish are different from those of other vertebrates. This difference is compounded by the fact that fish live in the water, making it more difficult to observe them and to obtain even rough estimates of the basic genetic parameters usually available for planning the breeding and management of nonaquatic species.
- No other major food resource of man is captured from wild populations. Geneticists perhaps have not been inclined to develop management schemes for the conservation of wild stocks that have difficult-to-define natural boundaries.
- Few breeders of other vertebrates had a similarly wide range of inter- and intraspecific variability from which to select. Many of the progenitors of chickens, cattle, and pigs faded into oblivion before serious selective breeding programs were initiated.
- Genetic results often appeared to contradict previous ecological or ethological understandings of the species, to be inconsistent with typological systematic conceptions, or to conflict with "biological intuition."

In the present section, we discuss some of the characteristics of fish that make them unique from a genetic perspective and have resulted in the delayed application of basic genetic principles to fisheries management.

Phenotypic Variation

Although morphological characters in different organisms cannot be compared directly, it appears that fish are phenotypically more variable than other vertebrates (Mayr 1969, p. 170). In particular, conspicuous and economically important differences in growth rate and body size are frequently observed within, as well as between, populations in many species of fish. "Dwarf" and "normal" forms occur within a large number of species such as brown trout, whitefish, perch, and Arctic char (e.g., Alm 1939, 1946, 1949, 1959, Svärdson 1979, Johnson 1980, Lindsey 1981). These forms also typically differ in size and color at sexual maturity, and they are frequently different with respect to a number of ecological charactersitics. Conspecific resident and migratory forms are known for most salmonid species, and the migratory forms commonly grow considerably larger than their resident counterparts. Similar size differences are found among individuals of the same population. Variation in age and body size of males at maturity constitutes an example frequently observed in anadromous populations of both Atlantic and Pacific salmon species (e.g., Gross 1985).

Tables 1.1 and 1.2 provide comparisons of the relative amount of variability in fish and other vertebrates. It appears that fish typically exhibit higher levels of variation both between and within populations (recognizing the limitations of comparing characters from different organisms). The coefficients of variation (SD/mean $\times$ 100, Table 1.1) indicate a considerably larger variation between individuals of the same population in fish than in other vertebrates. For length measurements the coefficients of variation are generally smaller than 10% in other vertebrates, whereas they exceed that number for most such characters in fish and range to over 20% for body length in Atlantic salmon. Similarly, coefficients of variation for body weight in nonfish species are typically smaller than 15%, a figure that is exceeded in all fishes and by more than 75% in Atlantic salmon (cf. Mayr 1969, p. 170).

The variation among conspecific populations of fish is also generally much larger than that observed in other vertebrates (Table 1.2). For instance, the intraspecific range of weight differences in Arctic char is over 4000%, i.e., more than 15 times as great as the corresponding differences between *species* of Darwin's finches. Similarly, for length measurements the range observed between *populations* of fish is two times or more that typically observed in other vertebrates at the *species* level.

Genetic Variation Within Populations

The larger phenotypic variation observed in fish species is not necessarily associated with greater genetic variability. Data suggest that the genetic-phenetic relationship in fishes may be somewhat different from what it is in other vertebrates. Heritability is the proportion of the total phenotypic variation within a population that is due to genetic differences between individuals. Heritability ranges from 0 (when all variation is completely environmental) to 1 (when all variation is due to genetic differences). Heritabilities for similar traits such as body length and weight are generally much lower within fish popula-

Table 1.1 Coefficients of variation (within populations) for select characters in various vertebrate species. The references are (1) Wright 1978, (2) Falconer 1981, (3) Schnell and Selander 1981, (4) Soulé 1976, (5) Gjedrem 1983, and (6) Ryman 1972.

Animal/character	Coefficient of variation (%)	Reference
Man		
Stature	4	1
Span	5	1
Forearm length	5	1
Mouse (*Mus*)		
Body weight (60 days)	11–14	2
Body weight (6 weeks, unselected lines)	6	2
Tail length	8	3
Hind foot length	4	3
Deer mouse (*Peromyscus*)		
Body length	3	3
Tail length	5–6	3
Ear length	3–5	3
Mammals (mean)		
Total length	5	3
Greatest length of skull	3	3
Lizards (*Anolis*)		
Number of toe scales	2–7	4
Rainbow trout		
Body length (140 days– 2.5 years)	10–19	5
Body weight (150 days– 2.5 years)	17–56	5
Atlantic salmon		
Body length (190 days– 3.5 years)	7–23	5
Body weight (190 days– 3.5 years)	25–76	5
Common carp		
Body weight (adults)	22	5
Channel catfish		
Body weight (juveniles)	46	5
Body weight (adults)	27	5
Guppy		
Body length (28–63 days)	10–12	6
Body weight (42–63 days)	35–36	6

Table 1.2 Range of variation between systematic/taxonomic units for select character averages in various vertebrate species. The references are (1) Price and Boag 1984, (2) Cramp and Simmons 1980, (3) Lowe and Gardiner 1974, (4) Angerbjörn (in press), (5) Niethammer and Krapp 1978, (6) Johnson 1980, (7) Alm 1939, (8) Alm 1946, and (9) Lindsey 1981.

	Range of variation		
Species/grouping/character	Smallest value	Largest value in % of smallest	Reference
Darwin's ground finches			
4 *species*			
Weight	11.8 g	253	1
Beak length	8.2 mm	187	1
Sparrowhawk			
4 *subspecies;* Europe-Asia			
Wing length, males	193 mm	109	2
Wing length, females	218 mm	114	2
Red deer			
4 European *subspecies*			
Skull length	310 mm	116	3
Skull breadth	138 mm	108	3
Field mouse			
36 insular *populations* from all over Europe; Mediterranean Sea to Baltic Sea			
Body length	82.5 mm	135	4
14 *populations* from all over Europe; Pantellaria—Sweden			
Body length	92 mm	112	5
Hind foot length	20.8 mm	115	5
Arctic char			
15 *stocks* from the Palearctic			
Length at 2 years	25 mm	800	6
Length at 4 years	80 mm	398	6
Length at 10 years	145 mm	417	6
10 nonmigratory *stocks* from the Palearctic			
Weight at maturity	23 g	4,213	6
No. of eggs	48	6,465	6
Brown trout			
11 European *populations*			
Length at 3 years	11.1 cm	270	7
Length at 6 years	19.3 cm	337	7
European perch			
23 Swedish *populations*			
Length	9.8 cm	315	8
Lake whitefish			
2 sympatric *forms*			
No. of gillrakers	23	150	9

tions than within populations of other vertebrates (Table 1.3; see also Purdom 1979 and Kirpichnikov 1981).

At first the heritability estimates obtained for fish may seem inconsistent with the conspicuously large phenotypic variation observed (cf. Table 1.3). However, the higher levels of phenotypic variation, coupled with the lower heritabilities, indicate greater susceptibility to environmental factors. This is not surprising considering some of the unique physiological characteristics of fishes. First, the indeterminate growth capacity of most fishes permits greater phenotypic adjustment to environmental factors, such as food availability and crowding, than is possible for most other vertebrates (cf. Purdom 1979). Second, fish are more sensitive than homoisothermic birds and mammals to variations in temperature directly affecting metabolic processes. Third, age and size at sexual maturity are interrelated in fish in a way that permits great flexibility without loss of reproductive success (e.g., Alm 1959, Jonsson et al. 1984, Gross 1985). We must mention, however, that heritabilities of meristic traits in fish are generally quite high (Leary et al. 1985a). This is not inconsistent with the preceding discussion, because meristic traits are often established early in development and, therefore, are not affected by such factors as indeterminate growth.

Genetic Divergence Between Populations

The classical and most frequently encountered genetic problem in fisheries management is the identification of genetically meaningful management units (e.g., Berst and Simon 1981 and references therein). In many species there is an apparent structure of more or less distinct and discrete subpopulations. Species restricted to fresh water are divided into a large number of subpopulations that are reproductively isolated from one another in unconnected lakes and drainages.

Both the anadromous and the freshwater resident salmonids have a strong tendency to home back to their natal streams for spawning. Marine species, such as cod and herring, are frequently divided into groups that spawn at different times and places. This structuring is associated with morphological or ecological differences and is also reflected in the terminology of fish biology ("stock," "run," "spring spawning," "anadromous," "piscivorous," etc.). The nomenclature implies genetic homogeneity within groups and stresses differences between them.

A considerable part of the variation observed *between* populations is likely due to environmental differences because of the comparatively large contribution of environmental factors to the phenotypic variation *within* populations. It has often been assumed that at least part of the variation observed between populations is determined genetically. Unfortunately, there is little information available concerning the relative importance of environmental and genetic factors affecting the phenotypic differences observed between natural populations of fish. The large phenotypic variation observed among populations

Table 1.3 Heritability estimates (within populations) for select characters in various vertebrate species. The references are (1) Falconer 1981, (2) Cavalli-Sforza and Bodmer 1971, (3) Pirchner 1969, (4) Robertson 1955, and (5) Gjedrem 1983.

Species/character	Heritability (%)	Reference
Man		
Stature	65	1
Systolic blood pressure	48	2
Cattle		
Body weight (adult)	65	1
Milk yield	20–40	3
Butterfat	40	1
Pig		
Back fat thickness	70	1
Weight gain per day	40	1
Poultry		
Body weight (at 32 weeks)	55	1
Egg weight (at 32 weeks)	50	1
Sheep		
Fiber diameter	20–50	3
Body weight	20–40	3
Mouse		
Tail length (at 6 weeks)	40	1
Body weight (at 6 weeks)	35	1
Drosophila melanogaster		
Body size	40	1
Thorax length	25–50	4
Rainbow trout		
Body weight (juveniles)	12	5
Body weight (adults)	17	5
Body length (juveniles)	24	5
Body length (adults)	17	5
Meatiness	14	5
Atlantic salmon		
Body weight (juveniles)	8	5
Body weight (adults)	36	5
Body length (juveniles)	14	5
Body length (adults)	41	5
Meatiness	16	5
Common carp		
Body weight (juveniles)	15	5
Body weight (adults)	36	5

has contributed to the development of a confusing picture of the role of genetics in fisheries management.

Questions concerning genetic population structure cannot be approached successfully without the analysis of distinct alleles at defined loci. Such genetic markers, the basic tools of empirical population genetics, were not available until the 1970s, when techniques were developed for electrophoretic examination of large numbers of protein variants controlled by individual genetic loci (Chapter 2). Prior to the development of protein electrophoresis, the amount of genetic divergence between populations was often assessed on the basis of the amount of divergence for quantitative characters that were sometimes known to be heritable within populations. Although reliable heritability estimates are necessary for designing efficient breeding programs, they provide almost no information concerning population structure and the amount of genetic differentiation between populations (Feldman and Lewontin 1975). The classical techniques of animal breeding refer to "statistical" genes rather than to distinct alleles at particular loci; and these techniques are generally of little use for stock identification, quantitative estimates of genetic divergence, or assessment of evolutionary relationships (Ryman and Ståhl 1981). It must be stressed that this limitation is true even for meristic characters, which generally have high heritabilities in fishes (e.g., number of vertebrae, gill rakers, and fin rays; Kirpichnikov 1981, Leary et al. 1985a).

There is another important misunderstanding concerning the use of morphologic characters for the assessment of genetic divergence among populations. It is often argued that such characters provide a large amount of genetic information on population structure because they are polygenic and therefore reflect allelic variation at more loci than can be examined with protein. However, a polygenic trait with 100% heritability (no environmental effects) contains only the same amount of genetic information about population relationships as a single electrophoretic locus (Rogers and Harpending 1983). The information provided by polygenic traits is greatly reduced when heritability is less than 100%. As mentioned by Lewontin (1984), there is "no way to escape the loss of information we suffer when we do not have detailed information on gene-phenotype relations."

The morphologically based Linnean principles for systematic classification have generally proven valid for genera over a wide phylogenetic range (Mayr 1969, Avise 1974). However, these principles have sometimes failed when used to delineate relationships in fishes (Table 1.4). The large phenotypic variation, coupled with the lack of information concerning the genetic basis for this variation, often resulted in oversplitting (e.g., Jordan and Evermann 1902, see also Behnke 1972). Many established concepts concerning genetic relationships within and between species were modified when electrophoretic data were collected and analyzed for fish. Some examples of revisions that emerged from biochemical genetic studies are listed in Table 1.4. A number of previously unknown species and subspecific groupings have been identified.

Table 1.4 Examples of biochemical genetic studies modifying previous assumptions of the genetic structure of fish species.

Relationship Indicated by Biochemical Genetic Data	Reference
Identification of previously unrecognized systematic groups at the:	
Intra-specific level — Major groups of rainbow trout corresponding to geographic region (coastal-inland) rather than drainage or life history pattern.	Allendorf & Utter 1979
Major population units of Pacific herring on each side of the Alaskan Peninsula.	Grant & Utter 1984
Sharp discontinuity of populations of several freshwater species east and west of the Apalachicola River in the southeastern U.S.A.	Bermingham & Avise 1984 Avise et al. 1984
Reproductivity isolated sympatric populations of brown trout.	Ryman et al. 1979 Ferguson & Mason 1981
Inter-specific level — Identification of previously unrecognized species of rockfish.	Westrheim & Tsuyuki 1967 Tsuyuki & Westrheim 1970 L. Seeb 1986
Identification of previously unrecognized species of bonefish.	Shaklee & Tamaru 1981
Inconsistencies with previous assumptions of genetic divergence based on:	
Residency vs. anadromy — Conspecificity of anadromous and landlocked forms of char of eastern North America.	Kornfield et al. 1981a
Lack of genetic divergence between anadromous and resident populations in rainbow trout, Atlantic salmon, and brown trout (less than 0.2% and 0.5% of the total gene diversity in Atlantic salmon and brown trout, respectively).	Allendorf & Utter 1979 Ryman 1983 Ståhl 1983
Time of spawning — No apparent genetic divergence between fall and spring spawning Atlantic herring.	Ryman et al. 1984
Major groups of chinook salmon corresponding to geographic region rather than time of spawning.	Utter et al., in prep.
Morphology — Little genetic divergence among morphologically distinct forms of cutthroat trout.	Busack & Gall 1981 Loudenslager & Kitchin 1979
Little genetic divergence among morphologically distinct species of pupfish.	Turner 1974
Little genetic differentiation between minnow species from two genera.	Avise et al. 1975
Conspecificity (and local random breeding) of distinct morphological types of *Ilyodon* previously considered separate species.	Turner & Grosse 1980
Lack of apparent genetic divergence between arid adapted (redband) and adjacent anadromous (steelhead) populations of rainbow trout.	Wishard et al. 1984
Lack of genetic divergence between two sturgeon species with overlapping geographic distribution.	Phelps & Allendorf 1984
Conspecificity of sympatric but trophically specialized forms of Mexican cichlids.	Kornfield et al. 1982b Sage & Selander 1975

Vague Management Goals

The lack of clear objectives in management programs is also a factor that has frequently confused the application of genetics to management. The distinction is rarely made between programs aimed variously at the following:

- Conservation of genetic variation between and within natural populations
- Maintaining the genetic characteristics of a stock that is artificially propagated in a hatchery
- Stock enhancement
- Selective breeding for production traits.

Such goals in most cases are genetically incompatible and cannot be achieved simultaneously. Nevertheless, only a few management programs acknowledge this incompatibility.

It should be obvious, for instance, that it is impossible both to maintain the genetic composition of a hatchery stock and to change it genetically through selective breeding. In addition to the intended genetic change that may be achieved, the restriction of the number of parental fish that is inherent to a selective breeding program will inevitably result in a loss of genetic variability. Nevertheless, some kind of selective breeding is usually practiced, intentionally or inadvertently, in most hatcheries (e.g., Calaprice 1969, Millenbach 1973, Hershberger and Iwamoto 1981). Similarly, a stock enhancement program cannot be combined with the conservation of the natural genetic population structure unless considerable care is taken to prevent unwanted interactions between the fish being stocked and the natural populations with which they come in contact.

In summary, we believe there are several characteristics of fish that distinguish them from other vertebrates that have been extensively managed. These include poikilothermy, indeterminate growth capacity, and a large phenotypic variation that has a considerable environmental component. These characteristics, coupled with the apparent intraspecific structuring of many fish species, produce a confusing picture of the genetic-phenetic relationships in managed fish species that has inhibited and delayed the application of adequate genetic tools and theory to fisheries management problems.

THE ROLE OF GENETICS IN FISHERY MANAGEMENT

Differential Harvesting Among Populations

As mentioned in the preceding section, the existence of a presumed genetic population structuring has long been recognized in the description of races, stocks, and subpopulations of fish species. This structure has also played an important role in the management of many species. For example, it has long been an accepted premise of management of Pacific salmon that "each population should be harvested separately" (Larkin 1981). Mixed harvesting of populations can lead to extirpation of local stocks through differential harvest rates because of unequal productivity of spawning adults (Ricker 1958), different average size of returning adults (Todd and Larkin 1971), or different time of return (Hartman and Raleigh 1964).

There are two basic requirements for harvesting populations separately. First, the population structure of the species must be understood. Second, the fishery must be regulated so that the harvesting of each population can be controlled individually. The advent of electrophoretic detection of genetic variation at protein loci has made it relatively inexpensive and easy to describe the genetic population structure of fish species (e.g., Allendorf and Utter 1979, Ryman 1983, Shaklee 1983). Returning to our example of Pacific salmon, a large number of studies have been published in the last ten years describing the genetic population structure of all five species of North American salmon from Alaska to California (e.g., Kristiansson and McIntyre 1976, Grant et al. 1980, Utter et al. 1980, Okazaki 1982, Ryman 1983, Utter et al. 1984).

The second step in the process is more difficult. Pacific salmon usually congregate in mixtures of populations except on the spawning grounds. Because of their low food value at sexual maturity, which usually precedes their arrival at the spawning grounds, they (and other species with similar life histories) are often harvested in mixed-fisheries. Harvests of population mixtures create uncertainty about the location, time, and number of fish to be caught in order to allow optimum escapement of each population. Techniques are therefore required to determine the contribution of individual populations to such mixed-fisheries.

Description of the components of mixed-fisheries is not limited to using genetic differences among populations since we are not trying to describe reproductive relationships but simply to identify different groups of fish. For example, a stable morphological difference between two populations is equally useful for identifying individuals whether the difference is genetically or environmentally based. The use of genetic differences between populations to distribute the composition of mixed-fisheries among the appropriate contributing populations is discussed in Chapter 10. The use of genetic and other information to discriminate populations of Pacific salmon in mixed-fisheries has been presented recently by Fournier et al. (1984).

Differential Harvesting Within Populations

The differential harvesting of fish within a population is perhaps the area of fishery management in which genetic considerations are most important, because the potential effects are the most pervasive. All populations of fish that are included in a sport or commercial fishery will inevitably be genetically changed by harvesting. If hatchery procedures produce a genetically undesirable hatchery stock, that stock can often be abandoned and a new stock brought into the hatchery from nature. However, natural populations with harmful genetic changes caused by harvesting cannot be discarded. Such induced genetic changes may take many generations, of fish and people, to correct. Some examples of these changes are examined below, and this topic is treated in detail in Chapter 15.

In spite of their importance, genetic effects have been largely ignored in this area for several reasons. It is difficult to prove genetic changes in natural populations. Observed changes in such important characteristics as survival, growth rate, and age at sexual maturity may be caused by either environmental or genetic changes. It is also difficult to predict what genetic changes are expected to occur with different harvesting regimes. In order to predict genetic changes, we must know how fish with different phenotypes or traits are being differentially harvested. We must also understand the underlying genetic basis of the trait being selected, as well as the phenotypic and genotypic correlations between traits of importance. All of these relationships are almost always unknown for natural populations of fish.

Thus, applying genetic principles to this important problem requires information that is at best difficult to obtain from natural populations. The effect of the fishery on the size of Pacific salmon is one of the best documented cases of changes in a fisheries resource caused by the selective effects of harvesting. Ricker (1981) has reviewed the evidence for genetically based differences in size and age of return for all five species of Pacific salmon on the west coast of North America. Over the past 60 years, the average size of chinook salmon *(Oncorhynchus tshawytscha)* caught along the Pacific coast of North America has declined by more than 50% and the average age of maturity has declined by approximately two years. Ricker (1981) has concluded "continuation of the present troll fishery, or selection of large fish by the gillnet fishery, can lead to a situation where practically the only chinooks left in the catch will be jacks and age 2 individuals maturing at ages 2 and 3 respectively; and chinooks weighing more than 15 pounds will be unheard of."

Similar effects have been reported with Atlantic salmon *(Salmo salar)*. Many anadromous populations of Atlantic salmon historically have had a small proportion of males (grilse) that become sexually mature after only one year at sea. In addition, some juvenile males become precociously sexually mature while still resident in freshwater. The harvesting of these populations in the ocean has reduced the probability of survival during the marine phase of the life cycle. This differential harvest has increased the reproductive success of grilse and precociously mature juveniles relative to other males. The result has been an increase in the proportion of grilse males and precociously mature juvenile males. In some populations, the majority of males are apparently no longer anadromous as the result of this selective process (W.L. Montgomery, personal communication).

Hatchery Populations

There are a variety of well-known genetic principles from animal husbandry that have been frequently ignored in the founding of hatchery populations of fish. The harmful effects of the loss of genetic variation in small populations is well documented in both fish and a variety of other organisms (Ralls and Ballou, 1983). It is often difficult to acquire a large number of individuals to found new hatchery populations. Nevertheless, ignoring the expected effects of small population size (bottlenecks) will increase the probability of failing to achieve the objectives of the hatchery program.

Fishery managers have long recognized the importance of genetic differences between populations. Populations are often chosen for use in a hatchery program for certain desirable characteristics that they possess. Founding a hatchery stock with a small number of individuals from the desired population will at best have the effect of ensuring that the hatchery fish are not genetically representative of the chosen population (Ryman and Ståhl 1980). A severe bottleneck may even cause the expected effects of inbreeding depression in the hatchery fish (Allendorf and Phelps 1980, Leary et al. 1985b). There are a growing number of examples from the literature that show detectable changes and loss of variation associated with the founding of hatchery populations (see Chapter 6).

Maintenance of hatchery populations presents additional genetic problems. As in the founding of a population, a sufficently large population size must be maintained to minimize the effects of genetic drift. In addition, it is inevitable that those genotypes that are most suitable to hatchery conditions will increase in frequency. These "hatchery adapted" genotypes are not likely to survive or grow well when stocked in nature. Thus, care must be taken to minimize any selective effects of adaptation to the hatchery environment.

Examples of such unintentional genetic changes in hatchery fishes are well documented (Hynes et al. 1981a). Vincent (1960) and Moyle (1969) have demonstrated that the more docile behavior of hatchery trout is genetically based. Several studies have shown a genetic basis to the common lower survival of hatchery fish than wild fish in natural habitats (Greene 1952, Reisenbichler and McIntyre 1977, Chilcote et al. 1981). A more detailed examination of population genetics and hatchery management is presented in Chapter 6.

Release of Hatchery Fish

When hatchery fish are released into areas containing a natural population, they may have an effect on the natural population that, as the literature shows, is neither intended nor beneficial. Certainly there is the danger of ecological competition between the introduced and the natural fish (Leider et al. 1984). There are also potentially harmful genetic effects: the increase in frequency of nonadaptive "hatchery" genes, the breakdown of natural systems of semi-isolated populations (Ryman 1981b), and interspecific hybridization (Behnke 1972, Leary et al. 1983a).

Often those who make management decisions do not fully appreciate the seriousness of those dangers. An editorial written by the President of the American Fisheries Society in 1985 advocated the use of hatcheries to augment striped bass populations in Chesapeake Bay and, in doing so, completely dismissed any potential genetic effects on existing natural populations because "we are a century too late to worry about genetic 'purity' of striped bass" (Sullivan 1985). It may be true that hatcheries are needed to supplement this fishery; however, more consideration must be given to genetic effects before such a decision is made. It seems safe to assume that genetic adaptation to local environments differs among striped bass populations and that individual populations depend on adequate levels of genetic variation to grow optimally and reproduce and to adjust to fluctuating environments. Any responsible hatchery program must consider such matters, regardless of the ancestral purity of existing stocks.

FUTURE DIRECTIONS FOR GENETICS IN FISHERY MANAGEMENT

The preceding sections of this chapter have explored the historical interface of genetics and fishery management and examined those segments of fishery management to which genetic considerations are particularly pertinent. This section links current interactions with what we perceive to be appropriate future directions for genetics and fishery management.

The Need for Education

We have examined those attributes of fishes, and of fishery biology and management, that have delayed the sound application of genetic principles to the understanding and management of fish populations. Some of this delay was inevitable. For instance, a generation ago the appropriate tools were not available to unravel the often diverse and confounding phenotypic expressions of similar genotypes influenced by different environments. On the other hand, inbreeding has only recently been identified as a persistent problem in some hatcheries (Ryman and Ståhl 1980, Allendorf and Phelps 1980, chap. 6) although its causes, effects, and remedies have been known for over half a century.

Errors based on ignorance of existing knowledge are destined to continue in the absence of adequate education of those who ultimately become responsible for managing fishery resources. This education must begin in the curricula of colleges and universities at the undergraduate level. Graduates destined for positions in fishery biology and management who are adequately equipped with the fundamentals of Mendelian and population genetics will not view different taxa in typological terms. They will also tend to be receptive to conceptual and technical developments in genetics as they occur. A mandate rather than an option for such training is therefore seen as a necessary starting point for adequately implementing genetic principles in fishery management. We urge educators to recognize this requirement, and to act accordingly.

Curriculum changes do not occur overnight, nor will they have an impact on the bulk of the present work force. What is being done, or can be done, to provide adequate genetic training to workers whose formal education has been completed and are already now in professional positions? There are some encouraging signs. Some management agencies (e.g., Washington State Department of Fisheries; see Hershberger and Iwamoto 1981) are promoting the preparation and use of materials outlining the background and applications of genetic principles. A proposed genetic policy being formulated by the Alaska Department of Fish and Game to address stock transport, protection of wild stocks, and maintenance of genetic variation reflects that agency's awareness of genetic concerns. These examples typify at least a general recognition among management agencies of the need for training in genetics and application of genetic principles in fishery management.

Adequate Application of Existing Techniques

The unique ability of electrophoretic methods to readily provide reliable information on genotypes has revolutionized the understanding of genetic structures of natural populations (see Chapter 2). The information in Table 1.4 confirms the new insights on genetic structures of fishes resulting from applications of biochemical genetic methods. However, the actual application of such information in fishery management remains minimal relative to its potential.

It has sometimes been concluded from the absence of revealed structuring that preferential emphasis should be diverted from electrophoretic procedures and toward developing and applying other genetic methods such as those that directly examine the composition of the genes. We agree that methods for directly examining the gene require development and application. Chapters 11 and 12 describe applications of restriction enzyme analyses of mitochondrial DNA in fishes. The data provided by this procedure and by protein electrophoresis are largely complementary, as pointed out later in this chapter. Only a few studies have been carried out whose data from mitochondrial DNA in fishes can be compared with adequate sets of electrophoretic data (Avise and Smith 1974, Avise et al. 1984, Berg and Ferris 1984, Ferguson and Fleming 1983, Graves et al. 1984, Sharp and Pirages 1978, Gyllensten et al. unpublished). In these instances the different approaches have arrived at similar conclusions regarding the genetic structures of the groups studied. Biological reality rather than procedural limitations may very likely be reflected in those comparable studies that indicate minimal structuring.

The use of genetic differences between populations to estimate compositions of stock mixtures has been mentioned in the preceding section and is examined in greater detail in Chapter 10 (see also Fournier et al. 1984). This approach has many distinct advantages for discriminating stock mixtures (see Milner et al. 1985) and warrants extended use in groups of fishes, such as salmonids, that tend to be structured into genetically distinct populations.

Allelic and genotypic information obtained by electrophoretic methods permits genetic monitoring of hatchery and wild stocks. This important application of electrophoretic data has received relatively little attention to date. Monitoring can provide managers with critical genetic insights related to a population. Some uses of genetic monitoring with respect to hatchery practices have already been mentioned in the preceding section of this chapter, and further elaboration is given in Chapter 6.

With monitoring, genetic stability or change can be measured directly. A need for monitoring exists within all hatchery populations, or in situations affected by hatchery operations (e.g., in areas where released hatchery fish may interbreed with wild populations). Information obtained from these measurements can validate or disprove assumptions that were previously untestable, such as those concerning genetic stability based on particular breeding practices, representative sampling of parent populations in founder stocks, and reproductive successes of hatchery fish in the wild. Monitoring of totally wild populations is equally feasible and desirable to measure their genetic stability in the absence of the influences of hatchery fish. Monitoring also can provide a valuable updating of genetic information on wild and hatchery populations if these populations are to be included in analyses of stock mixtures.

We reaffirm an advocacy that has persisted for more than a decade to include the collection of biochemical genetic data in all investigations dealing with questions of population structures or intrageneric relationships of fishes (e.g., Avise 1974, Utter et al. 1974). Indeed, a major goal of this book is to provide a bridge between fishery managers and population geneticists so that existing procedures can be applied much more broadly. It is encouraging to see some management agencies taking steps toward broader application of existing genetic technologies in fishery management. It is archaic to perform time-consuming and tedious computations manually when computers can do the job instantaneously and more effectively. It is similarly archaic not to collect sets of genetic data that are readily available and can provide valuable insights unobtainable by other procedures.

Development of New Techniques

The need for adequate applications of existing procedures should not be interpreted as a resistance to new techniques. For instance, less than one percent of the total genetic material of an organism would be examined even with full usage of genetic information detected by electrophoretic methods (see Chapter 2). Clearly, a vast amount of genetic information lies beyond the reach of this extremely useful procedure. In addition, new statistical procedures are continually being developed to more effectively use the genotypic and allelic data typically resulting from electrophoretic analyses, as well as to interpret genetic information from other sources such as DNA fragments and nucleotide sequences. It is the joint responsibility of geneticists and fishery managers to maintain an awareness of current technical developments and to assimilate these developments into management planning and operations.

Many technical developments are currently being made with regard to the direct examination of the genetic material DNA. Direct sequencing of nucleotides in DNA remains tedious. However, genetic information can be obtained much more readily through enzymatic fragmentation of mitochondrial DNA at specific recognition sites, as mentioned above. The potential usefulness of this procedure as a population genetic tool for fishery management is presented in detail in Chapters 11 and 12. The specific usefulness of data from mitochondrial DNA in studying particular cases of hybrid populations is also discussed in Chapter 7. As this procedure becomes more routinely applied, it appears destined to join protein electrophoresis as a major descriptive tool in population genetics because of the complementary types of information provided by the two methods (i.e., data from fragments of maternally inherited and rapidly evolving haploid DNA versus genotypic data from sexually inherited diploid loci and alleles).

Other procedures also appear destined for broader applications as tools for studying the population genetics of fishes. Recent developments in studying chromosomal morphologies and in manipulating sets of chromosomes, described in Chapter 13, permit induction of sterility and creation of new gene combinations as well as provide new kinds of genetic markers. Correlations observed between bilateral symmetry and heterozygosity (Leary et al. 1983b) open up a new sphere of investigation for relating morphological variation and fitness and suggest new approaches for measuring and monitoring genetic events affecting a population.

New statistical approaches are being developed that are applicable to fishery management. Procedures for measuring randomness of allelic associations at different loci are discussed in Chapter 4 (see also Chapter 7, Smouse et al. 1983, Chakraborty 1984). These procedures give valuable information on complexities of population mixtures, recent admixtures, and, in some instances, the action of natural selection in a particular sampling of individuals. A method for estimating gene flow based on the distribution of rare alleles has recently been developed (Slatkin 1981, 1985 a, b). The development of statistical procedures for quantifying individual stock contributions in stock mixtures (Chapter 10, Fournier et al. 1984) promises to see continued activity, based on use of both genetic and nongenetic characters.

SUMMARY

We have noted only minimal interactions of the disciplines of populations genetics and fishery management throughout the present century. An awareness of genetic concerns existing in fishery biology during this period has not been adequately reflected in management practices.

The thinness of this collaboration reflects, in part, several potentially confusing characteristics of fish that distinguish them from other vertebrates that have been managed extensively. Fish are largely invisible in their natural habitat, making location and enumeration often difficult or impossible. Attributes such as poikilothermy, indeterminate growth capacity, and extensive phenotypic variation with a considerable environmental component have tended to mask genetic-phenetic relationships. Such attributes coupled with either apparent or unobserved breeding units have suggested some invalid taxonomic divisions and yet failed to identify genetically distinct groups.

The application of genetics to fishery management has also been confused by vague management goals. Although it is clear that genetic conservation and (intentional or inadvertent) selection cannot be achieved simultaneously, this incompatibility has rarely been acknowledged.

Genetics and fishery management can interact in several ways. When the genetic population structure of a species is known, the distribution of subpopulations in mixed-fisheries can be estimated. Regulation of harvests to protect weaker populations can be made based on these distributions. Genetic changes within a population because of differential harvests are difficult to measure. Such changes are important to identify and regulate because of the drastic and long-term effect they may have on a population. Hatchery populations are particularly susceptible to genetic changes because of modified environments and controlled reproduction. Considerable attention must be given to genetic principles in the founding and maintenance of hatchery populations. Attention to the genetic interactions of hatchery and natural populations is also important to minimize breakdown of adapted gene pools and to prevent loss of irreplaceable reserves of genetic variation.

A healthy future for genetics and fishery management requires a continuation and expansion of recent trends toward greater interaction. A requirement rather than an option for at least some education in genetics is needed both for students and for active professional workers. Existing genetic techniques that have proven value in fishery management require broader applications in problem areas. Systematic genetic monitoring of hatchery and wild stocks is seen as a particularly important application of current technologies. Finally, the development of new genetic and statistical tools is required for an optimal interaction of genetics and fishery management.

2.
Interpreting Genetic Variation Detected By Electrophoresis

Fred Utter, Paul Aebersold, and Gary Winans

Fishery biologists generally recognize the need to identify differences among stocks and to monitor genetic changes. However, many biologists do not recognize the link between stock differences, protein variation, and genetic variation inherited in a Mendelian manner. This chapter is intended to provide that link.

Outlined are some of the basic genetic principles and procedures underlying the practical application of population genetics to fishery management problems. Some of the information is now fundamental to introductory undergraduate courses, but many fishery biologists completed their training before molecular biology and the molecular basis of hereditary variation were taught at the undergraduate level. However, most of this chapter deals with the understanding and interpretation of genetic data as revealed by electrophoresis. This tool continues to play a major role in understanding the levels and patterns of genetic variability within and among populations.

It was pointed out in Chapter 1 that contributions of individual genes usually cannot be identified from studies of quantitative traits such as length, fecundity, and number of gill rakers. Quantitative characters are therefore excluded in this chapter. The reader is referred to Ferguson (1980) for a general, and in some areas more detailed, examination of many topics approached in this chapter.

Mendel's studies (1866) first identified units of inheritance which were subsequently termed *genes* by Johannsen (1909). Considerable theory concerning the dynamics of Mendelian genes had developed by the early 1930s. A conceptual framework for the interactions of mutation, migration, selection, and drift in the creation, maintenance, and distribution of Mendelian genes in natural populations was established through the writings of Fisher (1930), Haldane (1932), and Wright (1932).

Very little empirical information was available to match this theoretical groundwork for over 30 years. An exception was the detection of sizable levels of deleterious recessive genes detected in *Drosophila* (Dubinin et al. 1937). Also, knowledge of simply inherited human blood groups was accumulating by the 1930s (reviewed in Boyd 1966). Indeed, human blood groups provided early and extensive observations on intraspecific structuring of populations based on information from Mendelian genes (Mourant 1954).

A revolutionary advancement in the ability to identify Mendelian genes arose from two developments in the 1950s. Watson and Crick (1953) deduced the structure of the DNA molecule, which ultimately clarified the direct relationship between genes and proteins. This knowledge was followed by the development of electrophoretic procedures which permitted rapid and reliable identification of protein variations reflecting simple genetic differences (Smithies 1955, Hunter and Markert 1957). The ease with which Mendelian variants could now be detected by electrophoresis (contrasted with the previously great difficulty in detecting such variation) resulted in a proliferation of descriptive studies of Mendelian variants of proteins in many organisms (see deLigny 1969, 1972 for reviews of early electrophoretic studies of fishes). Eventually, the classical studies of Lewontin and Hubby (1966) in *Drosophila pseudoobscura* and of Harris (1966) in man clearly suggested that substantially higher levels of genetic variation exist throughout all classes of organisms than had previously been known.

The simple inheritance observed for blood groups in man (cited above) and other higher vertebrates (e.g., cattle, Stormont et al. 1951; chickens, Briles et al. 1950) suggested the existence of similar Mendelian markers in fishes and led to studies of blood groups in such fishes as tuna (Cushing 1956) and salmonids (Ridgway 1957, Sanders and Wright 1962). The anticipated usefulness of these studies for identifying Mendelian variations was not fulfilled because of technical limitations such as fragility of fish erythrocytes and difficulties in producing and preserving discriminating antisera (reviewed in Hodgins 1972). However, information from electrophoretic studies has subsequently met and surpassed the expectations envisioned from blood groups. Population structures of fish species are being clearly defined with the use of purely genetic data (e.g. Allendorf and Utter 1979, Winans 1980, Shaklee 1983, Ferguson and Mason 1981, Ryman 1983). Using such data, statistical and data processing procedures have been developed for obtaining detailed, accurate, and timely estimates of mixed stock compositions (Grant et al. 1980, Fournier et al. 1984, Milner et al. 1985, Chapter 10).

In this chapter we describe the direct connection between the gene and its expression as a protein molecule, with the intention of making the remaining chapters more accessible to the uninitiated reader. Basic principles and terms are first introduced. An extended section on the most frequently applied technique for studying genetic variation in natural populations—protein electrophoresis—is presented because of the central role this technique has played in revealing Mendelian characters for population genetic studies. Illustrations focus on the complex patterns of the extensively studied salmonids. The chapter closes with observations concerning the limitations of current techniques, which examine only a small portion of the total amount of genetic material.

BASIC PRINCIPLES AND TERMS

Some basic principles and terms concerned with the molecular basis of genetic variation are introduced at this point to clarify their use in this chapter. Most genes in higher diploid organisms such as fishes and man are contained in structures of the cell nucleus called *chromosomes*. A much smaller fraction of genes are found outside the nucleus and include those found in the mitochondria (see Chapters 11 and 12). Chromosomes (and therefore genes) occur in pairs as a consequence of individual sets of chromosomes that are inherited from each parent. These individual sets are transmitted in germ cells called *gametes* (sperm and egg cells in animals). The process of gamete formation (gametogenesis) includes *meiosis* (outlined in any elementary genetics text), which allows chromosomes of each parent to assort independently to each gamete (Mendel's second law). Gametes unite through sexual processes; the egg is fertilized by an individual sperm cell to form a *zygote*. Fertilization, then, results in the pairing of individual sets of chromosomes.

The single-celled zygote soon develops into a highly differentiated collection of tissues and organs performing broadly diverse functions. This differentiation occurs because, although each cell has an identical complement of genes, very few of the total number of genes are active in a particular cell of higher organisms (e.g., see Alberts et al. 1983). Differentiation is the result of the interactions of *regulatory* genes (which determine at what time and in what tissue a particular gene is expressed) and *structural* genes (which contain coded information for proteins that are produced by the organsism). These complex and still poorly understood interactions lie beyond the scope of this chapter (but see McDonald 1983 for a recent review of progress in understanding these interactions). However, the direct relationship between structural genes and their protein products is well understood and is our primary focus.

The condition in which only a single set of chromosomes is present, such as in gametes, is called *haploid,* while *diploid* describes the paired chromosome complements in the zygote and subsequently formed tissue cells. Occasionally zygotes are formed with three or four sets of chromosomes *(triploidy* and *tetraploidy)*. The ability to induce the triploid condition is currently receiving considerable attention because of the general sterility of such individuals (see Chapter 13). The tetraploid ancestry of some families of fishes (e.g., Catastomidae and Salmonidae) has resulted in some special evolutionary opportunities for these species (see Ferris and Whitt 1979, Allendorf and Thorgaard 1984). Some consequences of tetraploid ancestry with respect to complexities of electrophoretic expressions are examined later in this chapter.

The location of a gene on a chromosome is called a *locus* (plural *loci*). The paired set of genes inherent in diploidy permits two different forms of a gene for a particular locus to exist in a single individual. Different forms of a gene are called *alleles*. Many alleles may exist for a particular locus in a species, but a single diploid individual can carry no more than two alleles at a locus. An individual is *homozygous* at a particular locus if the genes at that locus are identical, and *heterozygous* if they are different. A locus is said to be *monomorphic* if only one allele is known, and *polymorphic* if two or more alleles are known. The set of alleles possessed by an individual at a particular locus (or set of loci) is referred to as the individual's *genotype* at this locus. The *phenotype* is the observed character of an individual, and may be influenced by the environment as well as the genotype.

The fundamental chemical substance of the gene is *deoxyribonucleic acid*, or DNA. DNA is a giant molecule constructed in a so-called double helix

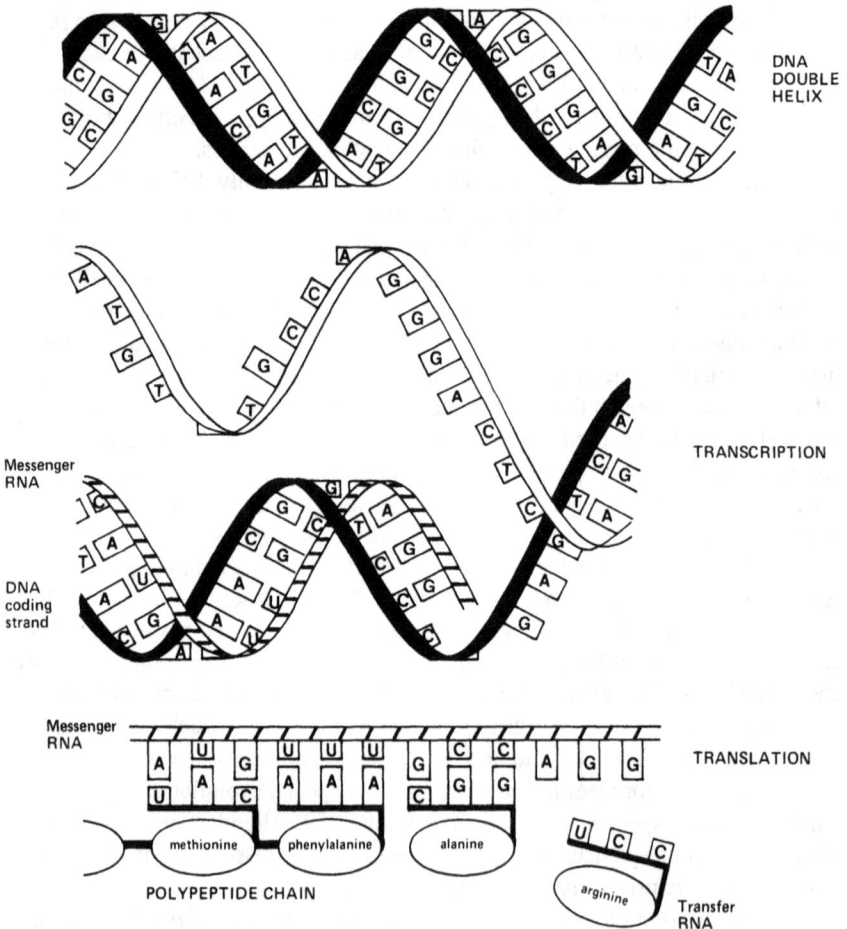

Fig. 2.1 Molecular processes relating base sequences of DNA to amino acid sequences of polypeptide chains (proteins). Messenger RNA, synthesized during transcription, provides a template for the synthesis of the polypeptide chain during translation. The bases in DNA are cytosine (C), guanine (G), adenine (A), and thymine (T). In RNA, the base uracil (U) replaces thymine (T) of DNA.

like a very long spiral ladder (see Fig. 2.1). The sides of the ladder are alternating sugar (called deoxyribose) and phosphate groups. The rungs of the ladder attached to the sugars are pairs of *bases:* adenine (A), guanine (G), thymine (T), and cytosine (C). These bases always combine as either $A-T$ (or $T-A$) or $C-G$ (or $G-C$) because of physical and chemical constraints that preclude other pairings. The genetic information is contained in different sequences of these four bases read from one side of the ladder, the DNA coding strand.

The base sequence of DNA has a direct linear relationship to the structure of proteins. Proteins are similar to DNA in that they are large molecules made up of different components called *amino acids*. There are 20 common amino acids in nature. Amino acids are connected by peptide bonds in a series to form *polypeptide* chains. Active proteins are made of polypeptide chains alone or in aggregate, depending upon the protein. Each polypeptide chain is called a *subunit*. Most proteins have at least 100 amino acids.

It has been found that different combinations of the four bases read in sequences of three (called *coding triplets* or *codons*) have information (or code) for different amino acids. This information, placed in a line on the DNA molecule, tells the cell which amino acids form a protein molecule and in what order they go. The four bases G, C, A, and T in various combinations of three can be arranged in 64 different ways. Some of the 64 triplets are used to tell the cell where one gene stops and another one starts on the long strands of DNA of a particular chromosome. Others represent a redundancy in the genetic code, more than three times as many combinations as are needed to code for all 20 amino acids; the sequences GGT, GGC, GGA, and GGG, for example, all code for the amino acid glycine.

The uncoding of the segment of triplets in the DNA molecule into the sequence of amino acids in a protein is a two-step process (Fig. 2.1). First, the genetic information from the DNA template is copied, or transcribed, into the nucleotide sequence of a second type of nucleic acid, *ribonucleic acid* (RNA). This RNA is called *messenger* RNA (mRNA) because it carries the coded information of the DNA molecule from the nucleus to the cytoplasm, where protein synthesis occurs. This process of synthesizing mRNA is appropriately called *transcription*. Unlike DNA, the mRNA is single stranded and very small, enabling it to pass from the nucleus to the cytoplasm through small pores in the nuclear membrane. The polypeptide chains are assembled in the cell cytoplasm on structures (organelles) called *ribosomes* through a process called *translation*. This process involves sequential base pairing of triplets on the mRNA with triplets on molecules of *transfer* RNA (tRNA), which, like mRNA, are encoded by a DNA template. Each transfer RNA carries a specific amino acid; there are different tRNAs for each of the 20 amino acids. Coupling between a triplet of mRNA and a specific tRNA occurs through recognition of complementary triplets on the two RNA molecules. Thus, through transcription and translation, the sequences of amino acids in proteins are direct reflections of the base sequences of DNA that constitute the genes. The sequence of DNA encoding on a single polypeptide chain is currently defined as the *unit gene* (Rieger et al. 1976).

Cell division and chromosome and DNA replication are complicated processes. Mistakes are made occasionally in the formation of gametes. Mispairings of bases can lead to amino acid substitutions in proteins (e.g., a base substitution of TTC for TTA results in an amino acid substitution of phenylalanine for leucine in a polypeptide chain) or to an actual discontinuation of chain building and very likely a nonfunctional protein. Such mispairings are a common source of *mutations* (others include actual structural changes in the chromosomes), which, when passed on to the next generation, are the ultimate origins of all genetic variation.

We can summarize to this point:

- Most genes in most diploid organisms occur in pairs located on individual sets of chromosomes contributed by each parent.
- DNA is the informative chemical substance of the gene.
- Amino acid sequences in proteins directly reflect base sequences in genes.
- Mispairings of bases in DNA are one source of mutations that can result in amino acid alterations in proteins.

GENOTYPIC DATA FROM ELECTROPHORESIS

The Process of Electrophoresis

Five of the 20 common amino acids which make up proteins are charged; the charges of lysine, arginine, and histidine are positive, while those of aspartic acid and glutamic acid are negative. Thus, different proteins tend to have different net electrical charges. Electrophoresis uses this physical chemical property of proteins to separate mixtures of proteins on the basis of charge. If allelic differences (i.e., different forms of a gene) occur at a protein coding locus, the net charge of the protein often changes. Gel electrophoresis makes it possible to identify such allelic differences.

The basic procedures of gel electrophoresis are outlined in Fig. 2.2. The process of electrophoresis includes a gel (commonly starch or polyacrylamide) in which introduced solutions of proteins are separated by passage of a direct electrical current through the gel. Initially, mixtures of proteins are extracted with water (or buffered aqueous solvents) from tissues such as skeletal muscle, heart, and liver, unless they are already contained in body fluids such as vitreous humor or blood serum. The water soluble protein mixtures are typically introduced to the gel on a piece of filter paper that is saturated with the mixture. Protein mixtures from 50 or more individuals are often introduced to a single gel, although only 10 individuals are pictured in Fig. 2.2.

A direct current is usually applied for 3–5 hours through the gel. The actual time is determined by such variables as composition of the buffer solution used to make the gel, its ionic strength, and the thickness of the gel. Pro-

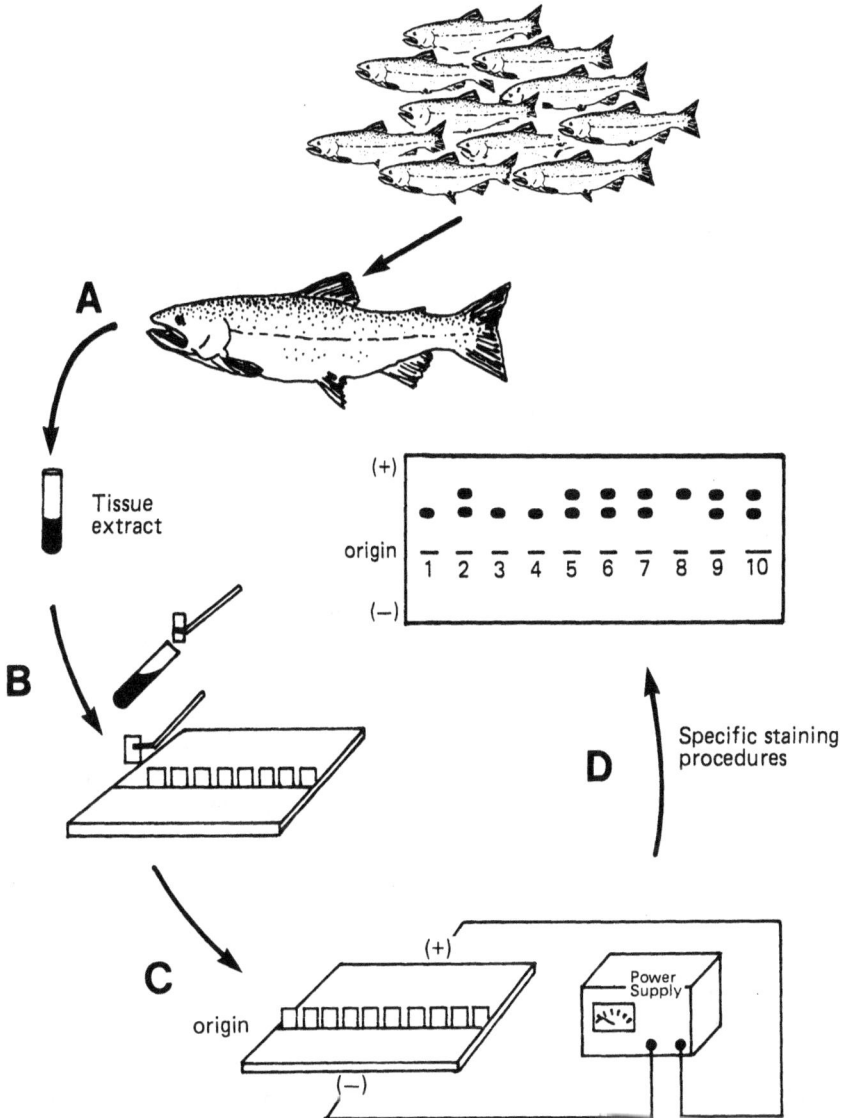

Fig. 2.2 Standard steps for obtaining genotypic data from electrophoresis (modified from Gharrett and Utter 1982). (A) Crude protein is extracted from tissue such as muscle or liver. (B) Extract from each fish is introduced individually to gel by filter paper inserts. (C) Different forms of a particular protein often move different distances from the point of applications when electric current is applied because they do not have identical electrical charges. (D) These forms are readily identified by a specific stain for each protein type. Specificity in staining permits identification of both the activity and the exact location of a particular protein for an individual fish from a complex mixture of proteins in each protein extract. (Intensities of banding patterns do not reflect differences of gene dosages in this depicition.)

teins with a positive net electrical charge move toward the negative (or cathodal) pole and negatively charged proteins move toward the positive (or anodal) pole. The rate of migration is determined by the absolute charge of the protein. A dye solution is added to the sample to mark the progress of electrophoresis. Following electrophoresis, the gel is sliced horizontally into multiple slabs and each slab is stained for the activity of a specific protein.

Most proteins that are studied by electrophoresis are *enzymes,* the catalytic molecules vital to all life, because it is easy to develop histochemical staining procedures to visualize activities of specific enzymes (Hunter and Markert 1957). A number of sources give detailed descriptions of many procedures for visualizing enzymatic activities following electrophoresis (e.g., Harris and Hopkinson 1976, Siciliano and Shaw 1976). Each procedure uses a product of the enzyme's specific activity to precisely locate that enzyme in the gel.

The localization of an enzyme's activity in a gel has been called the "isozyme method." Isozyme refers to different distinguishable molecules found in the same organism which catalyze the same reaction (see Markert and Moller 1959, Shaw 1964, Brewer 1970). *Allozyme* commonly refers to the electrophoretic expression of allelic proteins at a particular locus. The capability to visually localize an enzyme's activity has resulted in the detection of activities of dozens of enzymes reflecting 90 or more loci (e.g., Morizot and Siciliano 1984).

Specific staining for an enzyme's activity permits particular isozymes to be distinguished, one at a time, in a mixture of hundreds of proteins typically found in a tissue extract from an individual fish. The final result of the electrophoretic procedure is bands, such as those in Fig. 2.2, which identify the locations of various forms of a single type of protein on a gel. The banding pattern of an individual contains information on that individual's genotype with respect to the locus (loci) coding for that particular protein.

Expressions of Single Loci

The connection between DNA base sequences, protein amino acid sequences, and the electrophoretic expression of different genotypes is most easily illustrated for a *monomeric* protein. Monomeric proteins are proteins composed of single subunits (i.e., a single polypeptide chain). Let us assume that

- A locus is coded for a monomeric protein having two alleles designated A and A' (i.e., a polymorphic locus);
- These alleles produce subunits (the active protein for monomers), designated a and a' respectively, that are distinguishable by different electrophoretic mobilities; and
- The a' protein encoded by the A' allele migrates more slowly than the a protein encoded by the A allele.

Three different genotypes are possible for an individual at this locus: AA, AA', and $A'A'$. An individual with the AA genotype produces only the faster migrating protein form. This form appears at one location on the gel as a single band. Similarly, an individual with the *A'A'* genotype produces only the slower migrating form at a different location on the gel. The heterozygous *(AA')* genotype produces both protein forms and therefore is reflected as two bands on the gel. We assume that each allele results in the production of equal amounts of protein having the same levels of activity. Therefore, each of the two bands of heterozygous individuals is expected to reflect half the amount of protein that is reflected by either of the homozygous types of individuals; that is, each band of a heterozygous individual expresses half the *dosage* of the single band expressed by a homozygous individual. This pattern of genotypic expression of a monomeric protein encoded by a single locus with two alleles is pictured at the top of Fig. 2.3. Commonly studied monomeric proteins include the serum protein transferrin and such enzymes as phosphoglucomutase, mannosephosphate isomerase, and aconitate hydratase.

Banding patterns on a gel become more complicated when the active protein is *multimeric*, i.e., composed of two or more protein subunits. If we extend the above assumptions to a *dimeric* protein (i.e., one consisting of two subunits), the expected banding patterns are those pictured in the middle section of Fig. 2.3. An individual with the genotype *AA* is expressed as a single

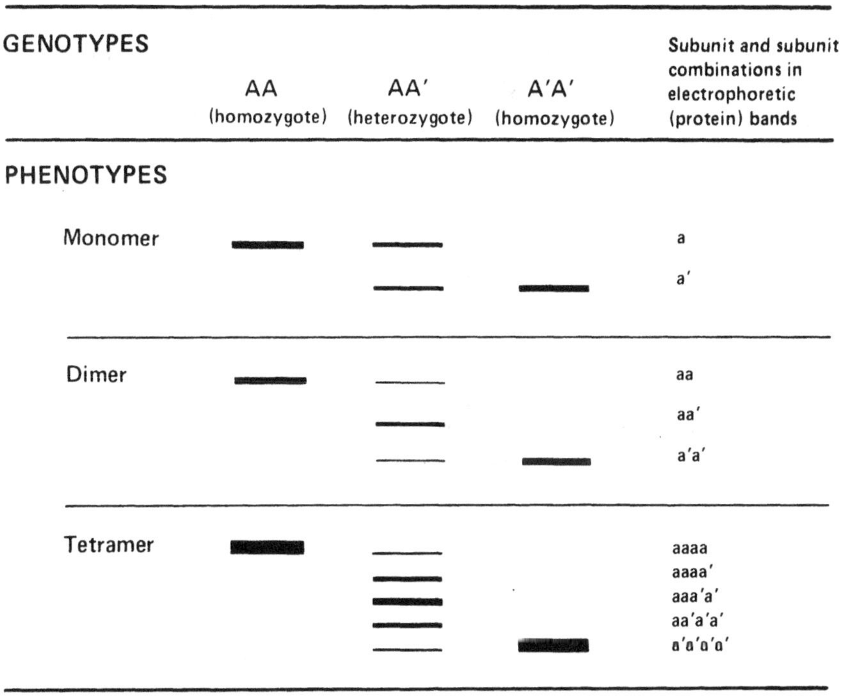

Fig. 2.3 Electrophoretic phenotypes when one locus is expressed. Individuals are homozygous and heterozygous at loci coding for monomeric, dimeric, and tetrameric proteins: the locus is polymorphic, with alleles *A* and *A'* resulting in subunits *a* and *a'*, respectively.

band reflecting identical molecules of a subunits combined in pairs. Similarly, the expression of an individual with the $A'A'$ genotype is another single band reflecting paired $a'a'$ subunits at a different location on the gel. An individual with the AA' genotype, however, is expressed by three bands reflecting the random combination, in pairs, of the two electrophoretically distinguishable types of subunits. Two of the bands are *homomeric* combinations of aa and $a'a'$ subunits. The third middle band is a *heteromeric* band reflecting combinations of a and a' subunits. (Note that monomers cannot form heteromeric bands because the single subunit is the active protein.) The sum of the intensity of the three bands expressed by heterozygous genotypes is expected to equal the intensity of single-banded homozygous expressions because the same number of subunits are produced by both heterozygous and homozygous individuals. Dimeric proteins commonly studied by electrophoresis include the enzymes malate dehydrogenase, isocitrate dehydrogenase, and aspartate aminotransferase.

The banding patterns of a protein having four subunits (a *tetramer*) are pictured in the lower portion of Fig. 2.3. Again, we assume a single locus polymorphic for two electrophoretically detectable alleles. The respective homozygous expressions are single-banded because of the identity of each of the four subunits. The heterozygous individual has five bands, representing random combinations of two allelic subunits in aggregates of four. The five bands include three heteromeric bands in addition to the two homomeric bands; again, their combined intensity is equivalent to the single band of the homozyous expressions. Commonly studied tetrameric proteins include lactate dehydrogenase (see Fig. 2.4), iditol dehydrogenase, and malate dehydrogenase (NADP dependent).

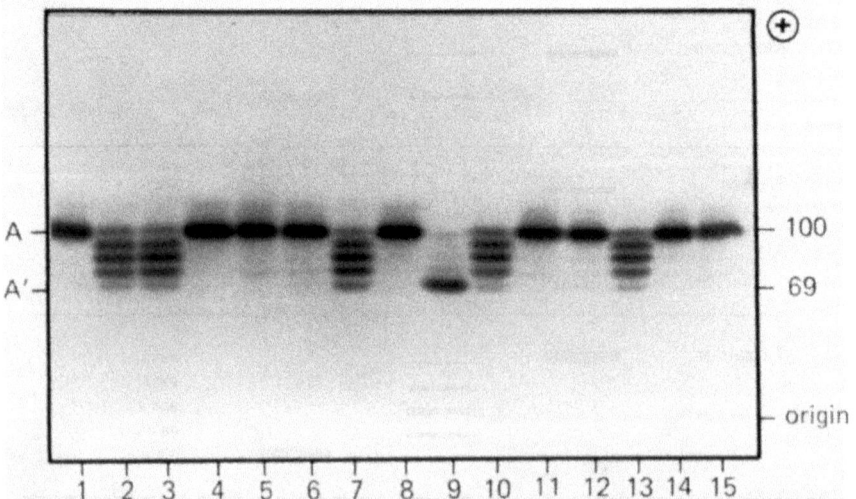

Fig. 2.4 Phenotypes of a two-allele polymorphism for a locus encoding a tetrameric enzyme, lactate dehydrogenase, from livers of 15 rainbow trout. Note five-banded expressions of heterozygous individuals (2, 3, 7, 10, 13).

The expected number of bands and their relative intensities for individuals heterozygous at a particular locus can be predicted assuming that subunits combine randomly following their synthesis. The basis for these expectations can be demonstrated through the randomness of flipping a coin, whose sides represent the allelic subunits a and a'. The coin is flipped in repeated series, with the number of flips in each series representing the number of subunits in the protein (one flip for a monomer, two flips in a series for a dimer, and four flips in a series for a tetramer). The sequential outcome of each series is recorded before going on to the next series.

For a dimeric protein, four combinations are possible when flipping the coin in a series of two: aa, aa', $a'a$, and $a'a'$. There is an equal probability of getting any of the four types. The sequences aa' and $a'a$ represent identical dimeric proteins which would form a single electrophoretic band; therefore, their probabilities can be pooled. Thus, the expression of the dimeric heterozygote is three-banded, with the combined aa'-$a'a$ band at twice the intensity of the respective aa and $a'a'$ bands. This coin flip analogy can be applied to proteins with other subunit structures.

The expected numbers of bands and their relative intensities for individuals heterozygous for protein coding loci can also be predicted from binomial expansion of the two categories of allelic subunits (a and a'). For a dimeric protein the expression would be

$$(a + a')^2 = a^2 + 2aa' + a'^2 \ .$$

In reference to the left-hand side of the binomial formula, the a and a' represent the actual protein subunits and the exponent 2 represents the number of subunits in the protein. In the expanded right-hand side of the formula, the three terms represent the number of bands, and their respective coefficients (1, 2, 1) represent their relative intensities. For a tetramer, the exponent becomes 4. Following expansion, then, the relative intensities of 1:4:6:4:1 would be expected from tetramers. Proteins are sometimes encountered whose subunit structures are something other than monomeric, dimeric, or tetrameric (e.g., the enzyme purine nucleoside phosphorylase has three subunits). The expected numbers and relative intensities of electrophoretic bands can be predicted for them in the same manner if the subunit structure of the protein is known.

The expected banding patterns are idealized configurations. It is important to recognize that some deviations from the expected numbers and relative intensities of bands are frequently seen. There are both genetic and nongenetic reasons for these deviations, some of which are discussed later in this chapter. However, understanding the basis of these idealized configurations is essential for properly interpreting the genotypic basis of electrophoretic patterns.

The electrophoretic banding patterns such as those pictured in Fig. 2.2 and Fig. 2.3 are phenotypes. They are expressions of the genotypes (i.e., the actual alleles) with possible—usually minimal—influences of the in vitro environment of the protein. Thus, genotypes can usually be deduced directly from such phenotypes when the subunit composition of the protein is known. For instance, consider that a group of individuals are subjected to electrophoresis and the resulting gel is stained for lactate dehydrogenase activity. Banding patterns would be observed like those in the lower portion of Fig. 2.3 and in Fig. 2.4. It is safe to assume that such phenotypic patterns reflect the respective homozygous and heterozygous genotypes of an *LDH* (lactate dehydrogenase) locus of that species because the phenotypes conform to the expected numbers and intensities of bands for a tetrameric protein.

Expressions of Additional Loci

More complicated electrophoretic patterns than those depicted in Fig. 2.3 frequently occur when the same type of protein is encoded by two or more loci. These complications include additional protein bands arising from combinations of subunits, encoded by different loci and having different electrophoretic mobilities, or electrophoretic patterns resulting from two (or more) loci whose protein bands have the same or overlapping mobilities.

The latter patterns are particularly frequent in salmonids because of their tetraploid ancestry. Salmonids have about 50% more protein loci expressed than teleosts of diploid ancestry (see Allendorf and Thorgaard 1984). The salmonids have undergone the most intensive electrophoretic examination of any group of fishes, and the complexities of their electrophoretic patterns have often been confusing; an examination of these complexities is therefore warranted.

Let us extend the assumptions of the genotypic expressions of Fig. 2.3 to include a second locus. Every individual is homozygous for the *B* allele at this locus (i.e., a monomorphic locus) which encodes electrophoretically identical *b* subunits. Both loci are expressed at equal levels. Homomeric bands of *b* subunits have electrophoretic mobilities that are distinct from those of the *a* or *a'* subunits encoded by the *A* or *A'* alleles of the first locus. The phenotypes of individuals homozygous at both loci (columns 1 and 3 in Fig. 2.5) resemble the phenotypes of heterozygous individuals in the single locus expression of Fig. 2.3. This resemblence is due to the similarity between the expression of two alleles of a single locus in Fig. 2.3 and of two loci in Fig. 2.5. In monomeric proteins, a single band is expressed for each allele (in this case the *A* and *B* alleles contrasted with the *A* and *A'* alleles in the single locus case). Multimeric proteins express the additional bands of random interactions of the individual subunits (in this case *a* and *b* subunits contrasted with the *a* and *a'* subunits in the single locus case). The expected relative intensities of the bands of dimeric and tetrameric phenotypes of homzygous individuals at two loci are also the same as those of the phenotypes of heterozygous individuals when only a single locus is expressed.

The expression of a heterozygote when one of two loci is polymorphic for a monomeric protein is again a single band for each allele. However, hetero-

GENOTYPES				Subunits and subunit combinations in electrophoretic (protein) bands
	AA (homozygote)	AA' (heterozygote)	A'A' (homozygote)	

PHENOTYPES

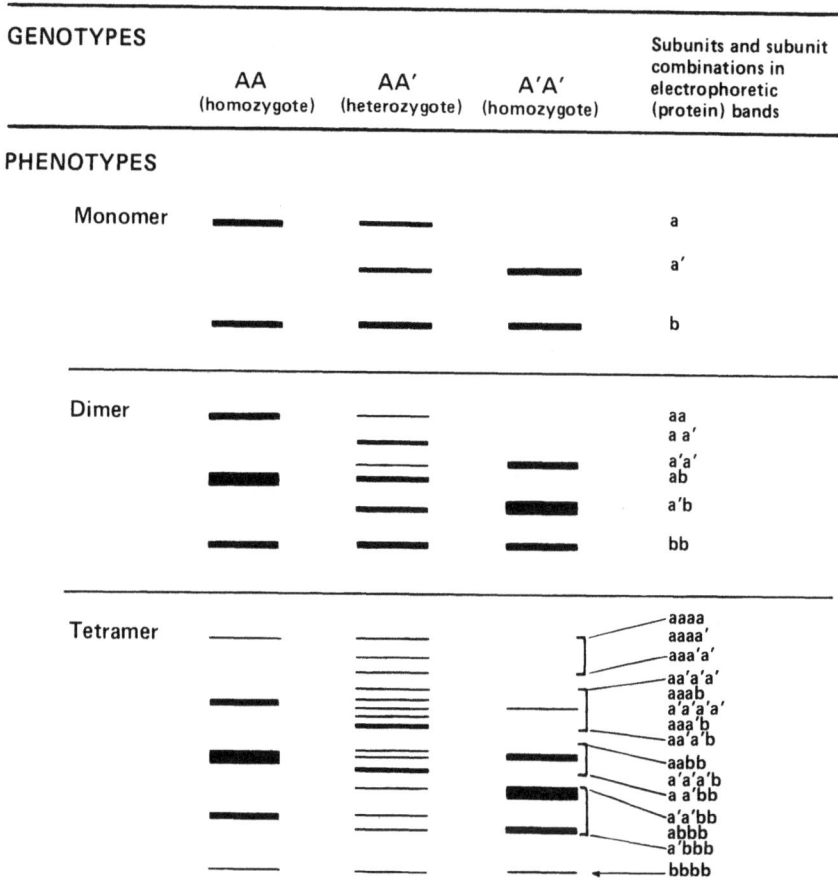

Fig. 2.5 Electrophoretic phenotypes when two loci are expressed. Individuals are homozygous and heterozygous at loci coding for monomeric, dimeric, and tetrameric proteins: one locus is polymorphic (with alleles A and A' resulting in subunits a and a', respectively); and a second is monomorphic, coding for a subunit (b) with an electrophoretic mobility that differs from subunits a and a'.

zygote phenotypes of the multimeric proteins in the situations given in Fig. 2.5 are complicated by combinations involving a third electrophoretically distinct subunit. The number of bands involving combinations of the b subunit from the second locus with a and a' subunits from the first locus are readily predicted for heterozygous phenotypic expressions by including this b subunit in the binomial expansion. For a dimer, with both doses of the single allele of the monomorphic second locus producing b subunits and the respective alleles of the first locus producing a and a' subunits, the squared expansion becomes

$$(a + a' + 2b)^2 = a^2 + 2aa' + a'^2 + 4ab + 4a'b + 4b^2$$

with six (three homodimeric and three heterodimeric) electrophoretically distinguishable subunit combinations (as pictured in the central portion of Fig. 2.5). The fourth power expansion for tetrameric proteins predicts the 15 bands depicted in Fig. 2.5. In practice, some of the predicted bands may not be seen because of overlapping unless the respective homomeric bands are adequately separated by electrophoresis. An actual gel showing a polymorphism for one of two loci encoding a monomeric protein is shown in Fig. 2.6.

Exceptions to Codominant Expression

The phenotypes of Figs. 2.3 and 2.5 are called *codominant* expressions of the respective genotypes because the contributions of all alleles can be identified. Codominant expression is an important attribute of electrophoresis because of the value of genotypic information at individual loci in population genetic studies. However, there are exceptions to codominant electrophoretic expressions that need to be considered.

The occurrence of electrophoretically identical subunits synthesized by two different loci is observed in some fishes (e.g., salmonids). The genetic and evolutionary basis for such *isoloci* (having *isoalleles* giving rise to electrophoretically identical gene products) has been reviewed by Wright et al. (1983) and Allendorf and Thorgaard (1984). One locus or both loci may be variable. In either case the electrophoretic expression of isoloci complicates the determination of genotypes.

A part of this complication is that it is often impossible to assign alleles

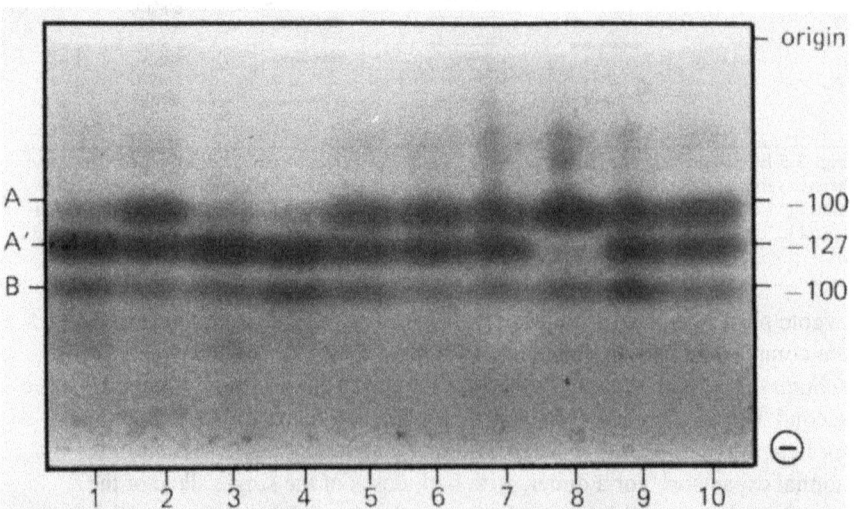

Fig. 2.6 Phenotypes of a two-allele polymorphism for a locus encoding a monomeric enzyme, phosphoglucomutase, from skeletal muscle of 10 sockeye salmon. A second monomorphic locus is expressed cathodal to the bands reflecting the different genotypes of the polymorphic locus. Note that the intensity of expression of the monomorphic locus is less than that of the polymorphic locus, suggesting different levels of synthesis. Note also distinct shadow banding of individuals 1, 3, and 4 that coincides with the mobility of the alternate allelic form.

to specific loci when two (or possibly more) loci code for electrophoretically identical subunits. The problem is apparent from the phenotypes of Fig. 2.7, which pictures the expressions of isoloci where one of the loci is polymorphic and the other monomorphic. The assumptions underlying the phenotypes of Fig. 2.5 are the same for Fig. 2.7 except that the products of the *B* alleles (i.e., *b* subunits) of the monomorphic locus are electrophoretically indistinguishable from those of the *A allele (i.e., a* subunits) of the polymorphic locus. This situation results in the inability to distinguish the contributions of the *a* and *b* subunits from the electrophoretic phenotype. Consequently, the numbers and mobilities of bands expressed by both *AA'* and *A'A'* genotypes are the same. Although no *a* subunits are produced by the *A'A'* genotype, the two doses of *b* subunits having the mobility of the *a* subunit mask this absence. Exactly the same situation would occur if the *B* locus instead of the *A* locus were polymorphic. There is no way to distinguish which locus is polymorphic if both loci are equally expressed in all tissues.

There are ways to deal with this problem. Phenotypes of isoloci are often recorded and analyzed as the summed contribution of four allelic doses for two loci (e.g., May et al. 1979). This procedure gives no information about the diploid genotypes of the individual loci (which has considerable value for determining whether individuals within a sampling represent a single random mating population). If only one locus is polymorphic (as in Fig. 2.7) it is con-

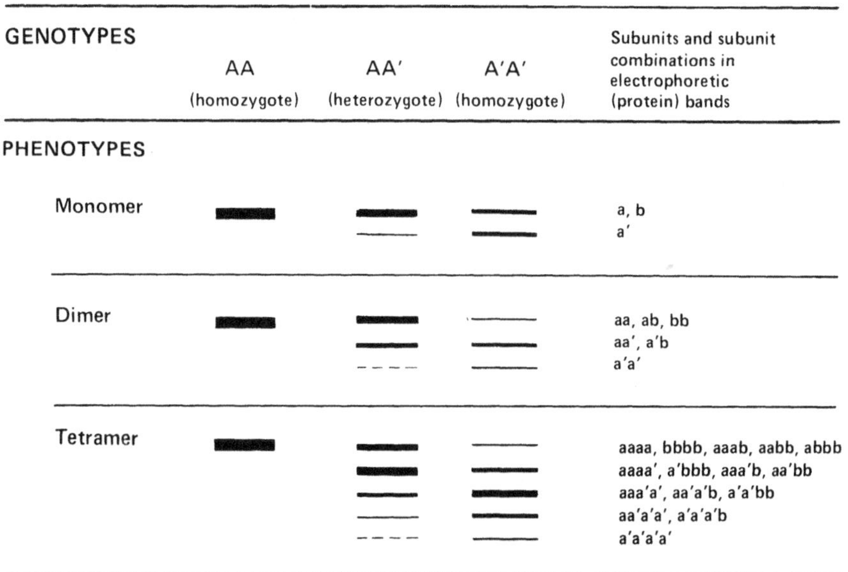

Fig. 2.7 Electrophoretic phenotypes when isoloci are expressed. Individuals are homozygous and heterozygous at loci coding for monomeric, dimeric, and tetrameric proteins: one locus is polymorphic (with alleles *A* and *A'* resulting from subunits *a* and *a'*, respectively); and a second locus is monomorphic, coding for a subunit (*b*) with an electrophoretic mobility identical to that of subunit (*a*).

venient to arbitrarily assign the variation to one of the loci. Such assignment gives information on diploid genotypes within a sampling of individuals but would be misleading in comparisons between samplings if different loci were polymorphic in the sampled groups. A gel showing a polymorphism at one of two isoloci encoding a dimeric protein is seen in Fig. 2.8.

The situation becomes more complicated if both loci are polymorphic. Different procedures have been used to assign alleles to one or the other locus (e.g., Imhoff et al. 1980, Gall and Bentley 1981). Accurate characterization of individual diploid genotypes is still precluded through these methods.

Unambiguous genotypic information can be obtained for isoloci whose encoded proteins are synthesized at different levels in different tissues (e.g., Allendorf and Thorgaard 1984). However, such differences have not often been found.

Codominant expression of isoloci can be masked even when it is known which of the two isoloci is the polymorphic one (as is assumed in Fig. 2.7). The problem here, as indicated above, is the potential ambiguity of phenotypic expression for heterozygous individuals (the second column of Fig. 2.7 and individuals 4–6 of Fig. 2.8) and individuals homozygous for the A' allele (the third column of Fig. 2.7 and individuals 7–9 of Fig. 2.8). The same number of bands is expressed in both instances, although the expected relative intensities of the bands differ between the two genotypic expressions. The gene dosages differ for the production of the respective bands of the heterozygote and the $A'A'$ homozygote (3:1 for $AA'BB$ versus 2:2 for $A'A'BB$). The heterozygous genotype is expressed by asymmetrical relative intensities of banding. Expected

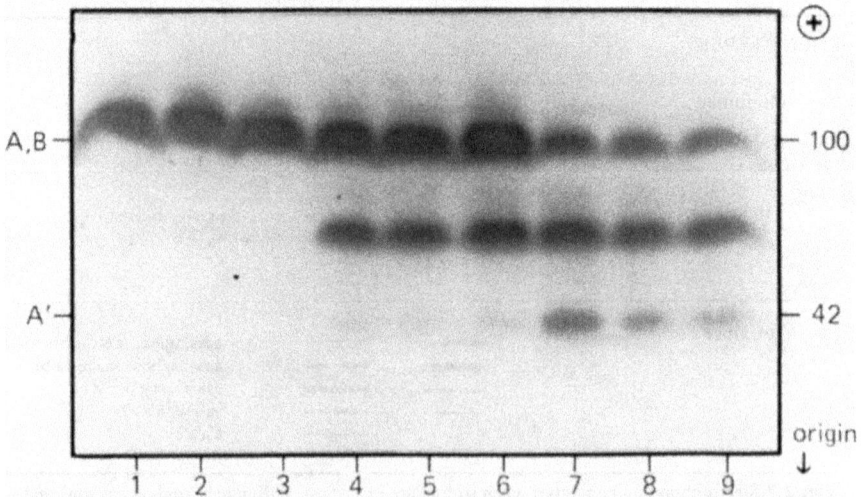

Fig. 2.8 Phenotypes of a two-allele polymorphism for one of two isoloci encoding a dimeric enzyme, isocitrate dehydrogenase, from livers of nine rainbow trout. Note that the same bands are expressed in individuals having single doses (4–6) and double doses (7–9) of the varying allele; only their relative intensities differ.

intensities based on binomial expansions are 3:1, 9:6:1, and 81:108:54:12:1 for the monomeric, dimeric, and tetrameric proteins, respectively. On the other hand, the homozygous $A'A'$ genotype has equivalent production of electrophoretically distinguishable subunits; and the respective relative intensities of bands for monomeric, dimeric, and tetrameric proteins are symmetrical 1:1, 1:2:1, and 1:4:6:4:1 ratios.

These differing expected intensities can be used to differentiate between homozygous and heterozygous individuals expressing the same bands. However, in practice, such distinctions are sometimes difficult or impossible to make. Insensitivities of the electrophoretic procedures coupled with possibly different activities or levels of synthesis of allelic products may prevent identifying genotypes among isoallelic phenotypes (e.g., Utter and Hodgins 1972, Allendorf et al. 1976).

Inactive, or *null* alleles, which result in no active protein being detected electrophoretically (e.g., Lim and Bailey 1977), are another category of variants whose genotypes are often difficult or impossible to determine from electrophoretic phenotypes. The expressions of genotypes having no null alleles and those heterozygous for a null allele are usually ambiguous and distinguishable only on the basis of different gene dosages and consequently differing intensities of the same banding patterns.

Heterozygous genotypes for null alleles are particularly difficult to detect when only a single locus is expressed. In such instances, the only clue to the correct genotype is a reduced intensity of the single band. The existence of the null allele is usually verified by the absence of any electrophoretic banding from individuals with homozygous genotypes for the null allele (see Utter et al. 1984). Such quantitative differences, particularly in the absence of individuals that are homozygous for the null allele, usually cannot be identified reliably. Heterozygotes for null alleles are more readily detected for loci encoding multimeric proteins whose expressions include interactions of subunits with those of the same protein encoded by other loci. The reduced synthesis of subunits caused by the null allele results in reduced intensities of multiple bands, providing more visual clues for genotypic recognition. However, correct identification of the heterozygous nature of such individuals by visual observation is still difficult (see Stoneking et al. 1981) and may require that reduced activity be verified by measuring the intensities of the banding patterns or the protein activities (e.g., Allendorf et al. 1984). Such measurements are also usually required to differentiate between a null allele and an allele (isoallele) encoding a homomeric protein of the same mobility as the homomer of the second locus.

Exceptions to the models for molecular aggregation outlined above are also known. A notable example is the electrophoretic expression of the dimeric enzyme creatine kinase extracted from the skeletal muscle of teleost fishes. These banding patterns do not reflect the subunit aggregations expected from a dimeric protein; in fact, they show no heteromeric bands at all. Fisher and

Whitt (1978) and Utter et al. (1979) have provided molecular and genetic interpretations of this anomalous electrophoretic expression.

Genotypic Nomenclature

The material introduced to this point has involved not more than two loci and, in most instances, not more than two alleles at each locus. A particular diploid species has thousands of loci, and a particular locus often has more than two alleles (although an individual of course, can express only one or two of them). The electrophoretic expressions of a monomeric protein in which four alleles are segregating are pictured in Fig. 2.9. The designations for loci and alleles require more explicit designations when multiple loci and alleles are simultaneously considered. A useful convention of allelic nomenclature that has been adopted by many workers is one outlined by Allendorf and Utter (1979), in which loci are identified by italicized lettered abbreviations reflecting the name of the protein (e.g., *LDH* for lactate dehydrogenase). If there is more than one locus coding for that protein in the species in question, a hyphenated number is added to the locus abbreviation (e.g., *LDH-1* and *LDH-2);* the sequence of these numbers starts with the least anodal homomeric band. Alleles are also numerically identified on the basis of electrophoretic mobilities of homomeric proteins. One allele is designated 100—(e.g., *LDH − 2(100)*—and other alleles are designated according to the percent mobility of their homomeric proteins relative to the distance migrated by the homomeric protein of the *100* allele. For instance, an allelic designation of *LDH − 2(80)* would indicate that the homomer encoded by this allele migrates 80% of the distance of the homomeric protein of the *LDH − 2(100)* allele.

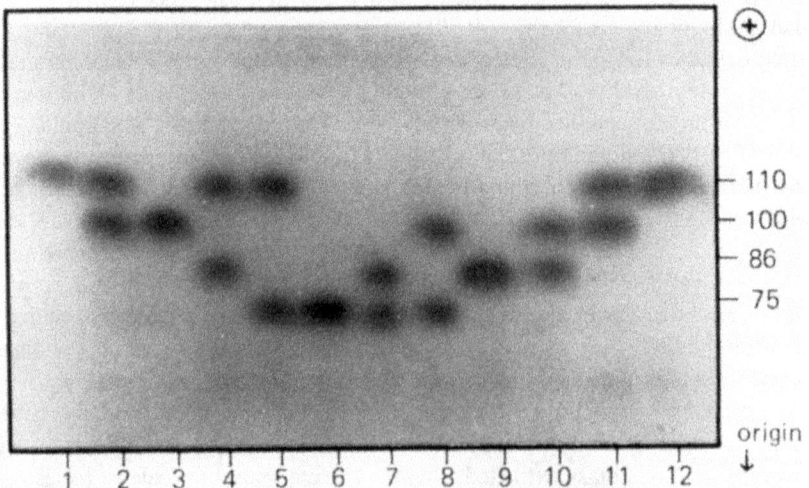

Fig. 2.9 The phenotypes of a four-allele polymorphism for a locus encoding a monomeric enzyme, aconitate hydratase, from livers of 12 chinook salmon.

STRENGTHS AND LIMITATIONS OF ELECTROPHORETIC DATA FOR STUDYING PROTEIN LOCI

The principles outlined above for obtaining genotypic data directly from electrophoretic patterns are widely applied and have resulted in electrophoresis being generally recognized as "the most useful procedure yet devised for revealing genetic variation" (Hartl 1980). The unmatched power of electrophoresis to detect allelic variation is enhanced by the large volumes of data that can be collected with a given amount of effort. Protein extracts can be prepared easily. Many samples can be run on a single gel, and multiple slices can be made from a gel, with each slice stained for different proteins to reveal different loci. For instance, a trained worker can run six gels per day, each gel containing 50 samples for a total of 300 individuals. Data for at least six loci can be obtained from each individual, because each gel can be sliced into six or more slabs and each slab can be stained for a different type of protein. Usually data from more than six loci per individual can be obtained because of the expression of more than one locus for particular types of proteins.

There are limitations to the information that can be obtained by electrophoresis at protein coding loci. The information needed in population genetics relates to base sequences of DNA studied either directly or indirectly. The amino acid substitutions of proteins detected by electrophoresis are indirect reflections of the actual base substitutions in base sequences. All base substitutions do not necessarily result in changes of amino acids. Furthermore, all amino acid substitutions do not result in protein changes that are electrophoretically detectable. It has been estimated that only about a third of the amino acid substitutions are detected under the conditions used to collect electrophoretic data in most laboratories (Lewontin 1974).

It is apparent, then, that electrophoretic identity does not necessarily mean identity of base sequences in DNA. Thus homozygosity is often a conditional concept with electrophoretic data, as it is with many other classes of genetic data whose alleles are inferred by phenotypes (see Allendorf 1977). The term *electromorph* (e.g., King and Ohta 1975) has been introduced to deal with this potential genetic heterogeneity within allelic classes. The distinction between alleles that are alike in state and identical by descent is fundamental in measurements of inbreeding (see Chapter 3). Although it is useful to treat homozygous phenotypes as expressions of the same allele, the possibility of unrevealed genetic heterogeneity must be kept in mind.

All electrophoretic expressions described to this point have had a genetic basis. It soon becomes apparent to any person collecting genetic data by electrophoresis that electrophoretic phenotypes can be affected by influences other than the individual genotypes. Electrophoretic expressions of proteins can be strongly modified by length and conditions of storage. A common reflection of such variables is extra bands expressed anodal or cathodal to the primary band of a particular protein. We have found that collection on dry ice ($-80°C$), storage at similarly low temperatures, and analysis within a few weeks of collection provide optimal conditions for minimizing the occurrence of these "shadow bands," or "conformational isozymes." These artifacts are more of a problem for some proteins, e.g., glucosephosphate isomerase, adenosine deaminase, and phosphoglucomutase , than for others (see Fig. 2.6). Harris and Hopkinson (1975) describe shadow bands and the use of thiol reagents for altering or reducing these spurious bands.

Shadow bands can present problems in recording genetic information from observed phenotypes on a gel. Several points should be kept in mind for accurate interpretations of gels. If a homomer of one allele is accompanied by a shadow band, homomers of each allele for the same locus will probably be accompanied by its own shadow band, in the same direction (cathodal or anodal) and at the same distance from the homomeric bands. It is therefore understandable why a monomer such as phosphoglucomutase can have heterozygous genotypes from a single locus expressing a phenotype of four bands (e.g., Winans 1980). Often, additional shadow bands occur beyond the initial shadow band, again with the same direction and spacing. They usually occur in progressively reduced intensities forming a serial pattern.

Knowing the subunit structure of the protein and, therefore, the expected number of bands and their relative intensities in the heterozygous phenotypes, helps in accurate gel interpretation. In gels in which more than one locus is expressed for a particular multimeric protein, heteromeric bands and shadow bands make correct gel interpretation even more difficult. Correct gel interpretation may initially be possible only when the frequency of a particular allele is high enough to detect a homozygote for that allele. Alternatively, such homozygous individuals may be generated if inheritance studies are feasible.

Criteria for Allelic Variation: Importance of Inheritance Studies

The ultimate test for an allelic basis of an electrophoretic variant is through inheritance studies. It is highly likely that progeny from crosses between parents expressing the phenotypes shown in Figs. 2.3 and 2.5 will verify the inferred parental genotypes through Mendelian ratios. These ratios are readily calculated from a "Punnett square," which plots the expected genotypic ratios of the progeny from the gametic contribution of each parent. Since two heterozygous parents produce two types of gametes, A and A', a Punnett square for a cross of these parents would look like the following:

Gametes of parent 1

		A	A'
Gametes of parent 2	A	AA	AA'
	A'	A'A	A'A'

Four different genotypes are produced in the progeny. However, since $A'A$ equals AA', the predicted genotypic ratio among the progeny would be $1AA:2AA':1A'A'$. Expected genotypic ratios from other crosses (e.g., a homozygote and a heterozygote) are calculated similarly with a Punnett square. Genotypic proportions conforming to these expected ratios validate the Mendelian inheritance of the observed variation. Statistically significant deviations from these genotypic ratios suggest, among other thing, nongenetic origins of observed variations.

Understanding the heritable basis of some electrophoretic phenotypes is sometimes impossible without inheritance data, particularly in species, such as salmonids, with a high frequency of isoloci. However, it is frequently impossible or impractical to carry out breeding studies, and alternative criteria for establishing an allelic basis to variability must be met in such cases. A demonstrated genetic basis through inheritance studies for a presumed homologous variant in a closely related species is a strong criterion for an allelic basis of a particular electrophoretic variant. Extensive inheritance studies such as those of May (1980) in salmonids and Kornfield et al. (1981b) in herring provide valuable information to verify the simple inheritance of similar phenotypic variations among closely related species. Another good criterion is consistent expression of a variant electrophoretic pattern in different tissues within an individual where the same locus is expressed. Criteria relating banding patterns to the known subunit structure of a protein have been mentioned above. A Mendelian basis can usually be determined by meeting such criteria (see Allendorf and Utter 1979).

The electrophoretic expressions of some proteins do not conform to the phenotypes that are expected based on the known subunit structure of the protein. For example, heteromeric bands are not observed for the dimeric enzyme creatine kinase in the skeletal muscle of teleost fishes (Ferris and Whitt 1978). These unusual expressions, when coupled with the general absence of homozygotes for infrequent alleles (this absence is explained later), can only be interpreted through inheritance studies. In our own studies, phenotypic expressions of individuals homozygous for an infrequent allele that are generated from crosses of heterozygous parents have clarified the genetic basis for phenotypic variation of muscle creatine kinase (mentioned above) and adenosine deaminase (Kobayashi et al. 1984) in salmonids. The studies of adenosine deaminase also clarified the nature of artifact bands that had previously confounded interpretations of electrophoretic phenotypes. Of course, the Mendelian ratios observed in these studies further confirmed the heritable basis of these phenotypes.

Inheritance studies are needed more often for verification of null and monomeric protein alleles than for variants of multimeric proteins such as those typified in Fig. 2.5. The heteromeric bands of multimeric proteins aid in distinguishing true genetic variation from banding patterns that can arise from nongenetic causes such as storage and bacterial contamination. Our own studies indicate that artifact bands of monomeric proteins are more readily confused with true allelic variation. Particularly deceptive are the occasional second bands seen above or below a band that appears to be monomorphic for other individuals. In such cases it is uncertain whether the phenotype reflects a heterozygous individual for an infrequent allele or a shadow band. As indicated above, heterozygous individuals for null alleles of loci coding for monomeric proteins or multimeric proteins that are singly expressed in a particular tissue are difficult to distinguish from expressions of genotypes lacking the null allele. The null allele would be expressed only as heterozygotes in most samplings from populations where the allele is infrequent. In such instances, the null homozygotes needed to verify the existence of the null allele would require intentional mating of two heterozygous individuals.

Inheritance data are also necessary to study *linkage* relationships among loci, i.e., the occurrence of two loci on the same chromosome (e.g., May et al. 1979, Morizot and Siciliano 1982). The extensive gene duplications of salmonids have resulted in complicated inheritance patterns that have only recently been clarified through inheritance studies involving allelic variants at many protein loci; they are reviewed in Wright et al. (1983) and Allendorf and Thorgaard (1984).

Describing Populations Through Information from Individual Genotypes

The principles outlined above can now be extended to describe a sample in terms of genotypic and allelic frequencies. Let us use the electrophoretic results pictured in Fig. 2.2 as an example of 10 individuals subjected to electrophoresis and stained for a locus segregating for two alleles in this sample. We arbitrarily designate the allele encoding the more anodal (or faster) protein as A and the other allele A'. Among the 10 individuals, 1 has the genotype AA, 6 have AA', and 3 have $A'A'$. The genotype frequency, then, is 1:6:3.

The allelic frequencies are calculated by counting the number of A and A' alleles. The 10 diploid individuals represent 20 alleles. The 1 AA homozygous individual represents 2 A alleles, and the 6 AA' heterozygous individuals each contribute 1 A allele. Thus, the number of A alleles in this sample is 2 + 6 = 8; and the frequency of the A allele is 8/20 = 0.40. Similarly, the frequency of the A' allele is 12/20 = 0.60.

These genotype frequencies (1:6:3) and allele frequencies ($p = 0.40$, $q = 0.60$) can be compared with genotype and allele frequencies of other samples for this locus. The more distinct the differences between two samples, the greater the "genetic distance" between them. Comparisons among samples are usually based on data from several loci. The amount of genetic distance is then averaged over all loci (see Chapters 4 and 8). Such genotype and allele frequencies, usually calculated for many loci (e.g., 30 or more), are the basic units of information for genetically describing a particular sample of individuals, and for making genetic comparisons between this sample and other samples (see Chapters 5 and 9).

Usually larger sample sizes than 10 individuals (and 20 genes) are required for accurate estimates of allele frequencies of a particular population because of the low precision of estimates from small sample sizes. For reasonable precision of estimates, a minimum of from 50 to 100 individuals is commonly required. For instance, approximate 95% confidence intervals on the estimates of allele frequencies from Fig. 2.2 with 10 samples are ± 0.23. For sample sizes of 50 and 100 the intervals are reduced to ± 0.10 and ± 0.07, respectively.

In addition to permitting genetic comparisons among groups, allele and genotypic data are extremely useful for genetically characterizing a sample of individuals. A common statistic of genetic variability is the frequency of heterozygotes which can be estimated either directly from counting heterozygous individuals over all loci examined or indirectly from allele frequencies (assuming Hardy-Weinberg genotypic proportions; see below). Two other estimates are the percentage of loci that are polymorphic and the average number of alleles per locus.

The Hardy-Weinberg Law (presented in all introductory texts of general and population genetics) is a particularly useful and broadly applied test for the random distribution of genotypes. This law predicts that binomial expansion of the allele frequencies of a polymorphic locus establishes the genotypic proportions of that locus under random mating. For a locus with two alleles *(A* and *A')* having respective frequencies of *p(A)* and *p(A')* this expansion is

$$[p(A) + p(A')]^2 = p(A)^2 + 2p(A)p(A') + p(A')^2 ,$$

where the expected proportions of the homozygous *(AA* and *A'A')* and the heterozygous *(AA')* genotypes are, respectively, the first, third, and second terms of the expanded expression. The Hardy-Weinberg Law can be extended to more than two alleles, and to two or more loci (see Chapters 4 and 7).

Genotypic proportions predicted by the Hardy-Weinberg Law provide a valuable first approximation for expectations in samplings of individuals. For instance, it has been stated earlier in this chapter that homozygous genotypes for a particular allele are not expected to occur in samplings of individuals when the frequency of that allele is low in the sampled population. This expectation becomes apparent from the binomial proportions of an allele occurring at low frequencies. Assume that two alleles *(A* and *A')* are present in a population for a protein locus, and 1 heterozygous individual *(AA')* and 49 homozygous individuals *(AA)* are seen out of a sample of 50 individuals following electrophoresis and staining for that locus. The frequency of the *A'* allele *p(A')* in this sample is $1/100 = 0.01$. Therefore, the probability of an *A'A'* genotype is

$$p(A')^2 = 0.0001,$$

or 1 in 10,000. Thus the $p(A')$ would have to be about 0.14 before a single $A'A'$ homozygote is expected in a sample of 50 individuals.

Deviations from expected Hardy-Weinberg proportions may result from forces including mutation, migration, selection, and genetic drift (chance fluctuations in allele frequencies operating particularly in small populations). In the absence of such forces, allele and genotype frequencies remain constant over successive generations. Data comparing the observed and expected genotypic frequencies provide valuable insights concerning the operation of such forces within the group of individuals from which a sample was drawn, as do data from repeated samplings of a group at different times.

CONCLUDING OBSERVATIONS

The material presented in this chapter has reviewed the fundamentals of Mendelian genetics that relate to electrophoretic procedures. The collection and interpretation of genotypic data by electrophoresis is currently the primary means for measuring genetic variability within and among species of diploid organisms. Genotypic data from one or more loci are the basis of much of the material presented in subsequent chapters. Among other things, such data permit quantitative estimates of the amount of genetic variation in a sample. Tests and analyses of genotypic distributions in a sample may give information concerning its genetic variation and insight into past and present actions of evolutionary processes (cf. Chapters 4 and 8). Allelic frequencies for many loci estimated from genotypic data collected for two or more populations permit quantitative estimates of the amounts and distributions of genetic variation between and among populations. Questions concerning relative levels of genetic variability and relationships among populations can be answered.

In spite of the unquestionable power of electrophoresis to reveal genetic variation, it must be kept in mind that an electrophoretic sample of 100 loci still represents substantially less than 1% of the total number of genes of a particular diploid organism (Crow 1976). It must also be remembered that electrophoresis detects only a part of the genetic variation of the loci studied. Thus, while electrophoretically detected differences among individuals and populations are positive indicators of genetic differences, the absence of differences cannot be equated to genetic identity at the DNA level.

The differences detected by electrophoresis of proteins encoded by different alleles at the same locus appear to have very little or no effect on the fitness of the individual (see Kimura 1968, Nei 1983). This situation is a disappointment to those who had envisioned electrophoretically detected alleles as "useful genes" for breeding programs and assumed that many such genes could be directly related to fitness (see Robertson 1972). However, the general absence of phenotypic effects on fitness of most allelic proteins enhances the value of this variation as a more or less neutral genetic marker. The primary value of such markers is for inferring the distribution and magnitude of genetic variation resulting from evolutionary processes at the vast remainder of the genome that has not been sampled electrophoretically. The value of this information is explored in many of the following chapters of this book.

Despite valuable yet largely preliminary sketches revealed by electrophoresis, considerable genetic variation remains to be detected. New procedures are continually being developed for examining previously unstudied proteins. Previously undetected alleles have also been revealed from proteins that are commonly studied through application of more refined techniques, such as modifications of buffer and gel concentrations and testing for different thermal stabilities of allelic proteins (Singh et al. 1976, Coyne et al. 1978, Coyne 1982).

Procedures for directly examining the genes are emerging as tools for a potentially broader and deeper examination of genetic variation than is possible with methods of protein electrophoresis. Nucleotide sequencing of nuclear genes provides the ultimate information on genetic variation. Methods are tedious and data (see Nei 1983) are as yet sparse and extremely complex. Nevertheless, such information will inevitably become more readily and widely collected.

Restriction enzyme analyses of DNA are seeing accelerated application, particularly for the small but accessible and informative mitochondrial genome. Applications and advantages of these procedures for studies of fish populations are described by Ferris and Berg in Chapter 11 and Gyllensten and Wilson in Chapter 12. Restriction enzyme studies of nuclear DNA that are currently in progress are also yielding promising preliminary results (Bruce J. Turner, University of Virginia, personal communication).

Until recently, studies of fish chromosomes have been largely unproductive in identifying Mendelian variants within species relative to electrophoretic methods (Ihssen et al. 1981). However, new procedures and refined techniques that permit more detailed examinations of chromosome morphology (see Chapter 13) indicate a previously unrecognized potential of cytogenetic studies for identifying Mendelian variations in fish populations.

These procedures appear certain to become and remain valuable tools for fish population genetics. It seems equally certain that electrophoresis will remain a leading procedure because it can readily generate large volumes of reliable genotypic and allele frequency data.

3.
Inbreeding

G. A. E. Gall

This chapter introduces the basic theory of inbreeding and the closely related concept of random genetic drift; it is not intended as a review of the consequences of inbreeding. The presentation assumes that information is available on the genetic structure and the mating system of the population, since knowledge of genotypic frequencies, pedigrees of individuals, the family structure of populations, and population sizes is essential to understanding the concepts presented.

The first section deals with the basic mechanism of how mating among related individuals results in an increase in homozygosity. This is followed by a discussion of the calculation of inbreeding coefficients to illustrate expected rates of inbreeding. Random genetic drift, migration among populations, and effective population size are addressed in the next three sections. Finally, inbreeding depression is discussed, together with the effects of selection during the inbreeding of populations. The reader needing a more detailed presentation is referred to appropriate references in population genetics and animal breeding.

The breeding system for populations of interest to fisheries managers is generally not under management control. This fact, along with the paucity of information about the genetic nature of the population, makes it very difficult to quantify levels of inbreeding. Hatchery populations do undergo controlled mating, so accurate estimates of inbreeding are theoretically possible. This is not the case in practice, however, since information on family structure, age composition, and survival of offspring generally are not known. Several authors (for examples, see Allendorf and Phelps 1980, Ståhl 1983) have attributed loss of electrophoretic variability in hatchery populations to increased inbreeding relative to natural populations. It is difficult to determine the actual cause of the loss of variation since no information was available on genotypic frequencies of founder individuals, selection practiced among potential parents, the survival of offspring, or the selection criteria used in sampling offspring to provide future parents. Busack et al. (1979) provided evidence that electrophoretic variation in hatchery populations can be as great as or greater than that observed in natural populations.

The concept of inbreeding and related ideas are frequently misunderstood among fisheries managers. The term *inbreeding* has come into common usage in reference to the "size of the gene pool" in the context that limiting the number of founder individuals will produce a population with reduced genetic variability. This concept of inbreeding is only partially true in a strict genetic sense. Founding a population with a finite number of breeding individuals sets a maximum on the number of genes contributed to the population; however, the founding sample of genes may represent a high or a low level of genotypic variability.

The genotypic structure of a population is determined not only by the frequency of genes and the forces that affect their frequencies, such as migration, mutation, selection, and genetic drift, but also by the system by which an individual finds a mate. A mating system may be *random,* a situation which exists when each individual in a population has an equal chance of mating with any other individual. For dioecious populations, random mating means that each individual of a particular sex has an equal probability of mating with any given individual of the other sex.

Mates can also be chosen according to (a) their genealogical relationship or (b) their phenotypic resemblence to each other. When mates are more closely related to each other than individuals chosen at random, the mating is referred to as *inbreeding;* when they are less closely related we refer to the mating as *outbreeding.* The choosing of mates according to phenotypic likeness is called *assortative* mating. If the mates are more similar than average for morphological or performance traits, mating is referred to as *positive assortative* mating, in contrast to *negative assortative* mating in which mates are more dissimilar than if they were chosen at random. An increase in homozygosity and a corresponding decrease in heterozygosity relative to that expected under true random mating result both from matings among relatives and from positive assortative mating when phenotypic likeness is based on heritable traits.

Mating between related individuals is inbreeding. It can occur because related individuals are chosen to mate, such as might be the case in a hatchery. It can also occur if the number of randomly mating parents leaving offspring *(population size)* is small, because in that case a high proportion of potential mates in the next generation will be related. This phenomenon is often referred to as random genetic drift.

Two related individuals have one or more *common ancestors.* Therefore, they may have genes in common because each received some of the same genes from their common ancestor(s). It follows, then, that for some loci, the two alleles possessed by an offspring from a mating between related individuals may be identical copies of a gene both its parents received from a common ancestor.

Level of inbreeding is usually measured by a quantity we call the *inbreeding coefficient,* denoted by the symbol F. It is the probability that the two alleles at a given locus are identical copies of a single ancestral allele. Jacquard (1975) attempted to clarify the multiple and often confusing uses of the term *inbreeding* by delineating five separate usages.

IDENTITY BY DESCENT

The inbreeding coefficient is a relative concept; it measures the increase in homozygosity due to inbreeding or, conversely, the loss of heterozygosity, relative to some ancestral, or *reference,* population. The minimum assumption concerning the reference population is that all the genes in the population are unique; such a population is *noninbred* ($F = 0$). The inbreeding coefficient for an individual in an existing population is the probability that the two alleles at a locus arose by replication of a single allele from the reference (or ancestral) population. The reference population need not be remote in time relative to the present one; in fact, it is assumed to exist just prior to the time we became interested in the level of inbreeding of the existing population. It is the time at which we assume the population was noninbred.

The process of gametogenesis, particularly meiosis, guarantees that alleles carried by gametes will be exact copies of alleles possessed by the parent; they will have the same function and the same nucleotide sequence, unless a mutation has occurred. Such alleles are *identical by descent* (IBD). Alleles that have the same function but are not recent copies of an ancestral allele are said to be *alike in state* (AIS).

If there are two alleles with different phenotypic effects, say, A_1 and A_2, at a locus in a diploid population, individuals may possess different alleles, in which case the individual is a *heterozygote* (A_1A_2), or they can have two similar alleles and thus be *homozygous* (A_1A_1 or A_2A_2).

We can visualize the formation of genotypes in the following way. The parents of a particular generation produce many copies of each allele through the production of gametes, thus forming a "pool of genes." Zygotes are formed through the random union of gametes. If the two alleles carried by the gametes forming the zygote are copies of a common allele, the zygote is an *IBD homozygote.* This could occur if the two gametes were produced by the same parent (self-fertilization) or if the two gametes were produced by related parents. If the two alleles are the same but not copies of a common allele, the zygote is an *AIS homozygote.* Finally, if the two alleles are different, the zygote is a *heterozygote.* Thus, a population undergoing some inbreeding can possess three different types of genotypes at a single locus.

In the reference population all homozygotes are assumed AIS, while inbred populations will contain both AIS and IBD homozygotes. The inbreeding coefficient, F, the probability that the two alleles at a locus are IBD, is a measure of the increase in homozygosity. However, the increase is due to the production of the special IBD type of homozygote. In contrast, the effect of positive assortative mating, which also increases homozygosity, is to increase the frequency of AIS homozygotes for those loci controlling the characteristic used to select mates.

Clearly, a high frequency of IBD alleles in the parental generation increases the likelihood their offspring will be homozygous-IBD. Since it is not possible for an IBD individual to be heterozygous, the offspring will have a reduced level of heterozygosity relative to the previous generation. We will examine, as an example, the mating of half-sib parents. Since half-sibs are related, they have a higher probability of possessing IBD alleles than two random individuals drawn from the reference population.

Half-Sib Mating

The idea of the two types of homozygotes and the way inbreeding occurs can be demonstrated by examining the possible origin of the two alleles at a single locus for an offspring produced by a mating between half-sibs. A hypo-

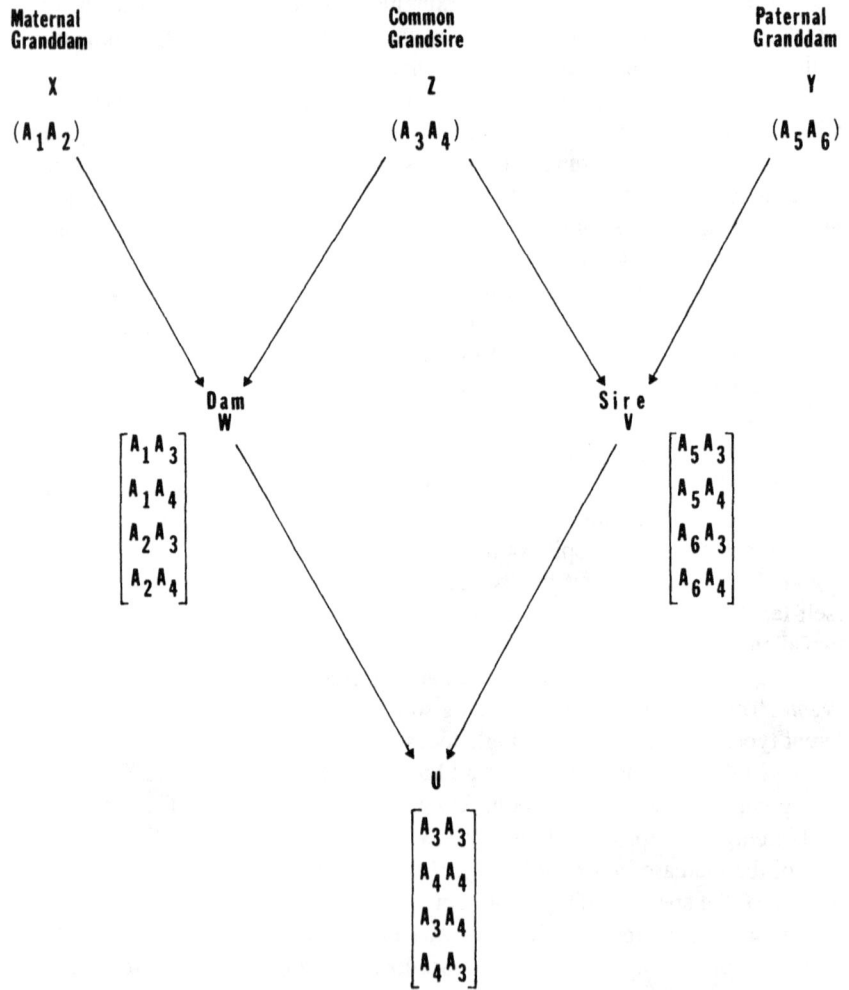

Figure 3.1 The inheritance of alleles for an offspring of a mating between half sibs. Grandparent alleles are designated as A_i and genotypes are listed parenthetically. Individual animals are represented with letters U through Z. Only four of the possible 16 genotypes are listed for individual U.

thetical pedigree is outlined in Fig. 3.1. We will derive F_U, the probability that individual U is an IBD homozygote at an arbitrary locus, A. The female parent (dam W) and the male parent (sire V) are the offspring of a single male (grandsire Z) and two females (granddams X and Y). Therefore, W and V are *half-sibs*. Initially, we will assume that all the alleles of the grandparents were independently derived from the reference population; i.e., the grandparents are unrelated and not inbred. The grandparents, individuals X, Y, and Z, are all ancestors of individual U. However, individual Z, the common male grandparent, is referred to as a *common ancestor* of individual U to denote that the parents of U may possess genes "in common," those derived from Z. The concept of the common ancestor is central to inbreeding since it is a necessary condition for a descendant, individual U in this example, to inherit alleles which are copies of a single ancestral allele.

To facilitate identification of the original six alleles, we will designate the grandparental genes as alleles A_1 through A_6. Genotypically, the grandparents may be heterozygotes or AIS homozygotes. Individual W can have one of four possible genotypes; she could have received either A_1 or A_2 from the granddam (X) and either A_3 or A_4 from the grandsire (Z). Individual V likewise can have one of four possible genotypes, receiving either A_5 or A_6 from the granddam (Y) and either A_3 or A_4 from the grandsire (Z).

Two important results should be noted at this point. First, it is possible for both parents, W and V, to inherit the same allele from the grandsire. However, it is *not* possible for the two parents to inherit the same allele from the granddams. Second, if the parents, W and V, did inherit the same allele, that is, A_3 or A_4, each would be carrying IBD copies of an ancestral allele. The alleles received from the granddams cannot be IBD unless the granddams are related.

The offspring, individual U, must inherit one allele from each parent. It will receive one of four possible alleles from the dam, W, and one of four possible alleles from the sire, V, so there are 16 possible offspring genotypes. Four of the 16 are listed in Fig. 3.1. (A_3A_4 differs from A_4A_3 only in the parental origin of the alleles.) To find the inbreeding level for U, we must ask the question, "What is the probability that individual U received two copies of an ancestral allele?"

In our example, the condition is satisfied by progeny genotypes A_3A_3 or A_4A_4. Since the two IBD genotypes represent 2 out of a possible 16, the probability of U inheriting identical-by-descent alleles is 2/16, or one-eighth. This probability can be interpreted by stating that of the 16 possible genotypes U could possess, only 2 represent IBD homozygotes. Some of the remaining 14 genotypes may be homozygous AIS and others may be heterozygous, but their occurrence will be a function of gene frequency, not relatedness among the parents.

To further understand the process of inbreeding, it is instructive to derive the probability of identity by descent using basic genetic principles. We can do this by tracing the likelihood that particular alleles, A_3 and A_4 in our example, are passed from a common ancestor to the individual of interest. First, we should note that the probability that an offspring inherits a particular allele possessed by a parent is 1/2 because of Mendelian segregation. Secondly, in a pathway of inheritance, the outcome of one Mendelian segregation is independent of the outcome of another segregation in the same or a previous generation. Finally, to be an IBD homozygote, individual U must inherit the same allele from both of its parents.

Consider the A_3 allele first. The probability that individual W inherits allele A_3 from Z and that U inherits it from W is $1/2 \times 1/2 = 1/4$. Similarly, the probability that U inherits allele A_3 through the *pathway* Z to V to U is $1/2 \times 1/2 = 1/4$. The probability that U inherits the A_3 allele from both parents is the product of the probabilities of these independent events. Thus, using the symbol P to mean the probability,

$$P(\text{U is } A_3A_3) = (1/2 \times 1/2) \times (1/2 \times 1/2) = (1/2)^4 = 1/16 .$$

Following exactly the same argument, it can be shown that the probability that U inherits the A_4 allele from both parents is

$$P(\text{U is } A_4A_4) = (1/2 \times 1/2) \times (1/2 \times 1/2) = (1/2)^4 = 1/16 .$$

Since two possible IBD genotypes are mutually exclusive events, that is, individual U can be IBD for either A_3 or A_4 but not both, the probability that U is an IBD homozygote is

$$P(\text{U is IBD homo}) = (1/2)^4 + (1/2)^4 = (1/2)^3 = 1/8 .$$

In our example, locus A represents any particular locus for these 6 individuals and the probability of IBD at the A locus is representative of what could occur at any locus. Therefore, the inbreeding coefficient F of individual U is $F_U = 1/8$ and represents the expected proportion of loci of U which are IBD, or "homozygous due to inbreeding."

It is worth noting the coincidental result, for future reference, that 1/8 is equal to 1/2 raised to the third power and that there are three ancestors in the pedigree of U involved in the transmission of potential IBD alleles: individuals W, Z, and V. The remaining individuals in the pedigree, X and Y, can be ignored because they cannot make a genetic contribution to both parents of U and thus cannot contribute to the inbreeding of U.

Inbred common ancestor. In our discussion to this point we have assumed that the two alleles carried by the common ancestor arose from independent origins. However, if the common ancestor is inbred, there is a probability, F_{anc}, that the alleles carried by the common ancestor are identical by descent. In our example (Fig. 3.1) if the grandsire, Z, was inbred, then, using the symbol $\equiv$ to mean IBD,

$$P(A_3 \equiv A_4) = F_Z .$$

Considering the 16 possible genotypes of U, two are composed of an A_3 and an A_4 allele. Let us first determine the probability of U receiving alleles A_3 and A_4. Following the rationale used to demonstrate the probability of U having genotype A_3A_3 and ignoring IBD for the moment, we see that

$$P(\text{U is } A_3A_4) = (1/2 \times 1/2) \times (1/2 \times 1/2) - (1/2)^4 = 1/16 ,$$

$$P(\text{U is } A_4A_3) = (1/2 \times 1/2) \times (1/2 \times 1/2) = (1/2)^4 = 1/16 .$$

Therefore,

$$P(\text{U is } A_3A_4 \text{ or } A_4A_3) = 1/16 + 1/16 = 1/8 .$$

As stated earlier, if the common ancestor, Z, is inbred, F_Z is the probability that A_3 and A_4 are identical by descent. Then it follows that the probability of U being an IBD homozygote for alleles A_3 and A_4 is equal to the product of the probability of the genotype being A_3A_4 and the probability that the alleles are IBD. That is,

$$P(\text{U} \equiv A_3A_4) = (1/16)F_Z + (1/16)F_Z = (1/8)F_Z .$$

We see that it is possible for the inbreeding of a common ancestor to be passed on to its descendants. This contribution is often referred to as "old" inbreeding since the copying of the alleles occurred at some earlier time. "New" inbreeding, by contrast, is that which occurs during the copying of alleles by individuals in the pedigree between the common ancestor and the individual of interest. The total inbreeding effect is the sum of the new and the old and the inbreeding coefficient for individual U is

$$P(\text{U is IBD homo}) = \frac{1}{8} + \frac{1}{8}F_Z$$

$$= (1 + F_Z)\frac{1}{8}$$

$$= (1 + F_Z)(1/2)^3 .$$

As an example, if Z was completely inbred, $F_Z = 1$, then

$$F_U = \frac{1}{8} + (1)\frac{1}{8} = (1 + 1)\frac{1}{8} = \frac{1}{4} ,$$

or twice what it would be if $F_Z = 0$.

For clarity, it must be pointed out that the inbreeding of the grandsire, Z, has no influence on the inbreeding of the parents, since neither W nor V can inherit two alleles from its sire. This result can be generalized to state that the inbreeding of a parent has no influence on the inbreeding of an immediate offspring. Thus, the mating of inbred but unrelated parents results in a noninbred offspring.

CALCULATING INBREEDING FROM PEDIGREES

General Case

The coefficient of inbreeding, F, is a probability value with limits of zero (0) and one (1). A value of zero is assumed for the reference population and indicates no IBD homozygotes, while a value of one represents complete IBD homozygosity. Values between the limits indicate the probability of occurrence of IBD homoyzygotes. The coefficient, F, has no meaning with regard to the initial level of heterozygosity or homozygosity but only to changes that may occur. For example, the founding individuals chosen to form a new inbred line could all be homozygous for the same allele at a particular locus even though they all were non-inbred individuals. Inbreeding would generate IBD homozygotes at the expense of AIS homozygotes with no apparent loss of heterozygosity. This concept is particularly important in differentiating between founder effects and inbreeding as the cause of low observed variability in populations.

The inbreeding coefficient can be applied in two independent ways depending on whether reference is being made to a particular individual or to a population of individuals (Franklin 1977). If the calculated F-value applies to a particular individual, then F is the expected proportion of all loci of that individual which are IBD homozygous. If the calculated F refers to a population, it is the expected proportion of individuals in the population that are IBD homozygotes at a particular locus. These two interpretations of the inbreeding coefficient cannot be applied simultaneously. This is because individuals with the same calculated F-value may be IBD homozygotes at different loci. However, the average of F-values calculated for all individuals in a population can be used as an estimate of inbreeding for the population. It is this latter idea of average F that is used when levels of inbreeding are estimated from observed levels of homozygosity.

Professor Sewall Wright developed a simple algorithm for the calculation of inbreeding from information on an individual's pedigree. In the calculation of inbreeding coefficients from analyses of pedigrees, the key elements are the parents and a pathway of inheritance, where *pathway* refers to an uninterrupted path of transmission of alleles from a common ancestor to the parents. If it is possible for each of the parents of the individual in question to receive copies of an allele from an ancestor, then

- the ancestor is referred to as a common ancestor,
- the parents will be related, and
- offspring from this pair of parents will be inbred.

The general formula for the coefficient of inbreeding of individual I is

$$F_I = \sum (1 + F_A)(1/2)^{m+n+1} \, ,$$

where

$m =$ number of generations (segregation steps) between the common ancestor and the sire,

$n =$ number of generations (segregation steps) between the common ancestor and the dam,

$m+n+1 =$ a simple count of the number of individuals in the pathway, including the common ancestor and the parents,

$1/2 =$ probability an allele is passed from one individual to the next in the pathway,

$F_A =$ inbreeding coefficient of any common ancestor A, and summation is over all independent pathways of transmission of alleles from the common ancestors to the parents.

Pedigree Analysis

The analysis of pedigrees to determine the coefficient of inbreeding of individuals has important applications in the management of hatchery populations when pedigree information is maintained. For our purposes, evaluating inbreeding from arbitrary pedigrees will be used to develop an appreciation of rates of inbreeding relative to the degree of relationship among the mates. The discussion will also demonstrate how past inbreeding is retained by a population unless a specific effort is made to outbreed. For populations managed in a hatchery, it becomes obvious that accurate pedigree records are necessary if level of inbreeding is going to be known with any degree of confidence.

In the examples discussed below, the pedigrees are depicted as arrow diagrams, with the arrows indicating the direction of inheritance. The diagrams include only those individuals contributing to the relationship between the parents of the individual of interest; all peripheral parents have been omitted. Pathways of inheritance are listed as strings of letters representing the individuals involved in the pathway, with the letter representing the common ancestor in boldface for purposes of bookkeeping.

One common ancestor, one pathway.

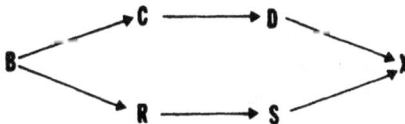

Determination of the pathway can be simplified if it is traced from one

parent backward to the common ancestor and then forward to the other parent. The common ancestor is B and the pathway is SRBCD. There are 5 individuals involved in the pathway, so $m + n + 1 = 5$, since $m = 2$ and $n = 2$. If we assume that B is not inbred,

$$F_X = (1/2)^5 = 1/32 = 0.03125 .$$

One common ancestor, two pathways.

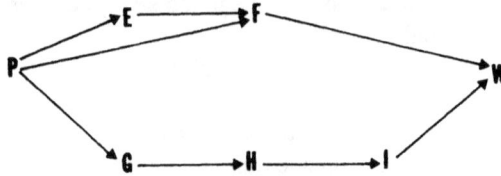

In this case the common ancestor, P, was used in a parent-offspring mating to produce individual F. Consequently, there were two independent opportunities for parent F to receive alleles identical by descent to an allele possessed by parent P. The two pathways and the contribution of each are

$$I\,H\,G\,P\,E\,F \qquad (1 + F_P)\,(1/2)^6$$

$$I\,H\,G\,P\,F \qquad (1 + F_P)\,(1/2)^5$$

If we assume P is not inbred, then the coefficient of inbreeding for individual W is

$$F_W = (1/2)^6 + (1/2)^5 = 3/64 = 0.0469$$

Of course individual F is also inbred due to the path PE, a parent-offspring mating. However, this fact does not affect the inbreeding of W since F can contribute only one allele to its offspring.

Two common ancestors, one of which is inbred.

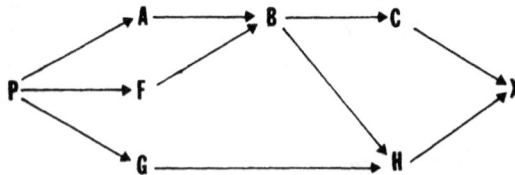

In this example, both individuals P and B are common ancestors of C and H, the parents of individual X. We also note that individual B is inbred due to a path to common ancestor P. Individual H is also inbred, but this fact will not contribute to the inbreeding of X.

The first step in the analysis of the inbreeding of individual X is to determine the inbreeding of the common ancestor B. The pathway of interest is FPA, so

$$F_B = (1/2)^3 = 0.125 \ .$$

There are three pathways that contribute to the genetic similarity between the parents, one associated with common ancestor B and two with common ancestor P. The pertinent pathways are

$$\text{H B C} \qquad (1 + F_B)\,(1/2)^3$$

$$\text{H G P A B C} \qquad (1 + F_P)\,(1/2)^6$$

$$\text{H G P F B C} \qquad (1 + F_P)\,(1/2)^6$$

The inbreeding of individual X is the total contribution of the three independent pathways. With $F_B = 1/8$ and $F_P = 0$,

$$F_X = (1 + 1/8)\,(1/2)^3 + (1/2)^6 + (1/2)^6 = 0.172 \ .$$

Rules of analysis. The use of algorithms such as the one above for F_X requires care and the application of a few rules. Note in the above example that a path, HBFPABC was not included in the calculation of the inbreeding of X; but note also that individual B occurs twice in the pathway because B is inbred. However, the inbreeding of B was included in the calculation of F_X. To assure that all paths are independent regarding inheritance of alleles, three simple rules should be applied:

- An individual can occur only once in a pathway.
- All paths must follow direct descendancy and not crisscross among generations. A good test is to make sure that all arrows in the diagram are traversed from head to tail when tracing the path from one parent to the common ancestor and from tail to head when tracing forward to the other parent.
- Inbreeding of all individuals is ignored except that of common ancestors. An inbred common ancestor will always occur as an intermediate individual if its parents are included in the pedigree. Thus, its inbreeding is considered only for paths to which it is the common ancestor, and ignored otherwise.

A number of computational methods have been developed for calculation of inbreeding coeficients for large populations based on pedigree information (Cruden 1949, Emik and Terrill 1949). Many of these methods are appropriate for computer applications.

Regular Systems of Inbreeding

It is of interest to consider examples of specific mating systems since they demonstrate expected rates of inbreeding and are often used for special breeding programs. The most frequent systems encountered are sib mating, parent-offspring mating, and repeated backcrossing. Self-fertilization is the most intense form of inbreeding, but an impossible one in most higher animals. Rates of inbreeding approaching selfing can be achieved when gynogenesis is employed; however, actual homozygosity is influenced by rates of crossing over (Thorgaard et al. 1983). The inbreeding coefficients realized from some regular systems of inbreeding are given in Table 3.1.

Selfing. In *selfing* the parent is always a common ancestor; thus, it is easy to derive a recurrence relation between individuals in consecutive generations. In the following pedigree, letters refer to individuals as usual and t refers to generation number:

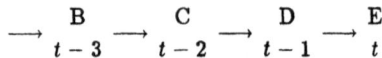

$$\xrightarrow{\quad} \underset{t-3}{B} \xrightarrow{\quad} \underset{t-2}{C} \xrightarrow{\quad} \underset{t-1}{D} \xrightarrow{\quad} \underset{t}{E}$$

If we apply the general formula, $F_I = 1/2(1 + F_A)$ and substitute F_t for the inbreeding of individual I and F_{t-1} for inbreeding of the common ancestor, then

$$F_t = \frac{1}{2}(1 + F_{t-1}) = \frac{1}{2} + \frac{1}{2}F_{t-1} \ .$$

If individual B, above, is noninbred, then $F_C = 1/2$, the inbreeding of individual D is $F_D = 1/2(1 + 1/2) = 3/4$, and $F_E = 1/2(1 + 3/4) = 7/8$.

Examination of these results demonstrates that the increase in inbreeding each generation is equal to one-half (the new inbreeding) plus one-half of the inbreeding level the previous generation. The equation for F_t can be rearranged to solve for $(1 - F_t)$. Multiplying both sides of the equation by (-1) and then adding (1) to each side yields

$$1 - F_t = 1 - \frac{1}{2}(1 + F_{t-1}) = \frac{1}{2}(1 - F_{t-1})$$

and

$$1 - F_t = (1/2)^t (1 - F_0) \ ,$$

where F_0 is the inbreeding in the initial generation.

The quantity $(1 - F_t)$ is sometimes called the *panmictic index*, a measure of the amount of heterozygosity in generation t relative to that of the reference, noninbred population. The interesting result demonstrated by this relationship between inbreeding coefficients in successive generations is that the loss of heterozygosity each generation is a function of $(1 - F)$, the hetero-

Table 3.1 Inbreeding coefficients for regular mating systems. Generation zero individuals are assumed to be noninbred and unrelated.

Inbreeding generation	Selfing	Full-sib	Parent-offspring	Repeated backcross
1	0.500	0.250	0.250	0.250
2	0.750	0.357	0.375	0.375
3	0.875	0.500	0.500	0.438
4	0.938	0.594	0.594	0.469
5	0.969	0.672	0.672	0.484
10	0.999	0.886	0.886	0.499
15	0.9999	0.961	0.961	0.4999
-	-	-	-	-
∞	1.000	1.000	1.000	0.500

zygosity remaining. For selfing, the loss in each generation is equal to one-half the heterozygosity present in the previous generation. It is also of interest to note that this recurrence equation does not consider loss of AIS homozygotes (see Inbreeding and Population Size).

Full-sib. Full-sib mating, also called brother-sister mating, and parent-offspring mating are the most powerful inbreeding systems available for higher animals. Full-sib mating is generally the system of choice for the production of inbred lines. The procedure involves repeatedly mating full brothers and sisters each generation as outlined in the following pedigree:

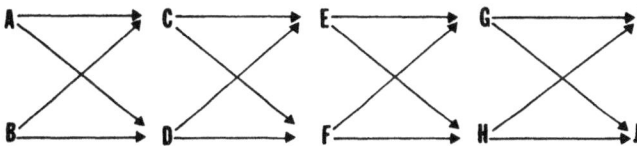

If we assume that generation zero individuals A and B are unrelated, individuals C and D are not inbred. However, their offspring, E and F, are equally inbred. The pertinent pathways and the contribution of each are

$$D\ A\ C \qquad (1/2)^3$$

$$D\ B\ C \qquad (1/2)^3$$

and

$$F_E = F_F = (1/2)^3 + (1/2)^3 = (1/2)^2 = 0.25.$$

The second generation of full-sib mating further increases the relationship between individuals and thus the inbreeding of individuals G and H. Since none of the common ancestors, A, B, C, and D, are inbred, the inbreeding coefficient for both G and H is equal to the sum of the contributions of the following paths:

$$F\ C\ E \qquad (1/2)^3$$

$$F\ D\ E \qquad (1/2)^3$$

$$F\ D\ A\ C\ E \qquad (1/2)^5$$

$$F\ C\ A\ D\ E \qquad (1/2)^5$$

$$F\ D\ B\ C\ E \qquad (1/2)^5$$

$$F\ C\ B\ D\ E \qquad (1/2)^5$$

$$\text{sum} = F_G = F_H = 3/8 = 0.375 .$$

The inbreeding coefficients for I and J, the offspring from mating individuals G and H, will be one-half (0.50). The calculation of F_I and F_J involves the contributions of a total of 14 pathways, 4 each for common ancestors A and B, 2 each for C and D, and 1 for each of the inbred common ancestors E and F. It can be shown algebraically that after a minimum of three generations of recurrent full-sib mating,

$$F_t = \frac{1}{4} + \frac{1}{2}F_{t-1} + \frac{1}{4}F_{t-2}$$
$$= \frac{1}{4}(1 + 2F_{t-1} + F_{t-2})$$

with $F_0 = F_1 = 0$. This recurrence equation demonstrates that after a lag of one generation relative to selfing, the inbreeding in generation t is a function of new inbreeding equal to one-quarter plus the old inbreeding accumulated through the previous two generations.

Parent-offspring. Parent-offspring matings are often used in breeding programs to "fix" a marker gene or a specific characteristic, such as a color morph, in a special population. When offspring are always mated to the younger of their two parents, the pedigree will have the following form:

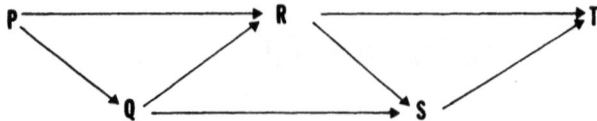

The first generation of parent-offspring mating yields inbred individuals with $F_R = 1/4$ due to the pathway **PQ**. Offspring of the second generation of parent-offspring mating have an inbreeding coefficient, $F_S = 3/8$, arising from common ancestors P and Q. In the third generation, one common ancestor, R, is inbred and the inbreeding of offspring from generation three matings is $F_T = 1/2$ from the pathways **RPQS**, **RQS** and **RS**.

From these results we see that the rate of inbreeding from repeated parent-offspring mating is identical to full-sib mating. The recurrence equation relating inbreeding over successive generations is also the same as given for full-sib mating. With $F_0 = F_1 = 0$ (represented by individuals P and Q, respectively) the new inbreeding is 1/4 each generation.

Repeated backcrossing. A second form of parent-offspring mating involves repeated backcrossing of an individual from a randomly mating population with $F = 0$. If B is the parent used for repeat mating, then a pedigree for repeat backcrossing is

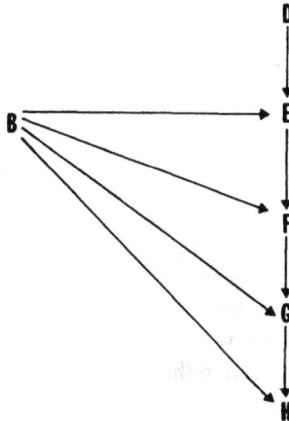

The recurrence equation for sequential generations of inbreeding using a non-inbred parent is

$$F_t = \frac{1}{4}(1 + 2F_{t-1}) .$$

The inbreeding coefficient can never exceed 1/2 since one parent is always non-inbred.

INBREEDING AND POPULATION SIZE

The availability of complete pedigree information allows one to calculate the inbreeding level for each individual in the population. Such data are invaluable in planning a breeding program for populations which are completely under the control of the fish culturist. However, there are many situations, involving both hatchery and naturally reproducing populations, where pedigree information is either not available or impossible to obtain. In these cases, we can assess level of inbreeding and its effect on genetic variability in terms of population size, that is, the number of breeding individuals each generation. In order to develop the relationship between population size, inbreeding, and the concept of random genetic drift, we must define a more generalized reference population. A suitable reference population is one which is infinitely large and randomly mating with no selection, migration, or mutation occurring.

The high degree of specificity of such assumptions in no way detracts from the practical value of the approach; it is merely the simplest, most predictable reference to use in assessing expected changes. The infinitely large, randomly mating population will have five important characteristics: every male will have the same probability of mating with every female, regardless of genotype; for every mating, the number of offspring surviving to breed will follow the same probability distribution; offspring surviving will represent a random sample of the genetic possibilities expected for the mating; offspring will be equally viable, regardless of genotype; and there will be no selection among parents or offspring, so the probability of a mating between parents of particular genotypes will be a direct function of the frequency of each genotype.

General Consideration
Of Finite Populations

All populations should be considered finite in size with some being larger than others, remembering that *population size* must be defined as the number of parents contributing genes to the next generation. It is often perceived that all populations of a given species should have about the same level of genetic variability. But when we consider that each individual possesses many, many loci, we must question the extent to which we should expect two populations to have the same genetic composition.

Consider as an example the possible number of combinations of alleles for some trait, say, body length or fecundity, controlled by n loci, and assume that each locus has two alleles. The following equalities express the *potential genetic variability:* 2^n = number of different gametes possible, 2^n = number of different homozygotes possible, 3^n = total number of different genotypes possible. The number of gametes and genotypes for various values of n are

Table 3.2 Number of different genotypes and gametes expected in a population with two alleles at each of n loci.

Loci	Genotypes	Gametes
1	3	2
5	243	32
10	59,049	1,024
20	3.5×10^9	1.0×10^6
50	7.2×10^{23}	1.1×10^{15}
100	5.2×10^{47}	1.3×10^{30}

given in Table 3.2. It soon becomes clear that even with only a moderate number of loci, say 20, the number of genotypes possible for a single trait far exceeds the number of individuals of any species in existence at any given time. Therefore, the individuals that do exist must represent a sample of the possible genotypes. It then follows for finitely small populations that since only a subsample of the possible gametes will be retained in the next generation, most populations will differ to some extent and random changes in gene and genotype frequencies will occur over time. In fact, it was an understanding of this enormous potential for genetic variability that gave rise to the phrase, "no two individuals are exactly alike."

Two separate but closely related approaches can be used to predict the effects of finite population size. The first is based on the variance in gene frequency expected from one generation to the next due to random sampling of gametes. A small population reproduces itself through a process of sampling a small number of alleles carried by gametes produced by a limited number of parents. The chance fluctuations in gene frequencies resulting from random sampling of gametes are called *random genetic drift*.

The other approach is to consider the increase in IBD homozygotes expected from randomly sampling a small number of alleles from a finite pool of genes. The rate of inbreeding will depend on the number of parents contributing genes to the pool of gametes. For example, if there are 6 parents, there can be no more than 12 different alleles at any one locus. Repeated random sampling among 12 alleles to provide 6 individuals as parents for the next generation is likely to result in some of the 6 individuals being IBD homozygotes. The effect of the process of random sampling is predictable in magnitude under certain assumptions, but the direction of the change is not predictable in any given generation.

The expectations of random change in gene frequency and the associated loss of heterozygosity outlined in the next two sections apply to closed, randomly mating populations not under the influence of selection, migration, or mutation. Mutation can be ignored because its effect over the short term will be negligible in small populations. However, non-random mating, selection, and migration can have a significant effect on both gene and genotype frequencies.

The frequency of genotypes depends on the frequency of the alleles and the manner in which mating occurs. As noted earlier, if mates tend to pair assortatively, we expect an increase in alike-in-state homozygosity. Levels of homozygosity observed for such populations would give an overestimate of the effects of inbreeding. Conversely, disassortative mating tends to maintain heterozygosity. Directional selection also has the effect of changing gene frequency and increasing homozygosity. Successful selection to change a specific phenotypic characteristic, such as may occur at a hatchery, results in an increase in the frequency of alleles favorable to the desired phenotype. Small to moderate selection pressures seem probable for all populations for which man controls at least part of the life cycle (Doyle 1983) and therefore would contribute to observed changes in gene frequency and heterozygosity during at least the early generations of artificial propagation.

Populations are seldom closed to the influence of breeding individuals from other populations. Migration among neighboring populations has the effect of restoring alleles lost through random genetic drift, since populations are as equally likely to drift apart in gene frequency as they are to become more similar. The circumstantial evidence for aquatic species suggests that migration is a general phenomenon even though there appear to be many spatial and temporal barriers limiting migration.

One exception relates to the current philosophy of managing hatchery populations of fish. Because of geographic restrictions or disease there is a general tendency for hatchery programs to restrict gene flow and maintain closed populations. Under these conditions, population size becomes a critical factor in determining the long-term genetic makeup of the populations. This problem is recognized by livestock breeders and has resulted in the adoption of breeding systems which encourage the exchange of breeding individuals among herds or flocks. It may be that hatchery programs based on closed populations will not be successful over the long term.

Sampling of Gametes

One way to visualize the process of random genetic drift is to consider a hypothetical situation in which a large population is subdivided into a large number of small subpopulations. We could imagine, as an example, using a base population, say, an infinitely large population of hatchery fish, to populate a large number of small streams that did not contain fish. Each subpopulation is then allowed to reproduce as an isolated unit. To eliminate the effects of other genetic forces, we must also apply the following idealized conditions for each subpopulation (Falconer 1981):

- Each generation, there are a constant number of breeding individuals in all subpopulations.
- There are equal numbers of males and females of a dioecious species.
- All breeding individuals are replaced each generation.
- All mating occurs at random.
- All breeders produce a random number of offspring with an average family size of two offspring per breeding pair.
- There is no migration, mutation, or selection.

Any deviation from these assumptions will change the expectations outlined below. We will consider deviations from three assumptions—constant population size, equal numbers of males and females, and random family size—in a later section on effective population size. Overlapping of generations is expected to reduce the rate of drift relative to the case of discrete, nonoverlapping generations when each breeding population is dominated by a single year-class (generation). Mating among individuals from different generations has the effect of increasing the number of contributing parents, thus reducing the likelihood of mating between relatives (Felsenstein 1971).

The sampling scheme is depicted in Fig. 3.2. The system assumes a gene frequency of p, at an arbitrary locus, for the base population. Through random sampling, k initial populations are established, each with N breeding individuals with gene frequency p_{i1}. In essence, generation 1 is represented by the genotypes formed from random sampling gametes from the base population. In subsequent generations, N breeding individuals result from random sampling gametes within each subpopulation and the random survival of genotypes maturing to produce offspring.

If the number of subpopulations (k streams) is large, then the average gene frequency over all streams will equal the gene frequency in the base population. That is, the overall gene frequency will not change and

$$\bar{p} = \Sigma p_i / k = p.$$

If, on the other hand, only one or a few subpopulations are considered, the overall or average gene frequency will depend on the chance outcome of the sampling of gametes. This is because we do not expect the gene frequencies p_{i1} to be equal in all streams. The variability expected among streams can be estimated as the binomial sampling variance, V. The variance among subpopulations in generation 1 is

$$V(p) = pq/2N,$$

since with N breeding individuals in each stream, there are $2N$ alleles and $q = (1 - p)$.

Assuming a large number of streams (subpopulations) and a gene frequency near one-half in the base population, the distribution of gene frequencies will be approximated by a normal distribution with a mean equal to p. We would then expect, for example, that only 68% of the streams would have gene frequencies within the bounds of p plus and minus one standard deviation, the square root of $V(p)$.

The first generation of sampling establishes k subpopulations with gene frequencies p_{i1}. Changes in gene frequency in the second and subsequent generations will be unique for each subpopulation. The magnitude of the change for any one subpopulation is only a function of gene frequency under the assumptions of constant population size and no migration, mutation, or selection. Thus, the sampling variance for subpopulation i in generation j is

$$V(p_{ij}) = p_{ij}q_{ij}/2N.$$

Since the sampling variance is the variance in gene frequency for all

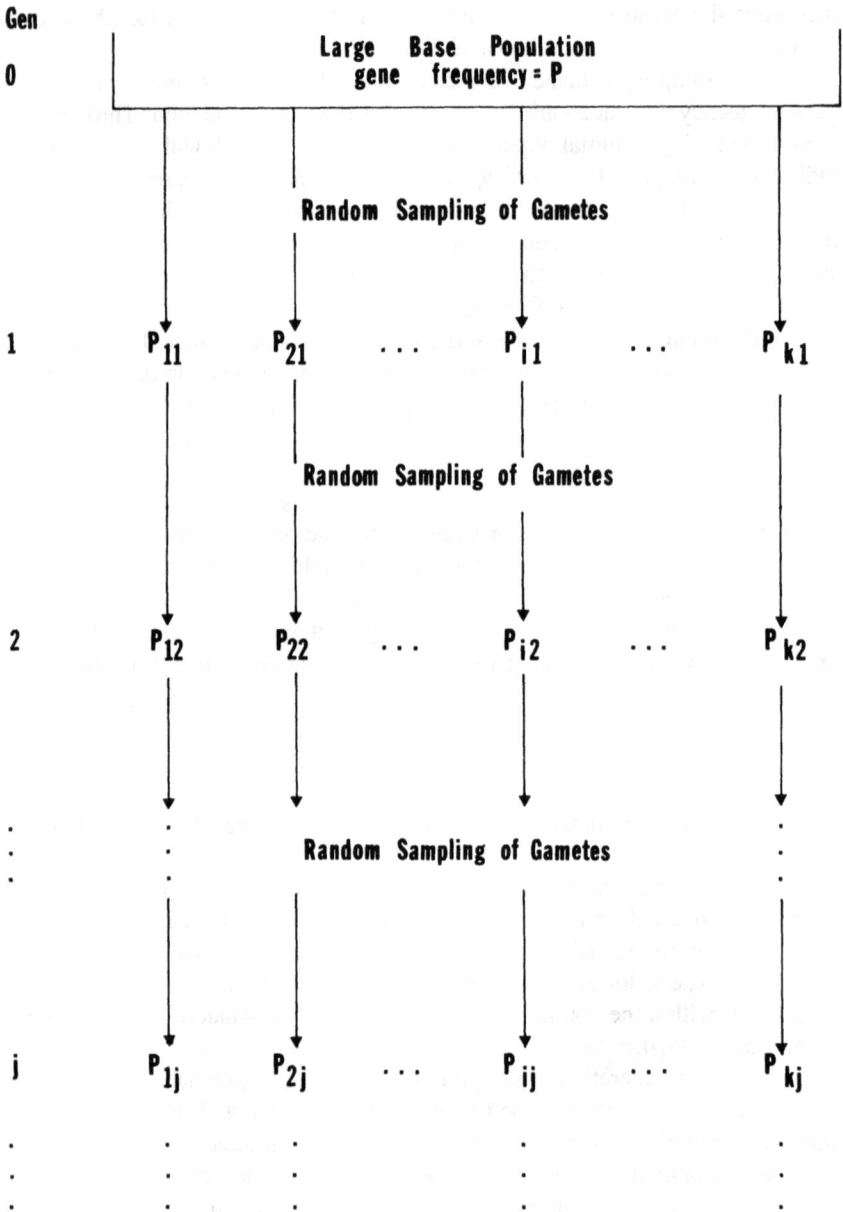

Figure 3.2 Hypothetical example of the division of a large population into K subpopulations. N_i represents population size of ith subpopulation; P_{ij} represents the gene frequency in the ith subpopulation and jth generation (Gen).

possible populations that could arise from sampling gametes, the expected variance of the change in gene frequency, $\Delta p = P_{ij+1} - P_{ij}$, in subpopulation i between generation j and $j+1$ is

$$V(\Delta p) = V(p_{ij+1}) = p_{ij}q_{ij}/2N,$$

where $V(P_{ij+1})$ is the expected variability for all possible outcomes of sampling in subpopulation i. Only one of the infinite number of possible outcomes of gamete sampling occurs in each stream, so each subpopulation may move to a new gene frequency each generation. We should note that if all subpopulations are not of the same size, for any given gene frequency those with smaller population sizes will show greater changes in gene frequency than larger ones. Regardless of the gene frequency or size of each subpopulation, if the number of subpopulations (k) remains large, the average or global gene frequency will also remain constant (see Falconer 1981, chap. 3).

The sampling variance is at a maximum at $p = 0.5$ assuming constant N. Therefore, the largest sampling effect is expected in those subpopulations with intermediate gene frequencies. As a corollary, we should expect those subpopulations with high or low gene frequencies, that is, those with low levels of heterozygosity, to change only slightly. Consequently, such populations will remain relatively homozygous and have a high probability of becoming completely homozygous simply by chance.

Using this idea of drift due to the random sampling of gametes, it is easy to see that if only one or a few subpopulations are established, the apparent effect of small population size would be to change gene frequency toward fixation for one allele or another. However, it should be noted that the change in average gene frequency would be due to the sampling of only a few gametes rather than a direct effect of inbreeding. It is also clear that the limited number of parental genotypes present in each subpopulation will result in increasing numbers of IBD alleles with increasing generations.

Inbreeding in Small Populations

The amount of inbreeding expected in one generation of random mating in a finite (small) population can be thought of as the probability of randomly sampling two gametes that contain IBD copies of an allele. To simplify the development of this concept, we will assume that self-fertilization can occur and examine the probability of IBD genotypes in a hypothetical population.

Assume a population containing 10 breeding individuals, all unrelated and equally likely to produce either one or both of the two gametes needed for zygote formation. Saying they are not related means that the 20 alleles possessed by the 10 individuals have all arisen from independent sources and none will be IBD. Therefore, the 20 alleles can be listed as independent items as

$$A_1, A_2, A_3, \cdots, A_i, \ldots, A_{20}.$$

Further assume that each individual contributes a large but equal number of gametes to a common pool and that fertilization occurs through the random union of any two gametes. The probability of identical gametes uniting then becomes a conditional sampling question: "Given that a gamete has been sampled, what is the probability of sampling a second gamete carrying the same allele as the first gamete?" The answer is quite simple. Since there are 20 alleles, the probability of sampling a particular allele is 1 out of 20. The probability of randomly sampling a second gamete that carries the same allele as the one carried by the first gamete is the probability of producing an IBD genotype. If, for example, an A_5 allele was sampled with the first gamete, the probability that the second gamete would also contain A_5 is 1 out of 20, so

$$P(A_5 A_5) = 1/20 .$$

Noting that $20 = 2N$ when $N = 10$ provides a general conclusion that the probability random sampling of gametes will result in a zygote that is an IBD homozygote is

$$P \text{ (IBD)} = P(\text{zygote is } A_i A_i) = 1/2N,$$

where N is the number of breeding individuals. The exclusion of self-fertilization changes the result only slightly so the value of $1/2N$ is a good approximation even for moderately small dioecious populations (Crow and Kimura 1970). Therefore, we will use this value as a sufficient approximation throughout the remainder of this presentation.

Since increased homozygosity through the production of IBD homozygotes represents what is referred to as inbreeding, the probability of IBD homozygotes occurring due to the random sampling of gametes must represent the expected increase in inbreeding. Therefore, the increase in inbreeding for one generation will be approximately

$$\Delta F = 1/2N.$$

It is clear then, that $1/2N$ does not represent the level of inbreeding in a population except in the first generation of selfing and as an approximation for inbreeding in the second generation for dioecious populations. Thereafter, the probability of IBD homozygotes will be increased by the probable existence of IBD alleles in the pool of gametes being sampled.

It is not obvious why random changes in gene frequency should result in inbreeding in small populations. Consider a population N of, say 10 parents, made up of an equal number of males and females. If each male mates with one female and if all 5 matings produce exactly 2 progeny, then the progeny generation (generation 1) will be made up of $N/2 = 5$ sets of full-sib families of 2 individuals each. In the next generation there will be a random chance of $1/5 =$

2/N that full sibs will mate with each other. We know that the progeny of full-sib matings have an inbreeding coefficient of $F = 1/4$. Thus, the inbreeding expected, on the average, for the progeny in generation 2 will be the product of the frequency of full-sib pairs and the inbreeding expected from a full-sib mating. Therefore, the increase in inbreeding, in terms of population size, would be

$$\Delta F = (2/N)(1/4) = 2/4N = 1/2N \ .$$

This is the same result as above for random sampling of gametes, except that the effect occurs one generation later. The one generation delay occurs because self-fertilization was not included in the example.

Let us now return to the general case. As we observed in the discussion of calculating inbreeding coefficients, after the first generation there are two sources of IBD alleles, those resulting from replication in the current generation and those replicated in previous generations.

Considering the idealized small subpopulations outlined in Fig. 3.2, we have shown that, since there are $2N$ different alleles, there is a probability of $1/2N$ that two gametes will carry identical alleles. Similarly, two gametes have a probability of $1 - (1/2N)$ of carrying different parental alleles. In the first case, the probability the alleles are IBD is one (1) in any generation. In the second case, the probability the different parental alleles are IBD is F_{t-1}, the inbreeding coefficient for the parental generation. This is true because, with random mating, the probability of any two gametes having IBD alleles is equal to the probability two alleles in the same zygote are IBD. Therefore, the inbreeding coefficient in generation t is equal to the sum

$$F_t = (1/2N) + [1 - (1/2N)]F_{t-1} \ .$$

It is possible to show that if H_0 is the frequency of heterozygotes in the base population, then the expected level of heterozygosity in generation t is

$$H_t = (1 - F_t)H_0 \ .$$

If we now replace F_t with the above equation for F_t, we find that

$$H_t = [1 - (1/2N)]H_{t-1}$$

$$= [1 - (1/2N)]^t H_0 \ .$$

This is an interesting result given the complexity and unpredictability of the random change in gene frequency. The result indicates that, on the average, heterozygosity in small populations will decrease by a fraction $1/2N$ each generation. When selfing is excluded, the approximate fraction is $1/(2N + 1)$. Thus, a population of two, one male and one female, will experience a decline in heterozygosity of about 1/5 each generation. Examination of the inbreeding values given for full-sib mating in Table 3.1 will show this result is approximately correct.

It is important to notice that heterozygotes do not decrease by a frequency of $1/2N$ each generation, but rather that a fraction $(1/2N)$ of the existing frequency (H_{t-1}) is lost. This idea can be developed further by returning to the equation for F_t. We have shown that the increment of inbreeding, that is, the new inbreeding each generation, is

$$\Delta F = 1/2N .$$

Therefore, the equation for the inbreeding coefficient in generation t can be rewritten as

$$F_t = \Delta F + (1 - \Delta F)F_{t-1}$$

when N is constant and selfing is included. Rearranging, we see

$$\Delta F = \frac{F_t - F_{t-1}}{1 - F_{t-1}} ,$$

which is not equal to $F_t - F_{t-1}$. This result demonstrates that the rate of inbreeding in any one generation is dependent on the inbreeding left to be achieved $(1 - F_{t-1})$ as well as on the population size. The equation $\Delta F = 1/2N$, therefore, measures the proportional increase in inbreeding per generation and not the absolute increase. This concept of *rate of inbreeding* is useful since it holds as a good approximation for any breeding system, not only for the idealized conditions assumed here. Falconer (1981) refers to ΔF as the rate of dispersion, and F as the accumulated effect of random genetic drift. We will examine this concept further under a discussion of effective population size.

Genotype Frequencies

The distribution of genotypes in individual populations is most often the parameter of concern to fisheries managers because it typifies the notion of population differentiation. It is necessary, then, to examine how changes in gene frequency and inbreeding resulting from division of a species into finite subpopulations produce variability among populations. Based on a discussion of sampling of gametes, we showed that the variance in gene frequency can be defined in terms of gene frequency and population size as follows:

$$V(p) = V(\Delta p) = pq/2N .$$

The important point to note is that this variance, representing the expected change in gene frequency due to sampling in any one subpopulation with gene frequency p as well as the variance among subpopulations after one generation of sampling from a base (reference) population with gene frequency p, is clearly dependent on frequency of heterozygotes ($2pq$) and is inversely proportional to population size.

Using a discussion of inbreeding in small populations, we showed that rate of inbreeding is also inversely proportional to subpopulation size ($\Delta F = 1/2N$). Therefore, for one generation of sampling, the variance of change in gene frequency within a subpopulation, $V(\Delta p)$, the variance among subpopulations, $V(p)$, and rate of inbreeding, ΔF, can all be equated as follows:

$$V(\Delta p) = V(p) = pq/2N = pq\Delta F .$$

As the sampling process continues, there is an increase in both the variance among subpopulations and the proportion of IBD alleles. It can be shown that the relationship between the variance in gene frequency and level of inbreeding for the population (species) as a whole at generation t is

$$V(p_t) = \overline{pq}[1 - (1 - 1/2N)^t] = \overline{pq}F_t ,$$

where $\bar{p}$ and $\bar{q}$ refer to the average gene frequencies for a large number of subpopulations (assumed to equal the frequencies in the base population).

This expression for $V(p_t)$ illustrates the somewhat confusing relationship between the notion of population differentiation and the concept of inbreeding. Gene frequencies change in each subpopulation and become distributed around the base (reference) population gene frequency due to random sampling of gametes. Small subpopulation size produces increasing numbers of IBD alleles (inbreeding), resulting in an increased frequency of homozygotes. Since the changes in gene frequency tend toward fixation for one allele in each subpopulation, and since more and more of the remaining alleles are becoming IBD, the subpopulations differentiate by possessing different arrays of homozygous genotypes.

We can develop an appreciation for what might happen in a single population and how it may differentiate from others by examining the expected genotype frequencies for a population subdivided into finite subpopulations. Although we will not address the question of the time required for differentiation to occur, it seems intuitively clear that very small subpopulations will change rapidly while relatively large ones will show rather slow changes.

Because of the accumulation of IBD alleles the frequency of homozygotes in a subdivided population will always be greater than expected for a large, randomly mating population with gene frequency equal to the average of the subpopulations. The increase in frequency of each homozygous type in the population as a whole will equal the accumulated variance in gene frequency at generation t. The frequency of the AA homozygote, for example, will be

$$p^2 + V(p) = p^2 + pqF_t.$$

This concept is called Wahlund's principle and is depicted in Table 3.3. The expected frequencies for a large population and a subdivided population with inbreeding of F are listed for the case of two alleles at one locus. The comparison presented could equally represent the case of many loci in a single line or subpopulation.

The first observation of note is that the subdivided population is expected to have only $(1 - F)$ as many heterozygotes as one large panmictic population since F measures loss of heterozygosity. We also expect the alleles lost as heterozygotes $(2pqF)$ to be redistributed equally to the homozygous classes along with a $(1 - F)$ reduction in the AIS homozygotes. As a result, each homozygous class is increased by an amount pqF above that expected in a single large population. Eventually, as F reaches a value of 1, all subpopulations become fixed for one allele or the other, and the frequencies of subpopulations fixed for each allele would be p and q, respectively. The result obviously assumes all alleles have equal viability.

One can visualize the possible outcome for many loci in a single small population, keeping in mind that the direction of change will be random; only the magnitude of the change can be predicted. Clearly, there will be a loss of heterozygotes. However, it is impossible for a single small population to retain both alleles in the homozygous state. Rather, the random sampling of gametes from a limited number of breeding individuals will cause the gene frequency to "drift." Therefore, we expect an increase in the frequency of one homozygote and a decrease in the other. The eventual outcome in a single small population is fixation at all loci; the allele at each locus will be a random allele unless selection is acting to cause preferential survival of one allele over another.

EQUILIBRIUM BETWEEN MIGRATION AND RANDOM DRIFT

If a population is divided into subpopulations and the subpopulations are small, we expect random changes in gene frequencies which will result in divergence among some subpopulations. Migration from one subpopulation to

Table 3.3 The frequency of genotypes expected under conditions of a single large random mating population and of a large random mating but subdivided population inbred to a level F. The case assumes one locus with two alleles at frequencies of p and q, respectively, in the base population.

Genotype	Random Mating Frequencies	Frequencies with Inbreeding of F	At $F = 1$
AA	p^2	$p^2 + pqF = p^2(1 - F) + pF$	p
Aa	$2pq$	$2pq - 2pqF = 2pq(1 - F)$	0
aa	q^2	$q^2 + pqF = p^2(1 - F) + qF$	q

another will counteract this effect by reintroducing alleles that have been lost. We will examine this in a very simplified manner (see Crow and Kimura 1970, chap. 6).

Let M be the amount of exchange each generation due to migration; M is the fraction of the parent population that are immigrants and is assumed constant each generation. If we assume equal subpopulation sizes, random mating, and no migration from any subpopulation to any other, then for any given subpopulation

$$F_t = 1/2N + [1 - (1/2N)]F_{t-1}$$

as before.

The IBD alleles will be maintained only if the individuals carrying them are not replaced by immigrants. The probability that neither of two uniting gametes has come from a migrant is $(1 - M)^2$, which is simply the probability that two individuals are not migrants. (This is not exact, because our model allows selfing.) Then an equation for F_t which corrects for random exchange of migrants is

$$F_t = [1/2N + (1 - 1/2N)F_{t-1}](1 - M)^2 .$$

If we simplify by assuming an equilibrium level of inbreeding as a result of a balance between loss of alleles from supopulations due to drift and their reintroduction due to migration, then $F_t = F_{t-1} = F$ and

$$F = \frac{(1 - M)^2}{2N - (2N - 1)(1 - M)^2} .$$

When the rate of migration, M, is small, so that the M^2 term can be neglected,

$$F \simeq \frac{1 - 2M}{4NM + 1 - 2M}$$

$$\simeq \frac{1}{4NM + 1} \quad \text{(for small } M) .$$

Although the above result is an oversimplification, it provides an indication of the importance of migration as a force counteracting genetic drift. If $M \lll 1/4N$, then F becomes large and we expect a considerable increase in homozygosity within subpopulations. If $M \ggg 1/4N$, then migration swamps the effect of genetic drift in local subpopulations and the subpopulations effectively behave as one large panmictic unit. A small amount of migration will have a marked damping effect on drift. For example, a fraction $1/N$ of the population, meaning one individual out of N, is considerably larger than $1/4N$; so if there is much more than one migrant per generation, we should expect little local differentiation.

The actual case for fish populations is poorly understood although Allendorf and Phelps (1981) and Busack (1983) have attempted to address this question. It appears that in some species, such as anadromous salmonids, migration may exceed $1/N$ per generation. Consequently, we should not expect extensive differentiation, at least not among neighboring populations. Allendorf and Utter (1979) provided evidence of differentiation, and thus little migration, among geographically distant populations but a lack of differentiation among local populations of rainbow trout. These results are in contrast to those of Ryman et al. (1979), who found a lack of migration between two groups of brown trout coexisting in a small lake.

EFFECTIVE POPULATION SIZE

The notion of population size as *effective number of breeders* is a concept used to evaluate the rate of inbreeding expected for small populations when there are particular deviations from the idealized conditions (Falconer 1981, chap 4). The most common deviations of interest in fisheries are
- unequal numbers of males and females,
- nonrandom distribution of family size,
- unequal numbers of breeders in successive generations.

The most convenient way to account for these deviations from the ideal is to express the actual number of breeders in terms of an *effective number*. This is the number of breeding individuals that would give rise to the same variance in gene frequency and thus the same rate of inbreeding as would be expected if the idealized condition had held. Therefore, the effective number, N_e, represents the number of breeders that would yield a rate of inbreeding equal to $1/2N_e$. The exact expressions for the conversion of the actual number to an effective number are often complex and difficult to evaluate (Crow and Kimura 1970). However, the approximations given below are sufficiently accurate to satisfy most situations. We should note, also, that population size refers to the number of breeding individuals in a single generation. The actual number of breeders is difficult to obtain directly from a population census unless the age classes are identified, the mating structure is known, and allowances are made for overlapping generations (Hill 1972).

Unequal Numbers of Males and Females

In many populations, particularly hatchery populations, the number of males contributing to the next generation is often smaller than the number of females. The two sexes, however, contribute equally to the allelic composition of the offspring, so having a limited number of one sex increases the probability of IBD alleles. Since the sampling variance is proportional to the reciprocal of the number of breeders, the harmonic mean of the numbers of alleles provided by each sex of parent yields an efficient estimate of the contribution of each sex to the next generation. This method of averaging has the effect of de-emphasizing large numbers.

Let N_m equal the number of males and N_f equal the number of females. Then, the harmonic mean number of breeding individuals is

$$\frac{1}{N_e} = \frac{1}{2}\left(\frac{1}{2N_m} + \frac{1}{2N_f}\right) = \frac{1}{4N_m} + \frac{1}{4N_f}$$

or

$$N_e = \frac{4N_m N_f}{N_m + N_f} .$$

The last formula shows, alternatively, that N_e is equal to twice the harmonic mean number of each sex. It is twice the harmonic mean because population size is males plus females. The one generation increase in inbreeding is

$$\Delta F = \frac{1}{2N_e} = \frac{1}{8N_m} + \frac{1}{8N_f} .$$

If the number of males equals the number of females, or $N/2$, then the effective number is the actual number. That is,

$$N_e = \frac{4(N/2)(N/2)}{2N/2} = \frac{4N^2/4}{N} = N .$$

The above equations demonstrate that effective population size (N_e) is determined primarily by the number of the less numerous sex. If, for example, a successful spawning was achieved with a single female and the eggs were fertilized with gametes from a large number of males, the effective number would be equal to only about 4, and the rate of inbreeding would be equal to that of a population of 2 males and 2 females.

In other situations, very few males are used relative to the number of females needed for reproduction. As an example, assume a closed population is maintained using 2 males and 50 females as the breeding population each generation. Then

$$\frac{1}{N_e} = \frac{1}{8} + \frac{1}{200} = 0.13$$

or

$$N_e = \frac{1}{0.13} = 7.7$$

and

$$\Delta F = \frac{1}{2(7.7)} = 0.0650 .$$

In terms of the increase in inbreeding, this population of 52 individuals is equivalent to a population of about 8, made up of 4 males and 4 females. Since the limiting sex will have the major impact on rate of inbreeding, an approximation of the effective number can be obtained if the number of one sex is known and the numerous sex is ignored. For the example above, ignoring number of females, which is equivalent to assuming an infinite number of females, and basing the calculation on 2 males gives the expected rate of inbreeding as

$$\frac{1}{N_e} = \frac{1}{4N_m} = \frac{1}{4(2)} = \frac{1}{8}$$

$$\Delta F = 1/16 = 0.0625$$

This result is close to the estimate of 0.0650 obtained above. Using a minimum of 4 males and a large number of females would produce an approximate inbreeding rate of 1/32, or 0.03125.

Non-Random Family Size

The concept of family size is probably the most important from the standpoint of understanding the effect of some parents leaving a large number of progeny while others leave only a few. The family size of interest is the number from each family in one generation that contribute progeny to the next generation of parents. We will first examine the distribution of family sizes expected under the ideal conditions outlined earlier and then consider two deviations from the idealized situation: increased and reduced variance. We will consider only the case of constant population size and initially assume equal numbers of males and females.

For the idealized populations, each parent has an equal opportunity to contribute gametes to the next generation, the contribution of progeny by individual parents is random but families vary in size, survival of progeny to breeding age is random, and all parents are replaced each generation. Since each parent must be replaced if population size is to remain constant, the mean number of progeny per parent must be one (1), yielding a mean family size of 2 per pair of parents. Although the potential number of progeny per pair of parents is large, an assumption of random survival of progeny with a low probability of survival of individual zygotes to breeding age is sufficient to meet the condition of random contribution of parents. By *random survival* of progeny we mean random chance of gametes forming a particular zygote and equal viability of zygotes regardless of genotype.

This determination of random family size follows the binomial distribution. However, if population size, N, is not extremely small and with low probability of survival, the distribution of family sizes is well approximated by the Poisson distribution. This distribution has the convenient characteristic that the variance is equal to the mean. Thus, for our case of random family size, the distribution will have a mean and variance of 2. The proportion of parents expected to contribute a given number of progeny is then as follows:

Number of Progeny	Proportion of Parents
0	0.135
1	0.271
2	0.271
3	0.180
4	0.090
5	0.036
6	0.012
7	0.003
8 or more	less than 0.001

The above distribution clearly shows that the assumption of parents having equal probability of contributing to the next generation does not imply that all parents will leave equal numbers of progeny.

The formulas given previously for rate of inbreeding and variance in gene frequencies were based on populations with family sizes following this distribution. However, when there is a large variance in family size, as is expected for most real populations, the rate of inbreeding and change in gene frequency will be greater than expected. It has been shown that, under assumptions of constant population size and random mating, the effective number, N_e, as a function of variance in family size can be approximated as follows:

$$N_e \simeq \frac{4N}{2 + V_k} \quad ,$$

where N is the number of parents and V_k is the variance in family size (Crow and Kimura 1970). The effective population size equals the actual size, N, when $V_k = 2$.

The relationship given above for effect of family size on effective population number holds approximately for monogamous, single pair matings when the variance in family size will be equal for the two sexes. If males mate with more than one female it is likely that the number of progeny and thus variance in family size will not be the same for the two sexes. Hill (1979) has shown that when the variance in family size differs for the two sexes, the effective population size can be approximated as

$$N_e = \frac{8N}{4 + V_{km} + V_{kf}} \quad ,$$

where V_{km} and V_{kf} are the variances in family size for males and females, respectively.

Variation in family size greater than that expected for the idealized population is the most important cause of the effective number being less than the actual number. It probably also gives rise to large departures from the ideal conditions in most fish populations. For example, large differences in parental contributions occur when selection is practiced in hatchery populations. Members of the same family tend to be similar in their performance, such as date of spawning, rate of hatch, or body size, so larger numbers are often selected from some families at the expense of other families. Large variability in family size under natural conditions is frequently due to chance predation. During early life, members of families tend to be localized, so a high proportion of the family is likely to be lost if predation occurs. For example, if a predator discovers a salmon spawn (a redd), it is likely most of the eggs or young will be destroyed; if they are not detected, there is a higher probability some will survive to breeding age.

Now consider an example with variance in family size $V_k = 6$ for both sexes. This is a typical value for hatchery-reared rainbow trout spawned as single pairs (Gall, unpublished data). Further assume a population size of 50 made up of 25 males and 25 females. Alternatively, this could be a population of about 18 males and 40 females if the variance in family size were equal in the two sexes. Then, the effective number of parents is

$$N_e = \frac{4(50)}{2+6} = \frac{200}{8} = 25 .$$

In terms of rate of inbreeding this actual population of 25 males and 25 females is equivalent to a population of about 12 males and 12 females producing random family sizes. The ratio of effective to actual population size is $N_e/N = 25/50 = 0.50$. Falconer (1981) reviewed the limited data available for natural populations and found ratios for N_e/N ranging from 0.49 to 0.94.

Different Numbers of Parents In Successive Generations

An approximation of the effective population size when the actual number of parents is not constant from generation to generation can be obtained from the harmonic mean number over all generations. That is, average effective number per generation is given by

$$\frac{1}{\overline{N}_e} = \frac{1}{t}\left[\frac{1}{N_1} + \frac{1}{N_2} + \ldots + \frac{1}{N_i} + \ldots + \frac{1}{N_t}\right] ,$$

where t is the number of generations and N_i is the effective number in the ith

generation, that is, the number corrected as necessary for unequal numbers of each sex and variance in family size.

Since the harmonic mean is a function of the reciprocal of the numbers each generation, small values of N_i will make the largest contribution to the average. Thus, having a large population during some generations will not eliminate the effect of small numbers in other generations. For example, if a population was reproduced for four generations with effective numbers of 10, 10, 50, and 10 parents, then the average effective number per generation is

$$\frac{1}{\overline{N_e}} = \frac{1}{4}\left[\frac{1}{10} + \frac{1}{10} + \frac{1}{50} + \frac{1}{10}\right] = \frac{1}{4}(0.32) = 0.08$$

and

$$\overline{N_e} \simeq 13 \ .$$

Thus, the rate of inbreeding expected over the four generations would be the same as would occur if 13 parents had been used in all generations. The effect of small numbers in some generations can be examined by considering the ideas of new and old inbreeding referred to earlier. Expanding the number in the population does not eliminate inbreeding from previous generations; it only reduces the amount of new inbreeding accumulated. Having a very small number of parents in one or a few generations is what is meant by the term *bottleneck* in population genetics.

Minimizing Inbreeding

There are situations in which it is desirable to keep rate of inbreeding as low as possible. The preservation of native stocks through hatchery propagation of captured wild breeders and maintenance of populations completely through hatchery management are examples. Increasing the number of breeders is not the only control available to the manager; available facilities and resources also limit the number of breeding individuals that can be maintained or procured.

If propagation is carried out using single pair matings, the number of each sex will be equal, thus maximizing effective population size. Another powerful tool is to reduce the variance in family size through careful choice of individuals to be parents in relation to family origin. If an equal number of parents are taken from each family, the variance in family size will be zero, and the resulting effective population size will be essentially twice the actual number. A more exact expression is $N_e = 2N - 1$ (Crow and Kimura 1970).

In the above expression the population is produced using equal numbers of males and females, and the equal family sizes are obtained by choosing two parents from each family. If the number of males and females is not equal, the variance in family size can be made zero by choosing one male parent from each sire's family and one female from each dam's family. Gowe et al. (1959), in their excellent discussion of controlling inbreeding, showed that the rate of inbreeding will then be approximately

$$\Delta F = \frac{3}{32N_m} + \frac{1}{32N_f} ,$$

where N_m and N_f are the actual numbers of males and females, respectively, and the number of females is greater than the number of males.

Avoiding matings among relatives, such as among sibs or cousins, is often considered a simple way of reducing rate of inbreeding. However, this procedure has little effect over the long term although it does reduce the initial rate. Falconer (1981) shows that whereas the increment of new inbreeding with completely random mating is $\Delta F = 1/2N$, the increment when both self-fertilization and full-sib matings are excluded is $\Delta F = 1/(2N + 4)$. The major advantage of avoiding matings among close relatives, as is often done for short-term research populations, is to make the rate of inbreeding more constant from generation to generation and the inbreeding coefficient similar for all individuals within generations.

Overlapping of generations (Hill 1979) is also expected to increase rate of inbreeding relative to equally large populations replaced each generation. However, if populations can be maintained as separate year classes for short periods, and each year-class maintained with numbers nearly equal to those which occur with mixed cohorts, some crossing between year classes at regular intervals, say every two or three generations, will restore genetic variability for the population as a whole. The effect of this crossing is equivalent to that for migration.

It should be obvious that all the effective methods of reducing rate of inbreeding require knowledge of the family structure of the population. At the present time such information is not generally available due to limitations in methods of identification or lack of concern by fisheries managers. It would appear that one of the most critical needs in fishery management is effort in the development of effective methods of fish identification and an awareness of its importance. Current technology is adequate for hatchery populations but is seldom utilized.

INBREEDING DEPRESSION

The increase in homozygosity resulting from inbreeding is generally accompanied by a loss of general vigor and fertility. In natural populations of outbreeding species, this reduction in average performance is closely connected with fitness. In hatchery populations, there is generally reduced performance for many of the important quantitative traits, particularly those associated with reproduction and physiological efficiency. The observed reduction in performance of inbred populations is referred to as *inbreeding depression*. Original levels of performance can often be restored by outcrossing inbred populations.

This opposite, or complementary, genetic phenomenon is called hybrid vigor and will be discussed briefly at the end of this section.

Observed levels of inbreeding depression from different experiments are summarized in Table 3.4. The values have been standardized to a base of 10% inbreeding assuming a linear relationship between F and phenotypic performance. Most of the characteristics studied relate to hatchery performance, an expected situation given the difficulty involved in data collection for natural populations. The variability in estimates from different experiments probably results from the diverse background and genetic history of the stocks studied,

Table 3.4 Typical levels of inbreeding depression estimated as the change per 10% breeding. A plus symbol (+) indicates an improvement in performance with inbreeding.

Trait (data source)	Depression	
	(Units)	(%)
Rainbow trout (Kincaid 1976, 1983)		
Egg mass at 2 years	7.2 g	7.1
Eggs hatched —F embryo	0.0 %	0.0
—F female	0.7 %	1.2
—F male	+ 6.7 %	+11.2
Abnormal fry	2.8 %	19.4
Survival to 147 days	4.8 %	6.7
Weight 147 days	121.5 mg	4.0
Weight 364 days	12.9 g	8.4
Weight at 2 years	31.5 g	4.7
Recovery to fishery (in 12 months)	2.3 %	5.3
Biomass recovered (per 100 g planted)	288.0 g	8.1
Rainbow trout (Gjerde et al. 1983)		
Egg mortality	——	2.5
Fry mortality	——	3.2
Fingerling growth	——	3.0
Saltwater growth to 18 months	——	5.1
Rainbow trout (Aulstad and Kittlesen 1971)		
Eggs hatch	10.6 %	13.7
Abnormal fry (per 1000)	114.0	11.1
Fry survival	12.0 %	12.2
Atlantic salmon (Ryman 1970)		
Recapture frequency	1.7 %	17.4
Brook trout (Cooper 1961)		
Weight at 7 months	——	11.1
Weight at 19 months	——	13.8
American oyster (Mallet and Haley 1983)		
Larval shell length, 12 days	+ 1.5 μm	0.9
Larval density, 12 days (per ml)	+ 0.3	+ 9.8
Spat surface area, 18 months	31.8 mm²	3.8
Spat weight, 18 months	0.2 g	2.8
Spat surface area 30 months	6.8 mm²	0.5
Spat weight, 30 months	0.1 g	0.9

but different reproductive strategies will affect experimental results as well. Some experiments have reported improved performance from inbreeding; these results may require further study.

Individual populations can differ greatly in the amount of change observed from inbreeding because separate subpopulations differ in gene frequency due to random drift. The past history, particularly selection in hatchery populations, also affects performance. As we will see, reducing the frequency of recessive deleterious alleles through selection has the effect of reducing the frequency of less desirable recessive homozygotes resulting from inbreeding. Therefore, to say that a trait exhibits inbreeding depression is to make a generalization about the average change in mean performance for a number of inbred lines from diverse sources.

Population Mean Under Inbreeding

Any change in average performance of a population can be attributed to changes in the frequency of genotypes, since inbreeding does not change gene frequency in the population as a whole. As we have seen, inbreeding does result in an increase in homozygotes and a decrease in heterozygotes. Therefore, changes in the performance of a population must be associated with differences in the genetic values of heterozygotes and homozygotes. What follows is a theoretical model used to compare the mean performance of random bred and inbred populations in terms of genotypic frequencies and genetic values. The model is outlined in Table 3.5.

Consider a population with average gene frequencies of p and $q = (1 - p)$ for two alleles, A_1 and A_2, at a single locus. We also assign arbitrary values to the genotypes in the following way. The two homozygotes are assigned values of $+a$ and $-a$ so that the homozygotes are an equal distance from their

Table 3.5 Theoretical model for the mean of a population in terms of genotypic values and level of inbreeding for a single locus with two alleles.

Model of genotypic values

A_2A_2		A_1A_2	A_1A_1
$-a$	0	d	$+a$

Mean for random mating population

Genotype	A_1A_2	A_1A_2	A_2A_2
Frequencies	p^2	$2pq$	q^2
Values	$+a$	d	$-a$

$$M_0 = a(p - q) + 2pqd$$

Mean for population with inbreeding, F

Genotype	A_1A_1	A_1A_2	A_2A_2
Frequency	$p^2 + pqF$	$2pq(1 - F)$	$q^2 + pqF$
Values	$+a$	d	$-a$

$$M_F = a(p - q) + 2pqd - 2pqF$$

average value, which we will define as zero (Table 3.5). The assigned values may be regarded as both genotypic and phenotypic values since they represent the average value of all individuals of that genotype. The heterozygote is assigned a value of d, representing the performance of the heterozygote relative to the average performance of the two homozygotes. The degree of dominance can be expressed as d/a. If $d = 0$, there is no dominance. If $d = a$, there is complete dominance. The population mean, calculated from the frequencies and values of the genotypes, represents the deviation of the mean from zero, the average of the two homozygotes.

The frequency of the genotypes and their values for a large random mating population are as listed in Table 3.5. The population mean is obtained by multiplying the frequency of each genotype by its value and adding the resulting three terms. Therefore, the mean for a large randomly mating population is

$$M_0 = p^2a + 2pqd - q^2a \ .$$

Noting that $(p^2 - q^2) = (p + q)(p - q) = (p - q)$,

$$M_0 = a(p - q) + 2pqd \ .$$

We see that the contribution of any one locus to the mean is made up of two terms: $a(p - q)$, attributable to the value of homozygotes, and $2pqd$, arising from the frequency of heterozygotes and dominance. If a population is fixed for the A_1 allele, then the population mean is a, the value of the A_1A_1 homozygote. If the two alleles are equally frequent ($p = 0.5$), then the average contribution of the homozygotes to the mean is zero.

The contribution of heterozygotes to the mean is of particular interest. If there is no dominance ($d = 0$), the heterozygotes have no effect on the mean because their value lies midway between the two homozygotes; under the case of no dominance, the mean is determined by the frequency of homozygotyes. On the other hand, if there is dominance ($d = 0$), the mean becomes dependent on the frequency of heterozygotes.

Subdivision of a population into small, finite subpopulations results in a decrease in the frequency of heterozygotes and proportional increases in each of the homozygotes. The genotype frequencies expected with an inbreeding coefficient or generation level of F are listed in Table 3.5. The mean for an inbred population, considering all subpopulations, is

$$M_F = (p^2 + pqF)a + 2pq(1 - F)d + (q^2 + pqF)(-a)$$

$$= a(p - q) + 2pqd - 2pqdF$$

$$= M_0 - 2pqdF \ .$$

The change in the mean due to inbreeding, called *inbreeding depression,* is clearly a function of the frequency of heterozygotes and level of dominance. If dominance is absent ($d = 0$) the mean is not expected to change with inbreeding. Lack of dominance is the probable reason some characteristics do not show inbreeding depression. With dominance, the change in the mean relative to a large random mating population is directly proportional to level of inbreeding. The magnitude of the reduction in the mean ($-2pqdF$) is a function of the original frequency of heterozygotes and the degree of dominance. It is also clear that the greatest reduction in the mean will occur for populations with intermediate gene frequencies since the term $2pq$ has its greatest value when $p = q = 0.5$.

Most phenotypic characteristics of importance are influenced by alleles at more than one locus. If the effects of the loci act completely additively, then inbreeding depression is expected to be a function of the sum of the effect, $-2pqd$, at each locus and the population mean is

$$M_F = M_0 - 2F\Sigma p_i q_i d_i ,$$

where summation is over all loci affecting the characteristic. We should note at this time that crosses between inbred lines that differ markedly in gene frequency will result in a high frequency of heterozygotes and high performance if dominance is important.

If effects of loci do not act completely additively but express epistatic interactions between loci, then an expression for the change in the mean is complex and involves higher order terms (Crow and Kimura 1970). The expectation is that the reduction in the mean will not follow a straight line when plotted against F. There is little evidence from animal populations for nonlinear rates of inbreeding depression; however, its detection is complicated by the effects of selection present in most populations studied.

Some general conclusions can be drawn concerning the circumstances leading to inbreeding depression. The change in mean is predominantly an effect of dominance, and the change is expected to be in the direction of the recessive alleles. The magnitude of the change will be dependent on initial gene frequency, with populations initially at intermediate frequencies showing the greatest inbreeding depression. The possibility does exist for the direction of dominance to be opposite at different loci for a given character. Under these circumstances the sum of effects over all loci may cancel each other, with the result that there is no apparent change in the mean. Thus, the lack of observed inbreeding depression is not a sufficient basis for assuming the absence of dominance.

Effect of Selection

It is not realistic to ignore the presence of selection during the process of inbreeding. Inbreeding theory, as outlined above, assumes equal survival of all genotypes so that all alleles are retained in the population; the gene frequency averaged over all subpopulations (e.g., inbred lines) is assumed to remain constant (see Pirchner 1969, chap. 13).

Since inbreeding often results in reduced fitness, natural selection is likely to oppose equal survival of both homozygotes in favor of either the most heterozygous individuals or those with the least deleterious homozygous genotypes. This fact also poses difficulties in assessing the effects of inbreeding in experimental populations. As inbreeding proceeds and reproductive capacity declines, some families or sublines are unavoidably lost. The survivors are then a select group which will not be representative of the genotype frequencies expected from the theory. Consequently, some genotypes (alleles) will be lost due to their relatively deleterious effects. Other, more favorable, alleles may be lost simply because they were present in sublines or individuals lost due to the effects of deleterious alleles.

The effect of selection among lines or subpopulations on the process of random genetic drift is also an important consideration. The principal effect of random genetic drift is to produce differences in allele frequencies (called genetic divergence) between subpopulations. Level of divergence for extant populations is often estimated with the so-called F-statistics (see Hartl 1980), since random genetic drift theory predicts that subdivided populations will be more homozygous than large random breeding populations. If selection is also acting to eliminate some subpopulations, due to poor reproductive performance, for example, the array of genotypes among surviving subpopulations will show less variability than expected from only random forces.

A general guideline for assessing the importance of selection versus random genetic drift as the cause of genetic divergence can be obtained from a consideration of subpopulation size (Crow and Kimura 1970). If the selection coefficient associated with the least desirable homozygote exceeds 1 over 4 times the effective population size ($s = 1/4N_e$), selection becomes a stronger force than random drift in determining change in gene frequencies. For example, if the effective number is 50, then the critical value for the selection coefficient is 0.005. This is equivalent to a fitness of the least favorable homozygote of 99.5% of the most favorable homozygote.

Unfortunately, little is known about the magnitude of selection coefficients in natural populations. It seems that random genetic drift is the major force determining gene frequency in very small subpopulations. However, selection could play a significant role in moderate sized populations, say an effective size above 50 or 100, because selection coefficients would not need to be very large to oppose random forces. In nature, genetic divergence is a function of migration and mutation as well as selection and random drift. Thus, data intended to measure divergence must be interpreted carefully with regard to the forces which may have produced any observed divergence (Allendorf and Phelps 1981, Busack 1983). Measures of genetic distance presented in later chapters of this volume have been designed with these complications in mind.

There is one other complicating factor regarding rate of inbreeding and the likely distribution of gene frequencies in the presence of selection. The coefficient of inbreeding, determined either from pedigrees or effective population size, measures the state of population differentiation only in the absence of selection, migration, and mutation. However, inbreeding depression is always a function of the actual genotype frequencies. Thus, with selection or migration, the coefficient of inbreeding will overestimate the actual loss of heterozygosity and the assumed random increases in particular homozygotes. The greatest discrepancy between the actual state of differentiation and that expected from F occurs when the rate of inbreeding is slow due to the accumulated effects of selection over time. Thus, experimental evaluations of inbreeding should always use rapid rates of inbreeding, whereas development of inbred lines can be enhanced by slow rates of inbreeding.

The effects of selection on the expected outcome of inbreeding and population subdivision present a number of difficulties in interpreting observed results. Three will be discussed briefly here: estimating inbreeding depression, genotype frequencies expected due to random genetic drift, and crossing inbred lines to restore performance (hybrid vigor).

To study the effects of inbreeding, a number of inbred lines are initiated from a base population, usually through full brother-sister mating. During the early stages of the program, each brother-sister mating produces a new line, so the number of sublines soon becomes very large if more than one mating is made from each full-sib family. Practical considerations of space and resources generally make it impossible to maintain large numbers of lines, and so some fixed number is usually established with the first generation of full-sib mating. Most animal experiments are restricted to 10 or 20 initial lines. The other usual procedure is to maintain the initial number of lines by replacing lines lost due to poor performance with offspring from duplicate matings from surviving full-sib lines.

What array of genotypes should we expect for the population of inbred lines after a number of generations of inbreeding? We should not expect the gene frequencies for the inbred lines to equal that of the original base population. Therefore, the proportion of each type of homozygote present among the inbred lines will not be that expected from the original gene frequency. The types of homozygotes that are present will be determined by two forces. Initial sampling of breeding pairs (referred to as the founder effect) will determine the alleles present and their frequencies. Identical-by-descent homozygotes will be derived from these alleles. Selection, acting through loss of lines, will eliminate some of these alleles. If lost lines are replaced from surviving lines, as is the usual procedure, the lost alleles will be replaced by alleles with a high probability of being identical by descent to surviving alleles, because relatives from the surviving lines are used to replace lost alleles. Consequently, we should expect a higher degree of genetic similarity among surviving inbred lines than theory would predict. If only a few of the original lines survive, the genetic composition of lines as a group will represent only a sample of the original alleles.

The performance of outcrossed, or "hybrid," populations derived from surviving inbred lines depends on the genetic composition of the inbred lines. If the inbred lines available for crossing possess similar arrays of alleles, the performance of the hybrid will be similar to the average performance of the inbred lines. On the other hand, if the inbred lines possess quite different arrays of alleles, high heterozygosity will be realized, with the performance of the hybrid exceeding the average performance of the parental inbred lines if the loci affecting the traits in question are characterized by significant dominance effects ($d \neq 0$). If the lines possess a representative sample of all the alleles of the original noninbred population, average performance of the hybrid population, after one generation of random mating, will equal that of the original base population.

The final possibility is that the lines crossed happen to be homozygous, or nearly so, for alternate alleles at many loci. Under these conditions—and only under these conditions—the average performance of the hybrid population may exceed the performance of original base population. The latter outcome occurs because the hybrids will be heterozygous at all loci for which different inbred lines were fixed for different alleles.

The unpredictability of the outcome of crossing inbred lines is primarily a result of the random nature of the inbreeding process: lines become fixed for a random allele at each locus. In addition, inbreeding depression usually results in the loss of many lines during the development of the inbred lines. Although depression in the performance responsible for the loss of a line may be due to the effects of only a few loci, the alleles carried at all loci are lost when the line is lost. Consequently, the restoration of original performance or levels of performance above the original level can be expected only if large numbers of inbred lines are produced and the surviving lines carry a substantial portion of the original genetic variability. Even then, only crosses among a few of the lines are expected to produce exceptional levels of performance (see Pirchner 1969, chap. 14).

4.
Genetic Variation Within A Subdivided Population
Ranajit Chakraborty and Olof Leimar

As early studies on managed species of salmonids will attest (e.g., Alm 1949, Svärdson 1945), fisheries biologists have long been concerned with intraspecific genetic differentiation. During the last two decades, large amounts of genotype and allele frequency data have been obtained from a large number of species, including many species of fish, primarily through the means of protein electrophoresis. These studies have shown that most species are subdivided into more or less distinct units that differ genetically from each other. The existence of such structure is a matter of major importance both for management and for the conservation of genetic resources. An appreciation of the forces leading to appearance or disappearance of substructure and the time scales over which these phenomena take place is needed for interpreting data and estimating the impact of man's activities on existing population structure.

In this chapter we present some basic elements of population genetics as applied to subdivided populations. The main emphasis will be on methods and results that are important for the analysis and interpretation of allele frequency data. Considerable discussion will be devoted to measures that are used to quantify the amount and distribution of genetic variation in a subdivided population and to the statistical problems encountered in attempts to infer such measures from data.

At present, the most commonly used procedure for obtaining genetic data, especially for natural populations, is protein electrophoresis. With this method, large numbers of individuals can be analyzed for many protein loci; thus, the frequencies of alleles at these loci can be estimated reliably. Having reliable estimates of the frequencies of detected allelic variants is important since, as far as is known, genetic differences between populations at the intraspecific level is due in large part to frequency differences of alleles existing in many or most populations, rather than to alleles unique to different populations. In recent years, other methods have come into use, such as DNA sequencing and the determination of restriction sites, making it possible to study genetic variation directly at the DNA level (see Chapters 11 and 12).

Population surveys on genetic variation based on these techniques are still relatively scanty. For this reason, our presentation focuses on electrophoretic data, but the concepts discussed here are also applicable to data gathered by restriction enzyme and DNA sequencing techniques (Nei and Li 1979, Kaplan and Langley 1979, Nei and Tajima 1981).

BASIC CONCEPTS

To gain an overview of the concepts to be dealt with, consider a total population consisting of either an entire species or some segment of a species. If the average genetic constitution of individuals from different parts of this total population varies, it is said to be genetically structured. A convenient way to visualize the structure is to think of the total population as consisting of a number of subpopulations. A subpopulation is a group of individuals that constitute a reproductive unit, e.g., a randomly mating group of individuals (*deme* and *population* are also often used to denote a reproductive unit). An important property of a subpopulation is that further subdivision should not reveal additional genetic differences between subgroups. It should be kept in mind that this concept of subpopulations can correspond more or less well with biological reality. In some cases, such as freshwater organisms isolated from conspecifics in other lakes or rivers, it may be possible to divide the population unambiguously into discrete units, but in other cases (e.g., a marine species) the subunits may be less clearly defined.

In practice, the most common procedure used to study population structure is to collect a number of samples from different localities. Data collected in this way give information about the structure, but it is worth pointing out that the picture obtained can be an oversimplification of the real situation. Extensive sampling over the species range is required if one wants a good picture of the existing structure.

Genetic differences between subpopulations will evolve in the course of time if there is little or no gene flow between them. The amount of gene flow needed to prevent differentiation depends on the strength of the evolutionary forces responsible for differentiation among subunits. These differentiating forces are selection, genetic drift, and mutation. Selection pressure from the varying conditions under which the individuals of different subunits live results in adaptation to local conditions. The differences in allele frequencies and in the average values of quantitative characters accompanying such local adaptations constitute the genetic variation among subpopulations that is of primary importance from the point of view of management and conservation.

Due to the finite size of subpopulations, allele frequencies at polymorphic loci tend to drift randomly. If the selective differences between alleles are sufficiently small, this drift will be the main factor causing genetic differentiation among subpopulations. For protein loci studied by electrophoresis, the question of the relative contribution of drift and selection to observed allele frequency variation is not yet resolved (see e.g., Koehn et al. 1983, Nei 1983). However, when interpreting allele frequency data, an assumption of no selection is a useful starting point. If selection dominates, a different explanation for the allele frequency variation is required for each polymorphic locus, whereas drift will affect all loci in the same way.

In many cases an attempt to postulate various selective influences would be highly speculative. As discussed by Allendorf and Phelps (1981b), a good strategy is then to try to interpret electrophoretic data without invoking selection so long as no information to the contrary is available. Regardless of whether observed differences are due to selection, the information obtained about population structure is valuable. If subpopulations have been sufficiently isolated to allow allele frequencies to drift apart, then local adaptation is to be expected as well. An exception to this may be when the differentiation observed is due to very small subpopulations being isolated for a short time only. In such a case, the variation is not likely to reflect any major biological adaptations.

The rate at which mutation creates new alleles at a locus is quite low, at least when one considers alleles that are not strongly detrimental. The time required before there is a substantial probability of observing new alleles at appreciable frequency at a locus can often be longer than either the time of existence of a subpopulation as a unit or the time in which gene flow will transport alleles over most or all of the species' range. The picture obtained from electrophoretic surveys is that alleles are generally shared between widely separated subpopulations of a species. Thus, it would seem that mutation is not a major force in intraspecific differentiation. On the other hand, some parts of a species may have been isolated from each other for a considerable time and can have unique alleles at least for a few loci. While these loci may not be among those studied in an electrophoretic survey, new mutations can still be an important factor in allowing adaptation to local conditions. Note in this connection that the presence of unique alleles may also result from loss of alternative alleles by drift or selection after isolation of populations.

QUANTITATIVE MEASURES OF DIFFERENTIATION

Some kind of quantification of the amount of differentiation in a structured population is desirable. As a basis for comparison, we take a large, randomly mating population and consider a locus at which there is no selection. If the locus has two segregating alleles, A and a with respective frequencies p and q, the expected proportions of the genotypes AA, Aa, and aa will be the Hardy-Weinberg proportions p^2, $2pq$, and q^2. If, instead, the population is subdivided and allele frequencies vary among subpopulations, then, as was first pointed out by Wahlund (1928), the proportion of homozygotes in the total population will be greater than the Hardy-Weinberg expectations. Let p_i be the frequency of A in subpopulation i ($i = 1, \ldots, n$), and assume that in each subpopulation genotypes occur in the Hardy-Weinberg proportions given by the local allele frequencies. The frequency of AA in subpopulation i is then p_i^2, and if subpopulations have equal size the proportion of AA in the entire population is

$$\frac{\Sigma p_i^2}{n} = \bar{p}^2 + V_p , \qquad (1)$$

where $\bar{p} = \Sigma p_i/n$ is the average frequency of A in the total population and V_p is the variance of p_i over subpopulations. Thus, V_p is equal to the excess of homozygotes over the Hardy-Weinberg expectation.

Wright (1921) introduced a parameter F, called the fixation index, to characterize the genotypic distribution at a two-allele locus. For our case, the proportions of AA, Aa, and aa in the total population can be written, respectively, as $\bar{p}^2 + F\bar{p}\bar{q}$, $2\bar{p}\bar{q}(1-F)$, and $\bar{q}^2 + F\bar{p}\bar{q}$; and comparing with (1) we see that $F = V_p/\bar{p}\bar{q}$. Wright (1943, 1951, 1965) developed this concept further by introducing the three F-statistics, F_{IS}, F_{ST}, and F_{IT}, whereby the overall deviation from Hardy-Weinberg proportions can be split up into deviation caused by subpopulation differentiation and deviation from local Hardy-Weinberg proportions.

For subpopulation differentiation, the relevant F-statistic is F_{ST}, which was defined by Wright as the correlation between random gametes, drawn from the same subpopulation, relative to the total. If the total is taken to be the currently existing population, this definition leads to F_{ST}, $= V_p/\bar{p}\bar{q}$, which for local Hardy-Weinberg proportions is the fixation index defined above. Note, however, that F_{ST} is given by $V_p/\bar{p}\bar{q}$ regardless of whether subpopulations are in Hardy-Weinberg proportions or not. A convenient formula for F_{ST} can be obtained if we introduce $H_S = \Sigma H_i/n$, where $H_i = 1 - (p_i^2 + q_i^2)$ is the Hardy-Weinberg expectation of heterozygosity in subpopulation i, and $H_T = 1 - (\bar{p}^2 + \bar{q}^2)$. A simple calculation yields (Nei 1973a)

$$F_{ST} = \frac{V_p}{\bar{p}\bar{q}} = 1 - \frac{H_S}{H_T} . \qquad (2)$$

In some cases, F_{ST} defined according to (2) is an estimate of the average inbreeding of individuals in the population relative to some founder population. This holds, for example, if a large, randomly mating population is split up into many isolated subpopulations, and with no selection operating at the locus in question (cf. Chapter 3).

In practice, there are a number of difficulties with this interpretation. An existing population structure is often the result of a complex and unknown history, with gene flow and subpopulation sizes varying over time. Furthermore, selection may have influenced the allele frequencies, or the population average allele frequencies may have drifted from their original values. For reasons such as these, it will in general not be possible to estimate inbreeding from current allele frequencies. Regardless of this, F_{ST} is a useful measure of the degree of differentiation among subpopulations at a two-allele locus.

For a locus with more than two alleles, F_{ST} can be defined for each allele by combining the frequencies of all other alleles at the locus. The value of F_{ST} will in general differ among alleles, but at a two-allele locus F_{ST} will be the same for both alleles.

Gene Diversity Analysis

Nei (1973a, 1977) extended F_{ST} by providing a measure of differentiation, called G_{ST}, based on allele frequencies at several multiallelic loci. For a given locus, let p_{ik} be the frequency of allele k in subpopulation i and define H_i and H_S as follows:

$$H_i = 1 - \Sigma_k p_{ik}^2 \, ,$$

$$H_S = \frac{\Sigma_i H_i}{n} \, . \tag{3}$$

As in the the case with two alleles, H_i is the Hardy-Weinberg expectation of heterozygosity in subpopulation i. H_i can also be interpreted as the probability of nonidentity of two independently sampled alleles from subpopulation i, and H_S is then the average probability of nonidentity of two alleles sampled from the same subpopulation. H_T is defined as the Hardy-Weinberg expectation of heterozygosity obtained with population average allele frequencies:

$$\bar{p}_k = \frac{\Sigma_i p_{ik}}{n} \, ,$$

$$H_T = 1 - \Sigma_k \bar{p}_k^2 \, . \tag{4}$$

H_T can be interpreted as the probability of nonidentity of two alleles sampled from the total population. With more than one locus, $\bar{H}_S$ and $\bar{H}_T$ are defined as averages of H_S and H_T over loci. G_{ST} is then defined as

$$G_{ST} = 1 - \frac{\bar{H}_S}{\bar{H}_T} \, . \tag{5}$$

For a single locus with two alleles, this becomes the same as (2). Nei introduced the term gene diversity for a quantity such as $\bar{H}_S$ or $\bar{H}_T$. $\bar{H}_i$, the average of H_i over loci, is also often referred to as the average heterozygosity of subpopulation i, and $\bar{H}_S$ is then called the mean average heterozygosity of the total population.

G_{ST} defined in (5) is closely connected to F_{ST}. As mentioned above, F_{ST} can be computed for each allele. Let F_{STkl} be the value of F_{ST} for allele k at locus l, and $\bar{p}_{kl}$ the frequency of this allele averaged over the total population. Wright (1978) defined $\bar{F}_{ST}$ as the weighted average of F_{STkl} over all alleles, with weights proportional $\bar{p}_{kl} (1 - \bar{p}_{kl})$. This $\bar{F}_{ST}$ is identical to G_{ST} in (5) (see Wright

1978, chap. 3). For convenience, we will use the notation G_{ST} regardless of the number of alleles and loci considered.

Some comments are required concerning the averaging performed in the computation of H_S and H_T. Subpopulations might vary in size, and we could compute H_S as $\Sigma w_i H_i$, where the weights w_i are proportional to the sizes. Similarly, to get H_T we need allele frequencies averaged over the entire population. Apart from the fact that subpopulation sizes are very difficult to estimate in practice, they may also vary considerably over time. There is thus no strong theoretical reason to prefer these weights. Due to this, Nei (1977) recommended that equal weights be given to subpopulations ($w_i = 1/n$). We have followed this recommendation in the definitions (3) and (4) above. The average over loci could be taken either over all loci or over some class, such as electrophoretically detectable loci.

Hierarchical Structure

A population may be structured into groups larger than subpopulations, with individuals within the same group tending to be more similar than individuals from different groups. This kind of grouping can be continued with successively larger groups, and in this way a hierarchical structure with several levels is formed. For instance, local populations of Atlantic salmon could be grouped into drainages, and then the drainages could be grouped into major geographical regions. The case which considers only division into subpopulations can be viewed as a hierarchy with one level, namely subpopulations within the total. For this case, H_T can be partitioned (Nei 1973a) as $H_T = H_S + D_{ST}$, where $D_{ST} = H_T - H_S$ is the component of gene diversity due to variation among subpopulations within the total (for notational convenience we drop the bars on H_S and H_T). Then $G_{ST} = D_{ST}/H_T$ can be viewed as the proportion of gene diversity due to variation between subpopulations and H_S/H_T as the proportion due to variation within subpopulations. A natural extension of this to cases with more levels in the hierarchy has been suggested by Nei (1973a). Considering one more level, referred to as groups of subpopulations, H_T can be partitioned as

$$H_T = H_S + D_{SG} + D_{GT} , \qquad (6)$$

where $D_{SG} = H_G - H_S$ and $D_{GT} = H_T - H_G$. H_G is the average gene diversity of groups, defined as the average over groups of H_{Gj}, where H_{Gj} is the gene diversity of group j. In other words, H_{Gj} is the Hardy-Weinberg expectation of heterozygosity obtained from the average allele frequencies in group j. For consistency with the assumption of equal weights to subpopulations, the average over groups should be made with weights proportional to the number of subpopulations in the group. (See also Chakraborty 1980 and Chakraborty et al. 1982 for more details on the computational procedures.)

Dividing with H_T on both sides of (6) we get

$$1 = \frac{H_S}{H_T} + G_{SG(T)} + G_{GT} \, , \qquad (7)$$

and $G_{ST} = G_{SG(T)} + G_{GT}$ has been split into two components due to variation between subpopulations within groups and between groups within the total. H_S, D_{SG}, and D_{GT} are called *absolute* gene diversity components; and H_S/H_T, $G_{SG(T)}$, and G_{GT} are called *relative* gene diversity components. Note that we have defined $G_{SG(T)}$ as $D_{SG}/H_T = (H_G - H_S)/H_T$. Nei (1973a) used this notation to emphasize that H_T is used to normalize D_{SG}, and he reserved the notation G_{SG} for the quantity $D_{SG}/H_G = 1 - H_S/H_G$. The gene diversity components D_{SG} and D_{GT} can never become negative; and if there is differentiation, D_{GT} will usually be larger than zero even if the grouping chosen does not correspond to biological reality. Thus some care is needed when interpreting these components.

Equations (6) and (7) describe population structure by an additive partitioning of H_T into components. An alternative approach is to write H_S/H_T as the product of H_S/H_G and H_G/H_T, or equivalently: $(1 - G_{ST}) = (1 - G_{SG}) (1 - G_{GT})$ (see Nei 1973a and Wright 1978, chap. 3).

Components of Variance

Equations (3)–(5) above define H_S, H_T, and G_{ST} as functions of current allele frequencies in subpopulations. A somewhat different approach has been taken by Cockerham (1969, 1973). In Cockerham's approach, the current allele frequencies are regarded as random variables, and a statistical model for these random variables is postulated. Parameters in the statistical model are then estimated from the allele frequencies observed in samples from subpopulations. The method of analysis is similar to a model II (random effects) analysis of variance. The parameters in the model are called components of variance and are related to gene diversity components, such as H_S and H_T. Cockerham (1969, 1973) considered a single two-allele locus; Weir and Cockerham (1984) generalized the method to deal with many loci and alleles.

In practice, it matters little whether one describes population structure with a statistical model or with some functions of allele frequencies. The reason is that usually the only information that is available about a statistical parameter characterizing a structured population is the current genetic constitution of the population. Any attempt to estimate such a parameter can then be regarded as an attempt to estimate some function of the current allele or genotypic frequencies.

To clarify the connection between Nei's gene diversity analysis and Cockerham's components of variance, we apply this argument to the components of variances that correspond to H_S and H_T. We begin by expressing H_T in (4) in a different way. Define H_{ij} and $H_{SS'}$ by

$$H_{ij} = 1 - \Sigma_k p_{ik} p_{jk} ,$$

$$H_{SS'} = \frac{\Sigma_{i \neq j} H_{ij}}{n(n-1)} . \tag{8}$$

H_{ij} is the probability of nonidentity of an allele from subpopulation i with one from subpopulation j, and $H_{SS'}$ is the average of H_{ij} over all pairs of different subpopulations. Expanding the square of $\bar{p}_k$ in (4) now gives

$$H_T = \frac{H_S}{n} + \frac{H_{SS'}(n-1)}{n} , \tag{9}$$

where n, as before, is the number of subpopulations. For large n, H_T is approximately equal to $H_{SS'}$. Weir and Cockerham (1984) present variance component estimators, denoted by a_k, b_k, and c_k, for each allele (labeled by k) at a multiallelic locus. These estimators are then summed over alleles. The following expectations hold:

$$E\Sigma_k(b_k + c_k) = H_S ,$$

$$E\Sigma_k(a_k + b_k + c_k) = H_{SS'} . \tag{10}$$

The expectations in (10) are computed for given allele frequencies in the subpopulations. These relations are most simply obtained from the estimators of Weir and Cockerham (1984) by considering the case in which large samples of equal size are taken from all subpopulations. For varying sample sizes and not all subpopulations sampled, (10) assumes that sample sizes and subpopulations are chosen independently of the allele frequencies in the subpopulations.

With more than one locus, Weir and Cockerham (1984) recommend that variance component estimators be averaged over loci, which is the same procedure as is used with gene diversity components. In analogy with the definition (5) for $G_{ST'}$, we can define a measure θ as

$$\theta = 1 - \frac{H_S}{H_{SS'}} . \tag{11}$$

When there are many subpopulations (large n in Eq. 9), θ and G_{ST} will be close to each other. Weir and Cockerham (1984) consider a parameter, called coancestry, that is defined as a ratio of expected components of variance. The estimation procedure for coancestry that they consider can be regarded as an attempt to estimate θ in (11).

Gene diversity components, being functions of the current allele frequencies, require no assumption about absence of selection or equality of subpopulation sizes for their definition. Weir and Cockerham motivate their estimation procedure by considering a large randomly mating population that is split into isolated subpopulations of equal sizes, and with no selection operating at the loci in question. In this case, θ in (11) will estimate the probability of identity by descent of two alleles drawn from the same subpopulation (coancestry).

This discussion has dealt with the theoretical definition of gene diversity components as compared with variance components. Further comments on the statistical properties of the estimation procedure for variance components recommended by Weir and Cockerham (1984) will be given in the section on estimation below.

Other Measures of Diversity

Many ways of quantifying variation among subpopulations have been suggested in the literature, and we give a few examples. Lewontin (1972) presented an approach based on the Shannon information index, and a similar method was used by Smouse and Ward (1978); Mitton (1977) used an analysis based on Hedrick's (1971) measure of genotypic identity; Latter (1980) utilized the proportion of shared genes between genotypes; Avise and Felley (1979) performed an analysis of variance on arcsin transformed allele frequencies. A general discussion of measures of diversity can be found in Rao (1982a,b). Without a precise statement of what constitutes a good measure, it is impossible to evaluate different methods. We choose to present Nei's gene diversity analysis because of its close connection with the concept of heterozygosity, which gives it a certain intuitive appeal. Many of the suggested methods will result in similar conclusions when applied to the same data (Latter 1980, Rao 1982b, Ryman et al. 1983).

Partitioning genetic variation into components is not the only way to analyze data. If the population structure is due to splitting of subpopulations which then remain isolated, it is natural to attempt to estimate an evolutionary tree, either from pairwise genetic distances or with some other method (see Chapter 8). However, regardless of the true history of the population, a genetic distance matrix and a dendrogram computed from the matrix will provide some information about population structure. One can also use multivariate statistical methods, such as principal component analysis, to obtain a picture of the structure (see Chapter 5 for an example). To provide the investigator with a comprehensive view of the population structure observed, the simultaneous use of several methods of analysis can well be recommended.

ESTIMATION OF GENE DIVERSITY COMPONENTS

The genotypic data obtained in an electrophoretic survey can be used to estimate the various gene diversity components defined above. The gene diversity components are functions of the true allele frequencies in the subpopulations. In practice, one will rarely be able to observe the genotypes of all individuals in all subpopulations, and this introduces a certain amount of statistical uncertainty. One can distinguish two types of statistical errors in an estimate of a quantity such as G_{ST}. First, the expected value of the estimate might differ from the true value of the quantity, and in that case the estimate is said to be biased. The bias (B) is defined as the difference between the expected value of the estimate and the true value. Second, the estimate will vary randomly around its expected value, and the magnitude of this variation can be described by the standard deviation (SD) of the estimate. The standard error of the estimate is then defined as $SE = (B^2 + SD^2)^{1/2}$.

It is important to keep in mind how the true value of the quantity that one attempts to estimate is defined and what the cause of the statistical uncertainty is. For instance, the true value of G_{ST} could be defined from the allele frequencies at the particular loci scored in a survey and in the particular subpopulations sampled. In that case, the only cause of statistical uncertainty is that not all individuals in the subpopulations have been analyzed.

However, in many cases one wants to estimate quantities that characterize not only the particular loci scored and the particular subpopulations sampled but also the entire species (or some segment of the species) and perhaps a larger class of loci. Whenever an interspecific comparison is made of the magnitudes of absolute and relative gene diversity components, one assumes that species parameters are compared (see Ryman 1983 for such a comparison in salmonid species). Other factors causing statistical uncertainty must then be taken into account in addition to the sampling of individuals from subpopulations. The particular loci scored could be regarded as a sample from all loci that in principle could be scored electrophoretically. Also, the particular subpopulations investigated could be regarded as a sample from all existing subpopulations. The true values of gene diversity components are then defined from the allele frequencies at all electrophoretic loci and in all subpopulations of the species. In the following we use notation such as H_S, H_T, and G_{ST} for the true values of gene diversity components and $\hat{H}_S$, $\hat{H}_T$, and $\hat{G}_{ST}$ for their estimates. The loci and subpopulations to which H_S, H_T, and G_{ST} apply depend on the context.

Sampling of Individuals

Consider first the estimation of H_S, H_T, and G_{ST} for the loci and subpopulations that have been studied in a survey. The simplest and most commonly used estimation procedure is to use the observed allele frequencies to compute estimates $\hat{H}_S(S)$ and $\hat{H}_T(S)$ (the label (S) stands for simple) by replacing the true allele frequencies with the observed frequencies in the expressions (3) and (4) for H_S and H_T. An estimate of G_{ST} is then obtained as $\hat{G}_{ST}(S) = 1 - \hat{H}_S(S)/\hat{H}_T(S)$. These estimates will not be unbiased, nor will they have the smallest possible standard deviation.

Nei and Chesser (1983) derived unbiased estimators of H_S and H_T. Consider a single locus and K samples with sizes N_i, $i = 1, ..., K$. Let $\hat{H}_{0i}$ be the observed proportion of heterozygotes in sample i, and $\hat{H}_0$ the unweighted average of $\hat{H}_{0i}$ over samples. Nei and Chesser's estimators are then

$$\hat{H}_S(NC) = [\hat{H}_S(S) - \hat{H}_0/2\tilde{N}]\tilde{N}/(\tilde{N} - 1) ,$$

$$\hat{H}_T(NC) = \hat{H}_T(S) + [\hat{H}_S(NC) - \hat{H}_0/2]/\tilde{N}K , \tag{12}$$

where $\tilde{N}$ is the harmonic mean of the N_i, i.e. $\tilde{N} = K/\Sigma(1/N_i)$. When sample sizes vary, they must be chosen independently of the allele frequencies in the sampled populations for the estimates to be unbiased. With more loci, the estimates given in (12) are averaged over loci.

Unbiased estimates of H_S and H_T can also be obtained from Weir and Cockerham's (1984) variance component estimators. The first relation in (10) shows how to estimate H_S and the second how to estimate $H_{SS'}$. With (9) (putting $n = K$ for the case considered here) an estimate of H_T can then be obtained. The formulae given by Weir and Cockerham (1984) for variance component estimators $(a_k, b_k,$ and $c_k)$ are rather complex and will not be reproduced here.

With equal sample sizes (all N_i equal), Nei and Chesser's estimates and those derived from Weir and Cockerham's variance component estimates are the same. When samples sizes vary, they differ in the respect that Nei and Chesser weight all samples equally, whereas Weir and Cockerham use weights that are proportional to sample size. Ideally, the weights should be chosen so that the variances of the estimates are minimized (note here that giving weights to samples depending on sample sizes in order to reduce the variances of the estimates of H_S and H_T is a distinct matter from the weights given to subpopulations in the definition of H_S and H_T). The best weighting scheme is not known. Cockerham (1973) pointed out that the best method will depend on the amount of differentiation (G_{ST}). For small G_{ST}, it is advantageous to weight with sample size, but for large G_{ST} it is better to use equal weights (see the simulation results presented below).

Estimating G_{ST} using unbiased estimates of H_S and H_T [e.g., $\hat{G}_{ST}(NC)$ $= 1 - \hat{H}_S(NC)/\hat{H}_T(NC)$] will remove some of the bias incurred in the simplest procedure (i.e., when observed allele frequencies are used as true frequencies). Since most G_{ST} and H_S values given in the literature have been computed without corrections for bias, we give a short discussion of the magnitude of the error that is introduced.

Amount of Bias in Sample Estimates

Both $\hat{H}_S(S)$ and $\hat{H}_T(S)$ will tend to be underestimates. With many samples, the bias in $\hat{H}_T(S)$ will be much smaller than the bias in $\hat{H}_S(S)$, and thus $\hat{G}_{ST}(S) = 1 - \hat{H}_S(S)/\hat{H}_T(S)$ will overestimate G_{ST}. The bias in $\hat{H}_S(S)$ is approximately $H_S/2\tilde{N}$ (obtained by taking the expectation in (12) and putting $H_S = H_0$). Ignoring the error in $\hat{H}_T(S)$ and noting that H_S/H_T is usually rather close to one, we get $1/2\tilde{N}$ as a rough approximation of the bias in $\hat{G}_{ST}(S)$. With 50 individuals in each sample, $\hat{G}_{ST}(S)$ will thus tend to overestimate G_{ST} by about 0.01.

In view of the uncertainties due to other factors, such as the choice of subpopulations and loci, we consider that an error of this magnitude may well be neglected. If sample sizes are as small as 10, the bias in $\hat{G}_{ST}(S)$ will be about 0.05 and in this case the use of unbiased estimates of H_S and H_T will result in a substantial improvement. When considering the importance of a bias correction, the magnitude of the standard deviation must be taken into account. If the standard deviation is large compared to the bias, there is little reason to correct for the bias. Nei and Roychoudhury (1974a) have computed the sampling variance of average heterozygosity. Chakraborty (1974) gave an approximate formula for the standard deviation of $\hat{G}_{ST}(S)$. Increasing the number of loci will reduce the standard deviation but not the bias, and corrections for bias become more important. If one wants more detailed information on the error in an estimate of H_S, H_T or G_{ST} that is due to small sample sizes, one possibility is to simulate multinomial sampling, using computer generated random numbers and the observed allele or genotype frequencies as parameters in the multinomial distribution.

Simulation Study

In order to compare the performance of the different estimation procedures that we have discussed, we have made such a simulation. The assumed true allele frequencies and sample sizes are given in Table 4.1. Two cases are presented: one with moderate amount of differentiation among subpopulations and rather small sample sizes, and one with strong differentiation and larger sample sizes. For the genotypic distributions in the subpopulations, local Hardy-Weinberg proportions and independence between loci are assumed. The results of the simulation for the two cases are presented in Table 4.2. Before each of the 1,000 sampling replicates, the samples in Table 4.1 have been reordered in a random fashion to assure statistical independence between sample sizes and allele frequencies.

Three ways of estimating H_S and H_T are compared in Table 4.2:
• the simple procedure (S) where observed allele frequencies are entered in expressions (3) and (4);
• the estimates of Nei and Chesser (NC) given in (12);
• and estimates derived from the variance component estimators given by Weir and Cockerham (WC).

In all three cases, G_{ST} has been computed from the estimates of H_S and H_T as $\hat{G}_{ST} = 1 - \hat{H}_S/\hat{H}_T$. For case 1, all three procedures do equally well (as judged from the standard error) with single locus data. With five loci, the bias in $\hat{G}_{ST}(S)$ leads to a somewhat greater standard error for this estimate than the others. For case 2, (S) and (NC) do equally well and are somewhat better than (WC). As mentioned, using weights proportional to sample size for averages over samples will be advantageous when G_{ST} is small. From additional simula-

Table 4.1 Allele frequencies and sample sizes used for simulation study of sampling properties of estimators of gene diversity components.

Subpopulation	Sample size	Locus				
		1	2	3	4	5
Case 1. Moderate differentiation						
1	11	0.574	0.955	0.603	0.818	1.000
2	60	0.342	0.975	0.658	0.833	0.922
3	74	0.535	0.926	0.688	0.743	0.878
4	58	0.644	0.960	0.770	0.853	0.842
5	18	0.667	0.917	0.471	1.000	0.707
6	22	0.699	0.909	0.707	0.932	0.905
7	15	0.742	0.900	0.577	0.733	0.683
8	13	0.723	0.923	0.785	0.923	1.000
9	17	0.657	0.941	0.767	0.941	0.939
Case 2. Strong differentiation						
1	51	1.000	1.000	0.610	0.804	
2	76	0.513	0.953	1.000	0.559	
3	100	0.654	0.420	0.945	1.000	
4	106	0.084	1.000	0.000	1.000	
5	48	0.902	1.000	0.906	1.000	
6	51	1.000	0.990	0.500	0.980	

Note: The allele frequencies and sample sizes (number of individuals) are from two surveys of Swedish brown trout populations.

Case 1: Observed frequencies of the common allele at five loci in nine samples from a restricted area. Reported in Chakraborty et al. (1982). Harmonic average sample size: 20.4.

Case 2: Observed frequencies of the common allele at four loci in six samples from different river systems. Reported in Ryman (1983). Harmonic average sample size: 64.8.

tions, we have found that (WC) does somewhat better than (NC) for very small G_{ST} ($G_{ST} < 0.01$).

Both Nei and Chesser (1983) and Weir and Cockerham (1984) give simplified versions of their estimators that depend only on observed allele frequencies (and not on observed proportions of heterozygotes). These estimators require local Hardy-Weinberg proportions to be unbiased. They have the advantage that only allele frequency data are needed for their computation. From simulations we have found that the simplified estimators have nearly the same standard deviation as the more general ones.

The difference in performance between the different estimation procedures is not very great. Overall, the estimators given by Nei and Chesser appear slightly better than the others. Preferably, the general version (12), requiring heterozygote frequencies, should be used. These comments apply also to estimation of θ in (11).

Table 4.2 Simulation study showing average (AV), standard deviation (SD), and standard error (SE) of estimates of H_S, H_T, and G_{ST}.

Estimator	Case 1				Case 2			
	TV	AV	SD	SE	TV	AV	SD	SE
Locus 1 only								
$\hat{H}_S(S)$	0.444	0.433	0.017	0.020	0.214	0.213	0.011	0.011
$\hat{H}_S(NC)$	.444	.444	.017	.017	.214	.215	.011	.011
$\hat{H}_S(WC)$	.444	.443	.016	.016	.214	.215	.031	.031
$\hat{H}_T(S)$	.471	.470	.012	.012	.426	.426	.009	.009
$\hat{H}_T(NC)$	.471	.471	.012	.012	.426	.426	.009	.009
$\hat{H}_T(WC)$	.471	.470	.017	.017	.426	.426	.039	.039
$\hat{G}_{ST}(S)$	.058	.079	.027	.034	.498	.500	.024	.024
$\hat{G}_{ST}(NC)$	.058	.058	.027	.027	.498	.496	.024	.025
$\hat{G}_{ST}(WC)$	.058	.057	.032	.032	.498	.493	.066	.066
All loci								
$\hat{H}_S(S)$	.281	.274	.009	.011	.166	.164	.004	.005
$\hat{H}_S(NC)$	.281	.281	.009	.009	.166	.165	.004	.004
$\hat{H}_S(WC)$	.281	.281	.015	.015	.166	.165	.013	.013
$\hat{H}_T(S)$	.298	.297	.009	.009	.315	.314	.006	.006
$\hat{H}_T(NC)$	.298	.298	.009	.009	.315	.315	.006	.006
$\hat{H}_T(WC)$	.298	.298	.016	.016	.315	.314	.018	.018
$\hat{G}_{ST}(S)$	.058	.078	.013	.024	.474	.478	.013	.014
$\hat{G}_{ST}(NC)$	.058	.058	.013	.013	.474	.474	.013	.013
$\hat{G}_{ST}(WC)$	0.058	0.058	0.014	0.014	0.474	0.474	0.045	0.045

Note: The first column (TV) contains the true value of the estimated quantity. The simulation has been performed with allele frequencies and sample sizes from the two cases in Table 4.1, assuming two alleles at each locus and local Hardy-Weinberg proportions. Number of replicates: 1000. For each replicate, the sample sizes from Table 4.1 have been reordered in a random fashion.

Sampling of Loci

The sampling associated with choice of loci is statistically less well defined than the sampling of individuals from subpopulations, but will nevertheless lead to random variation in one's estimates of H_S, H_T, and G_{ST}. If the loci studied can be considered to be randomly drawn from a large set of loci, then approximate standard deviations for gene diversity components due to variation among loci can be computed with the method given by Chakraborty (1974). Another possibility is to use the jackknife method (see Miller 1974 for a review). It should be kept in mind that unless a very large number of loci have been analyzed, say 100 or more, the standard deviations due to variation among loci in estimates of gene diversity components will be quite substantial and will usually contribute more to uncertainty in estimates than the sampling of individuals from subpopulations.

As an example, case 1 of the simulation study presented in Tables 4.1

and 4.2 was based on data from Chakraborty et al. (1982). With five loci, the standard error of $\hat{H}_S(S)$ due to sampling of individuals was 0.011 (see Table 4.2), as compared to the six times larger standard error of 0.064 that Chakraborty et al. (1982) estimated from variation among loci.

Sampling of Subpopulations

If gene diversity components are to be considered as species parameters, then the statistical uncertainty that is due to choice of subpopulations should be taken into account. Assume that the sampled subpopulations can be considered as a random selection from a large number *(n)* of existing subpopulations. Let H_S, H_T, and G_{ST} be computed from the allele frequencies in all subpopulations and $\hat{H}_S$, $\hat{H}_T$, and $\hat{G}_{ST}$ from the allele frequencies in the sampled subpopulations. (For the sake of the argument we assume here that large samples of individuals are taken, so that the allele frequencies are known.) Since $\hat{H}_S$ is the average of H_i for the selected subpopulations, $\hat{H}_S$ will be an unbiased estimate of H_S. $\hat{H}_T$ will, however, tend to underestimate H_T. From (9) we obtain that $H_T = H_{SS'}$ (*n* is assumed very large), and from (11) we get $G_{ST} = \theta$. Now, $\hat{H}_{SS'}$ computed from the allele frequencies in the sampled subpopulations will be an unbiased estimate of $H_{SS'}$, so that $\hat{H}_{SS'}$ should be used to estimate H_T. If K subpopulations are sampled, (9) gives

$$\hat{H}_T = \hat{H}_S/K + \hat{H}_{SS'}(K - 1)/K = \hat{H}_{SS'} - (\hat{H}_{SS'} - \hat{H}_S)/K .$$

Taking the expectation (with respect to choice of subpopulations) of this expression yields the expected value of $\hat{H}_T$ as equal to $H_T - (H_T - H_S)/K = H_T - D_{ST}/K$. Thus, $\hat{H}_T$ will tend to underestimate H_T by an amount D_{ST}/K. It follows that $\hat{G}_{ST} = 1 - \hat{H}_S/\hat{H}_T$ will tend to underestimate G_{ST}. A better estimate of G_{ST} (compare Eq. 11) will then be

$$\hat{\theta} = 1 - \frac{\hat{H}_S}{\hat{H}_{SS'}} .$$

We see here an advantage of using θ in (11) as a measure differentiation, namely that $\hat{H}_S$ and $\hat{H}_{SS'}$, will be unbiased estimates of H_S and $H_{SS'}$ regardless of the total number of subpopulations *(n)*.

For a single two-allele locus, an approximation for the standard deviation of an estimate of G_{ST} based on allele frequencies from a limited number of subpopulations can be obtained if one assumes that allele frequencies have a normal distribution among subpopulations (Lewontin and Krakauer 1973). For empirical purposes, what one wants to know is the statistical uncertainty caused by all sampling processes. Probably the best method is to use variation among loci to get an overall estimate of the statistical uncertainty in estimates of gene diversity components, i.e., to use the method given by Chakraborty (1974) or to jackknife over loci. Variation in estimates of H_S and H_T among loci will be due partly to variation in the true values of H_S and H_T among loci, but the statistical effects of small samples of individuals and subpopulations will also cause variation of the estimates among loci.

In practice, an assumption that the localities represent a random set of subpopulations is often unjustified, and this introduces further uncertainty and bias. For instance, if the total population is divided into major groups and if most samples available are from one of these groups, then H_T and G_{ST} will be underestimated. This type of statistical error will not be incorporated into a standard error obtained from variation of estimates among loci. See the discussion below for ways of coping with this bias.

Hierarchical Structure

The presentation above has been confined to estimation of H_S, H_T, and G_{ST}. If H_T is partitioned into additional components corresponding to a hierarchical population structure (see equations (6) and (7)), then statistical errors in these components should also be considered. Removing the bias that is due to the sampling of small numbers of individuals is straightforward. The formula (12) for $\hat{H}_T(NC)$ can be applied to each group of subpopulations separately, and an unbiased estimate $\hat{H}_G(NC)$ of H_G is then obtained by averaging over groups.

The statistical errors associated with the sampling of a limited number of subpopulations from various levels in the hierarchy present a more difficult problem. Let us consider two contrasting points of view regarding what one attempts to estimate. First, if one regards the total population as consisting of the sampled subpopulations only, then there will be no statistical errors due to choice of subpopulations. This point of view is implicit in most hierarchical gene diversity analyses that have been reported in the literature.

Second, one may regard the total population as consisting of a large number of subunits at each level of the hierarchy. For instance, with two levels in the hierarchy, the total population would consist of a large number of groups, and each group would contain a large number of subpopulations. In some cases, these assumptions may be valid for the currently existing population. In other cases, one will be dealing with a conceptual total population. This second point of view corresponds to the variance component analysis recommended by Weir and Cockerham (1984). One can also consider intermediates between these two approaches. Wright (1978) suggested that the sampled subunits should be regarded as a random selection from a large number of subunits at all levels of a hierarchy except for the most major subdivision.

To illustrate the quantitative effects of different assumptions regarding the relation between the samples taken and the structure of the total population, we give some different estimates of gene diversity components computed from the data on Atlantic salmon presented in the study by Ståhl (Chapter 5), where allele frequencies in 29 samples from natural populations are reported. From the dendrogram given by Ståhl, it appears reasonable to divide the samples into two major geographic groups: a European group and an American group. Most of the samples (24) were from the European group. For simplicity, we will not consider any further subdivision of the samples.

In the three estimation procedures below, different assumptions are made as to what the "true" structure of the total population is. Note that none of these assumptions is likely to be strictly correct for the existing total population of Atlantic salmon. In all procedures, the gene diversity components that we attempt to estimate are defined by giving equal weight to each to each of the subpopulations assumed to exist in each particular case.

First, if the total population is regarded as consisting of the sampled subpopulations only, one obtains the following estimates: $\hat{H}_S = 0.026$, $\hat{D}_{SG} = 0.007$, $\hat{D}_{GT} = 0.008$, and $\hat{G}_{ST} = 0.36$. Second, if each of the two major groups are assumed to contain a large and, for want of further information, equal number of subpopulations, then the major groups should be given equal weights in the definition of gene diversity components. One possibility for estimation is now simply to increase the weight of the American samples, but since the European group has been sampled more extensively, it may be preferable to take this into account in the estimation by giving more weight to the diversity observed in that group. The estimate H_S will then remain the same as before, the estimate of D_{SG} will increase very slightly, and the estimate of D_{GT} will increase to $\hat{D}_{GT} = 0.013$. The estimate of G_{ST} then becomes $\hat{G}_{ST} = 0.43$. Finally, if the two major groups are regarded as a sample from a large number of potential major groups, each major group containing a large (and equal) number of subpopulations, then the estimate of D_{GT} is further increased to $\hat{D}_{GT} = 0.026$, leading to $\hat{G}_{ST} = 0.55$.

It is clear that the way the investigator regards the samples in relation to the true structure of the species may have a considerable effect on the estimates of gene diversity components. If all subunits at some hierarchical level of an existing total population have been sampled, it is purely a matter of taste to regard these as a selection from a large number of potential subunits. On the other hand, if few subunits have been sampled but many more exist, it is natural to try to take this fact into account. Thus, one must decide from case to case what procedure to use.

In practice, however, an attempt to estimate components of variance for a hierarchy with many levels will often encounter problems. The reason for this is that the hierarchy that is formed by the samples is often unbalanced and lacks replicates at some points, either because some subunits have been sampled more extensively than others or because the total population that actually exists forms an unbalanced hierarchy. As an example, we could extend the hierarchy in the Atlantic salmon example above by further subdivision of the major groups. The European group could be divided into samples from the Baltic and the remaining European samples. For the American group, all samples come from a restricted geographic region, and no similar subdivision can be made. It may well be the case that additional American samples will reveal some structuring that corresponds to the one found for the European group, but from the available data one cannot estimate how much gene diversity would be found.

Summary

This rather lengthy discussion of the statistical problems encountered in estimation of gene diversity components represents an attempt to put into perspective an area where several more or less divergent points of view have been put forward in the literature, possibly leading to confusion among empirical investigators as to which procedure to prefer. Such confusion is largely due to a general reluctance of many workers to bridge the gap between the clearcut and tidy world of theory and that of biological and empirical reality.

When attempting to characterize the amount and distribution of genetic variation in a species, by far the most important consideration is to obtain genetic information from a sufficiently wide range of subunits at different hierarchical levels. No method of analysis is universally best at describing the genetic structure of a species based on very limited data. A given estimation procedure always contains idealized assumptions concerning the true nature of the genetic structure; these assumptions may or may not be valid for a particular species. Conversely, most methods are acceptable if sufficient data are available. It is probably more important for an investigator to be aware of the limitations that are due to an incomplete sampling of the species than to use a particular method of analysis.

STATISTICAL TESTS OF HYPOTHESES

A basic biological question to which a population genetic survey can provide an answer is whether there are genetic differences between subunits of a species. Statistically this question can be approached by a test of the null hypothesis that the genotypic frequencies are the same in all subunits sampled. For reasons of statistical power, one usually tests the related null hypothesis that allele frequencies are the same in all subunits sampled (test of allele frequency homogeneity). This can be considered as a test of the null hypothesis $G_{ST} = 0$.

Estimates of gene diversity components could be used to test hypotheses relating either to the genetic structure within a species or to a comparison between species. For instance, one might want to test whether one species has a higher heterozygosity than another species or whether the differentiation among subpopulations (G_{ST}) is more pronounced in one species than in another. However, as should be clear from the previous section, the sampling distributions of gene diversity components are poorly known, so that the statistical framework will in most cases be too vague to allow tests of hypotheses. The best one can do at the present time is to obtain a rough idea of the statistical errors in estimates of gene diversity components.

Testing for Allele Frequency Homogeneity

When testing for allele frequency homogeneity, one usually does not consider any specific alternative to the null hypothesis of identical allele frequencies in all subpopulations sampled. The commonly used method is to test for independence in an $R \times C$ table (R is the number of samples; C the number of alleles at the locus in question) either with the contingency χ^2 statistic or with the log likelihood ratio statistic (see, e.g., Sokal and Rohlf 1981). The decision of whether to reject or accept the null hypothesis usually will be the same with either of these two statistics. The above procedure assumes that alleles are randomly drawn from subpopulations, which is tantamount to assuming Hardy-Weinberg proportions. It is not known what method of testing should be preferred when there are deviations from local Hardy-Weinberg proportions, but one possibility is to test for homogeneity of genotypic frequencies. Another factor that may cause nonrandomness is family structure in the sample. See Allendorf and Phelps (1981b) for a discussion of this point.

When an allele is so rare (or sample sizes so small) that only a few copies are expected in each sample, then the sampling distributions given by the null hypothesis for the contingency χ^2 statistic and the log likelihood ratio statistic do not approximate a χ^2 distribution very well. For a multiallelic locus one can then pool the frequencies of the rarest alleles and test for allele frequency homogeneity with the pooled frequencies. For a two-allele locus, Fisher's exact test for a 2×2 table (see Sokal and Rohlf 1981) can be used to test for homogeneity between two samples.

With multilocus data, the question arises as to whether the tests for homogeneity performed for each locus should be weighed together into an overall test of homogeneity. Such an overall test of homogeneity can be made on the sum of all single locus statistics (this assumes statistical independence between loci). If genetic drift is thought to be the factor causing heterogeneity, then the expectation is that all loci will show more or less than same heterogeneity, and the overall test described above can be recommended. On the other hand, if selection is the factor, the test for each locus should be considered as a separate matter.

As mentioned earlier, a test of allele frequency homogeneity can be considered as a test of the null hypothesis $G_{ST} = 0$. If the null hypothesis is rejected, one might want to explain some of the heterogeneity by a grouping of subpopulations into larger units. The log likelihood ratio statistic is well suited to such an analysis, since it can be partitioned according to the levels in a hierarchy (Lewontin 1972, Smouse and Ward 1978). For instance, assume that the total population has been divided into major groups of subpopulations. If no heterogeneity is found within the major groups, one can proceed to test for homogeneity among the groups, and the positive value of G_{ST} may be explained as the result of variation among groups. However, if heterogeneity is found within the groups, there is at present no method whereby one can test whether the grouping explains a significant amount of the heterogeneity.

Sample Sizes

There is a close connection between the contingency χ^2 statistic and the estimator $\hat{G}_{ST}(S)$ for a single locus (Workman and Niswander 1970). This connection can be used to obtain an indication of sample sizes needed to detect a certain degree of heterogeneity. With a two-allele locus and N individuals in each sample, the relation is $\chi^2 = 2NK\,\hat{G}_{ST}(S)$, where K is the number of samples. From this one can show that the condition $2N > 1/G_{ST}$ yields a fairly good approximation for the sample size needed in order to achieve appreciable power, say around 50%, with a test at the 5% level. (G_{ST} is here the value applicable to the particular locus and the particular subpopulations sampled, and K is assumed not to be very large.) Thus, with sample sizes of 50 individuals, one will usually detect a G_{ST} that is somewhat larger than 0.01. See also Allendorf and Phelps (1981b) for a discussion on sample sizes needed to detect a certain heterogeneity. Requirements for sample sizes will be approximately the same for a locus with more than two alleles. From this one would conclude that for a locus with a given value of G_{ST}, the power of a test for homogeneity will be approximately the same regardless of the number of alleles segregating and of the average allele frequencies. This ceases to be true when only a few copies of an allele are expected in a sample. As pointed out above, the sampling distribution of the statistic given by the null hypothesis does not approximate a χ^2 distribution very well in this case.

Mixed Populations

It has been assumed so far that each sample contains individuals from one subpopulation only. If genetically distinct subpopulations occur sympatrically, there might also be heterogeneity within samples. The problem then arises of how such a mixture of genetically differentiated subpopulations can be detected statistically.

We approach this problem by first considering an alternative method of testing for among sample homogeneity. As mentioned earlier, allele frequency differentiation among subpopulations will result in an excess of homozygotes over the Hardy-Weinberg expectation for the entire population. If one assumes local Hardy-Weinberg proportions, one could then consider the total genotypic distribution at a locus and test for overall Hardy-Weinberg proportions. For a two-allele locus, there is a relation between the χ^2 statistic commonly used for such a test and the fixation index F (Li and Horvitz 1953). If F is estimated as $\hat{F} = 1 - \hat{H}_0/\hat{H}_T(S)$, where $\hat{H}_0$ is the observed proportion of heterozygotes, then, assuming N individuals in each of K subsamples, $\chi^2 = NK\,\hat{F}^2$ (if sample sizes vary, NK should be replaced by N_T, the total sample size). $\hat{F}$ tends to underestimate F slightly, and the χ^2 has one degree of freedom, so that an indication of the sample size needed to detect a certain F at the 5% level is given by

$NK > 4/F^2$. With local Hardy-Weinberg proportions, F will be equal to G_{ST}. Unless K is much greater than N, this test will be much less powerful than the contingency χ^2 test and cannot be recommended.

Consider now a case in which a single sample consists of individuals from more than one subpopulation. Analogously with the situation of the total sample discussed above, the sample size needed to have an appreciable chance of detecting such a heterogeneity is given by $N > 4/F^2$, where F is the fixation index of the mixture of subpopulations. With sample sizes of 50 individuals, F needs to be about 0.30, which is much greater than the between sample heterogeneity (G_{ST}) that can be detected with sample sizes of this magnitude. For more details on tests of Hardy-Weinberg proportions see Ward and Sing (1970) and Emigh (1980).

With data for more than one locus, heterogeneity within a sample could also be detected with a test for statistical independence of genotypes at different loci. The statistic commonly used in such a test is an estimate of a parameter D, the linkage disequilibrium, computed for a pair of loci (Hill 1974). In a mixture of subpopulations, D is expected to differ from zero even if $D = 0$ for each subpopulation (Sinnock and Sing 1972, Prout 1973, Nei and Li 1973). As an example, consider a mixture of two subpopulations and a pair of two-allele loci, denoted respectively by A and B, for which Hardy-Weinberg proportions and linkage equilibrium pertain to each subpopulation. The linkage disequilibrium in the mixture is then $D = F_A\, p_A(1-p_A)\, F_B\, p_B(1-p_B)$, where F_A, p_A, F_B, p_B are the fixation indices and allele frequencies in the mixture for locus A and locus B, respectively. The sample size needed to have a 50% chance of detecting this D with a test at the 5% level is then (Brown 1975) $N > 4/(F_A F_B)$. For $F_A = F_B = F$, this is the same requirement as that for detecting deviation from Hardy-Weinberg proportions.

The conclusion is that, unless very large sample sizes are used, a moderate within-sample heterogeneity caused by the presence of a mixture of subpopulations in the sample usually will not be detected. It is important to note also that both deviation from Hardy-Weinberg proportions and linkage disequilibrium can occur for other reasons than those given here, e.g., because of selection. A few instances of genetically distinct sympatric subpopulations of fish have been described (e.g., Allendorf et al. 1976, Ryman et al. 1979, Kirkpatrick and Selander 1979, Ferguson and Mason 1981), but, for the statistical reasons given above, samples larger than those generally taken are needed to settle the question of whether this is a common occurrence.

In this connection, we must point out that for multiallelic loci many different parameters D can be estimated and used to test for linkage equilibrium. Based on a suggestion by Sved (1968), Brown et al. (1980) and Chakraborty (1981) proposed an alternative test procedure that uses the distribution of the number of heterozygous loci. This procedure results in an overall test for linkage equilibrium. Chakraborty (1984) compared the statistical power of this approach with that of the traditional method through extensive computer simulation.

THEORETICAL MODELS OF THE EVOLUTION OF STRUCTURED POPULATIONS AT NEUTRAL LOCI

For a few rather idealized situations, theoretical models of the evolution of a subdivided population have been developed. Although the evolution of a subdivided population in nature may be more complex than any of the theoretically studied cases, such models can be used as a reference point for the interpretation of data. The study of models also leads to an appreciation of the strength of evolutionary forces affecting population structure and of the time scales over which quantities such as H_S, H_T, and G_{ST} change. For management and genetic resource conservation, the question of how long it takes for a particular genetic structure to evolve is an important one. The longer the time, the more likely it is that the structure will reflect important biological adaptations.

In this section, some results will be given for models of the evolution of a subdivided population for neutral loci, i.e., loci at which there are no selective differences between genotypes. A situation with a one-level hierarchy (subpopulations within the total) will be described, and then a hierarchical model with one more level will be studied.

The Island Model

This model was introduced by Wright (1943), and the name is derived from an application considered by Wright, namely a population occupying a group of islands with a subpopulation on each island and gene flow between them. Wright assumed that all subpopulations are randomly mating groups of equal size and that an individual migrating to a subpopulation is equally likely to come from any of the other subpopulations. Subsequently, variations of the island model have been studied in great detail (e.g., Maynard Smith 1970, Maruyama 1970, Nei and Feldman 1972, Latter 1973a, Nei 1975, Li 1976, Nagylaki 1984, Crow and Aoki 1984).

Here we consider the following situation. The population consists of n subpopulations, each having the effective size N_e. A proportion m of the individuals in a subpopulation is replaced every generation by migrants taken at random from all other subpopulations. At each of a large number of loci, mutation to neutral alleles, not previously existing, occurs at the rate u per generation. Assuming additionally, that generations do not overlap, that reproduction in a subpopulation occurs through random mating (including selfing), and that gene flow takes place through dispersal of gametes instead of migration of individuals, an exact solution for the evolution of quantities such as H_S, H_T, G_{ST} and genetic distance between subpopulations can be obtained (e.g., Latter 1973a, Li 1976). These assumptions will not be fulfilled for populations of fish, but the solution can still be an acceptable approximation (Nagylaki 1984).

Examples from nature where the island model might be applied are an Atlantic salmon population straying between rivers and a population of brown trout inhabiting a lake with the subpopulations representing different spawning aggregations. In order to ease comparison with real populations, particularly populations of fish, the parameters entering into the model will be discussed briefly.

If generations overlap, the generation length can be taken as the average age of a reproductively active individual. N_e should be the effective size (see Chapter 3) that a subpopulation would have in the absence of gene flow. With overlapping generations, a rough approximation of N_e is given by the number of spawners of mean reproductive age each season, multiplied by the generation length. Due to such factors as skewed sex ratios and variation in reproductive success, N_e usually will be smaller than this number. See Felsenstein (1971), Hill (1979), and Pollak (1980) for more details on effective population size when generations overlap.

The parameter m is the relative rate of gene flow per generation into a subpopulation. In population genetics, m is often referred to as the migration rate. It will be equal to the probability that a reproductively active individual randomly selected from a subpopulation originated in another subpopulation. The product $N_e m$ plays an important role in the model. If most individuals spawn only once, $N_e m$ will approximate the absolute number of individuals that migrate into a subpopulation from other subpopulations each generation and that reproduce successfully. $N_e m$ is sometimes referred to as the effective number of migrants per generation. When m or $N_e m$ are compared with estimates of rates of straying from recapture data, one should keep in mind that m may well be smaller than the fraction of individuals in the subpopulation consisting of migrants, since straying individuals may be less likely to reproduce (see, e.g., Ehrlich and Raven 1969, Ståhl 1981).

Estimates of rates of mutation to electrophoretically detectable alleles at protein loci can be obtained, if one assumes neutrality, from attempts to calibrate the "electrophoretic clock" (see Chapters 8 and 9). Current calibrations would correspond to mutation rates somewhat smaller than 10^{-7} per locus per year.

Equilibrium Gene Diversity
Components for the Island Model

As mentioned above, under certain assumptions it is possible to obtain an exact solution for the island model (Latter 1973a, Li 1976). Here we will give some simple approximate relations that will be adequate for most practical purposes. First, note that the values of gene diversity components and genetic distance referred to below are those obtained from a large number of neutral loci. The approximations below assume that the effective size of a subpopulation (N_e) is large, the migration rate (m) is much smaller than unity, and the mutation rate (u) is much smaller than m. The last assumption will not be justified if m is very small, as for instance for completely isolated subpopulations ($m = 0$). If the number of subpopulations (n) is large, then the equilibrium G_{ST} will be approximately

$$G_{ST} = \frac{1}{1 + 4N_e m} .$$
(13)

This formula was given by Wright (1943). If n is small, a better approximation is that provided by Takahata (1983) and Takahata and Nei (1984):

$$G_{ST} = \frac{1}{1 + 4N_e m (\frac{n}{n-1})^2} .$$
(14)

Thus, the equilibrium G_{ST} is approximately independent of the mutation rate. When G_{ST} is estimated from data and the island model can be considered to represent the biological situation, either (13) or (14) can be used to obtain an estimate of $N_e m$. An approximation for the equilibrium value of H_S (see, e.g., Crow and Aoki 1984) is

$$H_S = \frac{4nN_e u}{1 + 4nN_e u} .$$
(15)

This approximation assumes that nu is much smaller than m and will not be accurate for very low migration rates. Note that the equilibrium H_S is approximately independent of the migration rate. Since $G_{ST} = 1 - H_S/H_T$, (13) or (14) and (15) lead to an approximation for the equilibrium H_T. Another quantity of interest for the island model is the genetic distance between subpopulations. We will use Nei's standard genetic distance D (see Chapter 8). If the migration rate m is written as $m = (n-1)m_1$, where m_1 is the rate of migration from one subpopulation to another, then an approximation for the equilibrium genetic identity (Nei and Feldman 1972; see also Nei, Chapter 8) is $I = m_1/[m_1 + u]$, and thus

$$D = -\ln \left(\frac{m_1}{m_1 + u} \right) .$$
(16)

Rate of Approach to Equilibrium For The Island Model

When considering the equilibrium values given above, it is important to keep in mind the time required for a deviation from the equilibrium to disappear. Such a time can be expressed conveniently by a rate of approach of the equilibrium. A given quantity is characterized by the rate r when the time required is obtained from the condition that $\exp(-rt)$ should be small. Thus, af-

ter a time that is given by t >> 1/r or, equivalently, 1/t <<r, the quantity will
be close to its equilibrium value. The time does not need to be very much
greater than 1/r; if $t > 5/r$, then, for all practical purposes, the quantity in question will have reached equilibrium. For the island model, there are two rates,
one faster rate that applies to G_{ST} and one slower that applies to H_S, H_T, and D.
The approximations given below for these rates can be obtained from the eigen-
values given by Li (1976) (see also Crow and Aoki 1984). Measuring time in
generations, G_{ST}, will be close to its equilibrium value after a time given by

$$\frac{1}{t} << \frac{1}{2N_e} + 2m \ . \tag{17}$$

The rate $1/2N_e + 2m$ consists of the two components $1/2N_e$ and $2m$, which can
be considered as measuring, respectively, the strength of genetic drift in a subpopulation and gene flow between subpopulations as evolutionary forces. The
quantity $4N_em$ appearing in (13) and (14) is the ratio of these components, so
that when $4N_em$ is large, gene flow will dominate drift, leading to little differentiation; if $4N_em$ is small, drift will dominate, and allele frequencies will
differ strongly among subpopulations. The equilibrium G_{ST} is thus determined
by a balance between the opposing forces of gene flow and genetic drift in a
subpopulation.

An interesting application of (17) is gene flow caused by human activities, such as transplantation of fish. In this case, one would expect gene flow to
dominate drift, so that $1/m$ will be an estimate of the time needed to severely
change the genetic composition in the receiving population. This can be directly converted to the number of individuals transferred, summed up over
time. If that number represents an appreciable fraction of the size of the receiving population, then considerable genetic change will take place.

The time needed for H_S, H_T, and D to come close to equilibrium is
given by

$$1/t << 2u + \frac{1}{2N_{Te}} \ , \tag{18}$$

where N_{Te} is the effective size of the total population. Due to the subdivision,
N_{Te} will be larger than nN_e, the sum of the local effective sizes. An approximation for N_{Te} involving the equilibrium value of G_{ST} is

$$N_{Te} = \frac{nN_e}{1 - G_{ST}} \ . \tag{19}$$

Unless G_{ST} is close to one, N_{Te} will not be very different from nN_e. The relation
(19) is not restricted to the island model, but holds for any migration scheme
(Maruyama 1972). The rate (18) will be high only if N_{Te} is small, and in this

case H_S will decay to a very small equilibrium value (see Eq. 15). Roughly speaking, the equilibrium heterozygosity is determined by a balance between the opposing forces of mutation and genetic drift in the entire population. If nN_e is large enough to give an appreciable equilibrium heterozygosity, then due to the smallness of u, (18) will be very small. With a mutation rate smaller than 10^{-7} per locus per year, the time needed to reach equilibrium heterozygosity will be of the order of from one to several million years, so that this time may well be longer than the time of existence of the species in question.

The equilibrium D given by (16) is independent of N_e. The ratio $m_1/[m_1 + u]$ in Eq. (16) expresses a balance between the rate of appearance of new alleles in a subpopulation and the rate at which these alleles are exported to another subpopulation. If u is smaller than m_1, then alleles usually will be shared among subpopulations.

Numerical Illustration

In order to illustrate the above discussion, some numerical values are given in Table 4.3. For the initial state of the population, H_T is assumed to be smaller than the equilibrium value, and G_{ST} is assumed to be zero. Such a state could appear if a population experiences a bottleneck in population size, leading to a reduction in H_T, followed by an expansion and subdivision. The comparatively fast approach to equilibrium of G_{ST} can be seen in Table 4.3. With smaller subpopulation effective sizes, the approach to equilibrium of G_{ST} will be faster. See Allendorf and Phelps (1981b) for some numerical results for an island model with small subpopulations. Comparing the case of completely isolated subpopulations ($m = 0$) with that of $N_e m = 0.1$ or that of $N_e m = 1.0$ in Table 4.3, one notes that for the first few thousand generations H_S, H_T, and D change in a very similar way; to some extent, this holds also for G_{ST}. Gene flow will have a major effect only over times long enough for the total number of migrants to make up an appreciable fraction of size of the subunits in question. The increase of D with time for isolated subpopulations is also of interest. In order for D to grow linearly with time, H_S must stay constant (see Chapter 8).

Extensions of The Island Model

The assumptions of equal subpopulation sizes and time uniformity of migration rates have been relaxed in a few studies with the island model (e.g., Levene 1953, Deakin 1966, Chakraborty and Nei 1974, Li 1976a.) Another assumption in the island model that often may not be realistic is that the rate of migration is the same between all pairs of subpopulations. In nature, migrants are likely to come predominantly from nearby subpopulations. Models that take this into account have been developed, e.g., Wright's isolation by distance model (Wright 1943, 1951), Malecot's migration model (Malecot 1948), and Kimura's stepping stone model (Kimura 1953, Kimura and Weiss 1964). See Jorde (1980), Karlin (1982), and Slatkin (1985) for recent reviews on migration models. Compared with the island model, G_{ST} will be larger for a given m if

Table 4.3 Approach to equilibrium of gene diversity components and genetic distance for neutral loci in an island model.

t	$N_e = 5000, n = 20$				$N_e = 50000, n = 2$			
	H_S	H_T	G_{ST}	D	H_S	H_T	G_{ST}	D
$N_e m = 0.0$								
0	0.0200	0.0200	0.0000	0.0000	0.0200	0.0200	0.0000	0.0000
10^3	.0185	.0203	.0900	.0020	.0202	.0203	.0050	.0002
10^4	.0099	.0232	.5755	.0143	.0218	.0229	.0457	.0021
10^5	.0040	.0557	.9285	.0562	.0319	.0452	.2932	.0277
equil.	.0040	.9502	.9958	∞	.0385	.5192	.9259	∞
$N_e m = 0.1$								
0	.0200	.0200	.0000	.0000	.0200	.0200	.0000	.0000
10^3	.0185	.0203	.0881	.0019	.0202	.0203	.0049	.0002
10^4	.0119	.0232	.4881	.0121	.0219	.0229	.0436	.0021
10^5	.0175	.0517	.6621	.0374	.0348	.0447	.2200	.0206
equil.	.0636	.2056	.6909	.1740	.0709	.1132	.3731	.0953
$N_e m = 1.0$								
0	.0200	.0200	.0000	.0000	.0200	.0200	.0000	.0000
10^3	.0188	.0203	.0737	.0016	.0202	.0203	.0048	.0002
10^4	.0189	.0230	.1784	.0044	.0221	.0229	.0318	.0015
10^5	.0368	.0450	.1823	.0090	.0408	.0432	.0555	.0050
equil.	.0729	.0893	.1839	.0188	.0737	.0783	.0585	.0099
$N_e m = 10.0$								
0	.0200	.0200	.0000	.0000	.0200	.0200	.0000	.0000
10^3	.0199	.0203	.0216	.0005	.0202	.0203	.0034	.0001
10^4	.0224	.0229	.0219	.0005	.0227	.0228	.0061	.0003
10^5	.0419	.0428	.0220	.0010	.0424	.0426	.0062	.0005
equil.	0.0740	0.0756	0.0220	0.0019	0.0740	0.0745	0.0062	0.0010

Note: D is Nei's standard genetic distance between subpopulations, and time (t) is measured in generations. The initial state represents a population expansion following a bottleneck. Mutation rate per generation (u) is 2×10^{-7}. N_e is the effective size of a subpopulation, n is the number of subpopulations, and m is the migration rate.

migration takes place only between nearby subpopulations. The effect will be strongest if subpopulations are arranged in a linear (one dimensional) fashion. The approximation (13) for G_{ST} will, however, often be reasonable for cases with only short-range migration (Crow and Aoki 1984).

Hierarchical Model

The island model, the stepping stone, and the isolation by distance models all assume a population structure with one level, subpopulations within the total. Migration models describing a population with a more complicated structure have also been proposed (Malecot 1951, 1959, Bodmer and Cavalli-Sforza 1968, Smith 1969, Carmelli and Cavalli-Sforza 1976). We will give some numerical results for a model suggested by Carmelli and Cavalli-Sforza (1976).

This model applies to a hierarchically structured population (compare the discussion of Eqs. (6) and (7) above). Assume that the total population consists of g major groups of subpopulations and each major group contains k subpopulations. Thus, the total number of subpopulations is $n = gk$. Each subpopulation has the effective size N_e. The rate of migration between a pair of subpopulations is m_1 if they belong to the same group and m_2 if they belong to different groups. The proportion of a subpopulation that is replaced each generation by migrants from the outside is then $m = (k-1)m_1 + (g-1)km_2$. Similarly, for a major group the proportion $m_G = (g-1)km_2$ will be replaced. Mutation to neutral alleles at the rate u per generation is assumed as before. For the special case of $g = 1$ (or $k = 1$) this model becomes an island model. The product $N_e m$ will have the same meaning as in the island model, and a similar quantity for a major group will be the product $kN_e m_G$. If N_e is equal to the actual subpopulation size, $kN_e m_G$ is the number of individuals migrating into a major group each generation.

In Tables 4.4 and 4.5, values of gene diversity components (cf. Eqs. (6) and (7)) and genetic distances between subpopulations are given for a few different cases. The initial state of the population is the same as in Table 4.3. Values of the parameters $g, k,$ and N_e have been chosen so that the major groups in Tables 4.4 and 4.5 correspond to the subpopulations in Table 4.3. Although this model is more complicated than the island model, some of the approxima-

Table 4.4 Approach to equilibrium of gene diversity components and genetic distance for neutral loci in a two-level hierarchical migration model.

t	H_S	H_T	G_{SG}	$G_{SG(T)}$	G_{GT}	G_{ST}	D_1	D_2
				$N_e = 250, k = 20, g = 20$				
$N_e m = 1.0$	$kN_e m_G = 1.0$							
0	0.0200	0.0200	0.0000	0.0000	0.0000	0.0000	0.0000	0.0000
10^3	.0155	.0203	.1841	.1728	.0616	.2344	.0038	.0049
10^4	.0161	.0231	.1833	.1557	.1503	.3061	.0039	.0074
10^5	.0323	.0468	.1834	.1552	.1539	.3091	.0079	.0155
equil.	.0717	.1039	.1835	.1550	.1554	.3104	.0184	.0363
$N_e m = 10.0$	$kN_e m_G = 1.0$							
0	.0200	.0200	.0000	.0000	.0000	.0000	.0000	.0000
10^3	.0184	.0203	.0207	.0192	.0722	.0915	.0004	.0020
10^4	.0186	.0230	.0207	.0171	.1753	.1924	.0004	.0047
10^5	.0363	.0452	.0207	.0170	.1792	.1962	.0008	.0097
equil.	0.0727	0.0907	0.0207	0.0170	0.1808	0.1977	0.0017	0.0205

Note: D_1 is the genetic distance between subpopulations from the same group and D_2 that between subpopulations from different groups. Time (t) is measured in generations. The initial state represents a population expansion following a bottleneck. Mutation rate per generation (u) is 2×10^{-7}. N_e is the effective size of a subpopulation; k is the number of subpopulations in a group; g is the number of groups; m is the rate of migration into a subpopulation; and m_G is the rate of migration into a group of subpopulations. See Eqs. (6) and (7) in the text for a definition of the gene diversity components.

Table 4.5 Approach to equilibrium of gene diversity components and genetic *D*istance for *N*eutral *Loci* in a two-level hierarchical migration model.

t	H_S	H_T	G_{SG}	$G_{SG(T)}$	G_{GT}	G_{ST}	D_1	D_2
			$N_e = 250, k = 200, g = 2$					
$N_e m = 1.0, kN_e m_G = 0.1$								
0	0.0200	0.0200	0.0000	0.0000	0.0000	0.0000	0.0000	0.0000
10^3	.0162	.0203	.1972	.1964	.0041	.2005	.0041	.0042
10^4	.0179	.0231	.1972	.1903	.0354	.2256	.0045	.0061
10^5	.0307	.0467	.1974	.1617	.1808	.3424	.0079	.0255
equil.	.0697	.1284	.1974	.1336	.3233	.4569	.0187	.1139
$N_e m = 1.0, kN_e m_G = 1.0$								
0	.0200	.0200	.0000	.0000	.0000	.0000	.0000	.0000
10^3	.0162	.0203	.1972	.1964	.0039	.2003	.0041	.0042
10^4	.0180	.0231	.1972	.1922	.0256	.2178	.0045	.0057
10^5	.0350	.0457	.1973	.1885	.0449	.2334	.0090	.0133
equil.	.0724	.0947	.1974	.1881	.0475	.2356	.0195	.0293
$N_e m = 1.0, kN_e m_G = 10.0$								
0	.0200	.0200	.0000	.0000	.0000	.0000	.0000	.0000
10^3	.0163	.0203	.1972	.1967	.0028	.1994	.0041	.0042
10^4	.0184	.0231	.1972	.1962	.0049	.2012	.0046	.0049
10^5	.0361	.0453	.1974	.1964	.0050	.2014	.0093	.0097
equil.	0.0727	0.0911	0.1975	0.1965	0.0050	0.2015	0.0196	0.0205

Note: D_1 is the genetic distance between subpopulations from the same group and D_2 that between subpopulations from different groups. Time (t) is measured in generations. The initial state represents a population expansion following a bottleneck. Mutation rate per generation (u) is 2×10^{-7}. N_e is the effective size of a subpopulation; k is the number of subpopulations in a group; g is the number of groups; m is the rate of migration into a subpopulation; and m_G is the rate of migration into a group of subpopulations. See Eqs. (6) and (7) in the text for a definition of the gene diversity components.

tions given above can still be used. First, note that (15) is still a reasonable approximation for the equilibrium H_S. The equilibrium G_{ST} is no longer given by (13); but the quantity $G_{SG} = 1 - H_S/H_G$, which corresponds to G_{ST} if a major group is taken as the total, has an equilibrium value that is given by $1/[1 + 4N_e m]$. Also, the equilibrium value of $G_{GT} = 1 - H_G/H_T$ is well approximated by

$$G_{GT} = \frac{1}{1 + 4N_{Ge}m_G\left(\frac{g}{g-1}\right)^2} , \qquad (20)$$

where N_{Ge} is the major group effective size, which can be obtained from (19) applied to a major group. N_{Ge} will be somewhat larger than kN_e, so that G_{GT} will be small if the number of migrants per generation into the group is much greater than one. This is illustrated by the last example in Table 4.5 ($kN_e m_G = 10$), where G_{GT} is very small and D between populations from the same group is very close to D between populations from different groups. In spite of this, only a small fraction (5%) of the migration into a subpopulation consists of individuals from other major groups.

Concerning the approach to equilibrium, there are three different rates for this model. The fastest one, which is approximately given by (17), applies to G_{SG}. G_{GT} will approach equilibrium somewhat more slowly; the rate can be approximated by (17) with N_e and m replaced by N_{Ge} and m_G. The slowest rate is given by (18), and it applies to heterozygosity and genetic distance.

Assuming that an observed allele frequency variation is not determined by selection, empirical estimates of gene diversity components can be used to estimate effective numbers of migrants between subunits. When doing this, both the appropriate hierarchy and the time needed for equilibrium to be reached should be taken into account. For instance, consider subpopulations of Atlantic salmon in a river. The effective size of these subpopulations is likely to be small, maybe even less than a hundred. Taking rivers as major groups, the component G_{SG} of within-river differentiation will approach $1/[1 + 4N_e m]$ quickly, perhaps in fewer than a hundred generations. The effective size of the collection of subpopulations in a river will be larger, say a few thousand. After a longer time, perhaps one to a few thousand generations, G_{GT} will be near its equilibrium value and can be used to estimate between-river gene flow. If the hierarchy is continued with rivers grouped into larger units, larger effective sizes and thus longer times, perhaps several tens of thousands of generations, will apply.

LOCAL ADAPTATION

Adding selection to the evolutionary forces discussed in the previous section results in formidable mathematical difficulties. For a review of this field see Karlin (1982). We will give some results for the simplest case of genic selection.

Consider a locus with two alleles, A and $a,$ and assume that A is selected for in some subunit of a species but selected against outside this subunit. The strength of the selection is measured by the selection coefficient s (s is defined from the fitnesses of the genotypes AA, Aa and $aa;$ there are several slightly different definitions, e.g., the fitnesses can be given as $1 + 2s, 1 + s,$ and 1). Selection will tend to increase the frequency of A where s is positive and decrease it where s is negative. In order for this to take place, s must be large enough for selection to dominate over gene flow and drift. Roughly, if N_e *and* m are respectively the effective size and migration rate into the subunit where A is selected for, then A will be held at high frequency in the subunit if s is considerably greater than both m and $1/N_e$. Thus, for large subunits, where m and $1/N_e$ are likely to be small, selection is more likely to dominate than for small subunits.

With short-range migration, the condition on s will be less stringent, since subpopulations will be relativley isolated from gene flow from distant parts of the species. An example is a linear chain of subpopulations, e.g., along a coastline or a river, with migration rate m between nearest neighbor subpopulations. If A is selected for in a region containing l subpopulations, then a cline in allele frequency will appear if s/m is greater than $1/l^2$ (see Slatkin 1973, Endler 1977, Nagylaki 1975). There are a few examples of empirically observed clines in allele frequency for fish populations that have been interpreted as being maintained by selection (e.g., Place and Powers 1979, Christiansen and Frydenberg 1974). The time needed for selection to produce substantial allele frequency differences depends on how strong the selection is. For genic selection, this time can be taken as $1/s$ generations. The conclusion is that if selection is strong enough to dominate gene flow and drift, then differentiation will appear more rapidly at loci where selection is operating than at neutral loci.

An important aspect of local adaptation is variation in quantitative characters among subunits. Data for quantitative characters are often used in addition to electrophoretic data in studies of population structure. Very little theory pertinent to the problem of variation of quantitative characters in a subdivided population has been developed. Regarding the detection of population structure using quantitative characters there are some results. For a neutral quantitative character, Rogers and Harpending (1983) have shown that the additive genetic component contains the same amount of information about population structure as one allele frequency. Taking into account the fact that environmental effects will result in nongenetic variation in the character, the conclusion is that such a character is of limited use in the study of population structure. On the other hand, if differences among subunits in the mean value of a quantitative character are due to selection, then such a character is potentially a very powerful indicator of structure (Lewontin 1984). There are, however, severe practical difficulties associated with the determination of genetic and environmental components of the variation in the character.

CONCLUDING REMARKS

The picture that emerges from population genetic theory is that the amount of genetic variation within and between subunits of an existing population is determined by the relative strengths of the forces of selection, gene flow, drift, and mutation and by the time that has been available for these forces to act. When one attempts to judge the importance of a certain degree of differentiation observed at electrophoretic loci, information available on such factors as population size, potential for gene flow, and time of separation should be taken into account. For instance, in comparing a marine species such as Atlantic herring or cod with an anadromous or freshwater salmonid, a reasonable assumption is that the marine species will be characterized by a larger population size and smaller barriers to gene flow. Assuming that selection has not played a major role in determining the allele frequency differentiation, the expectation would be that the marine species would have higher H_S and smaller G_{ST}. This expectation is consistent with empirical data (see Gyllensten 1985 for a survey of published data relating to such a comparison).

As a concrete example, compare Atlantic herring with Atlantic salmon. Population genetic surveys of Atlantic herring have shown a striking similarity of allele frequencies at electrophoretic loci in samples collected over large geographical areas (Andersson et al. 1981, Kornfield et al. 1982, Grant 1984, Ryman et al. 1984). With samples from the Scandinavian waters, Ryman et al. (1984) estimated a G_{ST} of around 0.01 and genetic distances between sampled subunits of the order of 0.001. Grant (1984) obtained G_{ST} and genetic distances of the same magnitudes with samples from North American and European populations. For Atlantic salmon, Ståhl (Chapter 5) estimated a G_{ST} of about 0.4 and genetic distances up to about 0.04 with samples from North American and European populations. If neutrality is assumed for electrophoretic loci, then the comparatively greater differentiation of Atlantic salmon should be due to smaller effective sizes of populations, less gene flow, or longer time of separation.

Estimating the time that has been available for differentiation of Atlantic herring and Atlantic salmon populations would be speculative. The similar geographic distribution of these two species, and thus possibly shared geological history, at least indicates a comparable timescale. For instance, it seems reasonable that populations of these species invaded the Baltic Sea at approximately the same time. This points to the conclusion that differences in effective sizes and/or gene flow among subunits of these species are the major factors leading to contrasting genetic profiles. In relation to the question of local adaptation, one might conclude that populations of Atlantic salmon, showing more differentiation at electrophoretic loci, have had a greater opportunity to adapt to local conditions than populations of Atlantic herring. Although this might well be the case, it should be kept in mind that little allele frequency differentiation at neutral loci is the result of a high value of N_em and/or a large N_e and limited time of separation (N_e and m are here, respectively, the effective size and the rate of migration into a subunit). Local adaptation, on the other hand, will be possible if m is small enough. Thus, with very large effective sizes, local adaptation may take place with very little differentiation at neutral loci.

5.
Genetic Population Structure
Of Atlantic Salmon

Gunnar Ståhl

The Atlantic salmon *(Salmo salar)* is often perceived as an aristocrat of the Salmonidae (Netboy 1974). This status stems from an historical fascination with the size, strength, and pristine habitat requirements of the species, the angling skill that is frequently required for its capture, and the esthetic appeal of its physical appearance. Among others, these characteristics mark the Atlantic salmon as a primary target for sports and commercial harvest. Inevitably, numbers of salmon declined as a consequence of degradation of freshwater habitats as well as excessive harvesting throughout its natural range in north Atlantic drainages of Europe and North America. To supplement natural runs, hatcheries have existed on both sides of the Atlantic Ocean for over a century (MacCrimmon and Gots 1979).

The need for knowledge of the genetic structure of Atlantic salmon populations gradually became apparent. The intensification of high seas fisheries on stocks of unknown origin (Saunders 1966) heightened biological and political concerns about harvest allocations and preservation of less abundant populations. The increase in hatchery-reared Atlantic salmon has focused attention on the need to understand the genetic composition of natural populations and on the genetic effects induced by hatchery operations (Allendorf and Phelps 1980, Ryman and Ståhl 1980, Ryman 1981b, Ståhl 1983).

This chapter provides an overview of the genetic population structure of Atlantic salmon over the entire species range. The population genetic structure is elucidated by examining genetic differences among natural populations and hatchery stocks from various geographic regions. Evolutionary implications of the genetic structure of the species are discussed, and the use of genetic data in management practices is encouraged.

MATERIALS AND METHODS

Samples

The major part of the material used in the present study comes from analyses performed in our own laboratory. The electrophoretic results for parts of the material have previously been presented by Ståhl (1981, 1983), Ryman and Ståhl (1981), Ryman (1983), and Ståhl et al. (1983). In addition, the results from two electrophoretic surveys reported in the literature are included in the analysis of the intraspecific genetic structure.

Samples of anadromous Atlantic salmon were collected between 1979 and 1983 from 23 major river systems draining into the Western Atlantic Ocean, the Eastern Atlantic Ocean, and the Baltic Sea (Fig. 5.1, Table 5.1). In addition, landlocked Atlantic salmon populations in four European drainages were sampled. Samples were collected both from naturally reproducing populations (28 samples) and from hatchery stocks (21 samples, indicated by a "†").

All but three of the field collections were obtained by electrofishing a restricted area within each drainage. Specimens from electrofished samples represented under-yearling fry, 1+ parr, 2+ parr, and 3+ parr. At least two-year classes are represented in each collection, which guarantees that the sample represents more than a single family (cf. Allendorf and Phelps 1981b). No fish have been planted recently in any of the areas sampled by electrofishing, so the samples of young fish collected in the field should represent the progeny of natural reproduction.

Three of the Swedish samples (13a Lagan, 14 Fylleån, and 15a Ätran; Table 5.1) denoted as samples of naturally reproducing populations were collected from mature fish migrating upstream to spawn in watercourses where hatchery-reared smolt had been released. Thus, for these three samples it is not known whether the fish stem from naturally reproducing populations or from hatchery stocks.

Fish from hatchery stocks are produced through artificial spawning of annual catches of adult fish migrating upstream during the spawning season;

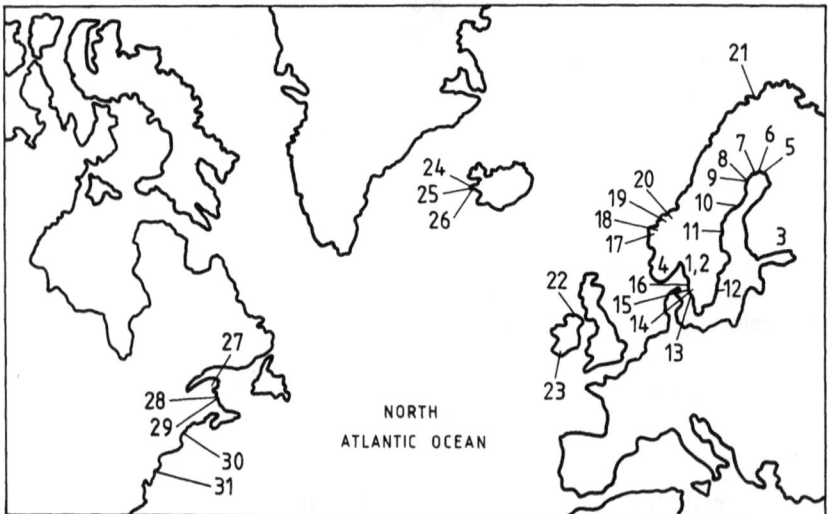

Fig. 5.1 Geographical distribution of major drainages that were sampled. The map codes (1–31) refer to those in Table 5.1.

Table 5.1 Frequency of the variant alleles at nine variable loci and estimates of average heterozygosity (*H*) in samples representing naturally reproducing populations and hatchery stocks of Atlantic salmon. The estimates of average heterozygosity are based on a total of 38 loci. For samples of hatchery stocks (indicated by †) the year class is also specified (e.g., †Gullspång-81 hatched in 1981). The map code refers to Fig. 5.1, where the numbers indicate major drainages; multiple samples from the same drainage are denoted by a letter in addition to the drainage number.

Map code	Sample	Number of fish	AAT-3 50	AGP-2 25	AGP-2 50	LDH-4 130	LDH-4 120	ME-2 125	MDH-1 -200	MDH-3 115	PGI-1 80	PGI-1 140	PGI-1 185	PGM-1 75	SDH-1 -50	H (38 loci) %
Landlocked																
1 a	†Gullspång-81	101	0	0	0	0	0	0.238	0	0	0	0	0	0	0.015	1.0
1 b	†Gullspång-80	50	0	0	0	0	0	0.040	0	0	0	0	0	0	1.000	0.2
1 c	†Anten-81	101	0	0	0	0	0	0.495	0	0	0	0	0	0	0	1.3
2 a	†Klarälven-81	225	0	0	0	0	0	0.199	0	0	0	0	0	0	0.742	1.8
2 b	†Klarälven-80	106	0	0	0	0	0	0.108	0	0	0	0	0	0	1.000	0.5
3	†Sairaa-79	63	0.088	0	0	0	0	0.008	0	0	0	0	0	0	0.258	1.5
4 a	†Byglandsfjorden-82	100	0.258	0	0	0	0	1.000	0	0	0	0	0	0	0.760	2.0
4 b	†Byglandsfjorden-80	26	0.481	0	0	0	0	1.000	0	0	0	0	0	0	0.750	2.3
Baltic Sea																
5 a	Laimio	14	0.571	0	0	0	0	0.071	0	0	0	0	0	0	0.244	2.6
5 b	Torne	100	0.280	0	0	0	0	0.050	0	0	0	0	0	0	0.219	2.2
5 c	†Kukkola-78	50	0.360	0	0	0	0	0.020	0	0	0.030	0	0	0.040	0.452	3.0
5 d	†Torne mynning-79	130	0.282	0	0	0	0	0.050	0	0	0	0	0	0	0.449	2.6
6 a	Kalix	98	0.378	0	0	0	0	0.082	0	0.005	0	0	0	0.020	0.394	3.0
6 b	Kaitum	98	0.378	0	0	0	0	0.071	0	0	0	0	0	0	0.447	2.9
6 c	Ängesån Satter	46	0	0.217	0	0	0	0.141	0	0	0	0	0	0	0.263	2.6
6 d	Ängesån Vettasjoki	79	0	0.373	0	0	0	0.095	0	0	0.019	0	0	0	0.181	2.6
6 e	Ängesån Valtiojoki	29	0	0.362	0	0	0	0.155	0	0	0	0	0	0	0.281	3.0
7	†Lule-älven-79	31	0.145	0	0	0.016	0	0.097	0	0	0.009	0	0	0	0.378	2.4
8 a	Byske	59	0.110	0	0	0	0	0.280	0	0	0	0	0	0.144	0.324	3.4
8 b	†Byske-80	40	0	0	0	0	0	0.088	0	0	0	0	0	0	0.500	1.7
9	†Skellefteälven-80	390	0.110	0	0	0	0	0.073	0	0	0	0	0	0	0.302	2.0

Table 5.1 (continued) Frequency of the variant alleles at nine variable loci and estimates of average heterozygosity (*H*) in samples representing naturally reproducing populations and hatchery stocks of Atlantic salmon. The estimates of average heterozygosity are based on a total of 38 loci. For samples of hatchery stocks (indicated by †) the year class is also specified (e.g., †Gullspång-81 hatched in 1981). The map code refers to Fig. 5.1, where the numbers indicate major drainages; multiple samples from the same drainage are denoted by a letter in addition to the drainage number.

Map code	Sample	Number of fish	AAT-3		AGP-2		LDH-4	ME-2	MDH-1		MDH-3	PGI-1		PGM-1	SDH-1	H (38 loci)
			50	25	50	130	120	125	-200	115	80	140	185	75	-50	%
Baltic Sea (cont.)																
10	Lögde	69	0.014	0	0	0	0	0.181	0	0	0	0	0	0	0.448	2.2
11 a	†Indalsälven-80	162	0.077	0	0	0	0	0.077	0	0	0	0	0	0	0.259	1.8
11 b	†Indalsälven-79	103	0.063	0	0	0	0	0.024	0	0	0	0	0	0	0.346	1.6
12	†Emån-80	82	0	0.500	0.500	0	0	0	0	0	0	0	0	0	0.651	2.5
Eastern Atlantic Ocean																
13 a	Lagan	27	0.111	0	0	0	0	0.759	0.019	0	0	0	0	0	0.306	2.7
13 b	†Lagan-81	140	0.043	0	0	0	0	0.486	0	0	0.022	0	0	0	0.684	2.8
13 c	†Lagan-80	115	0.113	0	0	0	0	0.543	0.056	0	0.022	0	0	0	0.563	3.2
14	Fylleån	9	0.056	0	0	0	0	0.778	0	0	0.118	0	0	0	0.423	3.3
15 a	Ätran	23	0	0	0	0	0	0.500	0	0	0	0	0	0	0.583	2.6
15 b	†Ätran-81	128	0.020	0	0	0	0	0.621	0	0	0	0	0	0	0.576	2.6
16	†Rolfsån-81	124	0.331	0	0	0	0	0.226	0	0	0	0	0	0	0.731	3.1
17	Örstaelva	54	0.056	0.009	0	0	0	0.611	0	0	0.009	0	0.019	0	0.528	3.0
18	Bondalselva	50	0.060	0.040	0.007	0	0	0.540	0	0	0	0	0.011	0	0.630	3.1
19	Sokna	72	0.069	0.139	0	0	0	0.493	0	0	0	0	0	0	0.435	3.6
20	Mosvikelva	14	0	0	0	0	0	0.429	0	0	0	0	0	0	0.465	2.6
21 a	Alta Moseslandet	121	0	0	0	0	0	0.682	0	0	0	0	0	0	0.672	2.3
21 b	Alta Svarrfossen	80	0	0	0	0	0	0.681	0	0	0	0	0	0	0.613	2.4
21 c	Alta Yli Sierra	104	0.058	0.005	0	0	0	0.788	0	0	0	0	0	0	0.660	2.4
21 d	Alta Eiby	86	0.006	0	0	0	0	0.669	0	0	0	0	0	0	0.736	2.2
22	†Bushmills-80	39	0.103	0	0	0	0	0.500	0	0	0	0	0	0	0.359	3.0

Table 5.1 (continued) Frequency of the variant alleles at nine variable loci and estimates of average heterozygosity (*H*) in samples representing naturally reproducing populations and hatchery stocks of Atlantic salmon. The estimates of average heterozygosity are based on a total of 38 loci. For samples of hatchery stocks (indicated by †) the year class is also specified (e.g., †Gullspång-81 hatched in 1981). The map code refers to Fig. 5.1, where the numbers indicate major drainages; multiple samples from the same drainage are denoted by a letter in addition to the drainage number.

Map code	Sample	Number of fish	AAT-3 50	AAT-3 25	AGP-2 50	AGP-2 130	LDH-4 120	ME-2 125	MDH-1 -200	MDH-3 115	MDH-3 80	PGI-1 140	PGI-1 185	PGM-1 75	SDH-1 -50	H (38 loci) %
Eastern Atlantic Ocean (cont.)																
24	Edlida	51	0	0	0	0	0	0.706	0	0	0	0	0	0	0.539	2.4
25	Langa	24	0	0	0	0	0	0.917	0	0	0	0	0	0	0.592	1.7
26	Urrida	12	0	0	0	0	0	0.750	0	0	0	0	0	0	1.000	1.1
Western Atlantic Ocean																
27	†Gaspe[2]	39	0	0.816	0	0	0	1.000	0	0.211	0	0	0	0	1.000	1.7
28 a	Sillikers Bridge	11	0	0.900	0	0	0	0.955	0	0.397	0	0	0	0	1.000	2.0
28 b	Otter Brook	50	0	0.837	0	0	0.020	0.880	0.060	0.151	0	0	0	0	1.000	2.3
29 a	Rocky Brook	100	0	0.790	0	0	0.005	0.960	0.080	0.158	0.005	0	0	0	0.859	2.8
29 b	Taxis River	50	0	0.875	0	0	0	0.910	0.011	0.265	0	0.011	0	0	1.000	2.1
29 c	Sabbies River	50	0	0.874	0	0	0	0.940	0.040	0.337	0	0	0	0	0.900	2.7
30	†Grand Lake[2]	38	0	0.803	0	0	0	1.000	0	0.263	0	0	0	0	1.000	1.9
31	†White River[2]	27	0	1.000	0	0	0	0.885	0	0.009	0.046	0	0	0	1.000	0.8

[1]Data from Cross and Ward (1980). This sample was also reported variable at *IHD-3* and *SDH-3*, with the common allele segregating at frequencies of 0.832 and 0.952, respectively.
[2]Data from Leary and Allendorf (1983).

therefore, no two year-classes will have the same parents. Each hatchery stock sample belongs to a specific year-class, and is so identified in Table 5.1.

Samples of liver, muscle, and eye tissue were dissected from each fish, were frozen immediately, and then were transported to the laboratory, where they were stored at $-60°$ C until they were analyzed electrophoretically. Tissue preparation, starch gel electrophoresis, staining procedures, interpretation of electrophoretic banding patterns, and designation of loci have been described previously by Utter et al. (1974), Allendorf et al. (1977), Cross and Ward (1980), and Ståhl (1981). All fish were analyzed for the following 19 enzymes encoded by 38 loci given in parentheses:

aspartate aminotransferase *(AAT-1,2,3)*
alcohol dehydrogenase *(ADH)*
alpha-glycerophosphate dehydrogenase
    (AGP-1,2)
adenylate kinase *(AK-3)*
creatine phosphokinase *(CPK-1,3)*
diaphorase *(DIA)*
glyceraldehyde-3-phosphate dehydrogenase
    (GAPDH-2)
glutamate dehydrogenase *(GDH)*
beta-glucoronidase *(GUS)*

isocitrate dehydrogenase *(IDH-1,2,3)*
lactate dehydrogenase *(LDH-1,2,3,4,5)*
malate dehydrogenase *(MDH-1,2,3,4)*
malic enzyme *(ME-1,2,3)*
6-phosphogluconate dehydrogenase
    (6-PGDH-1,2)
phosphoglucose isomerase *(PGI-1,2)*
phosphoglucomutase *(PGM-1,2)*
sorbitol dehydrogenase *(SDH-1,2)*
superoxide dismutase *(SOD)*
xanthine dehydrogenase *(XDH)*

The duplicated nature of the salmonid genome (Ohno 1970) often results in confusing electrophoretic patterns and interpretation problems (Allendorf et al. 1975). For two of the variable loci *(MDH-3* and *SDH-1)* it was not possible to score all different genotypes unambiguously, and these loci have subsequently been treated as loci segregating for dominant alleles.

Data Analysis

The phenotypic ratios observed at variable loci segregating for codominant alleles were analyzed for consistency with Hardy-Weinberg expectations in each sample. The commonly used chi-square test for testing homogeneity between observed and expected random mating phenotypic proportions would not be applicable for many of the samples; in those samples with fewer than 50 individuals and skewed allele frequency distributions, the standard statistic may deviate considerably from a chi-square distribution (Hedrick 1983). For this reason an extended version of the exact test described by Vithayasai (1973) was used. The genetic composition of each sample was further tested for deviations from linkage (gametic phase) equilibrium (Hill 1974) at each pairwise combination of genetically variable loci (including the two loci segregating for alleles which had to be treated as dominant ones).

Tests for homogeneity of allele frequencies among samples were performed with the contingency *G*-statistic, using Williams's correction for small sample sizes (Sokal and Rohlf 1981). Phenotypic counts were used as the basis for the comparisons at the *MDH-3* and *SDH-1* loci, whereas absolute allele frequencies were used for the remaining loci. When more than two alleles were segregated at a locus, the variant alleles were combined to avoid small expected numbers within certain cells of the contingency table.

Two graphical procedures were used to depict the genetic relationships among samples. A dendrogram was constructed from a matrix of genetic distances, (*D;* Nei 1972) with the unweighted pair group method of arithmetic averages (UPGMA; Sneath and Sokal 1973). A principal component analysis (Sneath and Sokal 1973) was used as an alternative method, and the first two principal components were visualized in a two-dimensional scatter plot. Both analyses were based on the allele frequency distribution at all 38 loci.

The relative amount of genetic variation among populations was analyzed using Nei's (1973a) measure of gene diversity; this method partitions the total amount of genetic variability into, within, and between population components. A hierarchical procedure was used, based on the geographical distribution of samples, to determine the relative contribution of different components to the overall gene diversity (Chakraborty et al. 1982). The hierarchical structure used was: the total data set, regions (landlocked, Baltic Sea, Eastern Atlantic, Western Atlantic; see Table 5.1), drainages within regions, and samples within drainages. Although samples from landlocked populations do not represent a common geographical area, their suggested subspecific status (see below), led them to be treated as a group of samples representing the same hierarchical level as a group of samples from a particular geographical region.

Additional Allele Frequency Data

In addition to the 49 samples analyzed in our laboratory, allele frequency data reported in the literature on another four samples of Atlantic salmon were included in the analysis of the genetic population structure. Three of the four samples were described by Leary and Allendorf (1983) and represent three hatchery stocks of North American Atlantic salmon (27 †Gaspe, 30 †Grand Lake, 31 †White River; Table 5.1). Allele frequency data on the fourth sample, representing a southern Irish Atlantic salmon population (23 River Blackwater; Table 5.1), were reported by Cross and Ward (1980).

Two assumptions had to be made to make it possible to combine the allele frequency data from the four samples reported by Cross and Ward (1980) and Leary and Allendorf (1983) with those from the 49 samples analyzed in our own laboratory. The first one was necessary because the sets of loci scored by Cross and Ward (1980) and by Leary and Allendorf (1983) are slightly different from those routinely scored in our own laboratory. Particularly, those two studies provide no information on the following six loci: *ADH, DIA, GDH, GUS, 6-PGDH-2,* and *XDH*. No genetic variation was observed at any of these loci in the remaining 49 samples, so for the purpose of the present study these six loci were also treated as monomorphic in the four samples reported by Cross and Ward (1980) and Leary and Allendorf (1983).

The second assumption was necessary because the electrophoretic patterns of specimens analyzed in different laboratories had not been compared on the same gel. Thus, it has not been possible to establish unambiguously the validity of presumed electrophoretic identity of allelic products observed in the different studies. The allelic designations used by Cross and Ward (1980) and by Leary and Allendorf (1983) were therefore transformed into the present nomenclature on the basis of the relative electrophoretic mobility reported for their products and the geographic distribution of alleles observed in the samples analyzed in our own laboratory. These assumptions concerning identity of alleles observed in different studies provide conservative estimates of the amount of genetic variation between as well as within populations.

The data from Cross and Ward (1980) refer to a single sample (23 River Blackwater; Table 5.1). For this sample genetic variation is reported at two loci *(IDH-3* and *SDH-2)* that were scored as monomorphic for all samples analyzed in our own laboratory and by Leary and Allendorf (1983). It is possible that the discrepancy reflects some unidentified methodological differences between laboratories, but this issue cannot be settled without exchanging tissue samples for analysis in different laboratories. At any rate, the inclusion (or exclusion) of the four samples reported by Cross and Ward (1980) and Leary and Allendorf (1983) has little or no effect on the results of the present analyses; those samples were included mainly to provide a picture as complete as possible of populations from all over the species range, and to illustrate the overall agreement between the results obtained in different studies.

RESULTS

Genetic Variation Within Samples

Nine of the 38 loci were found to be polymorphic in the 49 samples analyzed in our own laboratory (Table 5.1). Allelic variation at loci which previously have been reported monomorphic (Ståhl 1983) was found at the following three loci, with the alleles observed given in parentheses: *LDH-4 (100* and *120), MDH-1 (100* and *−200),* and *PGI-1 (100, 140,* and *185).* In addition to the generally most common *100* allele, the variant alleles at the other six variable loci are: *AAT-3 (50* and *25), AGP-2 (50* and *130), MDH-3 (115* and *80), ME-2 (125), PGM-1 (75),* and *SDH-1 (−50).* Allele frequencies at variable loci, given in Table 5.1, were calculated by "allele counting" for all loci except *MDH-3* and *SDH-1.* At those two loci, the frequency of the *100* alleles was estimated from the square root of the proportion of the *(100/100)* homozygotes, which were the only genotypes that could be unambiguously classified (cf. Ståhl et al. 1983).

Since *MDH-3* and *SDH-1* had to be treated as loci segregating for dominant alleles, tests for deviations from Hardy-Weinberg expectations could not be performed for them. For the remaining (codominant) loci, a total of 98 combinations of sample/variable locus permitted such tests. Five of these tests indicated a significant deviation of genotypic proportions from Hardy-Weinberg expectations, with an excess of heterozygotes in all five cases. Three of these tests have P-values slightly less than 0.05 and may well represent statistical type I errors (†Lagan-80, *AAT-3*, P = 0.035; †Bushmills-80, *ME-2*, P = 0.027; Alta Eiby, *ME-2*, P = 0.049). This is because approximately 5 out of the 98 tests performed are expected to result in P-values less than 0.05 by chance only. In contrast, the deviations in the two remaining tests have P-values much smaller than would be expected by chance considering the number of tests (†Emån-80, *AAT-3*, P<0.0001; †Rolfsån-81, *ME-2*, P = 0.005); they require other explanations. The observed deviations from random mating proportions in these two hatchery samples probably reflect the use of a small number of parents to propagate the stock (Johansson 1981). In the †Emån-80 sample all individuals were heterozygous at the *AAT-3* locus (cf. Ståhl 1983).

A majority of the observed statistically significant deviations from random associations of phenotypes among pairs of loci (gametic phase equilibrium) probably reflect statistical type I errors. Of a total of 201 such tests, only 12, representing samples from both naturally reproducing populations and hatchery stocks, showed significant deviations from random association of alleles between loci. The only highly statistically significant deviation was between the *AAT-3* and *SDH-1* loci in the †Emån-80 sample (χ^2 = 17.7, d.f. = 1, P<0.001). This significance most likely reflects the same events as those resulting in the significant deviation from random mating genotypic proportions at *AAT-3* in this sample. Although the statistical power of tests for deviations from gametic phase equilibrium is usually poor (Brown 1975), there appears to be no compelling evidence for such deviations in the populations sampled.

Genetic Variation Among Samples

It appears that Atlantic salmon are naturally substructured into multiple genetically differentiated and more or less reproductively isolated populations within as well as between major drainages. Statistically significant allele frequency differences were obtained between samples from naturally reproducing populations collected from different localities within the same river in four out of five river systems where such a comparison was possible (comparison "a" in Table 5.2). All the variable loci contributed to these differences.

Large allele frequency differences were observed between year classes of the same hatchery stock (comparison "b" in Table 5.2). The differences observed in the present data probably reflect the genetic drift caused by using a small number of parents in the hatchery to propagate the stocks. The differences may also be due to the use of parental fish from genetically different populations that spawn within a river.

It might seem that a mixing of parental fish from genetically different populations would increase the genetic variation in the offspring generation because of hybridization. However, there are no such indications of a general increase in average heterozygosity *(H)* in samples from hatchery stocks (Table 5.1). On the contrary, a *reduced* heterozygosity in hatchery stocks relative to naturally reproducing populations has been demonstrated for Atlantic salmon

Table 5.2 G-values (Sokal and Rholf 1981) of genetic homogeneity tests between samples within major drainages by comparisons of allele and phenotype counts at variable loci. Three types of comparisons were performed:

(a) between samples representing naturally reproducing populations collected at different localities within the same drainage.
(b) between samples representing different derivatives of the same hatchery stock.
(c) between samples from natural populations and hatchery stocks assumed to represent the Atlantic salmon from the same drainage system.

d.f. = degrees of freedom. Levels of significance are indicated thus: *, P < 0.05; **, P < 0.01; ***, P < 0.001.

Map code	Drainage	Number samples	d.f.	Type comparison	AAT-3	LDH-4	ME-2	MDH-1	MDH-3	PGI-1	PGM-1	SDH-1	Sum (G)	d.f. (sum)
1	Gullspång	3	2	b	—	—	81.5***	—	—	—	—	228.8***	310.3***	4
2	Klarälven	2	1	b	—	—	8.8***	—	—	—	—	11.5***	20.3***	2
4	Byglandsfjorden	2	1	b	6.0*	—	—	—	—	—	—	0.9	6.9*	2
5	Torne	4	3	a,b,c	10.9*	—	2.3	—	6.8	—	10.0*	25.5***	55.5***	15
6	Kalix	5	4	a	9.3	—	5.8	—	5.4	—	7.8	28.7***	57.0***	20
8	Byske	2	1	c	13.5***	—	11.8***	—	0.7	—	18.1***	4.4*	48.5***	5
11	Indalsälven	2	1	b	0.4	—	7.3**	—	—	—	—	3.7	11.4**	3
13	Lagan	3	2	b,c	9.6**	—	14.3***	2.2	2.0	—	—	19.3***	47.4***	10
15	Ätran	2	1	c	1.4	—	2.3	—	—	—	—	0.0	3.4	3
21	Alta	4	3	a	29.1***	—	9.5*	—	—	—	—	2.8	41.4***	9
28	L.S.W. Miramichi	2	1	a	0.5	0.5	1.2	2.1	4.6*	—	—	—	8.9	5
29	S.W. Miramichi	3	2	a	5.3	1.0	2.9	8.4*	14.1***	1.8	—	1.4	34.9**	14

from the Baltic Sea (Ståhl 1983). This result supports the conclusion that the observed heterogeneity of allele frequencies is partially the result of genetic drift effects, i.e., a small number of parents (Ståhl 1983).

The dendrogram in Fig. 5.2, constructed from the matrix of pairwise genetic distance values, reveals three clearly distinct clusters representing samples from the Western Atlantic, the Eastern Atlantic, and the Baltic Sea drainages. The European samples which represent landlocked Atlantic salmon populations are found within both of the Eastern Atlantic and the Baltic clusters, with two and six of the landlocked samples found in the Eastern Atlantic and the Baltic clusters, respectively.

GENETIC DISTANCE

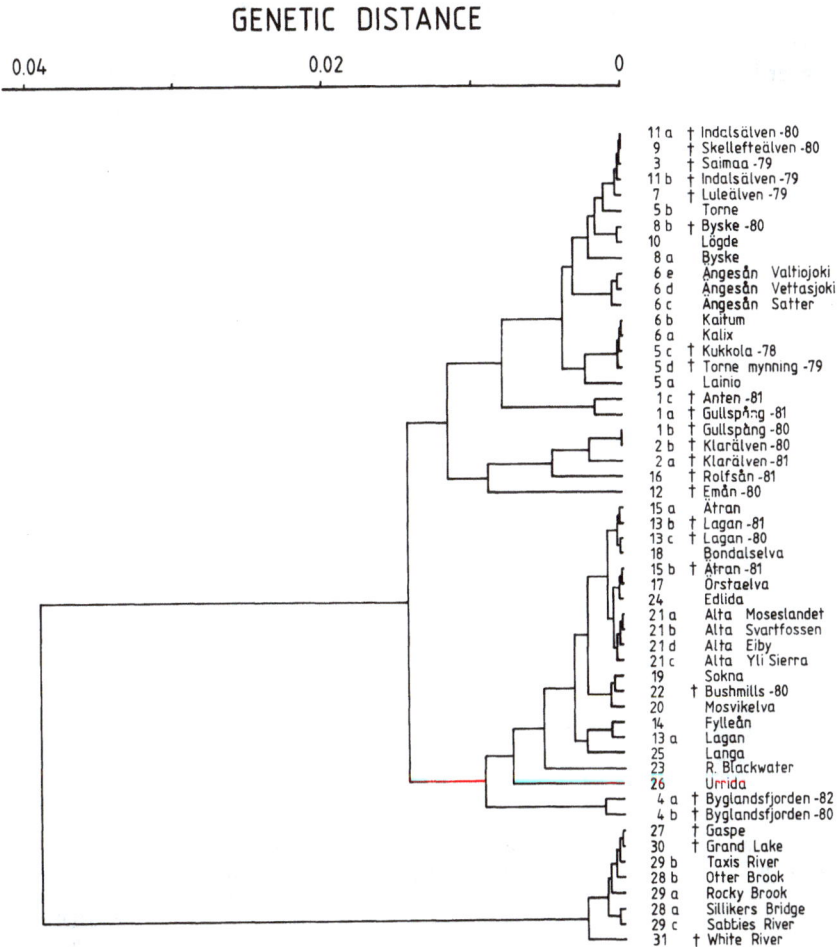

11 a	†	Indalsälven -80
9	†	Skellefteälven -80
3	†	Saimaa -79
11 b	†	Indalsälven -79
7	†	Luleälven -79
5 b		Torne
8 b	†	Byske -80
10		Lögde
8 a		Byske
6 e		Ångesån Valtiojoki
6 d		Ångesån Vettasjoki
6 c		Ångesån Satter
6 b		Kaitum
6 a		Kalix
5 c	†	Kukkola -78
5 d	†	Torne mynning -79
5 a		Lainio
1 c	†	Anten -81
1 a	†	Gullspång -81
1 b	†	Gullspång -80
2 b	†	Klarälven -80
2 a	†	Klarälven -81
16	†	Rolfsån -81
12	†	Emån -80
15 a		Ätran
13 b	†	Lagan -81
13 c	†	Lagan -80
18		Bondalselva
15 b	†	Ätran -81
17		Örstaelva
24		Edlida
21 a		Alta Moseslandet
21 b		Alta Svartfossen
21 d		Alta Eiby
21 c		Alta Yli Sierra
19		Sokna
22	†	Bushmills -80
20		Mosvikelva
14		Fylleån
13 a		Lagan
25		Langa
23		R. Blackwater
26		Urrida
4 a	†	Byglandsfjorden -82
4 b	†	Byglandsfjorden -80
27	†	Gaspe
30	†	Grand Lake
29 b		Taxis River
28 b		Otter Brook
29 a		Rocky Brook
28 a		Sillikers Bridge
29 c		Sabties River
31	†	White River

Figure 5.2 Dendrogram (UPGMA, Sneath and Sokal 1973) summarizing the genetic relationships among 29 samples representing naturally reproducing populations and 24 samples from hatchery derivatives (marked by †) of Atlantic salmon. The dendrogram is constructed from genetic distances (Nei 1972) between samples based on the allele frequencies at 38 loci. The map codes refer to those in Table 5.1.

A similar pattern of population structure as that depicted by the dendrogram is obtained in the principal component analysis shown in Fig. 5.3. The first two principal components explain 83.4 % of the total allelic variation among samples. The three easily identified groups and their relative degree of separation correspond to the clustering observed in the dendrogram. Both methods of analysis support the genetic distinctness of samples from the Western Atlantic, Eastern Atlantic, and Baltic Sea.

Allele frequency differences at the *AAT-3, ME-2, MDH-3,* and *SDH-1* loci (Table 5.1) largely explain the geographic pattern of genetic relationships. Samples in the Baltic cluster are distinguished from other regions by a somewhat lower frequency of the variant alleles at the *ME-2* and *SDH-1* loci (Table 5.1). The Western Atlantic populations differ from the others by having a much higher frequency of the variant alleles at the *AAT-3, ME-2, MDH-3,* and *SDH-1* loci. The Eastern Atlantic populations show frequencies of the variant alleles

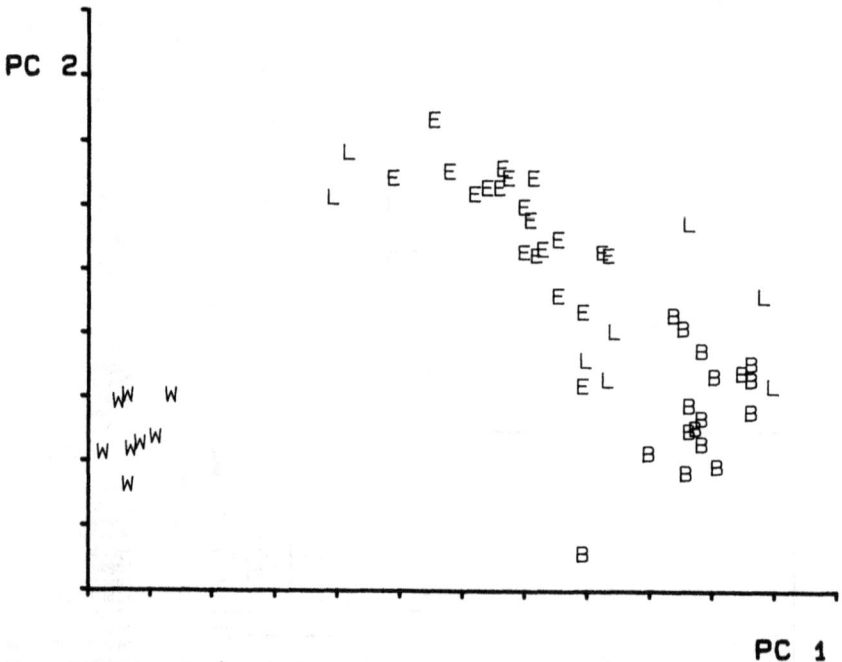

Figure 5.3 Principal component scatter plot derived from analyses of allele frequency estimates at 38 loci in 29 samples from naturally reproducing populations and 24 samples from hatchery stocks of Atlantic salmon. The first principal component (PC 1) accounts for about 65% of the total variance and the second component (PC 2) for 19%. Identical scales are used for the x- and the y-axis.

B: sample from Baltic Sea,

E: sample from Eastern Atlantic Ocean,

W: sample from Western Atlantic Ocean, and

L: samples from landlocked populations (cf. Table 5.1).

that are intermediate between those of the Baltic and the Western Atlantic populations at both *ME-2* and *SDH-1* loci.

The distribution of genetic variation at different hierarchical levels was examined in a gene diversity analysis (Table 5.3). Separate analyses were performed on samples from hatchery stocks, and from naturally reproducing populations, and on all the samples combined. There are differences in both the amount and the distribution of genetic variation between natural populations and hatchery stocks. The total absolute gene diversity is about 10% lower among the 24 samples from hatchery stocks (0.037) than that estimated from the 29 samples representing naturally reproducing populations (0.041). Because of the large standard errors obtained for the estimated values of absolute gene diversity (Table 5.3), this observed difference is not statistically significant.

The relative genetic divergence between samples within drainages is three times as high for hatchery stocks as it is for natural populations (5.1% versus 1.5%). Similarly, the relative genetic divergence between drainages within regions is almost three times as high for hatchery stocks as for natural populations (14.5% versus 4.9%). In contrast, there is about as much relative genetic divergence between hatchery stocks from different regions as there is for natural populations (26.5% versus 29.7%). A greater degree of genetic drift when fish are artificially propagated constitutes a possible explanation for the observed differences in the pattern of gene diversity among samples from hatchery stocks and samples from natural populations. This would result both in a loss of genetic variation and in an increased amount of relative gene diversity between hatchery stocks.

In the Atlantic salmon a remarkably large fraction of the total gene di-

Table 5.3 Gene diversity analyses (Nei 1973) based on 38 loci in geographically grouped samples from naturally reproducing populations and hatchery stocks of Atlantic salmon. Regions are: landlocked, Baltic Sea, Eastern Atlantic Ocean, and Western Atlantic Ocean (see Fig. 5.1 and Table 5.1).

Item	Hatchery stocks	Natural populations	Total
Number of			
Regions	4	3	4
Drainages	17	18	31
Samples	24	29	53
Fish	2,410	1,699	4,109
Absolute gene diversity			
Total	0.037	0.041	0.040
Standard error	(0.020)	(0.021)	(0.021)
Relative gene diversity (in percent)			
Between regions	26.5	29.7	28.4
Between drainages within regions	14.5	4.9	9.0
Between samples within drainages	5.1	1.5	3.6
Within samples	53.9	63.9	59.0

versity is found between samples (more than 40%; Table 5.3) in comparison with other fish species. As calculated by Gyllensten (1985) the average relative gene diversity between localities for six marine, four anadromous, and nine freshwater species are 1.6%, 3.7%, and 29.4%, respectively. The largest part of the between-population component for the Atlantic salmon is that observed among regions (28.4%). On the basis of European samples only, Ryman (1983) estimated that 12.3% of the total amount of genetic variability in hatchery stocks and natural populations when considered together was due to genetic divergence between the Baltic and the Eastern Atlantic regions. Thus, inclusion of the Western Atlantic region more than doubles the between-region component (Table 5.3).

It should be noted that estimates of the proportions of gene diversity reflect both the relative numbers of samples from different regions and the distribution of samples within and between drainages (Chapter 4). For instance, additional samples from Western Atlantic drainages would most likely result in a further increase of both the total gene diversity and the between-region component.

The pattern of genetic variation for individual loci varies considerably among the 11 polymorphic loci contributing to the total gene diversity (Table 5.4). Three of these loci, *AAT-3*, *ME-2*, and *SDH-1*, contribute more than 90% of the total absolute gene diversity. The component of relative gene diversity between regions varies among loci from less than 1% up to almost 40%. The

Table 5.4 Distribution of gene diversity at variable loci among 53 geographically grouped samples of naturally reproducing populations and hatchery stocks of Atlantic salmon (see Fig. 5.1). The average is based on a total of 38 loci.

	Absolute gene diversity		Relative gene diversity (in percent)			
Locus	Total	Within samples	Between regions	Between drainages within region	Between samples within drainages	Within samples
AAT-3	0.410	0.203	39.7	8.3	2.4	49.6
AGP-2	0.0009	0.0008	0.04	1.3	—	98.7
IDH-3	0.006	0.005	0.5	16.0	—	83.5
LDH-4	0.0009	0.0009	0.3	0.5	0.9	98.3
ME-2	0.497	0.256	35.2	11.7	1.5	51.6
MDH-1	0.010	0.009	1.3	2.2	1.7	94.7
MDH-3	0.076	0.059	17.0	3.0	2.3	77.7
PGI-1	0.001	0.001	0.1	1.2	0.2	98.6
PGM-1	0.008	0.007	0.8	4.1	5.9	89.2
SDH-1	0.492	0.343	15.5	7.9	6.9	69.7
SDH-2	0.002	0.002	0.1	4.6	—	95.3
Average	0.040	0.023	28.4	9.0	3.6	59.0
Standard error	0.021	0.012	5.8	1.1	1.3	5.2

major part of the overall divergence between regions is caused by differentiation at the three most polymorphic loci and at an additional fourth locus, *MDH-3*, which is highly polymorphic in the Western Atlantic samples only.

In summary, Atlantic salmon seems to be naturally substructured into more or less reproductively isolated and genetically differentiated populations within river systems. There are at least three major genetically distinct and geographically separated groups of Atlantic salmon within the natural range of the species. The greatest genetic differences are found between populations from the two continents of North America and Europe. The differentiation between the two continents is more than twice as great as that between the two European clusters, i.e., between the Eastern Atlantic and the Baltic populations. There are indications of both (1) a reduced amount of genetic variation and (2) genetic instability within the hatchery stocks as reflected by allele frequency differences between year classes and deviations in genetic equilibrium conditions.

Also, the pattern of genetic differentiation among hatchery stocks is different from that for natural populations. The major reason for these differences between hatchery stocks and natural populations is probably the use of too few parental fish when stocks are propagated in the hatcheries.

DISCUSSION

Evolutionary Implications

In absolute terms, i.e., as measured by genetic distance, relatively little overall genetic differentiation appears to have occurred among populations throughout the natural range of Atlantic salmon. The differentiation observed between the Western Atlantic and the European populations agrees qualitatively with the results of Payne et al. (1971), who identified contrasting distributions of transferrin alleles in European and Canadian collections. However, the average genetic distance between these two groups (D = 0.04) lies well within the range found between conspecific populations of other salmonids (Allendorf and Utter 1979, Ryman and Ståhl 1981, Ferguson and Fleming 1983, Ryman 1981). The limited divergence over all loci indicated by the present data supports Behnke's (1972b) contention that subspecific recognition on the basis of the single locus coding for transferrin (Payne et al. 1971) is unjustified.

The current data indicate that European populations are substructured into two major genetical groups corresponding to the geographical regions of the Eastern Atlantic and the Baltic Sea. Payne et al. (1971; see also Thorpe and Mitchell 1981) postulated the existence of a Boreal, a Celtic, and presumably additional "races" of Atlantic salmon within the European continent on the basis of the geographical distribution of allele frequencies at a transferrin locus According to their postulation, the Boreal race is distributed all over northern Europe including the Baltic Sea, while the Celtic race is native to the southern parts of the British Isles (Payne et al. 1971). Populations from southern Europe were presumed to represent a possible third race of the European Atlantic salmon. According to those authors, all but one of the present samples from the European continent represent the Boreal race.

The sample from River Blackwater, Ireland, was collected within the geographical range suggested for the Celtic race (Payne et al. 1971). Although there are two variant alleles observed exclusively in the sample from River Blackwater, in the dendrogram this sample clusters together with the other samples from the Eastern Atlantic region (Fig. 5.2). The geographical distributions suggested for the Boreal and the Celtic races (Payne et al. 1971) do not correspond with the major genetical grouping of European Atlantic salmon observed in the present study. A more extensive sampling of British and southern European populations is required to settle the issue of whether the Eastern Atlantic salmon is subdivided into more than one major genetic unit.

All localities sampled in the present study, except perhaps the Irish River Blackwater, were covered by an ice cap during the last glaciation (Sparks and West 1971). The amount of genetic differentiation observed among populations within each one of the three major geographical regions would therefore reflect a maximum time since divergence of approximately ten thousand years or less. That there is greater genetic differentiation between the two continental groups of Atlantic salmon than within continents suggests that North American and European populations diverged prior to or during the last glaciation. Each continent was then repopulated from refuge populations following the Pleistocene glacier recession.

The geological events leading to the formation of the present Baltic Sea (Lundqvist 1965, Gudelis and Königsson 1979) may explain the evolution of genetic differentiation observed between populations of Atlantic salmon native to rivers draining into the Baltic Sea on the one hand and populations native to rivers draining into the Eastern Atlantic Ocean on the other. Until ten thousand years ago a freshwater lake, the Baltic Ice Lake, was located south of the ice cap, approximately in the southern part of the present Baltic Sea. There was no water connection between this lake and the Atlantic Ocean to the west. As the ice melted and the glacier retreated north over the Scandinavian Peninsula, a connection opened between the Baltic Ice Lake and the Atlantic Ocean. During the following thousand years, the lake became a bay of the Atlantic Ocean with a marine environment. Several marine organisms, such as whales, seals, fishes and molluscs invaded the area at this time (Lundqvist 1965). It is likely that the Atlantic salmon was among the fish species migrating into this area. Due to an elevation of the land, the lake once again became isolated from the Atlantic Ocean and water salinity decreased. Some of the marine species were not able to survive this lacustrine period, which lasted about two thousand years. However, it is likely that the Atlantic salmon was able to survive because of its capacity to evolve a wide range of environmental adaptations, such as lacustrine and anadromous forms (see below).

The present outlet of the Baltic Sea originated about 7,500 years ago as a strait between Denmark and Sweden. The migratory behavior of the Baltic

Atlantic salmon of today is characterized by a nearly complete absence of individuals that migrate into the Atlantic Ocean (Christensen and Larsson 1979). This behavior might be due to their previous isolation in the Baltic freshwater lake. Accordingly, the Eastern Atlantic and the Baltic populations would have been reproductively isolated for about 9,000 years.

The concept of a nonanadromous subspecies of landlocked Atlantic salmon *(S. s. sebago),* existing on the continents of North America and Europe, persists in the literature (e.g., Christensen and Larsson 1979, Vuorinen 1982). The present data do not support this concept. In the dendrogram (Fig. 5.2), two of eight landlocked samples (†Byglandsfjorden-82 and †Byglandsfjorden-80) cluster with the anadromous populations of the Eastern Atlantic region, and the remaining six samples cluster with the anadromous samples from the Baltic Sea. A similar picture is revealed by the principal component scatter plot (Fig. 5.3). The results of both cluster analyses support Behnke's (1972b) hypothesis that nonanadromous populations of Atlantic salmon were derived from anadromous stocks in postglacial times. It should be noted that the present genetic data from the landlocked populations are obtained from hatchery derivatives. Thus, it appears that these stocks may not quite represent the genetics of the corresponding native populations (cf. comparison "c" in Table 5.2). However, it is not likely that this would obscure the classification with respect to their subspecific status.

Management Considerations

There is a general recommendation in fishery management to conserve as much intraspecific genetic diversity as possible in order to preserve future opportunities for management and use of populations (e.g. Ryman 1981c, Spangler et al. 1981, Chapter 15). Knowledge about the intraspecific genetic structure and a thorough understanding of the evolutionary relationships among naturally reproducing populations of Atlantic salmon are essential prerequisites for planning and carrying out biologically sound management programs. The heterogeneities of allele frequencies demonstrated between populations at electrophoretically detectable loci should be viewed as minimal estimates of genetic differences at the remainder of the genome (Lewontin 1974, 1984). Loss of genetically determined stock characteristics due to overexploitation of less productive stocks or by unintentional hybridization of different gene pools in hatchery operations will result in the disappearance of local adaptations reflected by population characteristics such as time of return and area of spawning. Considering the genetic instability observed within the hatchery stocks, it is not surprising that comparative studies of recovery ratios for hatchery stocks and naturally adapted Atlantic salmon populations have strongly favored natural populations (Stabell 1984).

The present picture of the genetic structure of Atlantic salmon has management implications that in some instances conflict with current practices. The practice of managing individual drainages as genetic units (e.g. Christensen and

Larsson 1979, Thorpe and Mitchell 1981) is not justified by the present data. Statistically significant allele frequency differences among samples from naturally reproducing populations within drainages (comparison "a" in Table 5.2) are indications of genetic differentiation that would ultimately vanish under such a management plan. In order to protect the present genetic structure among naturally reproducing Atlantic salmon populations, a discriminatory transplantation policy both within and between river systems should be implemented. By creating gene flow between previously reproductively isolated populations, a breaking up of adapted gene complexes may result from (an unintentional) hybridization of stocks. Eradication of locally adapted populations would then result in a decrease of the overall productivity of the species (Chapter 1). In an opposing argument, Larkin (1981) has suggested a more simplified approach, with a focus on similarities rather than differences among stocks as the basis for transplantation programs. However, similarities for many characters may mask undetected and highly important dissimilarities for others.

Transplantation of Atlantic salmon from the Baltic Sea to Norwegian waters is one example of a nondiscriminatory transplantation program which has been carried out for several years. During the last decade Norwegian hatcheries have not been able to produce enough Atlantic salmon smolts to supply the demand, and some fishery managers therefore took advantage of the contemporary overproduction of smolts in Swedish hatcheries. Although the Swedish rivers drain into the Baltic Sea and the Norwegian ones into the Atlantic, the Swedish and Norwegian populations of Atlantic salmon are undoubtedly similar in many respects. For instance, they are of the same latitudinal origin, and the geographic proximity of spawning sites (Fig. 5.1) implies that fish from these populations in many ways experience similar climatic and other environmental conditions during the freshwater stages of the life cycle.

Nevertheless, the transplantations from Sweden (Baltic populations) to Norway (Eastern Atlantic populations) have had disastrous effects on previously existing Norwegian populations of Atlantic salmon. A large number of the Norwegian drainages have now been found to be infected by a skin parasite, *Gyrodactylus salaris*, which attacks and kills the young of Atlantic salmon. Its presence has dramatically reduced the number of Atlantic salmon in many river systems (Heggberget and Johnsen 1981). Johnels (1984) suggested that the parasite was introduced into the Norwegian rivers by resistant Atlantic salmon from the Baltic Sea.

Thus, in this case a transplantation policy based on similarities (cf. Larkin 1981) has complicated rather than simplified the future management of Atlantic salmon in Norway. Preventing transplantation between drainages and especially between regions will greatly aid in preserving local adaptations and will prevent future disasters such as the reduction of many Norwegian populations.

The frequently observed genetic differences between hatchery and wild stock of the same river (comparison "c" in the Table 5.2) invalidate the generality that a hatchery stock is a random and genetically representative sample of the natural population. The allele frequency heterogeneities observed between year classes of the same hatchery stock (comparison "b" in Table 5.2) reflect a step toward further genetic impoverization. Although no studies have been performed regarding allele frequency variations among year classes in natural populations of Atlantic salmon, it appears that natural populations of salmonids seem to be characterized by an apparent stability of allele frequencies over the time spans so far examined (Ryman 1983 and references therein). The reason for the heterogeneities observed among year classes of the same hatchery stock appears to be either nonrandom sampling of parents or the use of too few parental fish to avoid genetic drift. Therefore, the common occurrence of such heterogeneities reflects a major need for remedial actions when they are detected, and for regular monitoring to reduce further losses of genetic variation (Chapter 1). To avoid inadvertent genetic changes no stock should be founded or perpetuated using fewer than 30 parents of the less numerous sex in any generation (Ryman and Ståhl 1980; cf. Chapter 6). The management implications of the genetic differences observed between hatchery and wild stock as well as between year classes of the same hatchery stock have been previously discussed for brown trout (Ryman and Ståhl 1980, Ryman 1981a, Vuorinen 1984), cutthroat trout (Allendorf and Phelps 1980), and Atlantic salmon (Cross and King 1983, Ståhl 1983). The present data extend earlier observations and emphasize the importance of their conclusions. The findings of pronounced genetic instability of many hatchery stocks strongly contradict the statement that "...electrophoretic analysis of populations was of little use in either identifying the source of past errors in broodstock management or predicting which stocks to select for propagation" (Gall 1983). Without electrophoretic data these findings would never have been detected. Electrophoretic analysis (Chapter 2) is the primary tool that is now available for identifying past errors and for pointing to counteractions.

The capability to use genetic data to estimate the proportions of contributing populations in a population mixture has applications to Atlantic salmon management, e.g., the high seas fishery off the West Greenland, around the Faroe Islands, and in the Baltic Sea. Biologically sound management, aimed at avoiding overexploitation of any of the contributing stocks in these fisheries, requires accurate estimates of the proportional contribution from each population to the total harvest. Conditions for obtaining genetic data for reliable estimates in a mixed fishery analysis include:
- known allele frequency differences among populations potentially contributing to a particular mixture, and
- a sufficient sampling of the mixture to provide estimates of adequate statistical precision (Miller et al. 1983, Fournier et al. 1984).

In the West Greenland fishery, Atlantic salmon from both North America and Europe are harvested (Saunders 1966, 1981). The present data suggest at least four loci *(AAT-3, ME-2, MDH-3, SDH-1)* that have the combined capability of identifying individuals to regional origins with high precision (Table 5.1 and 5.4).

The amount of genetic differentiation between populations within regions (Table 5.3) indicates sufficient differences for analysis of adequately sampled mixed populations within geographical regions as well. However, to obtain reliable estimates of the contribution from different stocks, it is necessary that the allele frequencies characterizing the contributing populations remain stable. This criterion is not fulfilled for the hatchery stocks (Table 5.2). If within-stock heterogeneities cannot be avoided in the hatchery programs, it will become impossible to estimate reliably the proportions of different contributing stocks in mixed fisheries. It should be pointed out that an intentional manipulation of allele frequencies of hatchery populations (genetic tagging) provides a method whereby the contribution of individual hatcheries may be much more readily detectable in population mixtures, and with an increased statistical precision for the obtained estimates. In this connection it is suggested that Atlantic salmon involved in both intentional and unintentional transplantation programs carry genetic tags. Such a tag would automatically be transferred to the offspring generation and possible interference with natural populations could then easily be traced.

The present knowledge of the genetic population structure of Atlantic salmon calls for an increased and immediate awareness of the genetic effects of various management activities as well as of the potential use of genetically determined population characteristics in aquaculture and fish breeding. Decisions about hatchery programs, transplantation policies, and harvesting strategies need to be based on information about the genetic structure of the populations concerned in order to avoid unintentional and unnecessary loss of genetic variation through genetic drift and inbreeding in hatchery stocks, inadvertent hybridization in hatcheries, or transplantation of fish into environments for which they are not adapted. Electrophoretic techniques provide an excellent tool for delineation of the genetic variation in natural populations and hatchery stocks, and such information is a prerequisite for the conservation of biological diversity in fish populations.

6.

Genetic Management of Hatchery Stocks

Fred W. Allendorf and Nils Ryman

The number of large-scale hatchery programs for artificial propagation of fish has increased dramatically during the last few decades. One group of hatchery projects reflects the growing global interest in aquaculture. Other projects are part of fishery management programs to produce fish for release into natural bodies of water. One purpose of these latter efforts is to augment natural reproduction or to reduce the impact of ecological damage caused by man's increasing alteration of the natural habitat of fishes. Another purpose is the conservation of wild populations whose natural habitat has been damaged or populations that are threatened by the introduction of exotic species.

Genetic variation is the basic resource of any successful animal breeding program. The goal for the management of hatchery populations of fish is dependent upon the purpose of the hatchery project. For instance, a selective breeding program aimed at producing fast-growing fish with high food conversion and docile behavior for aquaculture should start from a base population containing a large amount of genetic variability. Such a selection program will change the genetic composition of the base population by reducing its genetic variability; as selection continues, "positive" alleles replace "negative" ones. In contrast, programs that produce fish for release into the wild should not attempt to genetically alter populations to perform well under hatchery conditions, but, rather, should strive to maintain the genetic variation of the original wild population. The achievement of the latter goal is still more critical when the hatchery functions as a "gene bank" conserving the last remnant of an endangered population.

Some hatchery projects have failed to manage the genetic resources successfully. One reason is that their management goals sometimes have been vaguely defined. For instance, the genetic goals of aquacultural pen rearing and those pertinent to the raising of fish to supplement wild populations cannot be achieved simultaneously with the same stock. Furthermore, principles of population genetics and animal husbandry have frequently been ignored in the founding and maintenance of hatchery populations. It is often difficult, for example, to acquire a large number of individuals to found a new hatchery population, but to ignore the expected reduction in genetic variation due to small population size (bottlenecks) is to increase the probability that the hatchery program will fail to achieve its objectives. The effects of losing genetic variation in small populations is well documented in a variety of organisms (Ralls and Ballou 1983). Examples documenting such losses in hatchery populations of fish are listed in Table 6.1. However, this should not be taken to indicate that hatchery populations generally have reduced amounts of genetic variation. For example, most hatchery populations of rainbow trout *(Salmo gairdneri)* have approx-

Table 6.1 Examples of electrophoretic studies indicating genetic changes and loss of genetic variability in hatchery stocks.

Species	Genetic characteristics	Reference
Rainbow trout *(Salmo gairdneri)*	Conspicuously low level of genetic variation coupled with poor survival.	Allendorf and Utter 1979
Cutthroat trout *(Salmo clarki)*	Different from presumed source population. Differences between year-classes. Proportion of polymorphic loci reduced by 57%. Average number of alleles/locus reduced by 29%. Average heterozygosity reduced by 21%.	Allendorf and Phelps 1980 Leary et al. 1985c
Brown trout *(Salmo trutta)*	Different from presumed source population. Differences between year-classes. Reduced proportion of polymorphic loci. High mortality and atypical morphology.	Ryman and Ståhl 1980
Brown trout	Different from presumed source population. Differences between stock derivatives maintained in different hatcheries. Reduced levels of genetic variation.	Ryman 1981b
Brown trout	Differences between stock derivatives maintained in different hatcheries. Proportion of polymorphic loci reduced by 50%.	Ryman and Ståhl 1981
Brown trout	Different from presumed source population. Differences between cohorts. Average heterozygosity reduced at 33%.	Vuorinen 1984
Atlantic salmon *(Salmo salar)*	Different from presumed source populations. Differences between year-classes. Differences between stock derivatives maintained in different hatcheries. Linkage disequilibrium and strong deviations from Hardy-Weinberg expectations. Average heterozygosity within stocks reduced by 20%. Reduced divergence between stocks.	Ståhl 1983 and Chapter 5
Atlantic salmon	Different from presumed source population. Differences between year-classes. Reduced levels of genetic variability.	Cross and King 1983
Black seabream *(Acanthopagrus schlegeli)*	Different from presumed source population. Reduced levels of genetic variability. Effective number of parents only 15%-25% of actual number.	Taniguchi et al. 1983

imately the same amount of genetic variation as natural populations (Allendorf and Utter 1979, Busack et al. 1979).

The application of genetic principles to the founding and maintenance of hatchery populations is straightforward. The first step is to define the goals of the hatchery project. We are concerned here only with fishery management and therefore will not consider aquaculture. We assume that the purpose of the hatchery is to produce fish that will be placed into the wild to survive and at times to reproduce. The principal genetic goal of hatcheries with these objectives is to minimize any genetic changes caused by genetic drift or adaptation to hatchery conditions.

In this chapter, we discuss basic genetic principles relating to the founding, maintenance, and monitoring of hatchery stocks and recommend ways to minimize genetic alteration of hatchery stocks. Our discussion is based largely on our experience with salmonid fishes, but the genetic principles we consider apply to hatchery programs for all species of fishes.

FOUNDING OF HATCHERY POPULATIONS

Selection of Founding Populations

The first genetic decision to be made in establishing a hatchery population is the source of the founders. In many cases, a hatchery stock is established by taking individuals from an already existing hatchery population. There recently has been a major effort in the United States to compile information about the characteristics and relative performance of hatchery populations so that a suitable stock can be selected (Kincaid 1981). In the past, stocks have sometimes been chosen simply because they are readily available, without any consideration of genetic characteristics. It is important that the source of fish selected for founding be genetically suited to accomplish the goals of the management program.

Hatchery populations sometimes are derived directly from wild populations. This is often the case for anadromous salmonid populations, for hatcheries built to compensate for loss of spawning habitat, and for programs aimed at preserving the genetic resources of a particular species, subspecies, or population that is threatened in its native habitat. Hatchery programs starting with a wild stock must consider the importance of local adaptations. There is a well-documented tendency for salmonids to evolve genetically discrete, ecologically specialized populations by natural selection over thousands of generations of adaptation to local environmental conditions (Behnke 1972, Ricker 1972, Ryman et al. 1979). A number of recent studies with hatchery releases have indicated that hatchery fish derived from local populations perform much better in their native environment than hatchery fish from other populations (Bams 1976, Reisenbichler 1981, Altukhov and Salmenkova, Chapter 14).

Hatchery programs designed to preserve the genetic resources of species or subspecies through eventual reintroductions require additional considerations. The genetic variation of a species can be partitioned into genetic variation within local populations and genetic divergence between local populations. It is necessary to understand the amount and distribution of this genetic variation, as discussed by Chakraborty and Leimar (Chapter 4), in order to ensure preservation of genetic resources (cf. Chapter 5). Without this information, we act blindly in making decisions regarding the preservation of the genetic resources of a species.

No single strategy applies in selecting source populations for a hatchery program aimed at preserving the genetic resources of an entire species or subspecies. The best plan of action for a specific case must be based on the biology and distribution of genetic variation in a particular species, as well as practical considerations (e.g., how many separate hatchery strains can be maintained?). Vrijenhoek et al. (1985) conducted an analysis of genetic variation in the endangered Sonoran topminnow *(Poeciliopsis occidentalis)* and found that the genetic variation in the species is composed of diversity within localities (21%), between localities (26%), and between three major groups (53%). On this basis they concluded that it is vital to preserve representatives of all three major groups and less important to maintain multiple populations within these major groups.

In some cases it may be possible to maintain only one hatchery strain as a "gene bank." In this situation, a choice must be made between founding the hatchery stock from one local stock or combining many local populations into a single hatchery strain (Krueger et al. 1981). It may be desirable to combine many local populations when a significant proportion of the genetic variation is represented by between-population divergence, as long as this divergence is not so great as to suggest the possibility of more than one species. It must be noted, however, that local adaptations and the specific gene combinations of the local stocks will be lost when these stocks are combined into a single strain. Unfortunately, there is no way around this problem; the loss of local adaptations and gene combinations is the price to be paid if a major part of the gene diversity of a whole species or subspecies is to be conserved within a single population.

Number of Founders

A hatchery stock should be founded with a sufficient number of individuals to accurately reflect the genetic composition of the natural population from which it was derived. The number of founders is crucial because the alleles present in the founders represent the upper limit of allelic variation in subsequent generations, without the introduction of new alleles by mutation or from other populations. A single diploid individual can have a maximum of two alleles at any particular locus; it is therefore obvious that one or two individuals of both sexes do not adequately represent the genetic composition of a natural population. However, it can be shown from fundamental genetic principles that hundreds of individuals are not necessary to meet this requirement (cf. Chapter 3).

We can examine the genetic effects of the founding process by considering an extreme but not unrealistic example. Assume we found five separate hatchery stocks by taking one female and one male from a random mating natural population. Consider a single locus with two alleles, *B* and *b*, at frequencies of 0.6 and 0.4, respectively; the genotypic frequencies at this locus in the natural population will be in binomial (Hardy-Weinberg) proportions: 0.36*(BB)*, 0.48*(Bb)*, and 0.16*(bb)*. We assume there are 1,000 progeny produced by mating the one female and one male; these progeny are considered to be the first generation of the hatchery stock. The second generation of the hatchery stock is produced by randomly mating males and females from the first generation.

Figure 6.1 shows the results of simple Monte Carlo simulations of this situation. *P* is the frequency of the *B* allele and *H* is the proportion of heterozygotes. Two random numbers uniformly distributed between 0 and 1 were chosen to determine the genotypes of the two founders of each stock on the basis of the genotypic frequencies in the natural population. The genotype frequencies in the first generation are based on expected Mendelian proportions resulting from the founders. The genotypic frequencies in the second generation will be in Hardy-Weinberg proportions. We assume that the population sizes in

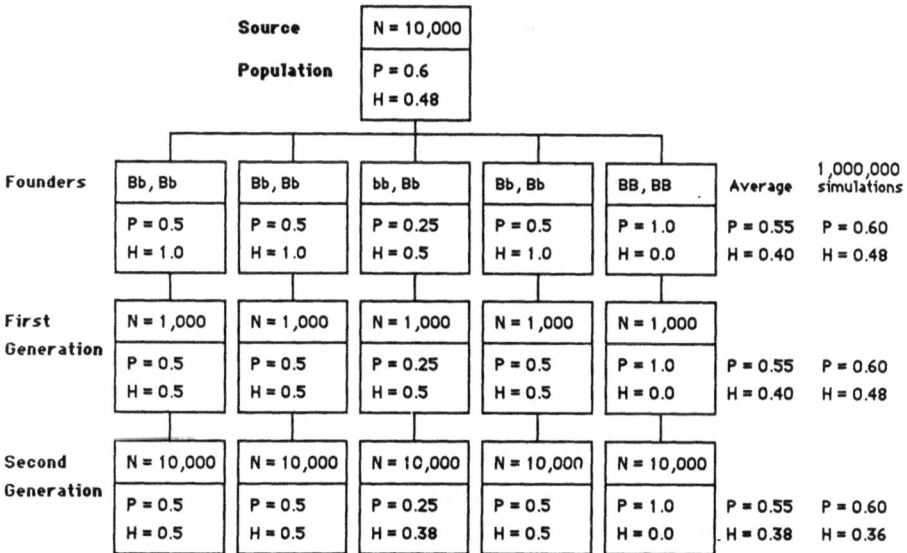

							Average	1,000,000 simulations
Source Population	N = 10,000 P = 0.6 H = 0.48							
Founders	Bb, Bb P = 0.5 H = 1.0	Bb, Bb P = 0.5 H = 1.0	bb, Bb P = 0.25 H = 0.5	Bb, Bb P = 0.5 H = 1.0	BB, BB P = 1.0 H = 0.0		P = 0.55 H = 0.40	P = 0.60 H = 0.48
First Generation	N = 1,000 P = 0.5 H = 0.5	N = 1,000 P = 0.5 H = 0.5	N = 1,000 P = 0.25 H = 0.5	N = 1,000 P = 0.5 H = 0.5	N = 1,000 P = 1.0 H = 0.0		P = 0.55 H = 0.40	P = 0.60 H = 0.48
Second Generation	N = 10,000 P = 0.5 H = 0.5	N = 10,000 P = 0.5 H = 0.5	N = 10,000 P = 0.25 H = 0.38	N = 10,000 P = 0.5 H = 0.5	N = 10,000 P = 1.0 H = 0.0		P = 0.55 H = 0.38	P = 0.60 H = 0.36

Figure 6.1 Monte Carlo simulations of the effects at a single locus with two alleles (*B* and *b*) of founding a hatchery stock from one male and one female from a source population with an allele frequency of 0.6 (P) of the *B* allele (see text for detailed explanation). *N* is population size and *H* is the proportion of heterozygotes present. The second column from the right indicates the average of the five diagrammed simulations. The column on the far right indicates the averages of these parameters for one million simulations.

the first and second hatchery generations (1,000 and 10,000) are large enough so that changes in allele frequencies in these generations can be ignored. Results are also presented from a more extensive series of identical simulations in which 1 million hatchery stocks were founded with one male and one female from this same natural population.

The individuals chosen as founders are random representatives of the natural population. Nevertheless, the allele frequency, *P,* within a pair of founders can only be 0, 0.25, 0.50, 0.75, or 1.00. Because of the large number of progeny produced, allele frequencies do not change within a stock in subsequent generations. Thus, sampling error associated with the founding event ensures that every stock will have a different allele frequency from the natural population. The average genotypic frequencies in the first hatchery generation are not expected to change since all individuals are simply the progeny of two individuals chosen at random from the natural population. Thus, there is no expected reduction in heterozygosity in this generation. However, all the individuals within a stock in the first generation are full-sibs. The "inbreeding" (Jacquard, 1975) caused by the limited size of the founding population results in reduced heterozygosity in the second hatchery generation. Allele and genotype frequencies are not expected to change in subsequent generations so long as a large number of individuals are randomly mated every generation to produce the next generation.

We can summarize these results as follows. A small number of founders results in allele frequency differences between the source population and the newly founded stock because of sampling error. Heterozygosity is expected to be reduced because of inbreeding in the second hatchery generation when the genotypic proportions in the hatchery stock reach Hardy-Weinberg proportions. The series of 1 million simulations can be interpreted as representing the expected effects at one locus in many stocks or at many loci within one stock. It is possible that heterozygosity at one locus in the newly founded stock is greater than in the source population. For example, if one of the founders happens to be heterozygous for an allele that is rare in the source population, this allele will occur in the hatchery stock with a frequency of 0.25 and *H* will be 0.375, as compared to an *H* of near zero in the source population. Nevertheless, the extended series of simulations shows that approximately 25% of the heterozygosity (0.48 versus 0.36) over the entire genome will be lost because of the use of only two founders.

The above results can be generalized for any size of founding population. Genetic drift is expected to change allele frequencies at variable loci. The *direction* of change in allele frequencies is random and therefore cannot be predicted. However, the expected *magnitude* of the changes in allele frequencies can be predicted. Consider a single gene locus with two alleles, *B* and *b,* in a population of size *N.* If *P* is the current frequency of the *B* allele, then 95% of the time the frequency of *B* in the next generation will be in the following interval:

$$P' = P \pm 2\sqrt{\frac{P(1 - P)}{2N}} \ . \tag{1}$$

When N is already large (e.g., 50 or greater), making it even larger will do little to reduce the expected change in allele frequency.

Minimizing the loss of genetic variation due to genetic drift is a primary concern in founding a hatchery population. The expected proportion of the original heterozygosity remaining after a bottleneck of size N for one generation is

$$1 - \frac{1}{2N} \ . \tag{2}$$

It is perhaps surprising that even very small population sizes will retain most of the heterozygosity in a population. For example, the progeny of one male and one female are expected to contain 75% of the total heterozygosity in a population. This expectation is realized in the simulations presented in Figure 6.1. Similarly, the use of five founders of each sex would result in an expected loss of only 5% of the heterozygosity in the source population. Nevertheless, it should be kept in mind that studies with a variety of agricultural species have shown that even a 10% loss of genetic variation has detectable harmful effects on such important traits as survival and growth rate (Falconer 1981). Kincaid (1976a,b) has shown in rainbow trout that a 25% loss of heterozygosity due to inbreeding is associated with an increase in morphological deformities (38%), decreased food conversion efficiency (6%), decreased fry survival (19%), and decreased weight at 147 days (11%) and 364 days (23%). Studies with a variety of fish species show that these results are general (reviewed by Kincaid 1983).

Equation (2) is potentially misleading because it uses the proportion of individuals that are heterozygous as the measure of genetic variation. Another important measure of genetic variation that can be used is the actual number of alleles present at a locus. Heterozygosity has been widely used to measure genetic variation because it lends itself readily to theoretical considerations of the effect of limited population size on genetic variation. Nevertheless, heterozygosity has the disadvantage of being relatively insensitive to the actual number of alleles at a locus. For example, consider two alternative situations. A particular locus in population X has 2 alleles at equal frequencies of 0.5. This same locus in population Y has 7 alleles, 1 with a frequency of 0.7 and the other 6 all at a frequency of 0.05. Which population has more genetic variation? The intuitive answer is that population Y has more genetic variation because of the much greater number of alleles present. However, population X has the greater amount of heterozygosity: 0.500 versus 0.495 in population Y.

Equation (2) provides an overly optimistic view of the effects of small numbers of individuals on the loss of genetic variation if we consider allelic diversity. The effect of a bottleneck on the number of alleles is more complicated than its effect on heterozygosity because it depends on the number and frequencies of alleles present in the original population. Consider a locus with n alleles at frequencies $P(1), P(2), P(3),...., P(n)$ in a large population. If a hatchery population is founded from this population, the expected number of alleles remaining (n') is (Denniston, 1978)

$$E(n') = n - \Sigma(1 - P_j)^{2N} , \qquad (3)$$

where N is the number of founders.

We can examine the effects of a founding event on the number of alleles present using Eq. (3) and assuming n equally frequent alleles in our initial population. We can use a quantity A (allelic diversity) as a measure of the proportion of genetic variation remaining based on the number of alleles retained at a polymorphic locus (n'), where

$$A = \frac{n' - 1}{n - 1} \qquad (4)$$

and n is the original number of alleles present. Allelic diversity ranges from 1, when all alleles are retained, to 0 when all alleles but one are lost. It can be seen in Fig. 6.2 that bottlenecks often have a greater effect on allelic diversity than on heterozygosity when multiple alleles are present. For example, although 75% of the heterozygosity is expected to be retained with a founding size of 2, a maximum of 4 alleles can be retained and less than half of the allelic diversity will be retained if there are 5 or more alleles.

In order to judge the potential importance of the loss of alleles in founding a stock we must know something about the number and frequency distributions of alleles throughout the genome of the species of interest. Unfortunately, we have very little information about this for any fish species. Almost all of our current information about genetic variation in fish species comes from electrophoretic examination of water-soluble proteins that either are enzymes or are present in large quantities. DNA-sequencing technology recently has provided

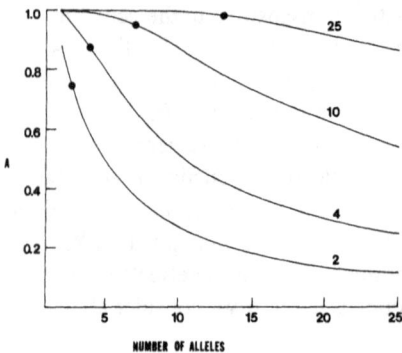

Figure 6.2 Allelic diversity (A) remaining versus original number of alleles after a bottleneck of a single generation of sizes 2, 4, 10, and 25 (Eq. 3). All alleles are assumed to be equally frequent. The solid circle represents the proportion of heterozygosity expected to be retained. $(1 - 1/2N)$.

NUMBER OF ALLELES

evidence that the variation detected by the examination of proteins greatly underestimates the actual amount of genetic variation present at a locus. Most of the genetic variation in DNA sequences at protein-coding loci that have been studied does not affect the amino acid sequence of the protein produced (Kreitman 1983, Ayala 1984). It should be noted, however, that just because a DNA substitution does not affect the amino acid sequence does not mean the substitution will not have a phenotypic effect. Evidence is accumulating that regulatory changes that affect the time and place of expression of protein-coding sequences may have greater phenotypic effects than changes in the protein itself (cf. Allendorf et al. 1983).

Immunological studies of vertebrates have revealed that blood group and other histocompatibility loci commonly have many alleles at substantial frequencies. For example, the HLA system in humans consists of several tightly linked genes that are highly polymorphic with many alleles at nearly equal frequencies (Bodmer et al. 1978). Different HLA genotypes are associated with resistance to a variety of diseases (Thompson 1981). Remarkably similar histocompatibility systems have been found in a wide variety of species (Zaleski et al. 1983). Bottlenecks may only slightly decrease heterozygosity, but they cause the loss of much allelic diversity at loci, such as *HLA*, with many alleles. The increased uniformity of individuals at such loci affecting disease resistance is likely to make a stock more susceptible to epizoötics.

Effective Population Size

The number of breeding individuals in a population usually does not indicate the actual rate of genetic drift. Such factors as the sex ratio and the number of offspring per individual also affect the rate of loss of genetic variation. It is desirable that all founding individuals contribute equally to a hatchery population to minimize the effects of genetic drift. This effect is seen most clearly by considering the number of male and female founders. The effective founding size is greatest when there are the same number of males as females so that each individual contributes equally. Consider a founding population of 100 individuals, consisting of 99 females and 1 male. One-half of the genes in the founded hatchery population will be contributed by the single male because each sex contributes equally to the next generation. The "effective" size of this population, with regard to the effects of genetic drift, will be much less than 100.

The effective population size can be defined as the size of an ideal population that would lose genetic variation at the same rate as the population under consideration (cf. Gall, Chapter 3). An ideal population is one that has a 1:1 sex ratio and in which all breeding individuals have an equal probability of being the parent of any progeny individual. The effective size with unequal numbers of males (N_m) and females (N_f) can be estimated by (Falconer 1981)

$$N_e = \frac{4(N_f)(N_m)}{N_f + N_m}.$$ (5)

Thus, a population of 99 females and one male will have an effective population size of only 4. In other words, genetic drift in this population is expected to be comparable to that in a population of 2 males and 2 females.

Fertility and fecundity differences must also be considered to ensure that all founders contribute equally to the next generation. For example, some females may have more eggs than others and some males may be more successful in fertilization because of hatchery spawning procedures. We can maximize effective population size by minimizing such differences between individuals. Such a procedure can actually increase the effective population size beyond the total number of parents used because the ideal population is expected to have random differences in contributions between individuals due to chance. Thus, if we can equalize the contribution of all individuals, the effective population size will approach twice the number of reproducing individuals (Denniston 1978).

As an example, consider a situation with 10 females, 7 of which have 1,000 eggs each, and each of the 3 remaining females have 10,000 eggs. Due to the large variation in fecundity the effective number of females is only about 4 (see Crow and Morton 1955 and Ryman et al. 1981 for computional procedures). Equalizing the contributions so that each female contributes 1,000 eggs will result in approximately 20 effective females. Thus, discarding the majority of the eggs greatly increases the effective number of females. It is fairly simple to equalize individual contributions with fishes that have external fertilization. Families from a single male and single female can be made and kept separately during the early period of high mortality. Equal numbers of fry from each family can then be placed together.

No ideal number of individuals needed to found a hatchery population that emerges from these theoretical considerations of genetic drift. The limiting consideration is probably the expected loss of infrequent alleles with small population sizes. For example, at least 30 founders have to be used to be 95% sure that an allele at a frequency of 0.05 will be present in the founders (Eq. 3). We suggest that a founding population of at least 25 females and 25 males is a reasonable absolute minimum (Ryman and Ståhl 1980, Shaklee 1983). We also strongly urge that efforts be taken to equalize the contribution of all founders. With salmonids, the procedure suggested above is practical and should be effective. Procedures with other species will depend upon the reproductive biology of the species and the hatchery facilities available.

MAINTENANCE OF HATCHERY POPULATIONS

Nonselective Changes

There are also general genetic principles that should be applied to achieve the objective of minimizing genetic change in hatchery stocks after the founding event. The principles of genetic drift that apply to maintenance apply equally to the founding of a hatchery population. Approximately $1/2N$ of the genetic variation is expected to be lost every generation (Eq. 2). Thus, after t generations

$$(1 - 1/2N)^t \qquad (6)$$

of the genetic variation is expected to be retained. Approximately 10% of the genetic variation will be lost in a stock with an effective population size of 50 after ten generations. Thus, a population size of considerably more than 50 is desirable after the initial founding event to achieve the objective of minimizing genetic changes induced by artifical propagation in a hatchery. Efforts should also be made within practical constraints to equalize the genetic contribution of individuals.

Selective Changes

Some selective changes in a hatchery stock are unavoidable (Hynes et al. 1981); for instance, those genotypes with poor survival in a hatchery will be selected against. Perhaps more important, however, are the mortalities that would occur in the wild that may not occur under hatchery conditions. Thus, physiologically or morphologically inferior genotypes that would be selected against in the wild (e.g., albinism) may contribute to subsequent generations in the hatchery and therefore may be found at higher frequencies in hatchery stocks. This "release" from natural selection may contribute substantially to the deterioration of hatchery stocks.

Selection for hatchery conditions must be minimized. For example, selection for larger or uniform size should be avoided. Such selection is common in hatcheries because of difficulties in maintaining fishes of different sizes. Selection for other traits, such as body shape or conformation, also should be eliminated. Only fish with obvious morphological deformities (e.g., missing fins or opercular flaps) should be selectively removed from the broodstock.

Time of spawning is also an important consideration. An effort should be made to save fish that are fertilized throughout the spawning period for inclusion in the broodstock. Use of fish spawned on only one day during the spawning season causes selection for a particular phenotype (time of spawning) and will have the same effects as other forms of selection. Individuals ripe for spawning on a particular day are, on the average, more likely to be related than two individuals chosen at random from the broodstock. This nonrandom sampling of the broodstock will decrease effective population size. Moreover, the population will respond to this selection, so that an increasing proportion of the population will be ripe at the time the brood fish are selected. This is likely to change the time and reduce the variability in time of spawning of the broodstock.

Reintroduction of eggs or sperm from wild fish may be desirable under some circumstances. This would minimize the divergence between the wild and hatchery stock and still allow some adaptation to hatchery conditions so that the population could be maintained successfully. There is no ideal frequency and proportion of such "infusions" from the wild. However, a 10% contribution of wild fish every second or third generation would be sufficient under most circumstances.

Recommendations for Population Maintenance

We recommend that a minimum population of at least 100 males and 100 females be used to sustain a hatchery population (see also Shaklee 1983). All types of selection should be avoided in order to minimize adaptation to hatchery conditions. We also recommend that efforts be made to equalize the genetic contribution of individuals by using equal numbers of males and females, by using approximately the same number of eggs from each female, and by using one male to fertilize the eggs from a single female. The importance of these actions depends upon the number of spawners used. If far more than 200 fish are used, these actions are probably not necessary. If the population is restricted to well below the recommended size, other actions can maximize effective population size. For example, marking full-sib groups would make it possible to equalize the contributions of each family to the next generation and thereby increase effective population size.

Our recommendations are based on the fact that a breeding program for hatchery-reared fish to be planted into the wild has goals very different from those of a breeding program for aquaculture. Fish planted in the wild must survive long enough to provide angling, and, in some cases, to reproduce. Thus, sufficient numbers of reproducing adults should be used to minimize the loss of genetic variation through the inbreeding effect of small populations sizes. In addition, we recommend that action be taken to minimize adaptation to the hatchery environment since such changes are likely to reduce the performance of fish planted in the wild.

The differences in the genetic goals of aquaculture and producing fish for release in the wild have not always been kept clear. For example, a recent book describing hatchery methods for salmon and trout culture (Leitritz and Lewis 1976) states the objective of the trout selective breeding program used by the California Department of Fish and Game: "to supply quality brood fish and at the same time to improve the economics of the production of fingerlings, subcatchable, and catchable fish through genetic means." Unfortunately, the criteria of quality used all relate to performance in the hatchery—size of eggs, number of eggs, percent egg mortality, size of fingerlings—and mortality of fingerlings yet only ten full-sib families are selected on the basis of these criteria to produce the next generation of the broodstock.

Such a program may be desirable for aquaculture but is likely to limit the performance of a broodstock designed to produce catchable fish. Five percent of the genetic variation is expected to be lost each generation because of the small number of parents used in this scheme (Eq. 2). Many studies with trout and other species have indicated that a loss of 10% percent of the genetic variation usually produces detectable harmful effects on a wide variety of important characteristics (e.g., fecundity, survival, and growth). In addition, the selection scheme is based only on hatchery performance with no consideration of the performance of the stock in the wild; this is likely to reduce further the performance of these fish when released from the hatchery.

MONITORING HATCHERY STOCKS

The recommendations outlined above are aimed at avoiding excessive genetic changes caused by genetic drift or adaptation to hatchery conditions. Unfortunately, following these recommendations provides no absolute guarantee that genetic changes will be kept at a low level. A mistake when handling fish or eggs may result in inadvertent mixing of stocks, for example, or reduced survival of some families in the artificial hatchery environment may drastically reduce the number of effective parents. A monitoring program is necessary to ensure that unwanted genetic changes do not occur, or at least that they do not go undetected (Hynes et al. 1981).

A number of factors should be considered when constructing a genetic monitoring program. The optimal design is dependent on the hatchery facilities available, the purpose and time scale of the hatchery program, and species characteristics (e.g., discrete or overlapping generations and generation interval). If the purpose is to conserve the last remnant of a unique population, it may be more important to identify inadvertent admixture with other stocks than to detect modest reductions in the level of genetic variability. The opposite may be true for a stock that is founded from a mixture of populations and that is perpetuated with the primary purpose of representing as much allelic diversity as possible.

A thorough treatment of the design aspects for efficient genetic monitoring programs for hatchery stocks would include a considerable amount of advanced population genetics and statistical estimation theory. Such a presentation is beyond the scope of the present volume, but an important question about the genetic management of many hatchery stocks follows: When perpetuating the stock, we have strived for a minimum number of 200 effective parents per generation. Do the gene frequency data collected over the last few years indicate that we have achieved this goal?

Severe reductions in the effective number of parents can be easily detected in most stocks, but the hypothesis testing is much more complicated for detection of minor differences between the assumed and real number of effective parents. For rigorous treatments of this and related problems, the reader is referred to texts such as Hill (1979), Pamilo and Varvio-Aho (1980), Nei and Tajima (1981b), and references therein. For the purpose of the present chapter, we confine the discussion to some basic aspects of genetic monitoring that should be applicable to most situations of genetic management of hatchery stocks.

A genetic monitoring program should ideally include many characteristics that directly or indirectly reflect the variation at a large part of the entire genome. However, the information needed for most practical purposes may be obtained by analyzing a sufficiently large number of electrophoretic loci and a carefully selected set of morphological characters.

Protein Electrophoresis

A properly designed monitoring program should be based on a detailed genetic description of the founders. The genes present in the founders constitute the basis of the stock, and considerable effort must be expended to depict the founders' genotypes in this initial phase of the program. We consider it a minimum requirement that some 30-50 electrophoretically detectable loci be scored for each founder contributing to the first generation of the stock. Analyzing only a random sample of the founders should not be considered unless the number of founders is large (e.g., exceeding 300 individuals of each sex). The loci analyzed should include as many as possible of those that have previously been shown to be polymorphic in the species. Frozen tissue samples from the founders may be stored to permit future analyses of additional loci.

The monitoring is quite straightforward once the genotypic frequencies of the founders have been determined. In each generation (or year-class) a random sample from the stock is analyzed for all the loci scored in the founders. It is imperative that loci found to be nonvariable in the founders also be included in the study; polymorphism at an initially monomorphic locus is the most obvious indication of mixing with another stock.

If the hatchery program runs satisfactorily, allele frequency variations between generations should not be larger than what may be explained by random genetic drift. However, the different steps in the various statistical sampling processes must be considered when evaluating observed shifts of allele frequencies. Preferably, the genotype should be determined for each fish used in breeding, that is, for all the fish chosen as parents for the next generation. The allele frequency observed in a particular generation of parents can then be regarded as a statistical parameter that is determined exactly without sampling error. This can be easily done in semelparous species such as the Pacific salmon, which die after spawning and may be sacrificed after stripping.

Consider, for example, a stock of a semelparous species that is maintained with a presumed minimum effective number of 200 parents. Assume that a particular allele is found to occur at the frequencies P and P' among the breeders from two consecutive generations, respectively. Then, the absolute value of the difference between P and P' is not expected (with 95% probability) to exceed $2\sqrt{[P(1-P)/400]}$ (Eq. 1). A larger difference would indicate some deviation from the genetic conditions desired for the stock. When information is available for multiple independent loci, the differences observed may be evaluated jointly. At each particular locus the quantity

$$\frac{(P' - P)^2}{\frac{P(1-P)}{2N_e}} \tag{7}$$

approximately follows a chi-square distribution with one degree of freedom. These quantities can be added over loci, and the sum should be chi-square distributed with degrees of freedom equal to the number of loci.

In iteroparous species the breeders are often spawned repeatedly, and they may not be available for electrophoretic analysis until years after producing their first offspring. In such cases, the magnitude of the allele frequency difference between two consecutive generations of breeders $(P - P')$ must be estimated indirectly from allele frequency estimates obtained from a sample of their offspring. In statistical terms this means that the parameters P and P' are estimated by the statistics p and p' representing the observed allele frequencies in the two samples of offspring, respectively. Similarly, the difference $(P - P')$ must be estimated from the difference $(p - p')$, the sampling error of which is considerably larger than a simple binomial one.

In situations like this it is imperative that the statistics p and p' be based on sample sizes large enough to provide an accurate picture of the true difference they are meant to estimate. This may be particularly important when generations are overlapping and the magnitude of temporal allele frequency changes must be inferred from differences between year classes rather than generations. Without going into the statistical details, we suggest the following rule of thumb, which should provide sufficient accuracy in most practical situations: If the goal is to maintain the stock with a minimum number of N_e effective parents per generation, each of the p and p' estimates should be based on no fewer than $2N_e$ offspring individuals.

In addition to the allele frequency data, the genotypic distributions may provide information of critical importance for understanding the genetic development of the stock. For instance, an excess of heterozygotes relative to binomial (Hardy-Weinberg) proportions is expected to occur in the offspring from a parental generation in which allele frequencies are different in males and females. Such a difference between sexes may result from stock mixture or from a severe reduction of the effective number of parents of either sex. A deficit of heterozygotes is expected if inbreeding is occurring; that is, if there is a tendency for related individuals to be mated together. Such a tendency could result from the fact that the individuals ripe on a particular spawning day have a greater than average probability of being related because of the genetic control of spawning time.

In contrast, an inadvertent stock hybridization that is not associated with gene frequency differences between sexes is not expected to result in deviations from Hardy-Weinberg proportions. Such a situation may occur if males and females from different stocks are mixed and mate at random within the mixture. However, if there are gene frequency differences between the stocks, a hybridization of this type will result in nonrandom associations between alleles at pairs of loci (linkage disequilibrium; cf. Campton, Chapter 7). Thus, the analysis of single and multiple locus genotypic distributions may provide extremely valuable information critical for the detection of genetic changes or for explaining the possible cause of unexpected shifts of allelic frequencies.

Some examples of the approximate sample sizes needed to detect allele frequency differences at a single locus with two alleles are given in Table 6.2. The sample sizes have been calculated (Sokal and Rohlf 1981, chap. 17) for a situation in which a sample from an uncontaminated stock is compared with one from a population representing a mixture between that stock and another one that has a different allele frequency. The statistical test is for allele frequency homogeneity; possible differences may be due to either hybridization or mixing of offspring from the two stocks. The 100% admixture may represent a situation in which a group of fish has been mislabeled. Fairly small samples are sufficient to detect admixture when allele frequency differences are large. However, mixing is difficult to detect when stocks have similar allele frequencies. Each locus that is examined may be considered an independent statistical test. Therefore, the ability to detect admixture increases with each additional locus that is analyzed. For example, the allele frequency differences typically ob-

Table 6.2 Approximate number of sampled fish required to provide a 90% probability of obtaining an allele frequency difference that is statistically significant at the 5% level. The admixed stock represents a mixture, or hybrids, between the uncontaminated stock that is being monitored and a foreign one characterized by a different allele frequency. Sample sizes are given for select combinations of foreign stock admixture rate and allele frequency for two cases with allele frequencies 0.5 and 0.9. See text for further explanation.

Case 1. Uncontaminated stock allele frequency = 0.5

Foreign stock admixture rate	Foreign stock allele frequency				
	0	0.1	0.2	0.3	0.4
10%	1,046	1,637	2,914	6,563	26,265
50%	37	61	112	258	1,046
100%	6	12	24	61	258

Case 2. Uncontaminated stock allele frequency = 0.9

Foreign stock admixture rate	Foreign stock allele frequency				
	0	0.2	0.4	0.6	0.8
10%	159	248	458	1,186	9,873
50%	9	15	27	65	458
100%	2	4	8	20	132

served in multiple locus studies of natural populations (e.g., Ståhl, Chapter 5) provide considerable statistical power even with moderate sample sizes.

Electrophoresis is the most efficient tool for determining the genetic constitution of large numbers of individuals (Chapter 2). It is, therefore, the basic and primary technique to be applied for direct genetic monitoring of hatchery stocks. However, in some special situations it may be necessary to apply additional techniques to depict the genetic characteristics of a stock. For instance, chromosomal markers may be sought in species or stocks which exhibit unusually low levels of electrophoretically detectable genic variation (Chapter 13). Restriction enzyme cleavage patterns of the maternally inherited mitochondrial DNA (mtDNA) may also be analyzed in such cases or when there is particular interest in determining the relative genetic contribution from different females (Chapters 11 and 12).

It must be remembered, though, that in the context of genetic monitoring of the hatchery stocks, mtDNA analyses can provide little direct genetic information other than the number of maternal mtDNA-lineages present in the stock. Analysis of mtDNA provides no information about either the genetic contribution of the males or the genes located on the chromosomes in the nucleus, which constitute the vast majority of the genome.

Morphological Characters

The loss of genetic variation has been found to have harmful effects on embryological development in fish. There are several reports of increased frequency of deformed individuals associated with the loss of heterozygosity caused by inbreeding in fish species: carp, *Cyprinus carpio* (Moav and Wohlfarth 1963, Kirpichnikov 1981), swordtail, *Xiphophorus helleri* (Baker-Cohen 1961), rainbow trout, *Salmo gairdneri* (Aulstad and Kittelsen 1971, Kincaid 1976a,b), and channel catfish, *Ictalurus punctatus* (Bondari 1983). Kirpichnikov (1981) has suggested that an increase in the frequency of morphological deformities could be used as an indicator of the loss of variation due to breeding practices.

Fluctuating asymmetry is potentially a much more sensitive morphological indicator of the loss of heterozygosity than the frequency of morphological deformities. Fluctuating asymmetry is said to occur when the difference between a character on the left and right sides of individuals is normally distributed about a mean of zero (Van Valen 1962). Phenotypic differences within individuals for bilateral traits exhibiting fluctuating asymmetry reflect the inability of an organism to develop precisely along predetermined pathways. Increased fluctuating asymmetry is an indication of increased developmental instability and may reflect reduced genetic variation.

Results with salmonids indicate that fluctuating asymmetry is sensitive enough to detect differences in developmental stability between individuals with different amounts of heterozygosity at isozyme loci in noninbred populations

(Leary et al. 1984b). Results with a hatchery strain of westslope cutthroat trout *(Salmo clarki lewisi)* known to have reduced genetic variation (Allendorf and Phelps 1980) and with gynogenetic diploid rainbow trout confirm that the loss of heterozygosity in inbred fish produces detectable decreased developmental stability as measured by fluctuating asymmetry (Leary et al. 1985a,b). Thus, fluctuating asymmetry provides a potentially valuable measure of the loss of heterozygosity in fish populations. In this section, we discuss the procedures for using this technique and its applications. The procedures can be outlined in three steps:

- Select a number of morphometric or meristic bilateral characters in which differences between the sides can be accurately detected. It is desirable to work with a suite of meristic traits that are all determined ("fixed") early in development. Thus, fishes of different ages and sizes can be directly compared. It is difficult to determine what correction should be used to compare fish for traits that continue to increase throughout the life of the fish (Valentine and Soulé 1973, Valentine et al. 1973, Soulé and Cuzin-Roudy 1982).
- Score a number of individuals from several populations for the characters chosen. Check for the existence of directional asymmetry or antisymmetry for each character (Van Valen 1962). Directional asymmetry occurs when there is normally a greater development of one side than of the other. It can be detected when the mean values differ systematically between the two sides. Directional and fluctuating asymmetry may occur together in the same character. The directional asymmetry can be corrected so that differences in fluctuating asymmetry can still be measured (Mather 1953). Antisymmetry occurs when asymmetry is normally present, but it is variable which side has greater development. An example of this is handedness in "major league baseball outfielders, who do not, however, form a Mendelian population" (Van Valen 1962). Antisymmetry results from a negative interaction between sides and can be detected by a bimodal distribution of the signed differences between the sides. This distribution may also result from extreme fluctuating asymmetry, as shown by Mather (1953) in populations where he selected for asymmetry.
- A measure of asymmetry must be used so that an overall value of asymmetry can be assigned to each individual and an average value of asymmetry can be estimated for the population. There is no single best estimator of overall individual asymmetry; it depends upon the distributions of the characters that are used (see Leary et al. 1984b, Valentine et al. 1973, Vrijenhoek and Lerman 1982, Felley 1980).

This procedure will provide a relative measure of the average amount of developmental stability in a population. We envision the most valuable use of this technique to be the monitoring of populations through time. A progressive increase in average asymmetry would indicate a loss of genetic variation through inbreeding or an increase in environmental stress. The ideal monitoring program would combine an examination of allele frequency changes at isozyme loci and changes in fluctuating asymmetry. Such a program would be able to both detect the loss of genetic variation and simultaneously evaluate the effects of such loss on the population.

The monitoring of fluctuating asymmetry is especially desirable in a program concerned primarily with detecting loss of genetic variation. Only a few of the many polymorphic loci throughout the entire genome can be detected with electrophoresis. Significant loss of genetic variation may go undetected by an examination of a few electrophoretically detectable loci. However, results with salmonids indicate that even slight reductions in genetic variation are detectable by an examination of fluctuating asymmetry (Leary et al. 1984b, 1985a,b). In addition, the generally high heritabilities of meristic traits (Leary et al. 1985c) makes them potentially suitable for detecting genetic changes caused by genetic drift or stock admixture.

7.
Natural Hybridization
And Introgression in Fishes
Methods of Detection and Genetic Interpretations
Donald E. Campton

> *The Jordan school of ichthyology held steadfastly to the view that*
> *the lines between fish species are almost never crossed. This*
> *mistaken idea was perhaps in part a holdover of the pre-Darwinian*
> *concept of the immutability of species, but was more pointedly, a*
> *reaction against the tendency of some European ichthyologists to*
> *explain as hybrids specimens that proved difficult to identify.*
>
> *Carl L. Hubbs (1955)*

The ability and propensity of taxonomically distinct fishes to interbreed
and produce viable hybrid offspring are now firmly established. Schwartz
(1972, 1981) compiled a total of 3,759 references dealing with the natural and
artificial hybridization of fishes. Most of these studies were published after
Hubbs (1955) summarized his extensive investigations of natural hybridization
among North American fishes.

Natural hybridization is believed to be more common in fishes than in
other groups of vertebrates. Several characteristics of fishes may account for
this distinction: external fertilization, weak ethological isolating mechanisms,
unequal abundance of the two parental species, competition for limited spawning habitat, and susceptibility to secondary contact between recently evolved
forms. These characteristics are affected to varying degrees by local habitat.

Natural and man-induced changes in environmental conditions are often
cited as causes of hybridization in fishes. For example, hybridization is relatively common among temperate freshwater fishes in areas where geologic and
climatic events since the Pleistocene have drastically altered aquatic environments, but hybridization appears to be rare among marine and tropical fishes
that inhabit more stable environments. Man-caused habitat changes in North
America have also been correlated with hybridization between both previously
allopatric and naturally sympatric pairs of species (Hubbs et al. 1953, Nelson
1966, 1973, Stevenson and Buchanon 1973). In addition, sympatric species that
rarely or never hybridize in nature often hybridize freely within the confines of
aquaria (Hubbs 1955). As Hubbs's (1955) extensive investigations led him to
conclude, "It is evident that the hybridization is conditioned by environmental
factors."

The ease with which fish hybridize in nature presents many problems to fishery biologists and management agencies. A major objective of fisheries management is to protect indigenous populations of fishes from overexploitation, habitat degradation, and exotic species that may interact detrimentally with native species through predation, competition, or hybridization. Unfortunately, the potential for hybridization has often been overlooked in many management programs. The widespread stocking of fishes outside their native geographic ranges for fishery enhancement or other management purposes has frequently resulted in hybridization between native and introduced forms. This has been true particularly in the western United States, where the indiscriminate stocking of rainbow trout *(Salmo gairdneri)* into nonnative regions has resulted in extensive hybridization with several indigenous trout species (references in Dangel et al. 1973, Behnke 1979). Similar hybridization has occurred between introduced smallmouth bass *(Micropterus dolomieui)* and endemic Guadelupe bass *(M. treculi)* in the southern United States (Edwards 1979, Whitmore and Butler 1982, Whitmore 1983).

Congeneric species of fishes are often interfertile, and hybrid swarms representing genetic admixtures of the two parental species may be produced following introductions of nonnative fishes. Repeated backcrossing of hybrid descendants with a parental species can further result in the *introgression* of genes from one species into the gene pool of another. This process of *introgressive hybridization* can cause the genetic loss of an entire species, subspecies, or unique population. In addition, small amounts of introgression may be very difficult to detect.

Fishery biologists and management agencies thus are confronted with the tasks of detecting hybridization and introgression in natural populations, distinguishing F_1 and later generation hybrids from parental species, and estimating the relative contributions of parental species to suspected genetic admixtures. These tasks are clearly problems in population genetics. In this context, hybridization can be viewed as a dynamic interaction between genetically differentiated groups of populations.

In this chapter, I review methods for detecting natural hybridization and introgression in fishes and describe some basic principles of population genetics that are useful for evaluating the extent and direction of these phenonomena. The chapter comprises two major sections. In the first, more general section, I review the empirical literature and discuss the capabilities and shortcomings of the various methods and approaches. The second section is largely theoretical and is intended as background material for those students and researchers who anticipate using the principles of population genetics to actually investigate natural hybridization in fishes. This second section also illustrates the expected behavior of gene and genotype frequencies in populations formed by hybridization. The approach is methodological. I use North American fishes for specific examples because natural hybridization has been most intensively studied in these fishes. This is due primarily to the pioneering investigations of Carl L. Hubbs and his associates from 1920 to 1960. The discussion is not restricted to hybridization between distinct taxa, however, but includes hybridization between populations within taxa. For example, the interbreeding of hatchery and wild fish can be viewed as the introgressive hybridization of two conspecific populations.

A review of natural hybridization in fishes should also include a special section on the family Poeciliidae, because interspecific hybridization within this family has resulted in several hybridogenic and gynogenetic forms. These hybridized biotypes have been the subject of several recent reviews (Schultz 1977, Turner 1982, Monaco et al. 1984, Moore 1984, Vrijenhoek 1984) and are not discussed here. Artificial hybridization is not discussed, but interested readers may consult the two bibliographies of Schwartz (1972, 1981) for references on specific fishes and crosses.

METHODS OF DETECTING HYBRIDIZATION

Detecting natural hybridization between fishes is often complicated because several situations are possible. Two closely related species, or conspecific populations, might coinhabit an area without ever interbreeding, yet hybridization may be suggested because of phenotypic overlap. On the other hand, hybridization might occur, but the hybrids themselves may never breed for one of several reasons, including infertility. If hybrids backcross with one or both parental species, introgression may have occurred or a hybrid swarm may be present. If hybridization is known to have occurred, one may want to distinguish individuals of mixed ancestry from those of the two parental species. However, as will be pointed out later, distinguishing individuals of mixed ancestry from parental species is often impossible if hybridization has proceeded past the F_1 generation.

This section is concerned primarily with detecting the past or present occurrence of hybridization between two species or populations. *Hybrid* here refers to any individual with a mixed ancestry.

Morphology

Until the mid-1960s, comparing morphological characters was essentially the only method available to ichthyologists and fishery biologists for detecting natural hybridization. In such studies, meristic and morphometric characters are counted and measured on the suspected hybrid individuals and on individuals representing the two hypothesized parental species. Specific procedures usually follow those described by Hubbs and Lagler (1970). The unknown fish are concluded to be interspecific hybrids if their meristic counts and morphometric measurements are, on the average, intermediate to the values for the two parental species.

One of the first significant reports of natural hybridization in fishes was by Hubbs (1920). In that study, Hubbs used morphological criteria to reinterpret the rare sunfish "species" *Lepomis euryarus* (family Centrarchidae) as a green sunfish *(L. cyanellus)* × pumpkinseed *(L. gibbosus)* hybrid (cited by Hubbs 1955). This and other suspected sunfish hybrids were verified by artificial matings between the parental species (Hubbs and Hubbs 1932). In all matings, the hybrid offspring expressed a combination of characteristics that were intermediate to the two parental species and virtually identical to those for the suspected natural hybrids. The sunfish hybrids, both natural and artificial, also tended to grow faster than either of the parental species and were predominantly sterile males (Hubbs and Hubbs 1931, 1932, 1933).

Morphological characters used to distinguish two species are usually not uniformly intermediate in known or suspected hybrids (e.g., Simon and Noble 1968, Ross and Cavender 1981, Leary et al. 1983). Hybrids often express a mosaic of morphological characters that separately resemble those for each parental species. The hybrid may appear morphologically intermediate for all characters combined, but for a specific character it may closely resemble one of the two parental species.

Hubbs and Kuronuma (1942) and Hubbs et al. (1943) defined a statistic called the *hybrid index* in order to measure the average morphological similarity of an individual fish to each of two species or other taxa. The index (I) is calculated separately for each character as $I = 100 \times [(u-X)/(Y-X)]$, where u is the value of the trait for the individual being evaluated and X and Y are the mean values of the trait for species X and Y. An individual fish with a value for the trait equal to X or Y will have an index value equal to 0 or 100, respectively. An index value of 50 indicates exact intermediacy for the character in question. The average value of the index for all diagnostic characters may be close to 50 in both known and suspected hybrids, but the individual values may vary widely from character to character (e.g., Ross and Cavender 1981). Hybrids may also express a value of the index for a particular character that is less than 0 or greater than 100 (Hubbs and Hubbs 1947). Average values of the index in the intermediate range of 30–70 for wild caught individuals are usually taken as evidence of hybridization (Hubbs et al. 1943, Mayhew 1983).

The primary advantage of the hybrid index method is that values of several characters for a large number of individuals can be summarized in the form of histograms to demonstrate the general morphological intermediacy of the suspected hybrid individuals. This allows the results to be easily evaluated quantitatively. Hubbs's hybrid index is still used to report natural hybridization in fishes when morphological criteria are applied (e.g., Gilbert 1978, Menzel 1978, Stauffer et al. 1979, Mayhew 1983). Misra et al. (1970) and Misra (1971, 1972) have made some sampling theory modifications of the index, but these modifications are rarely employed.

In recent years, use of Hubbs's hybrid index has lost ground to multivariate statistical methods (Smith 1973, Neff and Smith 1979). Two major objections to the hybrid index are that (1) it requires the *a priori* identification of individuals from the two parental species and (2) it fails to account for the variances and covariances of the discriminating traits. Highly correlated traits are often different measurements of the same biological phenomenon, reflecting the pleiotropic action of genes or the common response to environmental effects (Falconer 1981). The hybrid index essentially assumes, however, that all traits are uncorrelated. Furthermore, species-discriminating traits with low intra-specific variances should be given more weight than traits with high variances; however, the hybrid index weights all traits equally regardless of their discriminating power. Multivariate statistical methods can circumvent these objections by deriving weighting factors for each trait based on the variance-covariance structure of the data set.

Discriminant function analysis (DFA) is one multivariate statistical method that has frequently been used to analyze morphological data in studies of natural hybridization. DFA derives weighted linear functions of the measured traits so that two or more groups of individuals are maximally separated or dis-criminated in multivariate space. A common approach is to derive the discrimi-nant function using known individuals from the two parental species and then to calculate the discriminant function scores for each of the suspected hybrids. Histograms of discriminant function scores can then be used to demonstrate overall morphological intermediacy of the suspected hybrids (see Neff and Smith 1979, Fig. 2). If the hybrids can also be identified *a priori*, two-dimen-sional plots can be produced to demonstrate the ability of morphological char-acters to discriminate the two species and their hybrids (e.g., Whitmore 1983, Leary et al. 1983). Despite these apparent merits, DFA suffers from one major shortcoming which limits its usefulness in hybridization studies: DFA requires the *a priori* classification of individuals into the groups being discriminated (see Neff and Smith 1979). Individuals originally misclassified by the inves-tigator may distort the discriminant function and thereby cause the mis-classification of unknown individuals. Furthermore, DFA cannot detect individ-uals belonging to groups not specifically identified by the investigator. Consequently, DFA should be used only when the *a priori* classifications are known to be correct. This generally requires classification criteria independent of morphology for deriving the discriminant function (e.g., Leary et al. 1983).

Principal components analysis (PCA) is a second multivariate statistical method that is often used in hybridization studies. Unlike DFA, PCA does not require the *a priori* identification of groups or individuals. PCA only derives a new set of uncorrelated variables, the principal components, which are simple linear transformations of the original variables. Geometrically, this is nothing more than a multidimensional rotation of the original coordinate axes to yield a new coordinate axes system representing the principal components. The first principal component axis can be viewed in this new system as the least squares regression line through the cloud of points in the n-dimensional character space. The remaining axes are mutually orthogonal with the first axis and ori-ented sequentially in the directions of maximum variance. If several traits are correlated, the first two principal axes may account for a large proportion of the total variation among individuals. By projecting these individuals (in the least squares sense) from their positions in the *n*-dimensional character space onto the plane formed by the first two principal component axes, a visual approximation of the original data set can be obtained (e.g., Everitt 1978). A series of such projections for fish collected from areas with and without suspected hybrids can objectively summarize the morphological evidence for natural hybridization without making any *a priori* assumptions regarding the classification of individual fish. Dowling and Moore (1984), in an investigation of natural hybridization between two cyprinid fishes, performed a similar analysis by plotting histograms of the first principal component scores for individual fish from each of several locations (Fig. 7.1).

The technical details of DFA and PCA are beyond the scope of this

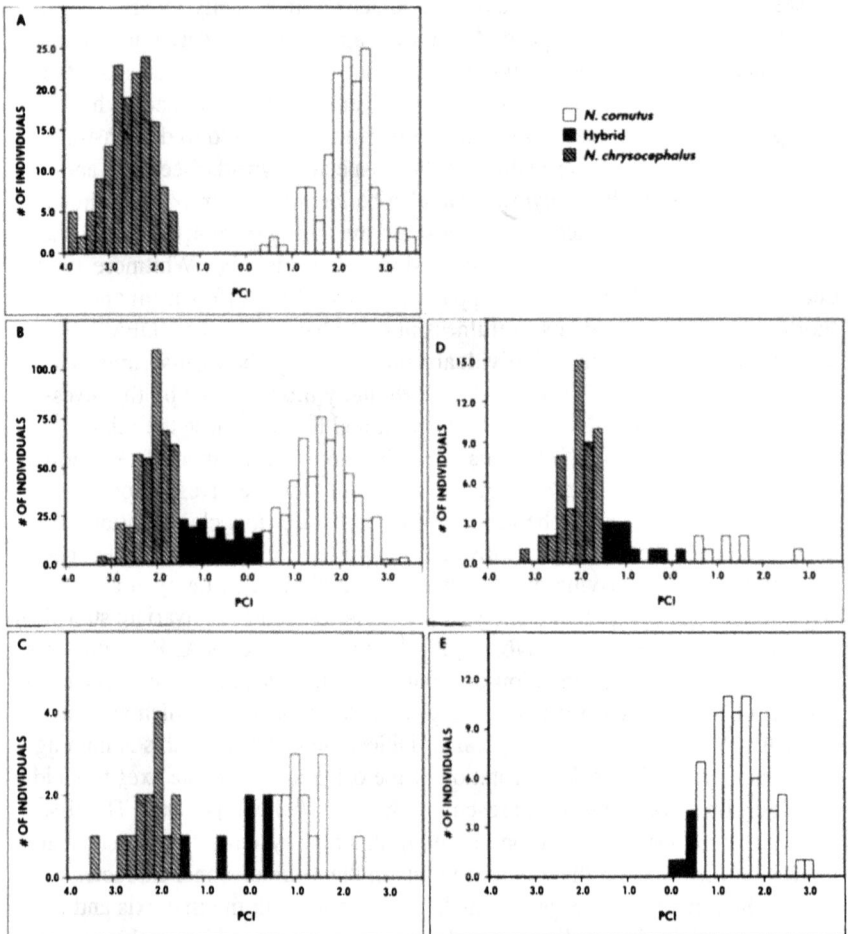

Fig. 7.1 Frequency distributions of first principal component scores for two cyprinid fishes, *Notropis cornutus* and *N. chrysocephalus*, and their suspected hybrids from each of five geographic locations. From Dowling and Moore (1984).

chapter. Interested readers should consult one of the currently available text-books (e.g., Johnson and Wichern 1982) as well as the articles by Smith (1973) and Neff and Smith (1979).

Detecting natural hybridization using morphological criteria suffers from many shortcomings. In general, morphological data can provide only circumstantial evidence for natural hybridization or introgression because hybrids are usually assumed *a priori* to be morphologically intermediate to the parental species. If morphologically intermediate fish are subsequently found, these fish are usually taken as evidence of hybridization. This often results in a circular argument. Morphological traits are generally polygenic, representing the phenotypic expression of a large number of genes and influenced to varying degrees by environmental effects (Barlow 1961, Ali and Lindsey 1974, Mac-Gregor and MacCrimmon 1977, Todd et al. 1981). As a result, the full range of intraspecific phenotypic variation possible for morphological traits, both collectively and individually, cannot be known precisely. For example, Leary et al. (1984) have shown that the ranges of meristic character values for rainbow trout, west-slope cutthroat trout *(Salmo clarki lewisi),* and introgressed populations between the two species all overlap.

In general, the *extent* of hybridization or introgression cannot be determined from morphological data because F_1 hybrids may not be individually distinguishable from F_2 or backcross (BC) hybrids. As previously mentioned, F_1 hybrids often express a mosaic of morphological characters that separately resemble those for one or the other parental species. Detecting introgression may also be extremely difficult because introgressed populations may appear morphologically identical to one of the parental species (e.g., Greenfield and Greenfield 1972, Busack and Gall 1981). These shortcomings have therefore motivated the use of more direct genetic methods for detecting natural hybridization.

Karyotyping

Only a few studies have used karyotypic methods (i.e., those involving chromosomal comparisons) to investigate natural hybridization in fishes (e.g., Setzer 1970, Greenfield and Greenfield 1972, Busack et al. 1980). It requires a great amount of labor to karyotype more than a few fish, and more efficient genetic methods are available (e.g., electrophoresis). In addition, a large number of congeneric species of fishes have identical karyotypes (Gold et al. 1980, Sola et al. 1981); chromosome comparisons in these fishes would be of little value for detecting hybridization. Compared with mammalian species, fish generally have a high number ($\geqslant 48$) of small chromosomes, and these small chromosomes further complicate karyotypic investigations.

Most attempts to differentially stain and thereby identify specific chromosomes in fishes by banding patterns have not been very successful. Some notable exceptions to this generalization are recent reports of replication and Q banding in chromosomes of salmonid fishes (Phillips and Zajicek 1982, Delaney and Bloom 1984, Phillips et al. 1985). Despite their inherent shortcomings, however, karyotypic analyses can provide an objective and independent method for verifying the assumptions associated with the morphological detection of natural fish hybrids (Greenfield et al. 1973).

Electrophoresis

Electrophoretic methods are commonly used to detect natural hybridization in fishes. Many studies have demonstrated electrophoretic techniques to be substantially more sensitive than morphological methods for detecting natural hybridization in fishes (Martin and Richmond 1973, McCleod et al. 1980, Pelzman 1980, Busack and Gall 1981, Whitmore 1983, Goodfellow et al. 1984, Leary et al. 1984). The many advantages of electrophoretic techniques for investigating genetic variation in natural populations are described by Utter elsewhere in this book (Chapter 1).

Detecting natural hybridization and introgression by electrophoretic methods is relatively straightforward when the two parental species are completely fixed for different alleles at two or more loci. Individuals of the two parental species will each be homozygous for different alleles, whereas F_1 hybrids will be heterozygous for these alleles at all diagnostic loci (e.g., Leary et al. 1983, Whitmore 1983). If hybridization has proceeded past the F_1 stage, however, hybrid descendants will express a broad mixture of recombinant types, including the two parental types (see Avise and Van Den Avyle 1984). Consequently, an individual with a composite phenotype identical to the phenotype for one of the parental species could be a hybrid descendant *if* F_1 hybrids have spawned among themselves or backcrossed with one or both parental species. The presence of recombinant types can therefore be taken as evidence of hybrid fertility and second-generation hybridization (F_2 or BC) if (and only if) the assumption of complete fixation of different alleles between species is true.

If the two species are fixed for different alleles at only one locus, one cannot distinguish first-generation hybrids from backcross or F_2 hybrids. All individuals in such a population will express either the F_1 hybrid phenotype or one of the two parental phenotypes regardless of the level of hybridization or backcrossing. In this situation, the presence of heterozygous individuals in a mixture of both parental species can only provide evidence that the two species have interbred; the extent of hybridization will be unknown. Many reports of natural hybridization have been based on the discovery of only a few heterozygous individuals (e.g., Soloman and Child 1978, Beland et al. 1981).

Detecting natural hybridization by electrophoretic methods is somewhat more difficult if the two species (or populations) are not fixed for different alleles at one or more loci. The two species may express different common alleles at several loci, but both alleles may be present in one or both species. In this situation, individual fish cannot be unambiguously classified as hybrid descendants (F_1, F_2, BC, etc.) because all possible combinations of electrophoretic phenotypes can occur within one or both species. For example, some fish may

be heterozygous at all distinguishing loci, whereas other fish may be homo-
gygous for the common alleles of one species at some loci but homozygous for
the common alleles of the other species at other loci. Hybridization must therefore be investigated *quantitatively* by determining whether the proportion of fish
with these "intermediate genotypic combinations" is significantly greater than
one would expect from the random mating within one or both species.

Campton and Utter (1985) addressed this problem in a report of natural
hybridization between steelhead trout (anadromous *Salmo gairdneri*) and
coastal cutthroat trout *(S. clarki clarki)*. These two species possess different
common alleles at four electrophoretically detectable loci, but one or both species are naturally polymorphic for both alleles at each locus. An electrophoretic
investigation of anadromous trout populations in the Puget Sound area of Washington state had revealed, however, a surprisingly high number of young-of-the-
year fish from two streams with intermediate genotypic combinations (Campton
and Utter 1985). To quantitatively demonstrate the likelihood of hybridization,
Campton and Utter devised a hybrid index measuring the relative probability
that the composite genotype for each fish arose by random mating within each
of the two species. The index (I_H) was defined as

$$I_H = 1.0 - \frac{\log_{10}(p_z)}{\log_{10}(p_z) + \log_{10}(p_y)} ,$$

where

$$p_x = \prod_{i=1}^{L} k_i \prod_{j=1}^{A_i} (X_{ij})^{m_{ij}} ,$$

$$p_y = \prod_{i=1}^{L} k_i \prod_{j=1}^{A_i} (Y_{ij})^{m_{ij}} .$$

Here X_{ij} and Y_{ij} are the average frequencies of the jth allele at the ith locus for
species X and Y, respectively; m_{ij} is the number of alleles of the jth type observed at the ith locus for the individual being evaluated; A_i is the total number
of known alleles at the ith locus for the two species combined; k_i is the binomial
sampling coefficient (e.g., $k_i = 2$ for Aa, $k_i = 1$ for AA or aa) associated with
the genotype of the individual at the ith locus; and L is the number of diagnostic loci used to distinguish the two species. The quantities p_x and p_y are the
conditional probabilities that the composite genotype at all loci for an individual could have arisen by random mating within species X and species Y, respectively, given the average allele frequencies for the two species and assuming gametic phase (linkage) equilibrium among all loci. The index can assume
any value between 0.0 and 1.0 and will be close to one of these two values
when individuals have a very high relative probability of belonging to species Y

or X, respectively. In our investigation of hybridization between steelhead trout and cutthroat trout (Campton and Utter 1985), histograms of hybrid index scores clearly revealed the presence of both species in an area of sympatry and the existence of a third, intermediate group at a separate sample site (Fig. 7.2). We interpret this genotypically intermediate group of fish to be the result of natural hybridization.

The power of electrophoretic methods to detect natural hybridization will increase as the number of distinguishing loci increases. This is easily demonstrated by the simple example given in Table 7.1. If the frequencies of alternate alleles (e.g., A and a) at a locus are 0.8 and 0.2 in one population and 0.2 and 0.8 in the second population, the probability of an individual being heterozygous in each population is simply $2pq = 0.32$. If the two populations interbreed, then the expected frequency of heterozygotes among the F_1 hybrids is 0.68, which is only 2.13 times the expected frequency of heterozygotes in each of the parental populations. As the number of distinguishing loci increases,

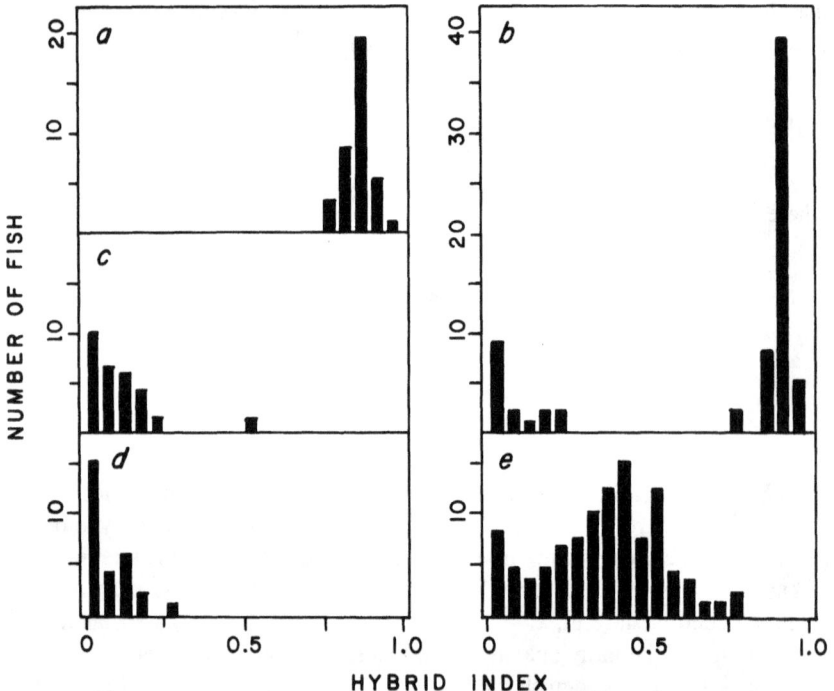

Fig. 7.2 Hybrid index scores for two sympatric species of trout, steelhead trout (anadromous *Salmo gairdneri*) and coastal cuttthroat trout *(Salmo clarki clarki)*, and their suspected hybrid descendants from three sample sites within a small stream. Individuals with values of the index close to 1.0 or 0.0 expressed composite electrophoretic phenotypes that have a high relative probability of occurring in *S. gairdneri* or *S. clarki clarki*, respectively. (a) Site 1, age 0+ fish (*gairdneri* only); (b) site 2, age 0+ fish (*clarki* and *gairdneri*, no hybrids); (c) site 2, age 1+ fish (*clarki* + 1 unknown or hybrid fish); (d) site 3, age 1+ fish (*clarki* only); (e) site 3, age 0+ fish (*clarki* + a large number of suspected hybrids). From Campton and Utter (1985).

Table 7.1 Probability of an individual being heterozygous simultaneously at one to six distinguishing loci in each of two parental populations (P) and among F_1 hybrids (P_h), where the frequencies of alternate alleles at each locus are 0.8 and 0.2 in one population and 0.2 and 0.8, in the other. The probabilities were calculated as $P = [2(0.8)(0.2)]^L$ for the parental populations and as $P_h = [(0.8)(0.8) + (0.2)(0.2)]^L$ for the F_1 hybrids, where L = number of loci. These expressions assume gametic equilibrium between all loci and Hardy-Weinberg genotypic proportions in each parental population.

Population	Number of distinguishing loci					
	1	2	3	4	5	6
Parental (P)	0.3200	0.1024	0.0328	0.0105	0.0034	0.0011
F_1 hybrid (P_h)	0.6800	0.4624	0.3144	0.2138	0.1454	0.0989
P_h/P	2.13	4.52	9.60	20.4	43.3	92.1

however, this ratio also increases. For example, with four distinguishing loci, the frequency of four-locus heterozygotes among F_1 hybrids is more than 20 times their expected frequency in each of the parental populations (Table 7.1). This ratio increases to 92.1 with six distinguishing loci.

In practice, a substantial number of multiple heterozygotes at several distinguishing loci, as well as other intermediate genotypic combinations (e.g., Aa, BB, cc, Dd, etc.), would provide strong evidence for natural hybridization as revealed by histograms of hybrid index scores (e.g., Fig. 7.2). However, even with four distinguishing loci, the expected frequency of four-locus hetero-zygotes within each of the parental populations in Table 7.1 is greater than 1%. Consequently, as previously pointed out, individuals cannot be unambiguously classified as hybrids if the two parental populations, or species, are not fixed for alternate alleles at one or more loci.

The hybrid index method used to evaluate hybridization between *S. gairdneri* and *S. clarki clarki* (Fig. 7.2) provides an objective method of summarizing the genotypic relatedness of many individuals to each of two established species. One advantage of the index method is that values can be calculated for individual fish even when data are missing at one or more loci. The major disadvantage of the index method is that it requires the *a priori* establishment of allele frequency profiles for each of the two parental species or populations. This could lead to some subjectivity depending on the amount of genetic differentiation among populations within each species and the particular populations used to establish the allele frequency profiles. Ideally, one would obtain allele frequency estimates for the hybridizing populations prior to interbreeding, but these estimates will probably be unavailable.

To overcome possible difficulties associated with establishing allele frequency profiles for the parental species (or populations), one could evaluate the likelihood of hybridization by a series of two-dimensional, principal component plots. The data set for fish collected from a particular sample site could be put in the form of an $n \times p$ matrix, where n is the number of fish and p is the total number of independent alleles for all diagnostic loci combined (number of independent alleles at a locus = total number of alleles − 1). The individual entries (x_{ij}) in the matrix would be the frequencies of the jth allele for the ith individual and would equal 1.0 if the individual was homozygous for the jth allele, 0.5 if it was heterozygous for the jth allele, or 0.0 if it was homozygous or heterozygous for alternate alleles. Each row of the matrix would represent a vector of genotypic scores for an individual fish (see Smouse and Neel 1977). A series of PCA plots for fish collected from areas with and without hybrid fish should respectively reveal the presence and absence of an intermediate group (Fig. 7.3). These plots can therefore provide a very objective method of presenting electrophoretic evidence for hybridization. However, the PCA approach has one

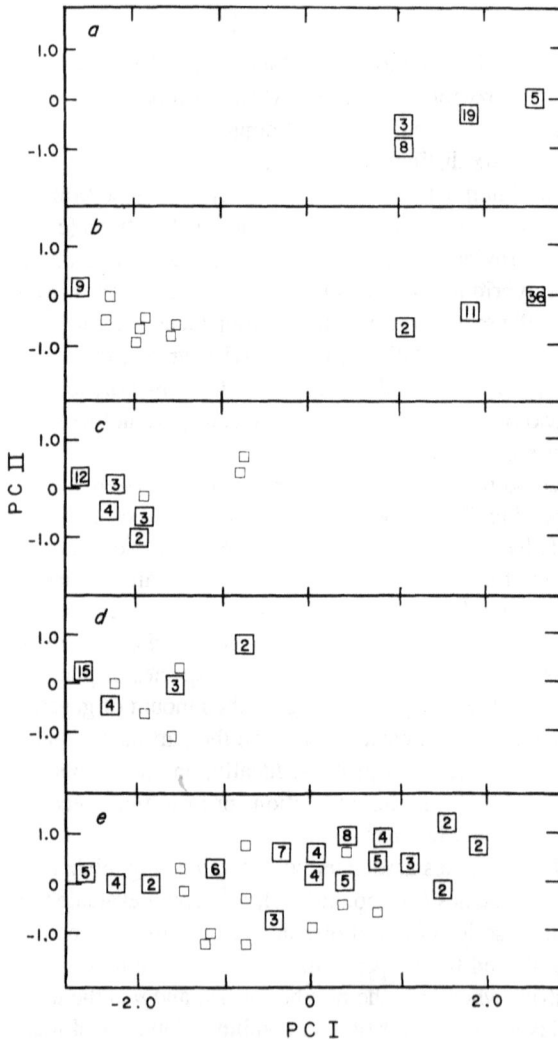

Fig. 7.3 Two-dimensional projections of the fish in Fig. 7.2 onto the first two principal component axes using genotypic scores (0, 1/2, or 1) at four distinguishing loci. Small squares represent individual fish and large squares represent the indicated number of fish. The projections in (e) and (b) clearly reveal the presence and absence, respectively, of a third, intermediate group. Data from Campton and Utter (1985).

disadvantage: fish with data missing for a particular locus must be excluded from the analysis because their positions in multivariate space along one or more axes are unknown.

Many species of fish are composed of two or more recently evolved but formerly allopatric forms. These forms, where they come in contact, may intergrade and hybridize over broad geographic areas. Electrophoretic techniques are especially useful for detecting such hybrid zones. For example, Avise and Smith (1974) describe an extensive hybrid zone between two morphologically distinct subspecies of bluegill sunfish, *Lepomis macrochirus macrochirus* and *L. m. purpurescens*. The two subspecies express nearly fixed allelic differences at two electrophoretic loci where they are allopatric in the southeastern United States, but an area of genetic intergradation and hybridization exists where the ranges of the two subspecies overlap (Fig. 7.4). Avise and Smith (1974) note that "gene exchange appears unrestricted and degrees of allelic introgression among loci are nearly equal." Phillip et al. (1983) report a similar zone of intergradation and hybridization in the southeastern United States between the northern and Florida subspecies of largemouth bass, *Micropterus salmoides salmoides* and *M. s. floridanus* (Fig. 7.5). The geographic distributions of the hybrid zones for the two subspecies of bluegill sunfish and the two subspecies of largemouth bass are amazingly similar (compare Figs. 7.4 and 7.5). Avise and Smith (1974) suggest that terrestrial islands formed during the Pleistocene in the southeastern United States isolated the freshwater fauna of this region prior to secondary contacts. The evolutionary significance of such hybrid zones

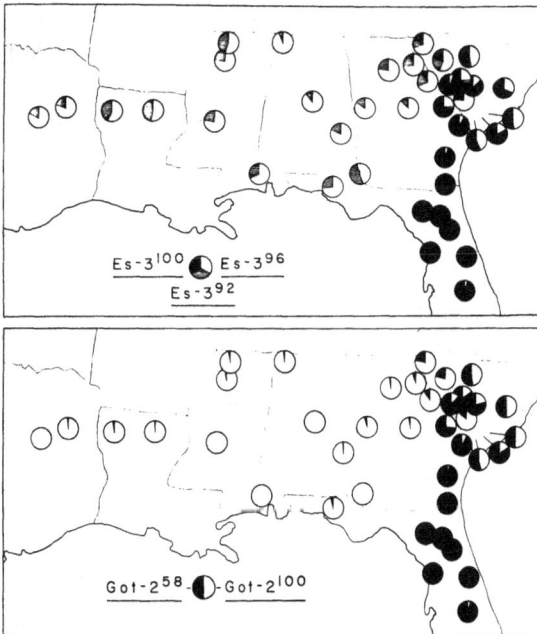

Figure 7.4 Geographic distributions of allele frequencies at two biochemical genetic loci for two subspecies of bluegill sunfish, *Lepomis macrochirus macrochirus* and *L. m. purpurescens*, in the southeastern United States. From Avise and Smith (1974).

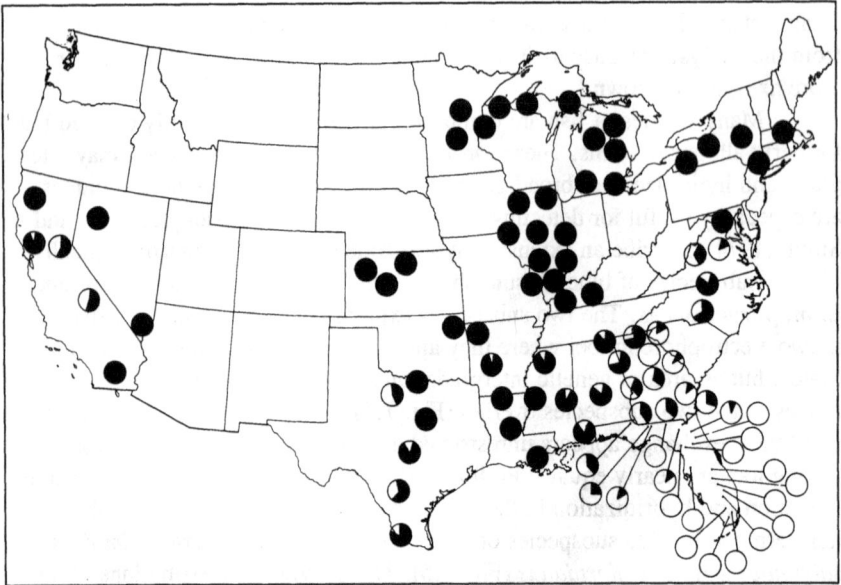

Fig. 7.5 Geographic distribution of allele frequencies at the *IDH-B* locus for two subspecies of largemouth bass, *Micropterus salmoides salmoides* and *M. s. floridanus,* in the United States. From Phillip et al. (1983).

is an important topic of research in evolutionary biology and has been discussed elsewhere (Remington 1968, Woodruff 1973, Moore 1977, Barton and Hewitt 1981, 1983).

Intraspecific hybridization between native and introduced populations of rainbow trout has also been investigated by electrophoretic methods (Allendorf et al. 1980, Wishard et al. 1984, Campton and Johnston 1985). Rainbow trout are native to the northwestern United States, but they form two geographic races that can be distinguished by divergent allele frequencies at two biochemical genetic loci (Allendorf and Utter 1979, Utter et al. 1980). These two races are separated geographically by the Cascade Mountains and are generally referred to as the coastal and inland races. Most hatchery strains of domesticated rainbow trout express allele frequencies consistent with being derived from the coastal race. As a result, Allendorf et al. (1980) were able to electrophoretically identify previously undescribed native populations of rainbow trout in the Kootenai River drainage of western Montana as well as a hybrid or introgressed population between native and introduced forms. Campton and Johnston (1985) similarly detected apparent introgression between native and introduced rainbow trout in the Yakima River of Washington (Fig. 7.6). Both studies indicated that native and introduced rainbow trout had randomly interbred to panmixia wherever the two forms came in contact. In contrast, Wishard et al. (1984) found no evidence of hatchery fish introgression among populations of rainbow trout in a desert region of southwestern Idaho, despite documented plantings of hatchery fish in this area. High stream temperatures (> 30

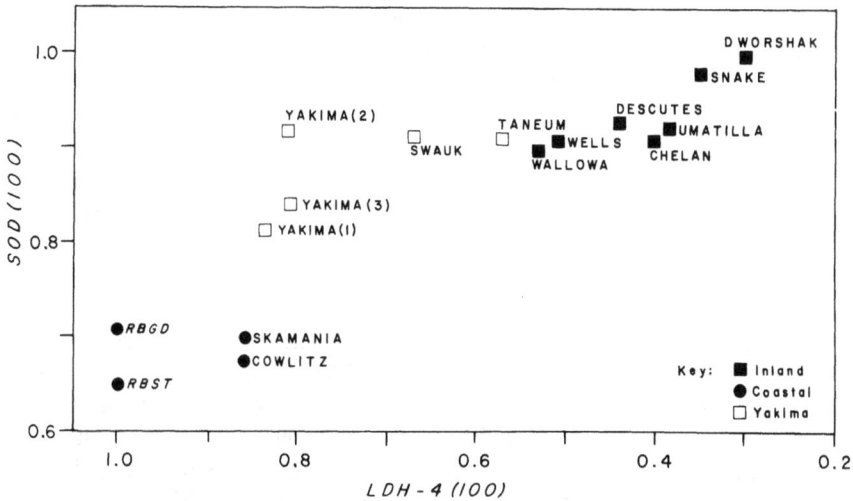

Figure 7.6 Frequencies of the common *(100)* allele at the *LDH-4* and *SOD* loci for populations of rainbow trout *(Salmo gairdneri)* in Washington state. The Yakima River is a major tributary of the Columbia River east of the Cascade Mountains, but rainbow trout inhabiting this river (sites 1, 2, and 3) and two tributary creeks (Swauk and Taneum) expressed allele frequencies intermediate to those for other inland populations and those for introduced coastal populations. RBGD and RBST represent two hatchery populations of nonanadromous rainbow trout. From Campton and Johnston (1985).

°C) in this region of Idaho are believed to have prevented nonnative rainbow trout from surviving or successfully reproducing, especially in competition with native rainbow trout, which are believed to be adaptively tolerant to the local conditions (Baake 1977, Behnke 1979). These studies exemplify the general finding that the likelihood of hybridization is very much dependent upon local habitat conditions.

Management agencies often want to know the extent to which hatchery fish spawn in the wild and contribute genetically to the wild populations. If the hatchery and wild populations are characterized by divergent allele frequencies at one or more loci, temporal shifts in allele frequencies for the wild population can be used to estimate the amount of genetic introgression that has occurred. On the other hand, conspecific populations are usually characterized by similar allele frequencies at electrophoretically detectable loci, and this prevents the detection of introgression between the hatchery and wild populations. In this situation, however, fish in the hatchery population can be selectively bred so that the frequency of one or more alleles among the released fish is significantly greater than the frequency of these alleles in the wild population (e.g., Schweigert et al. 1977). This "genetic marking" of hatchery stocks can thus provide management agencies with a *genetic* means of evaluating the success (or failure) of fish stocking programs. Genetic marking can also be used to estimate proportional contributions of hatchery and wild fish in a mixed population fishery (Chapter 10).

The major advantage of genetic marking is that allozyme "tags" are inherited; introgression into the wild populations can thus be detected, and hatchery fish do not have to be individually marked or handled once the breeding program is complete. The two potential disadvantages of genetic marking are natural selection associated with different allozyme phenotypes and inbreeding (or loss of genetic variation) resulting from intense artificial selection. The potential advantages and disadvantages of genetically marking hatchery stocks are further discussed elsewhere (Utter et al. 1976, Allendorf and Utter 1979).

Mitochondrial DNA

The use of restriction endonucleases for determining the nucleotide sequences of mitochondrial DNA (mtDNA) provides one of the newest and most direct methods for investigating genetic variation in natural populations. Of the many properties of mtDNA that distinguish it from nuclear DNA (see Chapter 13), the most important for studying natural hybridization are its maternal inheritance as a clonal genetic marker and the absence of recombination. The maternal species can therefore be identified in known or suspected F_1 hybrids if the two parental species are characterized by different mtDNAs. On the other hand, these same properties prevent mtDNA from being used as a stand-alone technique for detecting hybridization and introgression because each individual will have only one type of mtDNA, regardless of its parentage. The techniques of mtDNA must therefore be used in conjunction with other techniques in order to distinguish among species coexistence, first generation hybridization, and introgressive hybridization.

Mitochondrial DNA techniques, when used with electrophoretic methods, can detect the complete introgression of mtDNA from one species into the nuclear background of another species, provided female hybrids and their offspring are fertile. Such introgressive hybridization requires the successive backcrossing of female hybrid descendants with males of the paternal species. This introgression will be independent of the recombination and assortment events occurring within the nuclear genome if selection is not acting to maintain a nuclear-cytoplasmic compatibility (e.g., Takahata and Slatkin 1984). Historical hybridization events can therefore be detected when individuals of a particular species, as identified by morphologic, karyotypic, or electrophoretic methods, possess the mtDNA characteristic of a closely related species. Such interspecific transfers of mtDNA have recently been reported for congeneric species of *Drosophila, Mus,* and *Rana* (Powell 1983, Ferris et al. 1983, Spolsky and Uzzell 1984).

The use of mtDNA for investigating natural hybridization in fishes has been initiated only recently (Avise and Saunders 1984, Avise et al. 1984). Avise and Saunders (1984) used mtDNA in conjunction with allozymes to investigate the frequency of hybridization among nine species of sunfish (genus *Lepomis*) inhabiting two geographic locations in the southeastern United States. Their major findings can be summarized:

- Hybridization occurred at a relatively low frequency but involved five of the nine species examined.
- No mtDNA or electrophoretic evidence for introgression between species of *Lepomis* was detected; all hybrids appeared to be strictly F_1.
- Every detected hybrid represented a cross between a common (abundant) and a rare species.
- In six of seven possible hybrid crosses, the female parent was from the rare species as determined by mtDNA genotype. This result was attributed to intense competition among males for mating partners and the general promiscuity of females.

In the second study, Avise et al. (1984) investigated geographic variation in mtDNA for the two previously mentioned subspecies of bluegill sunfish, *L. m. macrochirus* and *L. m. purpurescens* (Avise and Smith 1974). One of their sampled populations was a known hybrid swarm. In this study, Avise et al. (1984) found

- Two distinct types of mtDNA, A and B, which separately characterized each of the two subspecies.
- Almost complete concordance between mtDNA genotype and the previously recognized geographic ranges of the two subspecies.
- One fish which electrophoretically and geographically belonged to subspecies A but expressed the B mtDNA genotype, suggesting that complete introgression had occurred.
- Random association of mtDNA genotype and allozyme genotype at each of two diagnostic loci in the hybrid swarm population but a nonrandom association with mtDNA genotype ($P < 0.05$) when the two isozyme loci were considered jointly.
- No evidence ($P > 0.95$) of paternal leakage or within-individual mtDNA polymorphism. In addition, the equal nuclear and mtDNA contributions by each subspecies to the hybrid swarm suggest that reciprocal matings between the two subspecies and among hybrid descendants were equally successful.

These two studies by Avise and his colleagues clearly demonstrate how the techniques of mtDNA can provide unique and previously unobtainable information regarding the dynamics of hybridization between fishes. Data from future mtDNA studies may reveal whether introgressive hybridization is responsible for maintaining identical allozyme polymorphisms in closely related but naturally sympatric species (e.g., Campton and Utter 1985). If such polymorphisms are maintained by introgressive hybridization, a redefinition of the "biological species concept" (Mayr 1970) may be in order. In this context, fishes might be particularly suitable subjects for studying the evolutionary significance of introgressive hybridization between naturally sympatric species.

GENETIC INTERPRETATIONS OF HYBRIDIZED POPULATIONS

Detecting natural hybridization requires the sampling of individuals from wild populations and their subsequent laboratory evaluation by one or more of the previously described methods. Prior to sampling, the investigator probably does not know whether hybridization has occurred. The investigator may sample a single random mating population with no history of hybridization or a mixture of two species (or populations) with some F_1 hybrids only. Several other situations are possible. If hybridization or introgression is detected, the next step is to quantitatively evaluate the extent and direction of these phenomena.

In this section, I review some theoretical models of population genetics that can be used to help distinguish the possible cases of hybridization. The theory of population genetics (e.g., Crow and Kimura 1970) is derived primarily in terms of simple Mendelian genes in which alleles are codominantly expressed at single loci. Genotypic data obtained by electrophoretic methods can be analyzed and interpreted directly in terms of these theoretical models. These models apply, however, to any data in which genotypes can be identified as the codominant expression of alleles at single loci. These models were initially developed for conspecific populations, but they are applicable to any situation where hybrids may be fertile (e.g., Avise and Smith 1974, Busack and Gall 1981, Campton and Johnston 1985). In these models, hybrid infertility can be treated simply as a special case where backcrossing and recombination do not occur. The problem with fishes, however, is that F_1 hybrids are often fertile and may backcross freely with one or both parental species. As a result, interspecific hybridization in fishes can be evaluated as a population genetic phenomenon. My purpose here is to provide a conceptual framework or guideline for interpreting allele frequency data in genetic studies of natural hybridization.

Deviations from Hardy-Weinberg Proportions

The first step in analyzing allele frequency data is to determine whether more than one population has been sampled. A single, randomly mating population is expected to conform to Hardy-Weinberg genotypic proportions at all loci. On the other hand, a mixture of two, non-interbreeding populations with different allele frequencies is expected to yield a deficiency of heterozygotes compared with these expected proportions (see Chakraborty and Leimar, Chapter 4). One can easily show that, for a particular allele at some locus, the expected deficiency of heterozygotes in a non-interbreeding mixture of two populations is

$$H_O - H_E = -2f_1 f_2 (p_1 - p_2)^2 , \qquad (1)$$

where H_O and H_E are the observed and expected frequencies of heterozygotes,

f_1 and f_2 are the proportional contributions of populations 1 and 2 to some mixture ($f_1 + f_2 = 1.0$), and p_1 and p_2 are the frequencies of the allele in populations 1 and 2, respectively (Appendix 7.1). The quantity in Eq. (1) will always be less than or equal to zero and will equal zero only if allele frequencies in the two populations are identical (i.e., $p_1 = p_2$). In addition, for a given set of allele frequencies (p_1 and p_2), $H_O - H_E$ is a maximum when the two populations are present in the mixture in equal proportions ($f_1 = f_2 = 0.5$). This deficit of heterozygotes is essentially an example of the *Wahlund effect* (Wahlund 1928, Hedrick 1983), in which a loss of heterozygosity will occur in a subdivided population if allele frequencies in the subpopulations have diverged from one another (see also Chakraborty and Leimar, Chapter 4).

A deficiency of heterozygotes will also occur if two populations interbreed to some extent (thus producing hybrids) but mate assortatively within populations. This can be illustrated by the preceding example assuming a fraction, *m*, of each population mate assortatively (i.e., within populations) while a fraction $1 - m$ randomly interbreed. The expected deficiency of heterozyotes under these conditions becomes

$$H_O - H_E = -2mf_1f_2(p_1 - p_2)^2 \ , \tag{2}$$

which is identical to Eq. (1) except for the factor *m* (Appendix 7.2). The deficiency of heterozygotes is expected to increase as a linear function of the amount of assortative mating within the two populations. If F_1 hybrids backcross with one or both parental populations, then this will have the effect of reducing the difference between p_1 and p_2 (because of gene flow) and thereby reduce the expected deficiency of heterozygotes in the population mixture. On the other hand, an excess of heterozygotes would be observed if the two populations (or species) had mated preferentially between populations, thereby causing F_1 hybrids to be present in greater numbers than one would expect from random mating. Such disassortative mating would not normally be expected under natural conditions, but would be important in artificial breeding programs if crossbreeding of strains or species was practiced.

Dowling and Moore (1984) describe an interesting case of hybridization between two nominal species of *Notropis* (Cyprinidae), *N. cornutus* and *N. chrysocephalus*, in which a consistent deficiency of heterozygotes was observed. They argue that these observations warrant recognition of *N. cornutus* and *N. chrysocephalus* as distinct species and not as subspecies, because "hybridization between subspecies should result in genotypic frequencies in Hardy-Weinberg equilibrium, while hybridization between species should result in a marked deficiency of heterozygotes, either because of assortative mating (pre mating isolation) or selection against hybrids (post-mating isolation)." In a subsequent study, the authors were able to attribute the deficiency in heterozygotes to natural selection against hybrids and not necessarily to assortative mating (Dowling and Moore 1985).

Detecting deficiencies of heterozygotes can reveal the existence of two (or more) populations or species that are maintaining some level of reproductive isolation (see Ryman et al. 1979 for one extreme example). On the other hand, measuring departures from single-locus, Hardy-Weinberg proportions is expected to be of only limited value for evaluating the dynamics of hybridization and introgression. In the absence of selection, a single generation of random mating will restore Hardy-Weinberg proportions in a mixture of two or more genetically differentiated populations. If males and females from each population do not contribute equally to the hybrid population, two generations will be required to attain Hardy-Weinberg proportions because allele frequencies will initially be different between the two sexes (see Hedrick 1983). In contrast, the random association of alleles *between* loci will only be approached asymptotically in a random mating population and may never be reached if the two populations (or species) are mating assortatively. As a result, one may be able to detect recent introgression or the residual effects of hybridization by estimating the amount of *gametic phase (linkage) disequilibrium*.

Gametic Phase Disequilibrium

Gametic phase disequilibrium (abbreviated to gametic disequilibrium) refers to the nonrandom association of alleles *between* loci. An extensive amount of theoretical literature has been published describing the conditions necessary for generating or maintaining gametic disequilibrium within or between populations (reviewed by Hedrick et al. 1978). The subdivision of a population into several subpopulations or the agglomeration of two or more previously isolated populations are two ways in which gametic disequilibrium can be generated (e.g., Ohta 1982). A theoretical assessment of gametic disequilibrium is therefore essential to understanding the dynamics of hybridization and introgression.

A population is in gametic equilibrium if the frequency of gametes carrying alleles from two (or more) loci is equal to the product of the respective allele frequencies. If p_1 and p_2 are the frequencies of two alleles (A_1 and A_2) at one locus and q_1 and q_2 are the frequencies of two alleles (B_1 and B_2) at a second locus, then gametic equilibrium is achieved in the population when $X_{11} = p_1 q_1$, $X_{12} = p_1 q_2$, $X_{21} = p_2 q_1$ and $X_{22} = p_2 q_2$, where X_{ij} is the frequency of gametes in the population carrying alleles A_i and B_j. If alleles are not randomly associated between loci (i.e., $X_{ij} \neq p_i q_j$), then one can show that

$$X_{11} = p_1 q_1 + D$$

$$X_{12} = p_1 q_2 - D$$

$$X_{21} = p_2 q_1 - D$$

$$X_{22} = p_2 q_2 + D \quad , \tag{3}$$

where D measures the amount of interlocus association between alleles within gametes (Hedrick 1983). D is termed the gametic (or linkage) disequilibrium coefficient and is defined as $D = X_{11} - p_1 q_1$ for a two-locus system (Lewontin and Kojima 1970). By substituting $X_{11} + X_{12}$ for p_1 and $X_{11} + X_{21}$ for q_1 in the preceding definition of D, one obtains

$$D = (X_{11} X_{22}) - (X_{12} X_{21}) . \tag{4}$$

Thus, the amount of gametic disequilibrium between two loci, each with two alleles, is simply the difference between the products of the coupled gametic frequencies (i.e., frequencies of $A_1 B_1$ and $A_2 B_2$) and the products of the repulsion gametic frequencies (i.e., frequencies of $A_1 B_2$ and $A_2 B_1$). Gametic disequilibria can also be defined in terms of three or more loci (e.g., third-order disequilibria, fourth-order disequilibria, etc.), but these expressions are considerably more complicated and are discussed elsewhere (e.g. Crow and Kimura 1970, p. 50, Bennett 1954).

The range of possible values for D depends upon the population allele frequencies at the two loci being considered. One can see from Eq. (4) that D attains a maximum value of 0.25 when $X_{11} = X_{22} = 0.5$ ($X_{12} = X_{21} = 0$) and a minimum value of -0.25 when $X_{12} = X_{21} = 0.5$ ($X_{11} = X_{22} = 0$). These maximum and minimum values of D are possible, however, only when the frequencies of both alleles at each locus are equal, i.e., when $p_1 = q_1 = p_2 = q_2 = 0.5$. The absolute value of D, therefore, may be difficult to interpret, especially when several D values are being compared. Consequently, Lewontin (1964) proposed the statistic $D' = D/D_{max}$ be used as a measure of gametic disequilibrium, where D_{max} is the maximum, absolute value of D for a particular set of allele frequencies (see Hedrick 1983). The value of D' ranges from -1.0 to 1.0, and it provides a relative measure of the nonrandom association among alleles between loci. An alternative measure of gametic disequilibrium is R, the correlation coefficient between alleles at different loci, which one can show is equal to

$$R = \frac{D}{(p_1 q_1 p_2 q_2)^{1/2}} \tag{5}$$

by assigning the value 1 to A_1 and B_1 and 0 to A_2 and B_2. R can also assume any value between -1.0 and 1.0 but only when allele frequencies at the two loci are equal, i.e., when $p_1 = p_2$ and $q_1 = q_2$; the range of possible values is somewhat less when allele frequencies at the two loci are not equal (Hedrick 1983). A major advantage of the correlation coefficient (R) is that it is a familiar statistic that can be interpreted easily by those not acquainted with the other measures of gametic disequilibrium.

As previously mentioned, a single generation of random mating will not restore gametic equilibrium once disequilibrium has been generated in a population. In the absence of selection, the disequilibrium expected between two loci after one generation of random mating is $D_1 = (1-r)D_0$, where r is the frequency of recombination between the two loci ($0 \leqslant r \leqslant 1/2$) and D_0 is the initial disequilibrium in the population (see Hedrick 1983, p. 342, for derivation). After two generations of random mating, the disequilibrium becomes $D_2 = (1-r)D_1 = (1-r)^2 D_0$, so that after t generations,

$$D_t = (1 - r)^t \ D_0 , \tag{6}$$

or, using the alternative measures of disequilibrium, $D_t' = (1-r)^t D_0'$ and $R_t = (1-r)^t R_0$. Gametic equilibrium is clearly approached only asymptotically in a random mating population *and* at a rate strictly proportional to the recombination rate between loci. For tightly linked loci (i.e., $r << 1/2$), the disequilibrium can persist almost indefinitely, especially in small populations where recombination will be counterbalanced by genetic drift.

In order to demonstrate the amount of disequilibrium expected in a mixture of two populations, let p_i and $1-p_i$ be the frequencies of two alleles (A and a) at locus 1 and let s_i and $1-s_i$ be the frequencies of two other alleles (B and b) at locus 2, where the subscript i refers to the ith population ($i = 1,2$). If the two populations are combined in the proportions f_1 and f_2 ($f_1 + f_2 = 1.0$), then the amount of disequilibrium expected in the mixture prior to random mating is

$$D_m = f_1 f_2 (p_1 - p_2) (s_1 - s_2) \tag{7}$$

where populations 1 and 2 are both assumed to be in gametic equilibrium (Cavalli-Sforza and Bodmer 1971, p. 69). If the two populations are not in gametic equilibrium, then the quantity $f_1 D_1 + f_2 D_2$ must be added to the right-hand side of Eq. (7), where D_1 and D_2 are the disequilibria in populations 1 and 2, respectively. Disequilibrium will thus be generated in the mixture even if the two contributing populations are each in gametic equilibrium, but only if allele frequencies differ between populations at *both* loci. One should note, however, that differences in allele frequencies between the two populations must be substantial for a significant amount of disequilibrium to be generated in the mixture. For example, if the frequencies of alleles differ between populations by 0.5 at both loci and the two populations are combined in equal proportions ($f_1 = f_2 = 0.5$), then D_m is equal to only 0.063. One should also note the high similarity between Eqs. (1) and (7); the former equation quantifies the nonrandom association among alleles within loci (e.g., departures from Hardy-Weinberg proportions), while the latter equation quantifies the nonrandom association among alleles between loci.

The results of Eq. (7) can be extended to a mixture of k populations, in which case the disequilibrium within the mixture (D_m) is given by

$$D_m = \sum_i f_i D_i + \text{Cov}(P, S) , \qquad (8)$$

where D_i is the disequilibrium between the two loci within the ith population and $\text{Cov}(P,S)$ is the covariance between p_i and s_i over all populations (Nei and Li 1973, Prout 1973). In this situation ($k>2$) it is no longer a sufficient condition for allele frequencies at both loci to vary among populations in order to generate disequilibrium; rather, the allele frequencies must be *correlated* among the populations. Nevertheless, the mixing and interbreeding of more than two populations will probably not occur in most instances of natural hybridization.

The treatment of gametic disequilibrium up to this point has been strictly theoretical. Evaluating gametic disequilibrium in natural populations, however, requires that the disequilibrium be estimated from samples of individuals. This is not a simple task, because the gametic genotypes (e.g., A_1B_1, A_1B_2) usually cannot be observed directly but must be inferred from the diploid genotypes at the loci being considered. For example, one can interpret $A_1A_1B_1B_2$ individuals as simply representing the union of A_1B_1 and A_1B_2 gametes. Similarly, gametic genotypes can be inferred for all individuals that are homozygous at one or both loci. In contrast, gametic genotypes cannot be inferred from double heterozygotes because the gametic phase for these individuals is usually not known; i.e., one cannot distinguish $A_1B_1/A_2 B_2$ from $A_1B_2/A_2 B_1$. Consequently, gametic disequilibrium cannot be explicitly calculated for most diploid organisms.

Several methods have been proposed for estimating gametic disequilibrium in natural populations when the two types of double heterozygotes cannot be distinguished. One approach is simply to ignore the double heterozygous class and to calculate D from the remaining individuals (e.g., Avise and Van Den Avyle 1984). One problem with this method is that double heterozygotes may constitute a substantial fraction of individuals in the sample, and their exclusion from the analysis can result in a significant loss of information. Alternatively, one can estimate D by the method of maximum likelihood *if* one assumes the population mates at random (Hill 1974). Under the assumption of random mating, the maximum likelihood estimate (MLE) of D is $\hat{X}_{11} - \hat{p}_1\hat{q}_1$, where

$$\hat{X}_{11} = (1/2N)\left[2N_{11} + N_{12} + N_{21}\right.$$

$$\left. + \frac{N_{22}\hat{X}_{11}(1 - \hat{p}_1 - \hat{q}_1 + \hat{X}_{11})}{\hat{X}_{11}(1 - \hat{p}_1 - \hat{q}_1 - \hat{X}_{11}) + (\hat{p}_1 - \hat{X}_{11})(\hat{q}_1 - \hat{X}_{11})}\right], \qquad (9)$$

$\hat{p}_1$ and $\hat{q}_1$ are the respective frequencies of A_1 and B_1 in a sample of N individuals, and the N_{ij} are the observed numbers of each composite genotype as given by the following table (Hedrick 1983):

Locus A

		A_1A_1	A_1A_2	A_2A_2
	B_1B_1	N_{11}	N_{12}	N_{13}
Locus B	B_1B_2	N_{21}	N_{22}	N_{23}
	B_2B_2	N_{31}	N_{32}	N_{33}

Equation (9) is a cubic function in $\hat{X}_{11}$ and can be solved explicitly for the three roots using one of several computer software packages. The root (solution) maximizing the likelihood function can then be identified according to the criteria given by Weir and Cockerham (1979). Hill (1974) had initially suggested that solutions to (9) be found by numerical iteration, but these solutions can converge to roots that do not represent the global maximum (Weir and Cockerham 1979). An example of Hill's (1974) likelihood method is given by Busack and Gall (1981), who found significant gametic disequilibrium in introgressed hybrid populations of rainbow and cutthroat trout.

Estimating gametic disequilibrium by the method of maximum likelihood is appropriate only for randomly mating populations. However, for many hybrid populations, the assumption of random mating would probably not be valid. Weir (1979) has argued against the general use of the likelihood method for estimating gametic disequilibrium in favor of an alternative procedure suggested by Burrows (unpublished, cited by Cockerham and Weir 1977) which does not make any assumptions regarding the mating structure of the population. In this method, the total disequilibrium D ($= X_{11} - p_1q_1$) is partitioned into two components, D_W and D_B ($D = D_W + D_B$), such that $D_W = X_{11} - Y_{11}$, $D_B = Y_{11} - p_1q_1$, and Y_{11} is the frequency that A_1 and B_1 occur together within zygotes (individuals) when contributed by different gametes, i.e., the frequency of A_1 and B_1 in the repulsion phase. If the coupling and repulsion phases of double heterozygotes cannot be distinguished, then X_{11} and Y_{11} cannot be calculated separately, but their *sum* is directly obtainable from the observed numbers or frequencies of each genotypic class. Using this fact, Burrows's composite measure of gametic disequilibrium (**D**) is defined (Weir 1979) as

$$\mathbf{D} = X_{11} + Y_{11} - 2p_1q_1$$

$$= (X_{11} - p_1q_1) + (Y_{11} - p_1q_1)$$

$$= D + D_B . \tag{10}$$

The composite measure $\mathbf{D}$ includes the normal measure of disequilibrium ($D = X_{11} - p_1q_1$) plus an added component ($D_B = Y_{11} - p_1q_1$) due to the nonrandom union of gametes. After one generation of random mating, however, Y_{11} (the frequency of A_1 and B_1 in repulsion) will equal p_1q_1 and D_B will go to zero. Hence, in a random mating population, $\mathbf{D}$ and D measure the same quantity, namely $X_{11} - p_1q_1$. However, estimating Burrow's composite measure ($\mathbf{D}$) does not require the assumption of random mating, and it is much simpler to calculate than the MLE of D. To show this, we first note that estimates of the coupling and repulsion frequencies of A_1, B_1 are given by

$$\hat{X}_{11} = \frac{2N_{11} + N_{12} + N_{21} + N'_{22}}{2N} ,$$

$$\hat{Y}_{11} = \frac{2N_{11} + N_{12} + N_{21} + N''_{22}}{2N} ,$$

where N_{22}' and N_{22}'' are the number of coupling and repulsion double heterozygotes in the sample, respectively ($N_{22} = N_{22}' + N_{22}''$). One should note that X_{11} and Y_{11} are not mutually exclusive; alleles are in both the coupling and repulsion phases for those individuals that are homozygous at one or both loci. Adding the expressions for $\hat{X}_{11}$ and $\hat{Y}_{11}$ and subtracting $2\hat{p}_1\hat{q}_1$ results in the following estimate ($\mathbf{D}^*$) of $\mathbf{D}$:

$$\mathbf{D}* = \frac{2N_{11} + N_{12} + N_{21} + N_{22}/2}{N} - 2\hat{p}_1\hat{q}_1 . \tag{11}$$

An unbiased estimate of $\mathbf{D}$ is further obtained as $\hat{\mathbf{D}} = [N/(N-1)]\,\mathbf{D}^*$ (Weir 1979). To test the hypothesis $H_0:\mathbf{D}\ (= D + D_B) = 0$, the following statistic can be used:

$$\chi^2 = N(\mathbf{R}*)^2 = \frac{N(\mathbf{D}*)^2}{[\hat{p}_1(1 - \hat{p}_1) + D_{.1}][\hat{q}_1(1 - \hat{q}_1) + D_{1.}]} \tag{12}$$

where $D_{.1} = N_{.1}/N$ (the observed frequency of A_1A_1) $- (\hat{p}_1)^2$; $D_{1.} (1)_1 = N_{1.}/N$ (the observed frequency of B_1B_1) $- (\hat{q}_1)^2$; $\mathbf{R}^*$ is the correlation between A and B adjusted for departures from Hardy-Weinberg proportions; and χ^2 is a chi-square distributed statistic with one degree of freedom (Weir 1979).

The statistical properties of the composite and maximum likelihood estimators have been compared by computer simulation (Weir 1979). The composite measure performs as well as, and sometimes better than, the MLE, even when random mating is assumed. When the population does not mate at random, the composite measure still estimates a definable quantity, namely D + D_B, but the quantity being estimated by the MLE is not defined.

An example illustrating the use of Hill's (1974) maximum likelihood method and Burrows's composite measure (Weir 1979) of gametic disequilibrium is given in Table 7.2. The data are for the same fish shown in Fig. 7.2e and represent estimates for a suspected mixture of coastal cutthroat trout and steelhead × cutthroat hybrids (data from Campton and Utter 1985). The first point to note from Table 7.2 is that all disequilibrium estimates are greater than zero, reflecting a positive correlation (R) between common alleles of the same species. However, the likelihood and composite estimates are not consistent. For example, the disequilibrium between *GLD-1* and *ME-4* is estimated as 0.105 by the method of maximum likelihood, but only as 0.026 by Burrows's composite measure. This discrepancy is due to the assumption of random mating by the likelihood method and the nonconformation of these fish to Hardy-Weinberg genotypic proportions (Campton and Utter 1985). Consequently, the composite measures are the only valid estimates of gametic disequilibria for these fish because a single, randomly mating population was clearly not sampled (Campton and Utter 1985). However, estimates obtained by the composite method must be interpreted with caution because this measure includes disequilibria generated by the nonrandom association of gametes (D_B). Unless two species have interbred randomly, such disequilibria would be expected in a mixture containing individuals of the parental species.

A second example of estimating gametic disequilibrium is presented in Table 7.3 (data from Campton and Johnston 1985). In this study, genotypes for rainbow trout from five localities in the Yakima River drainage (Washington state) conformed to Hardy-Weinberg proportions at virtually all loci, thereby indicating that random mating probably was occurring. If the intermediate allele frequencies observed for these fish at *LDH-4* and *SOD* (Fig. 7.6) had re-

Table 7.2 Estimates of gametic disequilibrium using Hill's (1974) maximum likelihood method and Burrows's composite measure (Weir 1979) for suspected hybrids of steelhead trout (*Salmo gairdneri*) and coastal cutthroat trout (*S. clarki clarki*). The statistic R is the estimated correlation between alleles, and χ^2 is the goodness of fit statistic ($= NR^2$) with 1 d.f. for testing $H_O:D = 0$ (MLE) or $D + D_B = 0$ (Weir 1979). The data and estimates are for the same fish as shown in Figs. 7.2e and 7.3e (From Campton and Utter 1985).

Loci	Maximum likelihood			Composite measure		
	D	R	χ^2	$D + D_B$	R	χ^2
GLD-1, ME-4	0.105	0.427	16.79***	0.026	0.239	5.26*
GLD-1, SDH-1	0.037	0.161	2.29	0.033	0.229	4.69*
ME-4, SDH-1	0.049	0.218	3.93*	0.062	0.248	5.09*

* P < 0.05; *** P < 0.001

Table 7.3 Estimates of gametic disequilibrium between *LDH-4* and *SOD* for populations of rainbow trout from five locations in the Yakima River drainage, Washington state. The statistic R is the estimated correlation between alleles, and χ^2 is the goodness of fit statistic ($= NR^2$) with 1 d.f. for testing $H_O:D = 0$ (MLE) or $D + D_B = 0$ (Weir 1979). (From Campton and Johnston 1985)

Location	Maximum likelihood			Composite measure		
	D	R	χ^2	$D + D_B$	R	χ^2
Yakima River (1)	−0.008	−0.052	0.19	−0.004	−0.026	0.05
Yakima River (2)	−0.006	−0.054	0.24	−0.009	−0.081	0.55
Yakima River (3)	0.040	0.282	6.27*	0.035	0.269	5.70*
Swauk Creek	−0.006	−0.047	0.08	−0.008	−0.048	0.09
Teneum Creek	−0.022	−0.158	1.25	−0.028	−0.206	2.12

*P < 0.05

sulted from the recent interbreeding of hatchery and wild fish, a negative correlation between alleles at the two loci would be expected. Only one of the five estimates of gametic disequilibrium was significant (P < 0.05), but the value of this estimate was greater than zero, not negative. However, the significance probability associated with this one positive result was not greater than one would expect by chance in one out of five independent comparisons (Cooper 1968). In contrast to the previous example, estimates of gametic disequilibrium obtained by the maximum likelihood and composite methods were very similar (Table 7.3). This is just as one would expect for randomly mating populations (Weir 1979). The conclusions to be drawn from Table 7.3 are that introgression from the hatchery to the wild populations, if it had occurred, was not a recent event and that hatchery fish and wild fish had randomly interbred to panmixia where they came in contact. On the other hand, relatively large sample sizes ($n > 200$) would be required to detect low to moderate levels of disequilibrium, given the observed allele frequencies in Fig. 7.6 In general, relatively large sample sizes are required to detect gametic disequilibrium in natural populations (Brown 1975; see also Chakraborty and Leimar, Chapter 4).

Estimating Admixture Proportions

Hybrid populations that have progressed past the F_1 stage, with little or no loss in viability or fertility, are often described as genetic (or population) admixtures. *Admixture* refers to the production of new genotypic combinations through recombination, and it was first used in this context to describe the interbreeding of human races following secondary contact in the western hemisphere (e.g., Glass and Li 1953, Glass 1955). Many human populations in North and South America essentially represent genetic admixtures of two or more previously isolated human races (Cavalli-Sforza and Bodmer 1971). If hybridization has been detected and if one can conclude that a single, randomly mating population has resulted (e.g., Campton and Johnston 1985), then one may wish to estimate the proportional contributions of the parental species (or populations) to the genetic admixture.

The relative contributions of two parental populations to a hybrid admixture can be estimated if allele frequencies for the hybrid population and the two parental populations are known or can be estimated. If p_1 and p_2 are the frequencies of some allele in populations 1 and 2, respectively, then the expected frequency of this allele in a simple mixture is $p_h = f_1 p_1 + f_2 p_2$, where f_1 and f_2 are the relative contributions of the two parental populations to the mixture. In the absence of selection, migration, and genetic drift (mutation being considered insignificant), the frequency (p_h) of the allele in the hybrid population will remain unchanged after one or more generations of random mating. Under these assumptions, the proportional contributions of the parental populations to the genetic admixture are obtained as

$$f_1 = \frac{p_h - p_2}{p_1 - p_2} \qquad (14)$$

where $f_2 = 1 - f_1$. Equation (14) was first proposed by Bernstein in 1931 (cited by Glass and Li 1953) and has been used somewhat frequently since then to estimate racial admixtures in human populations (e.g., Glass and Li 1953 and references therein, Reed 1969, Adams and Ward 1973).

If hybridization has occurred for several generations in a one-way direction such that the hybrid population is formed each generation from a fraction m of one parental population (say population 2) and a fraction $1 - m$ of the hybrid population from the previous generation, then the expected frequency of the allele in the hybrid (admixture) population at generation $t[p_{h(t)}]$ is given by

$$P_{h(t)} = (1 - m)^t p_1 + [1 - (1 - m)^t] p_2 .$$

If t is known, one can estimate m, the mean admixture rate per generation, as

$$m = 1 - \left[\frac{P_{h(t)} - p_2}{p_1 - p_2} \right]^{1/t} , \qquad (15)$$

which can be viewed as a simple one-way migration model, where $p_{h(0)} = p_1$. Equation (15) was first derived by Glass and Li (1953) to describe the dynamics of racial intermixture in the United States, where the population admixture in this case essentially represents the introgression of genes from Caucasians to American Blacks (see also Glass 1955, Roberts 1955, Saldanha 1957). Equation (15) has particular relevance to hybridization problems in fisheries because of the common practice of continuously stocking hatchery fish (e.g., steelhead trout and Pacific salmon) derived from one population into the native habitat of another population. The continuous stocking of an exotic species into the native habitat of a second species represents a similar problem at the interspecific level if hybrids are fertile (e.g., Busack and Gall 1981).

Equations (14) and (15) provide estimates of admixture proportions for two populations based on data for only a single allele or locus. In practice, data for many loci would be available and more than two populations could be contributing to the hybrid admixture. Elston (1971) describes the formal least squares and maximum likelihood estimators of admixture proportions for these more generalized situations. The details of these procedures are beyond the scope of this chapter, but interested readers are encouraged to consult Elston's (1971) excellent presentation.

The preceding methods of estimating admixture proportions (Eq. 14; Elston 1971) implicitly assume that allele frequencies for the parental populations have not changed since hybridization first occurred. They also assume that allele frequencies for the parental populations are estimated without error. These assumptions will not be true in most situations. If two populations are plotted in multidimensional space according to their allele frequencies at all loci, then a simple mixture of individuals from these two populations should fall somewhere on the interconnecting straight line segment (Fig. 7.7). However, allele frequencies of those individuals specifically contributing to the hybrid population may not be equal to those of the parental populations because of founder effects. In addition, allele frequencies would be expected to change in the parental and hybrid populations after several generations due to random genetic drift. Natural selection may further affect allele frequencies, especially in the hybrid population.

As a result, the hybrid population normally will not be collinear with the two parental populations in the multivariate allele frequency space, and estimates of admixture proportions using Eq. (14) will therefore vary among loci and alleles. Sampling error associated with the estimation of allele frequencies will be a further source of variability and departure from collinearity. Admixture proportions estimated by least squares (Elston 1971) will be represented by that point on the line segment minimizing the distance (squared deviations) be-

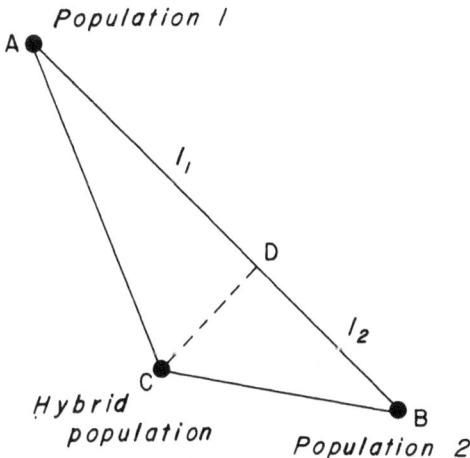

Fig. 7.7 Graphical representation of the relative positions of two hypothetical parental populations and a hybrid population in the n-dimensional sample space of estimated allele frequencies. The hybrid population is not collinear with parental populations because of founder effects, genetic drift, sampling error, and possibly natural selection. As a result, estimates of admixture proportions obtained by Eq. (14) will vary among loci. The least squares estimate projects the hybrid population to point D. If l_1 and l_2 are the lengths of line segments AD and BD, respectively, then the least squares estimates of f_1 and f_2 (the admixture proportions of populations 1 and 2, respectively) are $l_2/(l_1 + l_2)$ and $l_1/(l_1 + l_2)$, respectively. After Thompson (1973).

tween the line and the hybrid population (Fig. 7.7). The maximum likelihood estimate usually will be quite close to this least squares estimate (Elston 1971).

Thompson (1973) addressed some of these problems of noncollinearity by deriving maximum likelihood estimators of admixture proportions for two parental populations using a "drift and sampling model." In this model, the parental and hybrid populations are all assumed to have reproduced independently of one another for several generations following the hybridization event. Allele frequencies for present-day populations are therefore assumed to be different from those for the ancestral populations because of genetic drift. This model was used to estimate the proportions of Celtic and Norse genes in the human population of Iceland when data were available only on present-day Irish, Norwegian, and Icelandic populations. This study of Thompson (1973) is particularly interesting because it presents a rather sophisticated example of how admixture proportions can be estimated.

In addition to estimating admixture proportions for entire populations, one can estimate the proportions of an individual's genes derived from each of two parental populations (MacLean and Workman 1973a, 1973b). However, obtaining accurate estimates of admixture proportions for individuals may require a prohibitively high number of loci (Reed 1973, Cook and Weisberg 1974).

Estimating admixture proportions in hybrid populations of fishes may be of special importance in conservation programs. A common practice in the western United States is to poison (asphyxiate) local populations with rotenone when hybridization between native and introduced species is detected. However, electrophoretic and mtDNA techniques can detect very low levels of hybridization and introgression (e.g., Busack and Gall 1981, Powell 1983), and this raises some very important questions. Should a population of a native, perhaps threatened or endangered, species be considered "expendable" if only 5% of its genes were derived by hybridization with a nonnative species? What if the admixture proportion was only 1%? At what level of hybridization should fishery biologists stop poisoning these introgressed populations and preserve them as native forms?

These are clearly management questions with considerable political and social ramifications. The presence or absence of hybridization may therefore be a moot point, especially when the ability to detect hybridization is highly dependent upon the number of individuals sampled. Therefore, geneticists who are cooperating with fishery management agencies in attempting to detect natural hybridization should perhaps report estimates and standard errors of admixture proportions on a routine basis. From a management perspective, the answer "introgression was detected" or "hybrids were present" may be inadequate and unacceptable.

CONCLUDING REMARKS

Natural hybridization of fishes, both within and between species, presents many problems to fish management agencies. To a large extent, these problems have resulted from man's own activities. Habitat alterations, the transplantation of fish stocks, and the introduction of exotic species have all contributed to an increased incidence of hybridization. Our awareness of the extent of, and the factors leading to, natural hybridization in fishes is due in large part to the pioneering investigations of Carl L. Hubbs and his associates during the years 1920 to 1960.

In this chapter I have attempted to review a relatively diverse array of literature that is directly relevant to the application of population genetics to investigations of hybridization problems in fisheries management. Such a chapter would not be complete without discussing the use of morphological characters for detecting natural hybridization. Until the mid-1960s, measuring and counting morphological characters were essentially the only methods available for detecting hybridization in fishes. The advent of electrophoretic methods in the mid-1960s provided the first real opportunity to investigate natural hybridization in terms of population genetics. Mitochondrial DNA techniques offer hope for future understanding of the dynamics and evolutionary significance of introgressive hybridization in fishes. Future technologies will undoubtedly provide new answers to old questions and will certainly raise more questions as new discoveries are made.

From a population genetics perspective, hybridization is very much a dynamic process. It is the mixing and reorganization of genomes at the organismic level, and it is the mixing and reorganization of gene pools at the population level. It is a process whereby genetic disequilibrium is generated from genetic equilibrium simply by the mixing and interbreeding of two previously isolated gene pools. It is a process that confounds our desire to manage fishes as discrete populations or species. It is also a process that can exterminate rare and endangered fishes.

APPENDIX 7.1:

Derivation of Equation (1)

A mixture of two non-interbreeding populations with different allele frequencies will yield a deficiency of heterozygotes relative to the expected Hardy-Weinberg genotypic proportions. This can be illustrated easily by the following algebraic example. Let p_1 and $1-p_1$ be the respective frequencies of two alleles (e.g., A and a) at a particular locus in population 1, and let p_2 and $1-p_2$ be the frequencies of the corresponding alleles in population 2. In addition, let f_1 and f_2 be the relative proportions of the two populations in some non-interbreeding mixture ($f_1 + f_2 = 1.0$). If individuals mate at random *within* each population, then the frequency of heterozygotes in populations 1 and 2 will be $2p_1(1-p_1)$

and $2p_2(1-p_2)$, respectively. The observed frequency of heterozygotes (H_O) in some non-interbreeding mixture will therefore be $2p_1(1-p_1)f_1 + 2p_2(1-p_2)f_2$. However, the expected frequency of heterozygotes (H_E), assuming random mating and Hardy-Weinberg proportions, is $2\bar{p}(1-\bar{p})$, where $\bar{p} = p_1f_1 + p_2f_2$. The difference between the observed and the expected frequency of heterozygotes in the population mixture is

$$H_O - H_E = 2p_1(1 - p_1)f_1 + 2p_2(1 - p_2)f_2 - 2\bar{p}(1 - \bar{p}) \, ,$$

which, after some algebra, reduces to Eq. (1):

$$H_O - H_E = -2f_1f_2(p_1 - p_2)^2 \, .$$

APPENDIX 7.2:

Derivation of Equation(2)

A deficiency of heterozygotes will occur if two populations interbreed but mate assortatively, i.e., within populations. This can be illustrated by the example given in Appendix 7.1 except that in this case a fraction, m, of each population are assumed to mate assortatively while a fraction $1 - m$ randomly interbreed. The expected frequency of heterozygotes (H_E) in the population mixture remains as before if Hardy-Weinberg genotypic proportions are assumed, i.e., $H_E = 2\bar{p}(1-\bar{p})$, where $\bar{p} = p_1f_1 + p_2f_2$. On the other hand, the observed frequency of heterozygotes includes both those heterozygotes produced by random mating within each population and those produced by random mating between populations. If m is the same in both populations, then

$$H_O = [2p_1(1 - p_1)f_1 + 2p_2(1 - p_2)f_2]m + (1 - m)2\bar{p}(1 - \bar{p}) \, ,$$

where $\bar{p}$ is the same as before. The above expression for H_O is considerably more complicated if m differs between populations, but this situation will not be discussed here. Using the above expression for H_O and the results of Appendix 7.1, the observed deficiency of heterozygotes reduces to Eq. (2):

$$H_O - H_E = -2mf_1f_2(p_1 - p_2)^2 \, .$$

8.
Genetic Distance and Molecular Phylogeny

Masatoshi Nei

Genetic distance is the degree of gene difference (genomic difference) between species or populations that is measured by some numerical method. Thus, the average number of codon or nucleotide differences per gene is a measure of genetic distance. There are various kinds of molecular data that can be used for measuring genetic distance. When the two species to be compared are distantly related, data on amino acid or nucleotide sequences are used (e.g. Dayhoff 1972). In this case, the genetic polymorphism within species is usually ignored, since its effect on the total genetic distance is small. When two closely related species or populations are compared, however, the effect of polymorphism cannot be neglected, and one has to examine many proteins or genes from the same populations. Sequencing of amino acids or nucleotides for many proteins or genes is time-consuming and expensive, so that more efficient molecular techniques are needed for studying the genetic relationship of closely related organisms.

There are two such techniques available now. One is protein electrophoresis, which has been used extensively for the last fifteen years in evolutionary studies. This technique does not produce data on amino acid or nucleotide sequence differences, but the gene frequency data generated by this technique can be used to estimate genetic distance. The other is the restriction enzyme technique, which is now used by an increasing number of investigators (e.g., Brown et al. 1979, Avise et al. 1979, Shah and Langley 1979, Brown 1983, Avise and Lansman 1983). This technique does not produce data on nucleotide sequence differences, but these differences can be estimated by using some statistical methods (Nei and Li 1979, Kaplan and Langley 1979, Gotoh et al. 1979, Nei and Tajima 1983). Unfortunately, this technique is still time-consuming, and the accuracy of the estimates of genetic distance obtained is not necessarily high.

In this chapter we first consider statistical methods for estimating genetic distance for closely related organisms with special consideration of electrophoretic data. We then describe mathematical models that are important for understanding the process and mechanism of genetic differentiation of populations in terms of certain genetic distance measures. Finally, we discuss the empirical relationship between genetic distance and evolutionary time and outline several problems concerning the reconstruction of phylogenetic trees by using genetic distances.

GENETIC DISTANCE
Measures of Genetic Distance

In the past several decades various measures of genetic distance have been proposed. Some are direct applications of earlier measures of morphological distances which have been used in classical numerical taxonomy. For example, the measures proposed by Sanghvi (1953), Steinberg et al. (1967), Balakrishnan and Sanghvi (1968), and Siciliano et al. (1973) are all direct applications of Mahalanobis's (1936) D^2 statistic to gene frequency data. Bhattacharyya's (1946) measure, which is essentially the same as Cavalli-Sforza and Edwards's (1967), can also be regarded as an extension of Mahalanobis's D^2 statistic for the case of discrete characters.

In these theories populations are represented as points in multidimensional space and the genetic distance between two populations is measured by the geometric distance between the corresponding points in the space. Thus, the principle of triangle inequality is very important, but little attention is paid to the relationship between genetic distance and evolutionary change of populations. The absolute values of these measures do not have any particular biological meaning, and only the relative values are important for finding the genetic relationship among populations. In some distance measures such as Latter's (1973) ϕ^*, the distance is related to Wright's F_{ST}, which is in turn related to evolutionary time under the assumption of no mutation. However, these measures are not a direct measure of the amount of gene differences between populations.

Compared with these measures, Nei's (1972) distance measure is based on an entirely different concept. It is intended to measure the number of gene or codon substitutions per locus that have occurred after divergence of the two populations under consideration. Thus, the absolute value of this measure has a clear-cut biological meaning. Theoretically, Nei's method can be applied to any pair of taxa, whether they are local populations, species, or genera, if enough data are available. Of course, protein electrophoresis cannot detect all codon differences, so that we are forced to deal with only those codon differences that are detectable by the technique. Furthermore, there are some other statistical problems which make it difficult to estimate the exact number of codon differences. For these reasons, I have proposed three different measures of genetic distance: the minimum, standard, and maximum estimates of codon differences per locus (Nei 1973b).

Definition of Nei's Distance Measures

Consider two populations, X and Y, in which l alleles are segregating at a locus. Let x_i and y_i be the frequencies of the ith allele in X and Y, respectively. The probability of identity of two randomly chosen genes is $j_X = \Sigma x_i^2$ in population X and $j_Y = y_i^2$ in population Y. The probability of identity of two genes chosen at random, one from each of the two populations, is $j_{XY} = \Sigma x_i y_i$. Here, Σ indicates summation over all alleles. For example, $x_i^2 = x_1^2 + x_2^2 + \cdots + x_l^2$, and $\Sigma x_i y_i = x_1 y_1 + x_2 y_2 + \cdots + x_l y_l$. Note that the identity of genes defined in this way requires no assumptions about selection, mutation, and migration. We designate by J_X, J_Y, and J_{XY} the respective arithmetic means of j_X, j_Y, and j_{XY} over all loci in the genome, including monomorphic ones. Clearly, $D_{X(m)} \equiv 1 - J_X$, $D_{Y(m)} \equiv 1 - J_Y$, and $D_{XY(m)} \equiv 1 - J_{XY}$ are all equal to the proportion of different genes (alleles) between two randomly chosen genomes from the respective populations. In other words, $D_{X(m)}$ and $D_{Y(m)}$ are minimum estimates of codon differences per locus between two randomly chosen genomes from populations X and Y, respectively, whereas $D_{XY(m)}$ is a minimum estimate of codon differences per locus between two randomly chosen genomes, one from each of X and Y. ($D_{X(m)}$ and $D_{Y(m)}$ are equal to average heterozygosity.) Therefore,

$$D_m = D_{XY(m)} - \frac{D_{X(m)} + D_{Y(m)}}{2} \tag{1}$$

is a minimum estimate of net codon differences per locus between X and Y when the intrapopulational codon differences are subtracted. I have called D_m the *minimum genetic distance*. It is noted that if we denote by x_{ij} and y_{ij} the frequencies of the *i*th allele at the *j*th locus in populations X and Y, respectively, D_m can also be written as

$$D_m = \frac{1}{2R} \sum_{j=1}^{R} \sum_{i=1}^{l_j} (x_{ij} - y_{ij})^2$$

$$= \sum_{j=1}^{R} \frac{d_j}{R}, \tag{2}$$

where l_j and R are the number of alleles at the *j*th locus and the number of loci in the genome, respectively, and $d_j \equiv \Sigma_{i=1}^{l_j}(x_{ij} - y_{ij})^2/2$ is the distance at the *j*th locus.

The drawback of D_m is that $D_{X(m)}$, $D_{Y(m)}$, and $D_{XY(m)}$ are the proportions of different genes between two randomly chosen genomes, so that they are not proportional to the number of codon differences. Thus, D_m may be a gross underestimate of the number of net codon differences when $D_{XY(m)}$ is large. If individual codon changes are independent and follow a Poisson distribution, the mean number of net codon differences (substitutions) may be given by

$$D = -\ln I, \tag{3}$$

where

$$I = \frac{J_{XY}}{\sqrt{J_X J_Y}} \qquad (4)$$

is the normalized identity of genes (or genetic identity) between X and Y. I have called D the *standard genetic distance*. It is noted that D can be written as $D = D_{XY} - (D_X + D_Y)/2$, where $D_{XY} = -\ln J_{XY}$, $D_X = -\ln J_X$, and $D_Y = -\ln J_Y$. As will be seen later, if the rate of gene (codon) substitution per year is constant, it is linearly related to the time since divergence between the two populations. Also, in certain migration models it is linearly related to the geographical distance (Nei 1972).

If the rate of codon changes varies from locus to locus, D still may be an underestimate of codon differences. In this case the mean number of net codon differences may be estimated by

$$D' = -\ln I', \qquad (5)$$

where $I' = J'_{XY}/\sqrt{(J'_X J'_Y)}$, in which J'_{XY}, J'_X, and J'_Y are the geometric means of j_{XX}, j_X, and j_Y, respectively, over different loci. In practice, however, D' is affected considerably by sampling errors of gene frequencies at the time of population survey as well as by random genetic drift. These factors are expected generally to inflate the estimate of the mean number of net codon differences. Therefore, I call D' the *maximum genetic distance*. If any of the values of $j_{XY}/\sqrt{(j_X j_Y)}$ for individual loci is small, D' can be a gross overestimate. In fact, if there is a single locus at which there is no common allele between two populations, D' becomes infinite.

Nei at al. (1976) developed a somewhat different formula for this case, assuming that the rate of codon substitution varies among loci following the gamma distribution with coefficient of variation 1. It is given by

$$D_v = \frac{1 - I}{I} . \qquad (6)$$

The rationale of this formula will be discussed later. This distance measure seems to be superior to D', since it is not affected so strongly by sampling error.

Estimation of Genetic Distance

Theoretically, the genetic distance between two populations is defined in terms of the poplation gene frequencies for all loci in the genome. In practice, however, it is virtually impossible to examine all genes in the populations for all loci. Therefore, we must estimate the genetic distance by sampling a certain number of individuals from the populations and examining a certain number of loci. Let us now consider how to estimate genetic distance from actual data, following Nei and Roychoudhury (1974a) and Nei (1978a).

Clearly, there are two sampling processes involved in this case: sampling of loci from the genome and sampling of individuals (genes) from the population. In the following we assume that r loci are chosen at random and n individuals ($2n$ genes) are examined for each locus. Let $\hat{x}_i$ and $\hat{y}_i$ be the frequencies of the ith allele at a locus in samples of $2n$ genes from populations X and Y, respectively. The usual method of estimating genetic distance is to replace x_i and y_i in Eqs. (1), (4), or (5) by $\hat{x}_i$ and $\hat{y}_i$, respectively.

However, when sample size is small, this method gives a biased estimate (Nei 1973, 1978a). Unbiased (or less biased) estimates of D_m, D, D' and D_v may be obtained by replacing Σx_i^2, Σy_i^2, and $\Sigma x_i y_i$ in the formulae for genetic distance by the unbiased estimates of these quantities. The unbiased estimates of Σx_i^2, Σy_i^2, and $\Sigma x_i y_i$ are given by $\hat{j}_X = (2n\Sigma\hat{x}_i^2 - 1)/(2n - 1)$, $\hat{j}_Y = (2n\Sigma\hat{y}_i^2 - 1)/(2n - 1)$, and $\hat{j}_{XY} = \Sigma\hat{x}_i\hat{y}_i$, respectively, whereas the unbiased estimates $(\hat{J}_X, \hat{J}_Y,$ and $\hat{J}_{XY})$ of J_X, J_Y, and J_{XY} are the respective averages of j_X, j_Y, and j_{XY} over loci. For example, the unbiased estimates of D_m and D ($\hat{D}_m$ and $\hat{D}$, respectively) may be obtained by

$$\hat{D}_m = \left(\frac{\hat{J}_X + \hat{J}_Y}{2}\right) - \hat{J}_{XY} \tag{7}$$

and

$$\hat{D} = -\ln\left[\frac{\hat{J}_{XY}}{\sqrt{\hat{J}_X\hat{J}_Y}}\right]. \tag{8}$$

Obviously, Eq. (8) is valid only when the number of loci (r) is large (Li and Nei 1975).

The sampling variances of $\hat{D}_m$, $\hat{D}$, $\hat{D}_v$, and $\hat{I}$ can be computed by the methods given by Nei and Roychoudhury (1974a) and Nei (1978a, b). To compute the variance of $\hat{D}_m$, we first note that the unbiased estimate of d_j in (2) is given by

$$\hat{d}_j = \frac{2n_X\Sigma_i\hat{x}_{ij}^2 - 1}{2(2n_X - 1)} + \frac{2n_Y\Sigma_i\hat{y}_{ij}^2 - 1}{2(2n_Y - 1)} - \Sigma_i\hat{x}_{ij}\hat{y}_{ij} . \tag{9}$$

Therefore, the variance of $\hat{D}_m$ is

$$V(\hat{D}_m) = \frac{\Sigma_{j=1}^r(\hat{d}_j - \hat{D}_m)^2}{r(r - 1)} . \tag{10}$$

The variances of $\hat{D}$, $\hat{D}_m$, and $\hat{I}$ are more complicated, and usually a computer is required for the computation unless the populations are highly monomorphic. Such a computer program is available upon request.

When $\hat{I}$ is lower than 0.9 for all population pairs and average heterozygosity is low for all populations, the variances [$V(\hat{I})$ and $V(\hat{D})$] of $\hat{I}$ and $\hat{D}$ are approximated by

$$V(\hat{I}) = \frac{\hat{I}(1 - \hat{I})}{r} \tag{11}$$

$$V(\hat{D}) = \frac{1 - \hat{I}}{\hat{I}r} \tag{12}$$

(Nei 1971, 1978b). The reason for this is that in this case single-locus genetic identity $I_j = \hat{j}_{XY}/\sqrt{(\hat{j}_X\hat{j}_Y)}$ usually takes a value close to 1 or 0, so that $\hat{I}_j$ approximately follows the binomial distribution (Ayala et al. 1974).

In planning a survey of gene frequencies to estimate genetic distance it is important to know how many loci and how many individuals per locus should be examined when the total number of genes to be surveyed is fixed. This problem has been studied by Nei and Roychoudhury (1974a) and Nei (1978a) by decomposing the variance of genetic distance into the variance among loci and the variance due to sampling of genes within loci. The results obtained indicate that the interlocus variance is much larger than the intralocus variance unless n is extremely small, and thus it is important to study a large number of loci rather than a large number of individuals per locus to reduce the variance of the estimate of genetic distance.

Under certain assumptions genetic distance can be used to estimate the time after separation of two populations. In this case the standard error of the estimate of separation time may be computed from the variance of genetic distance considered above. The variance can also be used to test the difference between two estimates of genetic distances if independent sets of loci are used for computing the two distance estimates. In practice, however, it is customary to use the same set of loci for computing distance estimates for all pairs of populations. In this case, the variances obtained from (10) and its equivalent formulae for $\hat{D}$ and $\hat{D}_v$ are not appropriate for testing the difference between two distance estimates. This is because they include the variance resulting from the differences in the initial gene frequencies among loci at the time of population differentiation (Li and Nei 1975). However, the difference between a pair of D_m's can be tested in the following way. If we note $\hat{D}_m = \Sigma d_j/r$, the difference between a pair of $\hat{D}_m$'s, say $\hat{D}_{m1}$ and $\hat{D}_{m2}$, can be written as

$$\hat{D}_{m1} - \hat{D}_{m2} = \frac{\Sigma(\hat{d}_{j1} - \hat{d}_{j2})}{r}$$

$$= \frac{\Sigma \delta_j}{r},$$

(13)

where $\delta_j = \hat{d}_{j1} - \hat{d}_{j2}$ is the difference in $\hat{d}$ for the jth locus. Therefore, the difference between $\hat{D}_{m1}$ and $\hat{D}_{m2}$ is tested by the ordinary t-test for δ_j. Strictly speaking, δ_j is not normally distributed, but the above test would given an approximate significance level, since the t-test is known to be robust. It should be noted that a significant difference between $\hat{D}_{m1}$ and $\hat{D}_{m2}$ also implies a significant difference between the corresponding standard distances

So far we have been interested in the genetic distance defined as the number of codon differences per locus, so that a large number of loci are required for estimating this quantity. However, collection of gene frequency data is time-consuming, and under certain circumstances only a few loci are available for the study of gene differences. In this case the estimate of genetic distance may deviate considerably from the real value. When local populations within the same species are compared, this deviation is expected to be generally upward, since gene frequencies are studied more often with highly polymorphic loci than with less polymorphic loci, and monomorphic loci in these populations almost always have the same allele. However, if one is interested only in relative values of genetic distance among several populations, the estimate of distance based on a few polymorphic loci would still be useful though its variance inevitably becomes large.

MATHEMATICAL MODELS
OF POPULATION DIFFERENTIATION

The genetic differentiation of populations occurs only when the populations are partially or completely isolated from each other. Let us now consider the process of genetic differentiation of populations in terms of the genetic distance measures considered above.

Complete Isolation: General Case

When two populations are reproductively isolated, they tend to accumulate different genes due to mutation, selection, and genetic drift. With certain assumptions, this problem can be studied by a simple mathematical model. The assumptions we make are as follows:

- A population splits into two populations (X and Y) at a certain evolutionary time and thereafter no migration occurs between the two populations.
- Populations X and Y are in equilibrium with respect to the effects of mutation, selection, and random genetic drift, so that the average gene

identities (J_X and J_Y) within populations remain constant. This assumption seems to be satisfactory in many natural populations, since closely related populations or species generally show the same degree of heterozygosity. In some cases, of course, the bottleneck effect seems to be important, and this effect will be considered later.

● All new mutations are different from the alleles existing in the populations (infinite-allele model). This assumption seems to be satisfactory if alleles are identified at the codon (amino acid) level but probably not if they are studied by electrophoresis. I shall discuss the effect of violation of this assumption later.

● The rate of gene substitution per locus per year (α) remains constant and is the same for all loci. The first part of this assumption seems to be roughly correct at the amino acid level (e.g., Nei 1975, Fitch 1976, Wilson et al. 1977), but the second part is certainly incorrect. However, the effect of varying rates of gene substitution among loci can be corrected, as will be seen later. It can be shown that α is equal to the mutation rate per year (v) if all mutations are neutral, whereas it is equal to $4Nsv$ if mutant genes are advantageous and semidominant, where N is the effective population size and s is the selective advantage of a mutant gene (Kimura and Ohta 1971).

Under the above assumptions, Nei (1972, 1975) has shown that the genetic identity at the tth year is

$$I_A = I_0 e^{-2\alpha t} \, , \tag{14}$$

where I_0 is the value of I at time 0. Therefore, we have

$$D = 2\alpha t + D_0 \, , \tag{15}$$

where $D_0 = -\ln I_0$. In the present model $I_0 = 1$, so that $D_0 = 0$. It is clear from (15) that D measures the accumulated number of gene (codon) substitutions per locus between the two populations.

As mentioned earlier, however, the assumption that α is the same for all loci is incorrect. Nei et al. (1976a) have shown that the rate of amino acid substitution per polypeptide varies considerably with protein and is distributed roughly as a gamma distribution with coefficient of variation 1. They also showed that the subunit molecular weights of the proteins that are often used for electrophoresis also follow a gamma distribution. Furthermore, studies on the variances of single-locus heterozygosity and genetic distance in various organisms (more than one hundred different species) have suggested that the distribution of the rate of gene substitution or mutation rate roughly follows the gamma distribution with coefficient of variation 1 (Nei et al. 1976b, Fuerst et al. 1977, Chakraborty et al. 1978). Zouros's (1979) study on the relative mutation rates supports this conclusion. It is also noted that variation of the mutation rate is apparently related to the subunit molecular weight of protein (Koehn and Eanes 1977, Ward 1977, Nei et al. 1978).

Let us therefore assume that α has the following gamma distribution:

$$f(\alpha) = \frac{b^a}{\Gamma(a)} e^{-b\alpha} \alpha^{a-1} ,$$

where $a = \bar{\alpha}^2/V(\alpha)$ and $b = \bar{\alpha}/V(\alpha)$, in which $\bar{\alpha}$ and $V(\alpha)$ are the mean and variance of α. The expected genetic identity is then given by

$$\bar{I}_A = \frac{E(\Sigma j_{xy})}{\sqrt{E(\Sigma j_x)E(\Sigma j_y)}}$$

$$= \frac{\Sigma E(j_i)e^{-2\alpha_i t}}{\Sigma E(j_i)}$$

$$\simeq \int_0^\infty f(\alpha)e^{-2\alpha t}d\alpha = \left(\frac{a}{a + 2\bar{\alpha}t}\right)^a , \qquad (16)$$

where $E(j_i)$ is the expected homozygosity at the ith locus, and Σ stands for the summation for all loci in the genome. Equation (16) is expected to give an overestimate of the true value of expected genetic identity, since it is based on the assumption of no correlation between $E(j_i)$ and $\exp(-2\alpha_i t)$ though in practice there should be a positive correlation. In the case of neutral alleles, the effect of the above assumption on $\bar{I}_A$ can be evaluated, but unless $2\alpha_i$ is very large, the effect does not seem to be important (Griffiths 1980).

When the coefficient of variation $(a^{-1/2})$ is 1,

$$\bar{I}_A = \frac{1}{1 + 2\bar{\alpha}t} . \qquad (17)$$

Therefore, the mean number of gene substitutions per locus $(2\bar{\alpha}t)$ can be estimated by D_v in (6):

$$D_v \equiv 2\bar{\alpha}t = \frac{1 - \bar{I}_A}{\bar{I}_A} . \qquad (18)$$

Mathematically, $D_v > D$, but the difference between (14) and (17) is small when t is relatively small (see Table 8.1). However, note that, because of the assumption we have made above, Eq. (18) is expected to give an underestimate of $2\alpha t$ when $2\alpha t$ is large.

Table 8.1 Evolutionary time and genetic identity under the infinite-allele model $(I_A, \bar{I}_A)$ and the stepwise mutation model $(I_E, \bar{I}_E)$. I_A, $\bar{I}_A$, I_E, and $\bar{I}_E$ were obtained by Eqs. (14), (17), (20), and (22), respectively. In this computation the rate of gene substitution ($\alpha = \nu$) was assumed to be 10^{-7} per year (see text).

Time ($\times 10^3$ yrs)	I_A	$\bar{I}_A$	I_E	$\bar{I}_E$	Time ($\times 10^6$ yrs)	I_A	$\bar{I}_A$	I_E	$\bar{I}_E$
10	0.998	0.998	0.998	0.998	1	0.819	0.833	0.827	0.845
50	.990	.990	.990	.990	2	.670	.714	.697	.745
100	.980	.980	.980	.981	3	.549	.625	.599	.674
200	.961	.961	.961	.962	4	.449	.556	.524	.620
300	.942	.943	.943	.945	5	.368	.500	.466	.577
400	.923	.926	.925	.928	6	.301	.455	.420	.542
500	.905	.909	.907	.913	7	.247	.417	.383	.513
600	.887	.893	.890	.898	8	.202	.385	.353	.488
700	.869	.877	.874	.884	9	.165	.357	.329	.466
800	.852	.862	.858	.870	10	.135	.333	.309	.447
900	0.835	0.847	0.842	0.857	20	0.018	0.200	0.207	0.333

Recently, Hillis (1984) claimed that the effect of variation of substitution rate among loci can be taken care of if we redefine I in Eq. (3) as the mean of $j_{XY}/\sqrt{(j_X h_Y)}$ over loci. However, a theoretical basis for this claim has not been found.

Either formula (15) or (18) enables us to estimate the time after divergence between two populations if α is known. Using the average rate of amino acid substitution for 22 proteins that are often used for electrophoresis, Nei (1975) estimated α to be 10^{-7} for electrophoretic data (see the following section). Therefore, assuming $D_0 = 0$, t may be estimated by

$$t = 5 \times 10^6 \times D . \tag{19}$$

It should be emphasized, however, that the above value of α is based on a number of assumptions, and thus (19) gives only a very rough estimate of divergence time.

It should be mentioned that Eqs. (15) and (18) are valid only when a large number of loci are studied, since each event of gene substitution is subject to large stochastic errors. Nei and Tateno (1975) studied the distribution of single-locus gene identity $[I_j = j_{XY}/\sqrt{(j_X j_Y)}]$ under the assumption of neutral mutations by using computer simulation. The results obtained show that I_j shows an inverse J-shaped distribution when $2\bar{\alpha}t$ is small, whereas it shows a U-shaped distribution when $2\bar{\alpha}t$ is moderately large. Therefore, to obtain a reliable estimate of I, a large number of loci must be studied. This is true even if gene substitution is mediated by natural selection (Chakraborty et al. 1977). The mathematical formulae for obtaining the stochastic variance of genetic distance under the assumption of neutral mutations have been obtained by Li and Nei (1975).

Strictly speaking, Eq. (15) is not appropriate for electrophoretic data, even if the mutation rate is the same for all loci. This is because at the electrophoretic level the effect of back mutations becomes important as t increases. This problem can be studied by using Ohta and Kimura's (1973) stepwise model of neutral mutations, although some authors (Ramshaw et al. 1979, Fuerst and Ferrell 1980, McCommas 1983) have questioned the appropriateness of this model to electrophoretic data. Nei and Chakraborty (1973), Li (1976b), and Chakraborty and Nei (1976, 1977) have studied the expected genetic identity under the stepwise mutation model. The exact formula for the genetic identity for electrophoretic data (I_E) is rather complicated (Li 1976b), but for practical purposes we can use the following equation:

$$I_E = e^{-2vt} \sum_{r=0}^{\infty} \frac{(vt)^{2r}}{(r!)^2} , \tag{20}$$

where v is the mutation rate per generation (Nei 1978b). We note that $\alpha = v$ in this case since we are dealing with neutral mutations.

When v varies from locus to locus following the gamma distribution, the average value of I_E is given by

$$\bar{I}_E = \frac{b^a}{\Gamma(a)} \sum_{r=0}^{\infty} \frac{t^{2r}}{(r!)^2} \frac{\Gamma(a + 2r)}{(b + 2t)^{a+2r}} \tag{21}$$

approximately. At the present time, we do not know the a value for the stepwise mutation model very well. However, if we use $a = 1$ as before, we have

$$\bar{I}_E = \frac{1}{1 + 2\bar{v}t} \left[1 + \sum_{r=1}^{\infty} \frac{(2r)!}{(r!)^2} \left(\frac{\bar{v}t}{1 + 2\bar{v}t} \right)^{2r} \right] , \tag{22}$$

where $\bar{v}$ is the mean of v over all loci. This formula is expected to give an overestimate of the expected genetic identity when $2\bar{v}t$ is large, as in the case of Eq. (16).

Table 8.1 shows the values of genetic identity for the four different models, i.e., Eqs. (14), (17), (20), and (22). In this table, calendar year rather than generation is used as a unit of time, with $\bar{\alpha} = \bar{v} = 10^{-7}$. The relationship between genetic distance $D = -\log_e I$ and evolutionary time is also given in Figure 8.1 for four different models. It is clear that the genetic identity is virtually the same for all four models for the first one million years. Therefore, if the observed value of I is larger than about 0.82, Eq. (19) may be used for estimating divergence time. However, if the divergence time increases further, the difference between the models becomes pronounced. In this case, Eq. (19) should not be used for estimating divergence time, since the assumption of the same mutation rate for all loci is not necessarily valid. The formula for I_E is also expected to give an underestimate, since in this case the same mutation rate is assumed for all loci. Therefore, for estimating divergence time, Eq. (17) or (22) seems to be more appropriate, although the applicability of the stepwise mutation model used in (22) is yet to be confirmed. At any rate, the numerical values in Table 8.1 or Figure 8.1 can be used for getting a rough estimate of divergence time if the genetic identity value is available.

Complete Isolation: Short-term Evolution

In general the above theory does not apply to nonprotein loci such as those for blood groups, since the relationship between codon substitution and phenotypic change at these loci may not be so simple as that for protein loci (Nei 1975). However, if we consider a very short period of evolutionary time, all of our measures of genetic distance are approximately linearly related to evolutionary time. In this case we can neglect the effect of mutation. In the absence of selection, the values of J_X, J_Y, and J_{XY} in generation t [$J_X(t)$, $J_Y(t)$, and $J_{XY}(t)$, respectively] can be written as

$$J_X(t) = J_Y(t) = 1 - [1 - J(0)]\left(1 - \frac{1}{2N}\right)^t$$

$$\approx J(0) + [1 - J(0)]\left(\frac{t}{2N}\right), \tag{23}$$

$$J_{XY}(t) = J_{XY}(0) = J_X(0) = J_Y(0) = J(0) ,$$

where $t \ll 2N$ is assumed (see Nei 1975, p 124). Therefore, we have

$$D_m = 1 - J(0)\left(\frac{t}{2N}\right) . \tag{24a}$$

$$D \cong \frac{1 - J(0)}{J(0)}\left(\frac{t}{2N}\right) , \tag{24b}$$

$$D_v \cong \frac{1 - J(0)}{J(0)}\left(\frac{t}{2N}\right) . \tag{24c}$$

Thus, as long as $t \ll 2N$, our distance measures can be used even for nonprotein loci. In most human populations, $t \ll 2N$ appears to hold.

In a computer simulation, Reynolds et al. (1983) showed that in the absence of mutation, D can increase nonlinearly with time even for a relatively short evolutionary time. However, this happened because they started with unrealistic initial allele frequencies (200 different alleles in a population of 100 diploid individuals) and traced the genetic change of populations until a substantial amount (about 40 percent) of genetic variability was lost. When the mutation-drift balance is maintained with the infinite-allele model, the relation-

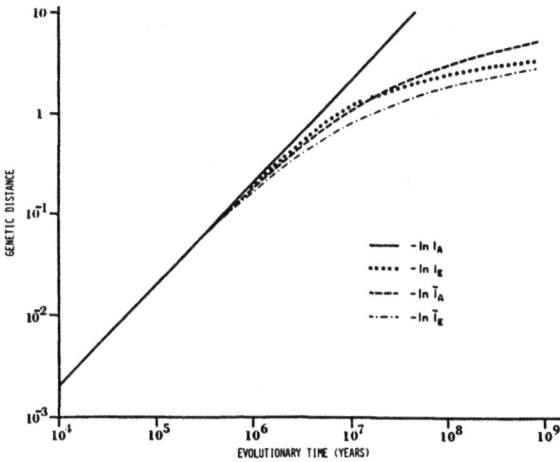

Fig. 8.1 Relationship between evolutionary time and genetic distance for four different models.

ship between D and t is approximately linear for a considerable period of evolutionary time, as seen from Fig. 8.1.

In this connection, it should be noted that the quantity that has a simple relationship with evolutionary time is the second moment of gene frequency (Wright 1931), and thus the genetic distance defined as a geometric distance in a multidimensional space is not proportional to evolutionary time. In my view, the linear relationship with evolutionary time is one of the most important properties a genetic distance measure should have. Unfortunately, many distance measures such as Manhattan distance, the d of Cavalli-Sforza and Edwards (1967), Roger's (1972) distance, and that of Siciliano et al. (1973) [which is identical with that of Thorpe (1979)] do not have this property even under the effects of genetic drift and mutation alone (Nei 1976, Nei et al. 1983). In the absence of mutation, Latter's (1973b) distance ϕ^* can be related to evolutionary time by $t = -2N\ln(1 - \phi^*)$. In the presence of mutation (infinite-allele model), the expectation of ϕ^* becomes

$$E(\phi*) = \frac{J(\infty) + [J(0) - J(\infty)]e^{-2v+(1/2N)t} - J(0)e^{-2vt}}{1 - J(0)e^{-2vt}},$$

where $J(\infty) = 1/(4Nv+1)$. When $J(\infty) = J(0)$, as is usually assumed, it reduces to

$$E(\phi*) = \frac{J(\infty)(1 - e^{-2vt})}{1 - J(\infty)e^{-2vt}}.$$

Therefore, $-2N\ln(1 - \phi^*)$ is no longer linear with evolutionary time. This limits the utility of ϕ^* and does not support the contention of Reynolds et al. (1983) that ϕ^* is preferable to D when evolutionary time is relatively short.

In our mathematical formulation, we assumed that the average hetero-zygosities of the two populations in question have remained constant throughout the entire evolutionary process. This assumption, however, may not always be satisfied. In fact, there are many cases in which one or both of the populations have gone through bottlenecks. The bottleneck effect on genetic distance has been studied in detail by Chakraborty and Nei (1974, 1977). They have shown that the genetic distance increases rapidly in the presence of bottlenecks and the rate of increase is higher when the bottleneck size is small than when it is large. However, if the population size returns to the original level, the bottleneck ef-fect gradually disappears, though it takes a long time for the effect to disappear completely (Fig. 8.2).

Under certain circumstances, it is possible to make a correction for the bottleneck effect. In the case where only one of the two populations has gone through a bottleneck, the following genetic identity may be computed:

$$I = \frac{J_{XY}}{J_X} , \qquad (25)$$

where J_X is the mean homozygosity (gene identity) for the population whose size has remained constant. If we use this I in (3) or (6), then D or D_v is linearly related to evolutionary time under the infinite-allele model (Chakraborty and Nei 1974). In the case where both populations have gone through bottlenecks, a similar correction can be made if there is a third population the size of which is known to have remained more or less the same as that of the foundation stock of

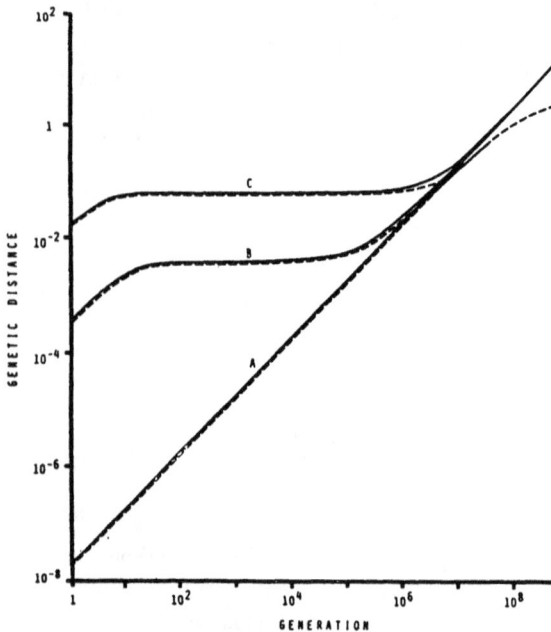

Fig. 8.2 Bottleneck effects on genetic distance. Solid lines represent the genetic distance for the infinite-allele model, whereas broken lines represent the distance for the stepwise mutation model. Computations have been made under the assumption that one isolated population (or species) is established through a bottleneck of size N_0 and thereafter population size increases to the level of the parental population following the sigmoid curve. The genetic distance in the ordinate represents the distance between this population and its parental population, which has undergone independent evolution. A: no bottlenecks. B: $N_0 = 100$. C: $N_0 = 10$.

the two populations under investigation. In this case, I may be computed by replacing J_X in (25) by the mean homozygosity for the third population.

Effects of Migration

In the early stage of population differentiation, gene migration usually occurs between populations. Migration retards gene differentiation considerably, and even a small amount of migration is sufficient to prevent any appreciable gene differentiation unless there is strong differential selection. The effect of migration on genetic distance has been studied by Nei and Feldman (1972), Chakraborty and Nei (1974), Slatkin and Maruyama (1975), and Li (1976a) under the assumption of no selection. Their main conclusions are as follows:

• If there is a constant rate of migration in every generation, the genetic identity (I) eventually reaches a steady-state value, which is given by

$$I = \frac{m_1 + m_2}{m_1 + m_2 + 2\alpha} \tag{26}$$

approximately, if $2\alpha << m_1 + m_2 << 1$. Here, α is the rate of gene substitution per locus per generation, and m_1 and m_2 stand for the migration rates between two populations (m_1 and m_2 may not be the same if the sizes of the two populations are not equal).

• The approach to the steady state value is generally very slow; the number of generations required is of the order of the reciprocal of the mutation rate. Formula (26) indicates that the genetic distance between populations cannot be large unless migration rates are very small.

• If we know α, Eq. (26) can be used for estimating the maximum amount of migration between two populations. That is, if we write $m = (m_1 + m_2)/2$, Eq.(26) becomes $I = m/(m + \alpha)$. Therefore, $m = \alpha$ $I/(1-I)$. This formula has been used by a number of authors (e.g., Chakraborty and Nei 1974, Nei 1975, Larson et al. 1984).

EMPIRICAL RELATIONSHIP BETWEEN GENETIC DISTANCE AND EVOLUTIONARY TIME

In the previous section I discussed the theoretical relationship between genetic distance and the time since divergence between two populations. This relationship can be used for estimating the divergence time from genetic distance data. In practice, however, the underlying assumptions are not always satisfied, so that the straightforward application of the theory can be misleading. Recently, Thorpe (1982) and Avise and Aquadro (1982) surveyed published data on the relationship between genetic distance and evolutionary time and concluded that there is no solid evidence for a universal electrophoretic clock. However, they did not examine the possible causes of seemingly nonuniversal molecular clocks. In reality, genetic distance is affected by various factors other than evolutionary time, and these factors should be considered when assessing the relationship between genetic distance and evolutionary time.

In recent years a number of authors estimated evolutionary times from genetic distances and compared them with the estimates from other sources such as fossil records, separation of lands and seas, and island formation. In these studies most authors used Eq. (15) or $t = kD$, where k is a proportionality constant and given by $k = (2\alpha)^{-1}$. Table 8.2 shows the results of these studies. It should first be emphasized that the estimates of evolutionary times in this table are not as certain as they might suggest; the numerical values are presented only to give a rough idea. Some results such as those of Highton and Larson (1979) are not included because of the uncertainty of these estimates. Despite this reservation, however, Table 8.2 shows that the agreement between the estimates from genetic distance and other sources is reasonably good in most studies, particularly if we consider the large standard error of genetic distances. However, there is one problem. That is, the proportionality constant used is not always the same, and indeed, there is a 20–fold difference between the largest and smallest k values. Therefore, the data in this table suggest that there is no molecular clock that is universal to all organisms. Before rushing to this conclusion, however, we must examine each set of data carefully, taking into account the other factors that might have affected genetic distance estimates.

Proteins Used

The first factor to be considered is the set of proteins used for electrophoresis. As discussed by Nei (1971, 1975), the rate of gene substitution per locus per year in Eq. (15) or (18) may be expressed as

$$\alpha = nc\lambda , \tag{27}$$

where n is the average number of amino acids per polypeptide, c is the proportion of amino acid substitutions that are detectable by electrophoresis, and λ is the average rate of amino acid substitution per year. The value of $\alpha = 10^{-7}$ in Eq. (19) is obtained by using $n = 400$, $c = 0.25$, and $\lambda = 1 \times 10^{-9}$ (see Nei 1975, p. 33, for the rationale). In practice, however, different investigators use different sets of proteins, though large-scale electrophoretic surveys usually contain many commonly used proteins (Avise and Aquandro 1982). For example, the n value (515) for the proteins used by Nevo et al. (1974) was somewhat larger than that (400) of Nei and Roychoudhury (1974b). They also assumed that $\lambda = 2.1 \times 10^{-9}$. Mainly because of these differences, Nevo et al.'s $k = 1.5 \times 10^6$ was about three times smaller than that of Nei and Roychoudhury (1974b).

Table 8.2 Estimates of divergence time from genetic distance and other sources. Proportionality constant k for $t = kD$ is also given.

Organism	Distance	Time (years) estimated from		No. of loci	k	Source
		Distance[a]	Other sources			
Mammals						
Negroid and Mongoloid (human)	0.024	$(1.2 \pm 0.6) \times 10^5$	$(5-20) \times 10^4$	35	5×10^6	Nei and Roychoudhury (1974b)
Man and chimpanzee	0.62[b]	$(4 \pm 1) \times 10^6$	5×10^6	44	(5×10^6)	Eq. (18)
Japanese and cont. macaques	0.11	5.4×10^5	$(4-5) \times 10^5$	28	5×10^6	Nozawa et al. (1977)
Pocket gophers (*Thomomys*)	0.08	$(1.2 \pm 0.8) \times 10^5$	$(1-2) \times 10^5$	31	1.5×10^6	Nevo et al. (1974)
Pocket gophers (*Geomys*)	?[c]	2.7×10^5	Ca 3×10^5	22	8.1×10^5	Penney and Zimmerman (1976)
Woodrats (*Neotoma*)	0.18	$(1.5 \pm 0.9) \times 10^5$	$(2-4) \times 10^5$	20	8.1×10^5	Zimmerman and Nejtek (1977)
		$(9 \pm 4.5) \times 10^5$			(5×10^6)	
Deer mice spp.	0.15	$(1.5 \pm 0.7) \times 10^5$	$(1-5) \times 10^5$	28	9.9×10^5	Gill (1976)
		$(7.5 \pm 3) \times 10^5$			(5×10^6)	Eq. (19)
Ground squirrels (spp.)	0.56	$(5 \pm 1) \times 10^6$	5×10^6	37	6.7×10^6	Smith and Coss (1984)
		$(4 \pm 1) \times 10^6$			(5×10^6)	Eq. (18)
Ground squirrels (subsp.)	0.10	$(6.9 \pm 0.3) \times 10^5$	7×10^5	37	6.7×10^6	Smith & Coss (1984)
		$(5.5 \pm 1) \times 10^5$			(5×10^6)	Eq. (18)
Birds						
Galapagos finch	0.12	$(6 \pm 3) \times 10^5$	$(5-40) \times 10^5$	27	5×10^6	Yang and Patton (1981)
Reptiles						
Bipes spp.	0.62	$(3.1 \pm 1) \times 10^6$	4×10^6	22	5×10^6	Kim et al. (1976)
		$(4 \pm 1) \times 10^6$			(5×10^6)	Eq. (18)
Lizards (*Uma*)	0.28	$(5 \pm 2.5) \times 10^6$	Ca 5×10^6	22	18×10^6	Adest (1977)
		$(1.6 \pm 0.8) \times 10^6$			(5×10^6)	Eq. (18)

Table 8.2 (continued) Estimates of divergence time from genetic distance and other sources. Proportionality constant k for $t = kD$ is also given.

| Organism | Distance | Time (years) estimated from | | No. of loci | k | Source |
		Distance[a]	Other sources			
Fishes						
Cave and surface fishes	0.14	$(7 \pm 4.6) \times 10^5$	$(3 - 20) \pm 10^5$	17	5×10^6	Chakraborty and Nei (1974)
Minnows	0.053	2.7×10^5	$(1 - 20) \times 10^5$	24	5×10^6	Avise et al. (1975)[d]
Panamanian fishes	0.32	5.8×10^6	$(2 - 5) \times 10^6$	28	18×10^6	Gorman & Kim (1977)
		$(1.9 \pm 0.6) \times 10^6$			(5×10^6)	Eq. (18)
Panamanian fishes	0.24[e]	3.5×10^6	$(2 - 5) \times 10^6$	31.4	18×10^6	Vawter et al. (1980)
		1.2×10^6			(5×10^6)	Eq. (18)
Echinoids						
Panamanian sea urchins	0.03 − 0.64	—	$(2 - 5) \times 10^6$	15 − 18	—	Lessios (1979)
	0.39f	2.4×10^6			(5×10^6)	Eq. (18)

[a]The standard error of D was not given in many papers. In this case it was computed by using Eq. (12), except for the case of small D.
[b]Data from King and Wilson (1975).
[c]It is not clear how the authors computed D.
[d]The authors used a different value of k, but $k = 5 \times 10^6$ gives a better result.
[e]Average for ten pairs of species.
[f]Average for four pairs of species.

Detectability of Protein Differences By Electrophoresis

The detectability of amino acid differences in proteins by electrophoresis depends on various biological and biochemical conditions of electrophoresis such as the tissue used and the pH and type of the gel. These conditions are not always the same for all organisms studied or for all laboratories where the experiments are done. The differences in these conditions are expected to cause some differences in the detectability of protein differences, c. Although Avise and Aquadro (1982) suggested that the technique of electrophoresis used is virtually the same for most laboratories, there is evidence that this is not necessarily the case. For example, King and Wilson (1975) reported a genetic distance of $D = 0.62$ between man and chimpanzee, whereas Bruce and Ayala (1979) and Nozawa et al. (1982) obtained $D = 0.39$ and 0.45, respectively, for the same pair of species. These differences are partly due to the differences in the proteins used. However, even when the same 15 protein loci common to the three studies were used, there were substantial differences: the D values for the data of King and Wilson, Bruce and Ayala, and Nozawa et al. were 0.83, 0.41, and 0.71, respectively (Nozawa et al. 1982). This indicates that c is not the same for all laboratories. In a study of ground squirrels, Smith and Coss (1984) speculated that the c value for their data was probably smaller than 0.25, and obtained $k = 6.7 \times 10^6$.

Sampling Errors

As mentioned earlier, a large number of loci should be used to obtain a reliable estimate of genetic distance, particularly when genetic distance is small. In practice, however, many investigators use a relatively small number of loci for various reasons. In these cases the results obtained could be misleading. For example, consider an extreme case in which two populations are virtually monomorphic and fixed for different alleles at 10% of the loci. In this case the expected genetic identity is 0.9 ($D = 0.11$), but the observed identity can be substantially larger or smaller than 0.9. Indeed, if r loci are examined, the observed value of I will be 1 (or D will be 0) with the probability of $(0.9)^r$. In the case of $r = 20$, this probability is 0.12, which is not very small. To make this probability smaller than 0.05, r must be equal to or larger than 29, whereas the minimum value of r that makes the probability less than 0.01 is $r = \log 0.01 / \log(1 - I) = 2/0.0458 = 44$, where $I = 0.9$. This is a substantial number of loci. In practice, of course, the number can be a little smaller than this, because usually some loci are polymorphic and this reduces the variance of I or D to some extent. Nevertheless, we must be cautious about the effect of sampling error when the number of loci examined is small.

In a study of the genetic distance between the sea urchins of the Pacific and those of the Atlantic coast of Panama, Lessios (1979) observed an unusually low genetic distance for one of the four species pairs examined, although the Panama Isthmus is known to have been above sea level for the last

2–5 million years. From this observation, he concluded that electrophoretic data cannot be used for dating evolutionary time. However, he studied only 18 protein loci, so that this unusual result could well be due to sampling error (Vawter et al. 1980). Furthermore, if we take the average of the genetic distances for all the four species pairs, it becomes 0.39, which corresponds to a divergence time of 2.4 million years (Table 8.2). This divergence time is consistent with the estimate from geological data.

Nonlinear Relationship of D With Time

When the time since divergence between two species is long, the relationship between D and t is no longer linear, as mentioned earlier. This factor has not been taken into account properly in most of the empirical studies cited in Table 8.2. In some cases, the correction for this nonlinearity seems to improve the agreement between the estimates of divergence times from genetic distance and other sources. For example, King and Wilson (1975) obtained $D = 0.62$ between man and chimpanzee using Eq. (3). If we use Eq. (19) with $\alpha = 10^{-7}$, this gives $t = 3.1 \times 10^6$ years, which seems to be too low when compared with the estimate (5×10^6 years) obtained from fossil records and Sarich and Wilson's (1967) immunological distances. In this case, however, Eq. (18) is expected to give a better estimate of genetic distance (D_v) than Eq. (3). If we use Eqs. (18) and (19), we have $t = (4 \pm 1) \times 10^6$ years, which is no longer incompatible with the estimate from other sources. Furthermore, if we use Eq. (22), t becomes even larger. Similarly, the estimate of t by Kim et al. (1976) can be improved by using Eq. (18), as shown in Table 8.2. The same property was noted by Smith and Coss (1984) in their study of ground squirrels, though they used a proportionality constant of 6.7×10^6.

The largest proportionality constant k in Table 8.2 is that of Gorman and Kim (1977), Adest (1977), and Vawter et al. (1980). This value was obtained by comparing Sarich and Wilson's (1967) immunological distance (d_I) with D (Maxson and Wilson 1974). In this comparison, however, many D values larger than 1 were used. Therefore, the k value obtained is expected to be an overestimate. In a later study of the regression coefficient (b) of d_I on D, Maxson and Maxson (1979) obtained $b = 26$. Since d_I is (stochastically) related to evolutionary time t by $t = 6 \times 10^5 d_I$ in mammals, reptiles, and amphibians (Prager et al. 1974), we obtain $t = 16 \times 10^6 D$. In a similar study Highton and Larson (1979) obtained $b = 24$ and $t = 14 \times 10^6 D$. The k values obtained in these studies are still about three times as large as that in Eq. (19), but they are again based on many D values larger than 1. If we use only the D values less than 0.5 in these studies, the k value seems to be substantially lower than 14×10^6 (see Fig. 4 of Highton and Larson 1979). (The regression coefficient should be computed by fitting $d_I = bD$ rather than $d_I = a + bD$, as in Highton and Larson.)

Nevertheless, the value of $k = 18 \times 10^6$ seems to be much better than $k = 5 \times 10^6$ in explaining Vawter et al.'s (1980) and Adest's (1977) electrophoretic data. In these cases one can use Eq. (18) to compute the expected evolutionary time under the assumption of $k = 5 \times 10^6$, but the values obtained are much smaller than the estimates obtained from other sources. The only data that can be accommodated with the value of $k = 5 \times 10^6$ are those of Gorman and Kim (1977) (Table 8.2). Since the data of Vawter et al. are based on 10 pairs of species, their results cannot be dismissed as a special case. This suggests either that the fishes and reptiles studied by Vawter et al. and Adest do not show the same evolutionary rates as those for many other organisms or that the electrophoretic technique used for these organisms did not detect protein differences so efficiently as in other organisms.

Bottleneck Effects

One of the troublesome problems in dating evolutionary time from genetic distance data is the bottleneck effect. As we have seen earlier, the bottleneck effect accelerates the increase of D temporarily. This acceleration occurs because the homozygosity [J_X or J_Y in Eq. (3)] in one or both populations increases under the bottleneck effect. The expectation of J_{XY} is not affected by the bottleneck effect. If the population size is restored to the original level after going through a bottleneck, J_X or J_Y also gradually returns to the original level. Once J_X and J_Y reach the original level, the bottleneck effect on D can no longer be detected. Usually, however, it takes a long time before this effect disappears. At any rate, in the presence of bottlenecks Eq. (19) is expected to give an overestimate of evolutionary time. On the other hand, if one tries to calibrate the evolutionary clock by using D values which are affected by a bottleneck, a relatively small value of k will be obtained.

In many cases, it is difficult to know whether a particular population or a particular pair of populations has undergone bottlenecks. In some cases, however, there is clear-cut evidence of a bottleneck effect, and we can make a correction for it. For example, in the case of cave and surface fishes of *Astyanax mexicanus* the cave populations are very small and clearly derived from the surface populations (Avise and Selander 1972). At the present time, the average homozygosities in the cave populations are virtually 1, and if we use Eq. (25) the bottleneck effect is eliminated. The genetic distance for this species in Table 8.2 has been estimated in this way (Chakraborty and Nei 1974).

In some other cases the bottleneck effect (or the effect of population size reduction) can be inferred from the level of heterozygosity or homozygosity even if we do not know the history of the populations. For example, in the case of pocket gophers *Thomomys talpoides* studied by Nevo et al. (1974), there are a number of subspecies which are fixed for different chromosome numbers. Apparently, these subspecies have been isolated from each other for a long time. Furthermore, average heterozygosity is quite low in these subspecies. Therefore, it is possible that the genetic distances among them are affected by bottlenecks. Possibly, for this reason the k value for these subspecies is lower than 5×10^6. Unfortunately, we cannot make a correction for D in this case. A similar bottleneck effect might have occurred in the other species (*Geomye* species) of pocket gophers listed in Table 8.2.

General Remarks

It is now clear that the conclusion that there is no solid evidence for a universal electrophoretic clock (Thorpe 1982, Avise and Aquadro 1982) is not so firm. Rather, if we take into account various factors that affect the relationship between D and t, electrophoretic data seem to be useful for getting a rough idea about evolutionary time. It should be noted that D is intended to measure the number of amino acid substitutions that are detectable by electrophoresis. Therefore, so long as amino acid substitution occurs at a constant rate, as seems to be the case, D should increase as evolutionary time increases. However, electrophoretic data give a less accurate estimate of evolutionary time than amino acid sequence data, since they are affected by several factors such as the detectability of protein differences and bottleneck effects. Therefore, when D is to be used as a molecular clock, it should be understood that it gives only a rough estimate of evolutionary time.

It should also be noted that the electrophoretic clock may not be the same for all groups of organisms. Even if the same electrophoretic technique is used, the detectability of protein differences may vary with the group of organisms used (Avise and Aquadro 1982). If this is the case, separate electrophoretic clocks must be used for different groups of organisms. This certainly reduces the utility of the molecular clock, but even under this restriction, the concept of a molecular clock is useful for clarifying the evolutionary relationships of closely related organisms, where no other methods for estimating evolutionary time are usually available. For some time, molecular evolutionists have thought that the rate of molecular evolution is significantly lower in birds than in mammals, reptiles, and amphibians (Prager et al. 1974, Avise and Aquadro 1982). Recent evidence, however, suggests that what is wrong may not be molecular data but fossil records (Wyles et al. 1983). If more studies are done on the relationship between genetic distance and evolutionary time, similar results may be obtained in some cases.

Finally it should be emphasized that the electrophoretic clock is mainly useful for closely related species. If D is too large (say $D \geq 1$), its variance becomes very large even if a substantial number of loci are studied, so that the reliability of dating declines. The high frequency of backward and parallel mutations at the electrophoretic level in the case of $D \geq 1$ makes the clock unreliable. Nozawa et al. (1982) noted that the estimate of the time since divergence between man and macaques was much smaller than that from fossil records (about 30 million years). I believe that man and macaques are too far distant for the electrophoretic clock to be applied.

RECONSTRUCTION OF PHYLOGENETIC TREES

Genetic distances can be used not only for estimating evolutionary times but also for constructing phylogenetic trees. For the latter purpose we do not have to know the exact relationship between genetic distance and evolutionary time. As long as D is approximately linear with time, the evolutionary relationship among organisms can be estimated by using genetic distances.

There are many methods for constructing a phylogenetic tree from genetic distance data, but the two most frequently used for electrophoretic data are the unweighted pair-group method (UPGMA) and Farris's (1972) distance Wagner method. UPGMA was originally proposed for phenetic classification by Sokal and Michener (1958), but it can be used for phylogeny construction as long as the distance measure used is (stochastically) proportional to evolutionary time. A simple explanation of this method is given by Nei (in press). When the topology of the tree to be made is known, this method gives least-squares estimates of branch lengths (Chakraborty 1977). On the other hand, Farris's method is intended to construct a minimum evolution tree (a tree requiring a minimum number of evolutionary changes) and supposedly requires a distance measure that satisfies the triangle inequality. A distance measure that is often used for this purpose is Rogers's (1972) distance. Tateno et al. (1982) and Nei et al. (1983) have shown that when genetic distance is subject to a large stochastic error, this method tends to give overestimates of gene substitutions, contrary to the original intention. To rectify this property, Tateno et al. (1982) modified the Farris method. A brief account of the Farris method and its modified version is given by Nei (in press).

In the last decade there has been a great deal of controversy over the relative merits of various tree-making methods. Some of the discussions are quite philosophical, whereas some others are based upon conjectures on the evolution of special groups of organisms. The main problem in this controversy is that in most cases we do not know the true evolutionary tree with which a reconstructed tree is to be compared. To address this problem Tateno et al. (1982) and Nei et al. (1983) conducted computer simulations to find out which method reconstructs the true tree with the highest probability. In simulation study one can set up a model tree, and a reconstructed tree can be compared with this model tree. It is therefore possible to know which method yields a better tree than others. In the following I present some of their results, particularly those of Nei et al. (1983).

Computer Simulation Studies

The method of computer simulation used in the study by Nei et al. (1983) can be summarized:

- The evolutionary change of populations was assumed to occur solely by mutation and genetic drift with $4Nv = 0.2$ following the infinite-allele model, where N and v are the effective population size and the mutation rate per locus per generation, respectively.

• An ancestral population, which was in equilibrium with respect to the effects of mutation and genetic drift, was split into two populations. At a later time, one of the two descendant populations was again split into two populations This process was continued until the model tree given by Figure 8.3a was completed.

• The expected number of gene substitutions per locus was proportional to evolutionary time. In Figure 8.3a the expected number (M) of gene substitutions for the shortest branch is 0.1, but we also studied the case of $M = 0.004$. $M = 0.1$ would represent a typical case of interspecific comparison, and $M = 0.004$ a case of comparison of local races or subspecies.

• Starting from the ancestral population, the gene frequency change in each population was followed, and at the end of evolution, the gene frequencies for all populations were recorded.

• Using the gene frequency data, five genetic distance measures were computed: standard genetic distance, D; minimum genetic distance, D_m; Rogers's (1972) distance, D_R; Cavalli-Sforza's (1969) distance, f_θ; and Nei et al.'s (1983) angular distance, D_A. These five distance measures were used, because the accuracy of a tree reconstructed was expected to depend on the distance measure used.

• To see the effect of the number of loci used, ten sets of distance matrices were computed for each distance measure by using the first 10 loci, first 20 loci, and so on, until all 100 loci were used.

• For each of these distance matrices, evolutionary trees were reconstructed by using UPGMA, the Farris method, and the modified Farris method.

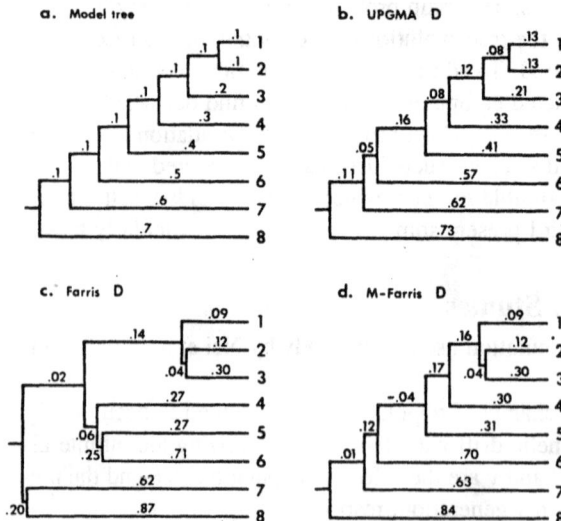

Fig. 8.3 Model tree (a) and reconstructed trees (b,c,d) by using D in one replication of computer simulation. The value given to each branch (or internode) in the model tree is the expected number of gene substitutions, whereas the corresponding number in the reconstructed tree is the estimate of branch length. $4Nv = 0.2$, $M = 0.1$, and the number of loci used is 50. (b) $d_T = 0$, $S_E = 0.073$; (c) $d_T = 8$, $S_E = 0.250$; (d) $d_T = 2$, $S_E = 0.138$.

• Each tree reconstructed was compared with the model (true) tree, and the extents of topological errors and errors in the estimates of branch length were evaluated.

• This simulation was repeated ten times for each value of $M = 0.1$ and $M = 0.004$.

There are two different criteria for measuring the deviation of a reconstructed tree from the model tree. One is the degree of distortion of the topology of a reconstructed tree, and the other is the amount of deviation of patristic (estimated) branch lengths from true lengths. The extent of the topological errors was measured by Robinson and Foulds's (1981) distortion index (d_T), which is roughly twice the number of interchanges of OTUs (operational taxonomic units) that are required for converting the topology of a reconstructed tree into that of the true tree. When the topology of a reconstructed tree is correct, d_T takes a value of 0. On the other hand, the amount of errors of the estimates of branch lengths was measured by the following average deviation of patristic distances from the expected distances.

$$S_E = \left[\frac{2\sum_{i>j}(D_{ij} - D'_{ij})}{n(n-1)} \right]^{1/2}, \tag{28}$$

where D_{ij} and D_{ij}' are the patristic distance and expected distance between OTUs i and j, respectively. The patristic distance between OTUs i and j is the sum of all branches connecting these two OTUs in the reconstructed tree, whereas the expected distance is the sum of branches connecting the same pair of OTUs in the model tree. The results obtained may be summarized as follows:

Topological errors. Figure 8.3 shows examples of reconstructed trees from one replication of computer simulation. In this case Nei's standard distances based on gene frequency data for 50 loci were used. The topology of the tree reconstructed by UPGMA is identical with that of the true tree, so that $d_T = 0$. On the other hand, the tree reconstructed by the Farris method has many topological errors, d_T being equal to 8, whereas d_T for the modified Farris tree is 2. Figure 8.4 shows the phylogenetic trees produced by using Rogers's distance and Cavalli-Sforza's f_Θ from the same set of gene frequency data. UPGMA again gives the correct topology ($d_T = 0$) for both D_R and f_Θ, whereas the other two methods give trees with several topological errors. Therefore, from the comparison of d_T alone, we can conclude that UPGMA is better than the other two evolutionary tree-making methods. However, this is the result from one replication of computer simulation, and to make a general conclusion we must consider the results from all ten replications.

The average distortion indices ($\bar{d}_T$) over ten replications for the case of $M = 0.1$ are given in A of Fig. 8.5 in relation to the number of loci used (r). It is seen that $\bar{d}_T$ is very large when $r = 10$ but rapidly declines as r increases. However, the decrease of $\bar{d}_T$ with increasing r is nonlinear, and the rate of decrease is not large when r is equal to or larger than 30. When 30 loci are used,

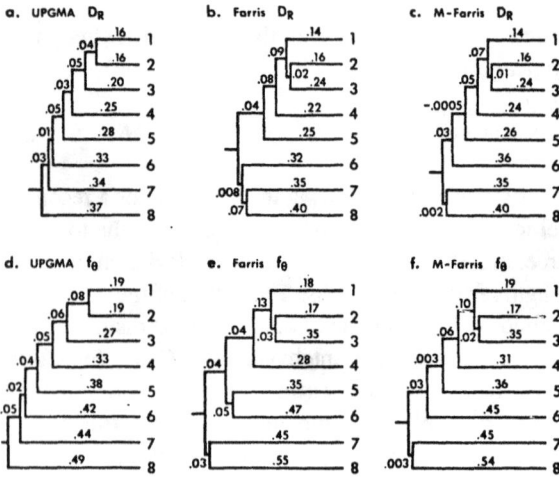

Fig. 8.4 Constructed trees by using D_R and f_Θ in the same replication of computer simulation as that of Fig. 8.3. The value given to each branch is the estimate of branch length. $4Nv = 0.2$, $M = 0.1$, and the number of loci used is 50.

(a) $d_T = 0$, $S_E = 0.448$;
(b) $d_T = 6$, $S_E = 0.388$;
(c) $d_T = 4$, $S_E = 0.449$;
(d) $d_T = 0$, $S_E = 0.270$;
(e) $d_T = 6$, $S_E = 0.243$;
(f) $d_T = 4$, $S_E = 0.278$.

d_T is already about 2 for UPGMA, which means that the amount of error of a reconstructed tree is about one interchange of OTUs from the true tree. As r increases further, d_T decreases very slowly, and even with $r = 100$, d_T is not 0, except in one case. This result suggests that in the construction of a phylogenetic tree for a group of species at least 30 loci should be used.

Figure 8.5 also shows that UPGMA and the modified Farris method generally give a smaller value of d_T than the Farris method for all distance measures and for all r values. This is true even if Rogers's distance (D_R), which satisfies the triangle inequality, is used. It is also noted that the differences in d_T among different distance measures are generally small, though D_A tends to give a better topology than other measures. The better performance of D_A compared with others seems to be due to the fact that D_A has a relatively small coefficient of variation.

The value of d_T for the case of $M = 0.004$ is generally larger than that for $M = 0.1$, as expected (Fig. 8.5.). To make d_T equal to 2, a large number of loci must be used. However, the rate of decrease of d_T with increasing r again declines around $r = 30$. In this case UPGMA always shows a smaller value of d_T than the other two methods, and the modified Farris method tends to show a smaller value than the Farris method. This is true irrespective of the distance measure used.

Errors of the estimates of branch lengths. Another important criterion of the accuracy of a reconstructed tree is the deviation of estimates of branch lengths from true branch lengths. We have seen that in the example trees of Figs. 8.3 and 8.4 the topology of the tree reconstructed by UPGMA is correct irrespective of the distance measure used. However, the estimates of branch lengths are considerably different from each other. Comparison of these trees with the true tree (Fig. 8.3a) indicates that D gives a better result for estimating branch lengths than the other distance measures. Indeed, the S_E value for D is

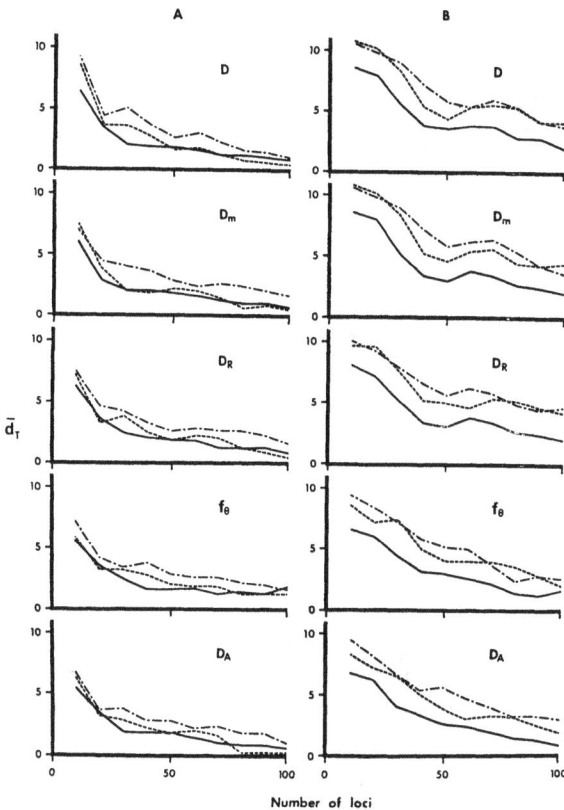

Fig. 8.5 Relationships between d_T and the number of loci used for the cases of $M = 0.1$ (A) and $M = 0.004$ (B). Solid line, UPGMA; chain line, Farris method; broken line, modified Farris method.

0.073, whereas the S_E for D_R and f_Θ is 0.448 and 0.270, respectively. Therefore, with this criterion the tree produced by UPGMA using D is the best among the three. One might think that the Farris method would give a good tree when D_R is used. That this is not the case can be seen from the comparison of Fig. 8.4b with Fig. 8.3c. Compared with D, D_R generally gives a tree in which the part near the root is condensed and the other part is elongated. This is because D_R is not proportional to the expected number of gene substitutions (Fig. 8.6). A similar pattern is observed for D_m, f_Θ, and D_A, although the results for D_m and D_A are not shown here.

The average values ($\bar{S}_E$) of S_E over all replications for the cases of 20 loci, 60 loci, and 100 loci examined are presented in Table 8.3. The value of $\bar{S}_E$ varies considerably with the tree-making method and the distance measure used. The smallest value is obtained when UPGMA with D is used. This supports our visual conclusion from Figs. 8.3 and 8.4. When D is used, the modified Farris method also shows a relatively small value of $\bar{S}_E$. In contrast, the $\bar{S}_E$ for the Farris method is nearly twice as large as that of UPGMA. In the case of D, $\bar{S}_E$ decreases as the number of loci used increases, as expected. When the other distance measures are used, UPGMA no longer gives the smallest value

Table 8.3 Means of average deviations of partristic distances from expected distances ($\bar{S}_E$). $4Nv = 0.2$ and $M = 0.1$. These results are based on 10 replications. All results should be divided by 10^3.

	D	D_m	D_R	f_θ	D_A
			20 loci		
UPGMA	252 ± 26	539 ± 12	457 ± 11	305 ± 17	426 ± 13
Farris	461 ± 66	497 ± 13	417 ± 13	272 ± 16	386 ± 14
Modified Farris	291 ± 31	539 ± 11	458 ± 10	309 ± 16	427 ± 12
			60 loci		
UPGMA	136 ± 10	540 ± 5	456 ± 4	295 ± 11	426 ± 5
Farris	225 ± 12	511 ± 7	430 ± 5	268 ± 12	397 ± 5
Modified Farris	161 ± 9	541 ± 5	457 ± 4	297 ± 10	427 ± 5
			100 loci		
UPGMA	122 ± 8	534 ± 5	452 ± 4	296 ± 8	420 ± 5
Farris	204 ± 13	510 ± 5	427 ± 5	271 ± 9	395 ± 6
Modified Farris	140 ± 6	535 ± 5	452 ± 4	297 ± 8	421 ± 5

of $\bar{S}_E$. This is because the other distance measures are not linearly related with time (Fig. 8.6). However, the $\bar{S}_E$ values for the other distance measures are always larger than those for D in UPGMA. This result indicates that UPGMA, in

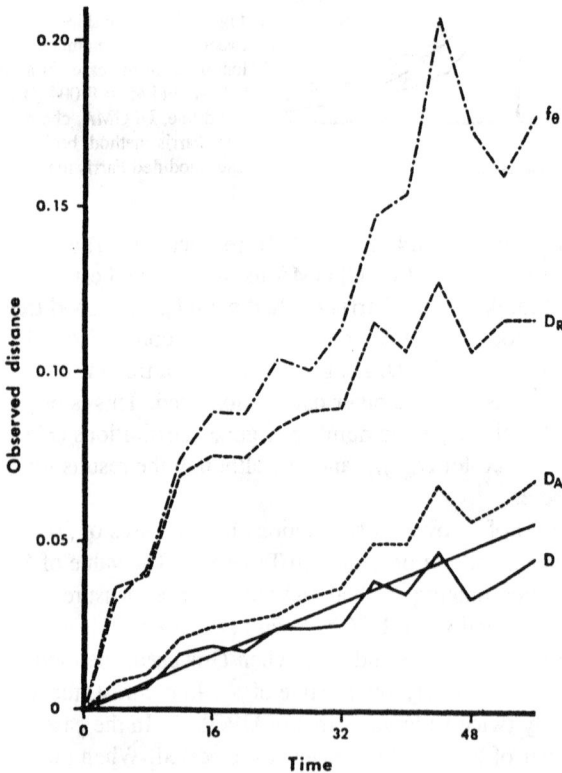

Fig. 8.6 Relationships between genetic distance and evolutionary time in one replication of computer simulation. $4Nv = 0.2$, $M = 0.004$, and the number of loci used is 20. The straight line represents the expected value of D. The expectations of D_R, f_θ and D_A are not linear with time. Time is measured in the unit of $2N$ generations.

combination with D, gives the best estimates of branch lengths. In the case of $M = 0.004$ the $\bar{S}_E$ values are considerably smaller than those for $M = 0.1$, but essentially the same conclusion has been obtained (see Nei et al. 1983).

General Remarks

We have seen that both the topology and branch lengths of a reconstructed tree are often quite wrong unless a large number of loci are used. In the study of phylogenetic relationships of related species many authors have used 20–40 genetic loci. The study by Nei et al. (1983) indicates that even if 30 loci are used and M is as large as 0.1, some parts of a reconstructed tree are incorrect with a high probability. In their study only 8 OTUs were used because of the limited computer time available, but the error in reconstructed trees is expected to increase disproportionately as the number of OTUs increases (Tateno et al. 1982).

One important factor for determining the accuracy of a reconstructed tree is the branch lengths of the true tree. If there are many branches of which the true distances are as small as 0.004, the reconstructed tree is usually incorrect even if 100 loci are used. This result is discouraging, but we must accept it since it is due to the stochastic nature of gene substitution. Clearly, we cannot be overconfident about evolutionary trees reconstructed from electrophoretic data. Nevertheless, a large part of the topology of a reconstructed tree seems to be correct if 30 or more loci are used. In many cases even this approximate phylogenetic tree is useful for studying various evolutionary problems.

The accuracy of a reconstructed tree also depends on a tree-making method and distance measure used. In general, UPGMA, in combination with D, is the best among the three tree-making methods examined. Some authors have used Fitch and Margoliash's (1967) method for tree-making. This method usually requires more computer time, yet the efficiency of recovering the correct tree is not so high as UPGMA (Tateno et al. 1982).

It is interesting to see that the simple UPGMA, which was originally proposed for phenetic taxonomy, shows the best performance. The reason for this seems to be that the genetic distance based on a relatively small number of loci is subject to a large stochastic error, and the procedure of distance-averaging used in UPGMA reduces this error to a considerable extent. It should be noted, however, that this conclusion is only for electrophoretic data and does not necessarily apply to other types of data such as amino acid sequences for distantly related organisms (Blanken et al. 1982, Nei in press).

In the past many authors have used the Farris method because in this method the unequal rates of gene substitution in different branches can be taken into account. However, a large part of the seemingly different rates of gene substitution are apparently caused by stochastic errors in gene frequency changes. The Farris method cannot distinguish these stochastic errors from the true variation in substitution rate and thus is susceptible to errors in tree-making. Furthermore, in the presence of stochastic errors the Farris method often gives overestimates of branch lengths (Tateno et al. 1982). Nevertheless, the Farris method seems to be superior to UPGMA in obtaining a correct unrooted topology when the rate of gene substitution varies substantially with evolutionary lineage and stochastic errors are relatively small (Tateno et al. 1982).

Some numerical taxonomists (e.g., Farris 1981) claim that the genetic distance measures used in phylogeny construction should satisfy the triangle inequality. They give two arguments for this. First, when one wants to represent the species or populations concerned in a multidimensional space and measure the geometric distances between them, it is necessary to use a measure that obeys this principle. Second, if every estimate of genetic distance between OTUs represents the sum of the actual number of gene substitutions for all relevant branches of the true tree, then the triangle inequality should hold. Representation of populations in a multidimensional space is mathematically interesting, but it is not necessary for tree-making. Furthermore, the *geometric* distance between populations measured in this way is not proportional to the number of gene substitutions, and thus it is inappropriate for measuring *genetic* distance (Nei 1978b). (Genetic distance, as defined earlier, is the extent of gene difference between two populations.)

Their second argument looks reasonable at first sight, but it is not realistic. Although D is not a metric in individual cases, its expectation is a metric. Thus, if a very large number of loci are used, the genetic distance between any pair of OTUs will represent the sum of the numbers of gene substitutions for all relevant branches, at least theoretically. In practice, it is virtually impossible to examine hundreds or thousands of loci for phylogeny construction at the present time. Therefore, we must estimate the genetic distance from a smaller number of loci, and in the process of this estimation the metricity of D is disturbed by statistical errors. Nevertheless, it is possible to construct a reasonably good phylogenetic tree by using D, as shown here and previously by Nei et al. (1983). Actually metricity is not really required for tree-making so long as a proper distance measure and a proper tree-making method are used. In this connection it should be noted that usual estimates of nucleotide or amino acid substitutions are not metrics either, because they are estimated statistically by taking into account back mutations, parallel mutations, and multiple mutations, and essentially the same argument as the above applies to these estimates (see Tateno et al. 1982).

Recently, Farris (1981) criticized Sarich and Wilson's (1967) immunological distance and Nei's (1972) distance for their nonmetricity and claimed that any nonmetric distance would not show clocklike behavior. However, he did not consider that the molecular clock is stochastic rather than deterministic and subject to errors due to backward and parallel mutations. Furthermore, his criticism of D is based on gene frequency differences for one locus. In practice, D is designed to be used for many loci and should not be used for one locus (Nei 1972). Note also that, contrary to Farris's assumption, the amount of gene frequency difference between two populations is not proportional to evolutionary time whether there is selection or not. Only when the dynamics of gene frequency changes in populations is taken into account properly can one develop a distance measure that is useful in evolutionary studies. D has been developed exactly in this way.

Rogers's (1972) distance (D_R) has often been used in conjunction with the Farris method, because it satisfies the triangle inequality. The fact that this distance does not give negative branches when the Farris method is used seems to have been attractive to some workers. However, metricity of distance itself does not give any advantage in tree-making, as mentioned above. Just like D, this distance may occasionally decrease in the evolutionary process because of stochastic errors (Fig. 8.6), and thus D_R does not necessarily show the true genetic relationship among OTUs. Furthermore, as mentioned earlier, it is not linear with evolutionary time.

In addition to its nonlinear relationship with evolutionary time, D_R also has theoretical defect: it is not necessarily 1 even when the two populations concerned have no shared alleles. This occurs when the populations are polymorphic. For example, when there are five nonshared alleles in each population and all allele frequencies are equal, i.e., 1/5, we have $D_R = 1/\sqrt{(5)} = 0.45$. On the other hand, if the two populations are fixed for different alleles, D_R becomes 1. From the evolutionary point of view, this is a poor property. A similar property is observed with D_m.

In this section we have been concerned with the reconstruction of phylogenetic trees from gene frequency data. In recent years evolutionists (e.g., Brown et al. 1979, Avise et al. 1979b, Shah and Langley 1979) have started to use the restriction endonuclease technique to study the genetic differences between species or populations. In this case the number of nucleotide differences per nucleotide site can be estimated by the statistical methods of Nei and Li (1979), Kaplan and Langley (1979), Kaplan and Risko (1981), and Nei and Tajima (1983). The estimates obtained by these methods have a statistical property similar to that of Nei's D. Therefore, the conclusions obtained in this paper seem to apply to these estimates as well. In this case, however, we must use a large number of restriction endonucleases to make a reliable phylogenetic tree (Li 1986).

9.

Genetic Divergence Between Congeneric Atlantic And Pacific Ocean Fishes

W. Stewart Grant

As on land, so in the waters of the sea, a slow southern migration of marine fauna, which during the Pliocene or even somewhat earlier period, was nearly uniform along the continuous shores of the Polar Circle, will account, on the theory of modification, for many closely allied forms now living in areas completely sundered.

Darwin (1859)

Two general factors are responsible for the complexity of modern world biota: (1) a continuing temporal sequence of vicariance events, and (2) subsequent dispersal modifying earlier vicariant patterns.

Croizat et al. (1974)

Fish systematists and marine biogeographers have long recognized the taxonomic affinities between many morphologically similar fishes in the Atlantic and Pacific oceans (Jordan 1885, 1908, Rosenblatt 1967), whose taxa include several closely related pairs that have been variously classified as conspecific populations, subspecies, or congeneric species. Two separate geologic events appear to be responsible for creating these pairs by subdividing ancestral populations or by permitting dispersal of founding populations from one ocean to the other. In the tropics, the rise of the Isthmus of Panama in the Pliocene Epoch (Fig. 9.1) terminated the faunal exchange between the Atlantic and Pacific oceans that had persisted since Mesozoic times. In the northern seas, the formation of the Bering Strait during the mid-Pliocene allowed reciprocal dispersal of North Atlantic and North Pacific fishes through a then warmer Arctic Ocean.

The theory of allopatric speciation postulates that after populations become reproductively isolated from one another by some barrier to migration or through isolation by distance, they begin to diverge genetically because of random genetic drift and natural selection (Mayr 1963). Divergence among fish taxa traditionally has been based on measurements of morphological variation. However, morphological classifications may not always reflect genetic relationships, because the rate of morphological change may not be the same in all lineages and because evolution may cause species to converge as well as diverge. Early fish classification schemes were largely typological (Simpson 1961), and

Epoch	Period	Era	Events
Pleistocene			← Bering Strait closed during 4 Pleistocene ice ages
Pliocene	Quaternary		← Bering Strait opening ← Panama Seaway closure
Miocene			
Oligocene	Tertiary	Cenozoic	← Opening between North Atlantic and Arctic Oceans
			← Formation of Euroasian Arctic Basin
Eocene			
Paleocene			← Kolyma Plate encloses Pacific arm of Arctic Ocean
			← Intrusion of Pacific Plate between North and South America
	Cretaceous	Mesozoic	
	Jurassic		← Beginning of formation of North Atlantic Ocean
	Triassic		

Millions of Years: 0, 10, 20, 30, 40, 50, 60, 135, 225

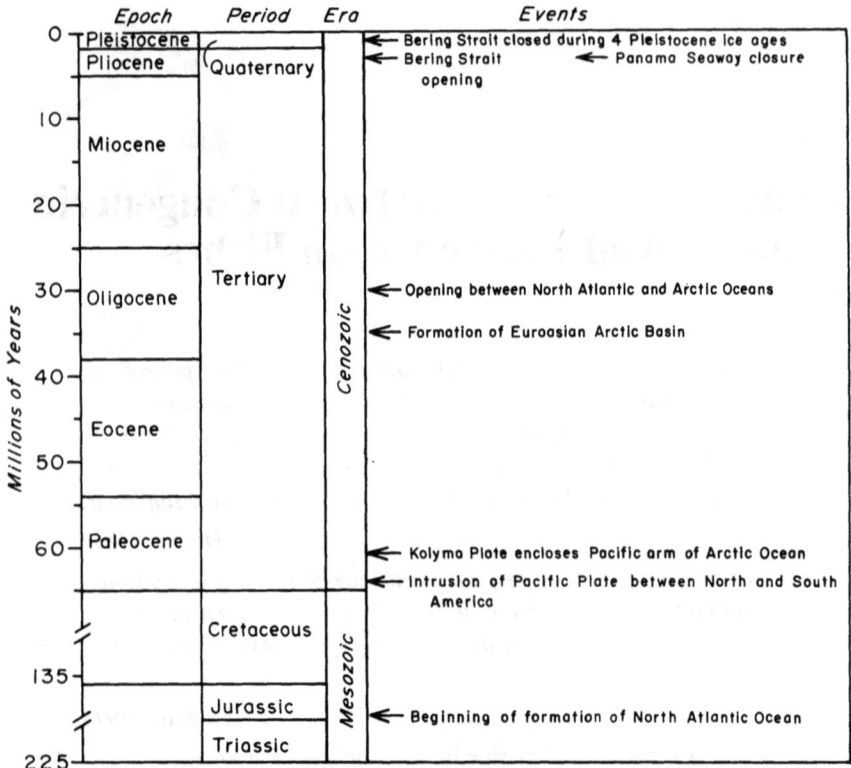

Fig. 9.1 Time scale showing approximate dates of geologic events discussed in this paper.

many higher taxa included fish from different evolutionary lineages (e.g., Jordan 1923). Recent arrangements have attempted to reflect phylogenetic relationships more clearly (Henning 1966, Greenwood et al. 1966).

The concept of time divergence between taxa is implicit in these recent classification schemes because closely related taxa are assumed to have diverged from one another more recently than distantly related taxa. Dates of cladogenetic events between related taxa traditionally have been inferred from fossil or geologic evidence. However, since Zuckerkandl and Pauling (1962) first postulated that biological macromolecules might contain temporal information, molecular methods—DNA hybridization, amino-acid sequencing, immunological methods, protein electrophoresis, and, more recently, DNA sequencing and restriction-enzyme mapping of mitochondrial DNA—have been used to study evolutionary relationships in a large array of organisms.

In this chapter we examine a practice frequently used in fish studies, of estimating divergence times from *genetic distance* as determined by protein electrophoresis (e.g., Ward and Galleguillos 1976, Smith et al. 1979, Fairbairn 1981). The emphasis here is on comparing molecular and geologic estimates of divergence times for several related fishes in the Atlantic and Pacific oceans. A more theoretical treatment of using genetic distance to estimate divergence times is presented by in Chapter 8 of this book. We also review the degrees of morphological and life history divergence between some of these pairs in light of their present taxonomic classifications.

THE ELECTROPHORETIC CLOCK

The electrophoretic analysis of proteins is at present the most widely used biochemical method for studying the genetic relationships of populations of closely related species, because it is a simple and cost-effective means for surveying genetic variation in a large number of individuals. This method measures only the amino-acid sequence divergence between homologous proteins that produce electrostatic charges or conformational differences affecting electrophoretic mobility; therefore, only about 25%–30% of the total number of amino-acid differences between two homologous proteins can be detected (Nei 1975, King and Wilson 1975). Details of this method and the statistical treatment of the resulting genotypic data are discussed in Chapter 1 of this book.

Nei's (1972) standard genetic distance, D, has been used the most in fisheries investigations to measure evolutionary divergence based on electrophoretic data and to estimate the length of time since two taxa shared a common gene pool. D is defined by

$$D = -\ln(I) ,$$

where I is the normalized probability of gene identity between two population samples:

$$I = \frac{J_{xy}}{(J_{xx}J_{yy})^{1/2}} .$$

Here J_{xy} is the probability that two randomly chosen genes, one each from taxa X and Y, are identical and J_{xx} and J_{yy} are the respective probabilities that the same genes within the taxa are identical. These probabilities include data for monomorphic loci as well and are averaged over all loci examined. With certain assumptions as to the size of the average protein being examined, D may be interpreted as the number of amino-acid differences in an average protein between two populations or taxa. The sample variance of D was derived by Nei and Roychoudhury (1974a) and comprises a large interlocus component resulting from the effects of mutation, selection, and genetic drift acting on each locus, and a smaller intralocus component influenced by sample size and allelic frequencies.

The Molecular Clock Hypothesis

The basis of the molecular clock hypothesis lies in the assumption that nucleotide-base substitutions in specific genes proceed at a more or less constant rate. These substitutions may be measured directly (as in DNA sequencing) or indirectly (e.g., amino-acid sequencing) by measuring some correlated attribute of the encoded proteins. Early proponents of the molecular clock hypothesis assumed that the rate of amino-acid replacement depended only upon the mutation rate (Kimura 1969, King and Jukes 1969), and that amino-acid replacement proceeded in a manner analogous to a Poisson process (Margoliash and Smith 1965, Nei 1969). However, Langley and Fitch (1974) and Fitch and Langley (1976) analyzed patterns of amino-acid replacement in seven proteins of 17 mammals and found that the observed variation was about twice that expected from a Poisson process. Nonetheless, apparent regularities in the rates of amino-acid replacement have been observed by some authors for globins (King and Jukes 1969), lysozymes (Prager and Wilson 1971), albumins (Maxson et al. 1975), and cytochrome c (Dickerson 1971) as well as several other proteins. Other authors however, maintain that the rates of change in some proteins have alternately accelerated and slowed over the course of biological evolution (Fitch and Langley 1976, Goodman et al. 1982).

One major criticism of the molecular clock hypothesis as it is generally applied to electrophoretic studies is that genetic distance may be nonlinear with time. Theoretical studies have shown that speciation itself may influence the initial genetic distance between two diverging taxa. For instance, Chakraborty and Nei (1974, 1977) show that genetic distance increases rapidly when a bottleneck in population size occurs, but the effects of the bottleneck eventually disappear with subsequent population growth. Nei (Chapter 8) suggests that when the heterozygosity of one of the diverging taxa is reduced by the bottleneck effect, the genetic distance may be corrected by the heterozygosity of the unaffected taxon. Templeton (1980) has shown that speciation in a highly subdivided ancestral population may produce a larger initial genetic distance than speciation in groups of panmictic populations. Korey (1981) further suggests that the interaction between population growth rate and generation time may influence genetic distance.

Another problem with using D to estimate evolutionary time is that amino-acid substitution rates vary from protein to protein and the value of D may depend to some extent on which proteins were included in an electrophoretic study. Authors have variously argued that an enzyme's metabolic function (Gillespie and Kojima 1968, Johnson 1974), its subunit number (Ward 1977, Harris et al. 1977), its subunit size (Eanes and Koehn 1978, Ward 1978), and its subcellular location (Gottlieb and Weeden 1981) may influence its heterozygosity and hence its evolutionary rate. Sarich (1977) argued that evolutionary rates of proteins are bimodal where enzymes such as transferrins, albumins, and esterases evolved an order of magnitude faster than other enzymes. However, in a survey of a large number of enzymes commonly included in electrophoretic studies, Skibinski and Ward (1982) did not find such a bimodality. Rather, single-locus genetic distance between related taxa was correlated in a continuous fashion with enzyme heterozygosity. They also suggested that, since a large proportion of monomorphic loci are fixed for different alleles in related species, heterozygosities for protein appear to change over time. Therefore, although it is desirable to calibrate each locus, present-day heterozygosity cannot be used as an absolute guide for calibration.

Theoretical Calibration

Nei's (1971) D can be stated in terms of the proportion of amino-acid differences detected by electrophoretic methods *(c)*, the number of amino-acids in the peptides being examined *(n)*, the rate of amino-acid replacement (λ), and the length of time since the two taxa separated *(t)*, such that

$$D = 2cn\lambda t .$$

When t is large and $cn\lambda t$ is larger than one, this relationship does not hold because of (1) the increased probability of convergence in the electrophoretic mobilities of alleles and (2) the inability to detect multiple replacements at single amino-acid sites. If $cn\lambda$ can be estimated from laboratory data, the length of time since separation may be estimated from empirically determined D using

$$t = \frac{D}{2cn\lambda} .$$

Nei's (1971) initial calibration was that a D of 1.0 represents 10^5 years of separation. However, as better estimates of $cn\lambda$ appeared, this calibration was revised to 5×10^6 years per unit of D (Nei, 1975).

Correlation of D
With Albumin Immunological Distance

Genetic distance, D may also be calibrated by correlating it with another molecular measure that has been previously calibrated with fossil or geologic estimates of divergence time. One such measure of protein difference is produced by the microcomplement fixation test (Levine and van Vunakis 1967), which is an immunological method based on antigen-antibody reactions between antibodies produced in rabbits and homologous proteins in related taxa. This procedure is often sensitive to small numbers of amino-acid differences (Cocks and Wilson 1969) and is most useful for comparing proteins having amino-acid differences less than 30%–35%, beyond which immunological recognition is lost. Albumins are frequently used as target proteins, and evolutionary distance is measured in albumin immunological distance (AID) units

(Sarich and Wilson 1966). Calibrations of D by correlation with AID have been frequently used in the electrophoretic literature.

There are, however, several problems associated with any attempt to calibrate D in this way. The first is that the errors of AID estimates are large. The percentage standard deviation generally averages from 5% to 15% per study, but may be as large as 30% or more for a single comparison (Wallace et al. 1973, Maxson and Wilson 1975, Maxson et al. 1979).

Another potential problem is that values of AID and D may not be linear over their useful ranges. Specifically, large values of D begin to asymptote with time because the probability of undetected multiple amino-acid replacements at a specific site increases with time. Furthermore, multiple replacements at a single site or simultaneous replacements at two or more sites may produce coincidental allelic mobilities (King 1973, Nei and Chakraborty 1973). Thorpe (1982) measured electrophoretic coincidence for 10 functionally similar proteins in pairs of unrelated organisms and found about a 5% occurrence of identical mobilities between presumably different proteins. Using this value, he suggested that the limiting value of D is $-\ln(0.05) = 3.0$. However, this limit may overestimate the useful range of D because homologous proteins in related taxa tend to concentrate in a smaller portion of a gel than functionally similar proteins of unrelated taxa.

Maxson and Maxson (1979) suggested that only about ten mobility states are resolvable for a protein because no more than about ten alleles are usually reported for any one locus. They found that D was linear with AID only to a value of about $D = 1.5$. This represents a period of time roughly extending back to the beginning of the Miocene Epoch, or about 25 million years. In contrast to D, which potentially measures only a single amino-acid replacement per protein, AID can measure the cumulative effects of about 200 amino-acid site differences in albumin (about 35% amino-acid sequence difference) between two taxa before immunological recognition is lost. Thus, the maximum length of time that can be measured is about 120 million years.

The first step in correlating D with AID is to correlate estimates of AID between pairs of taxa with fossil evidence or with geologic dates of continental drift for related taxa on different land masses. A rate of roughly one AID per 0.6 million years has been reported for frogs (Wallace et al. 1971, Maxson and Wilson 1975, Maxson et al. 1975), lizards (Gorman et al. 1971), marsupials (Maxson et al. 1975), carnivores (Sarich 1969) and primates (Sarich and Wilson 1967). Radinsky (1978), however, has charged that these studies tended to choose a "best" estimate from the fossil record to support a preconception. One example of a regression of AID values with fossil estimates of divergence time is shown in Figure 9.2. In this study Carlson et al. (1978) found that one AID unit yielded a divergence time of 0.54 million years.

A summary of several correlations of D with AID is presented in Table 9.1. These calibrations vary considerably from one taxonomic group to the next, but not in any consistent way, such as mammals versus reptiles. Estimates

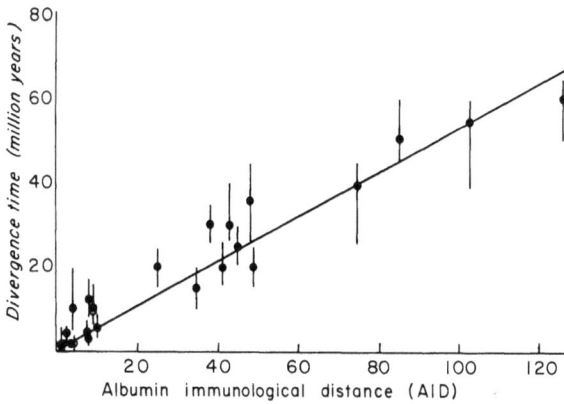

Fig. 9.2 Calibration of albumin immunological distance (AID) with paleontological estimates of divergence times between 25 pairs of ungulates and carnivores. The best estimates of divergence times are indicated by solid circles and the ranges of these estimates by vertical lines. Taken from Carlson et al. (1978).

range from 22 AID units per unit of D for rodents (Sarich 1977) to 55 AID units for frogs (Case 1975). The overall average of these calibrations is 35.8 AID units per unit of D.

Maxson and Maxson (1979) attempted to improve their calibrations by excluding values of D larger than 1.5 because of the saturation effect of multiple amino-acid replacements. A revised estimate was 24 AID units per unit of D, or about 14 million years of divergence time. In contrast, the best estimate of Wyles and Gorman (1980) for species of *Anolis* lizards was considerably

Table 9.1 Least-squares correlations of Nei's D and albumin immunological distance for several taxonomic groups. Modified from Wyles and Gorman 1980.

Group	Correlation coefficient (r)	Number species pairs	Slope	Millions of years[1]	Reference
Frog					
Hylidae	0.93	15	35	18.9	Case et al. 1975
Ranidae	0.66	22	55	29.7	Case 1975
Salamander					
Plethodontidae	0.70	79	26	14.0	Maxson and Maxson 1979
Lizard					
Iguanidae					
Anolis	0.83	70	38	20.5	Wyles and Gorman 1980
Sceloporus	0.26	55	22	11.9	Wyles and Gorman 1980
Sceloporine	0.73	44	37	20.0	Wyles and Gorman 1980
Iguanine	0.35	37	42	22.7	Wyles and Gorman 1980
Mammal					
Rodent	0.36	32	22	11.9	Wyles and Gorman 1980[2]
Primate	0.51	120	45	24.3	Wyles and Gorman 1980[3]

[1] Using fossil calibration of 0.54 million years per unit of AID (Carlson et al. 1978).
[2] From Sarich 1976.
[3] From Sarich 1967, Cronin 1975, Bruce 1967.

larger: 39 AID units per unit of *D,* or about 23 million years. Some authors have tried to refine their calibrations by attempting to account for the different evolutionary rates of the enzymes used to estimate *D.* This modification improved the agreement between taxa for some data (Sarich 1977, Maxson and Maxson 1979), but not for other data sets (Webster 1975, Wyles and Gorman 1980).

It is clear that large errors are associated with estimates of divergence time using electrophoretic genetic distance because of potential errors at each step of the calibration with time. Therefore, at best, Nei's *D* with its standard error can be used only to estimate an *interval* of time that most likely includes the actual date of cladogenesis between two related taxa. Choosing a specific calibration is somewhat arbitrary, but for heuristic purposes in this paper the calibration of 19 million years per unit of *D* (Carlson et al. 1978) is used because it is close to the average of several empirically derived estimates and because it has been the one most often used in fish literature (Fig. 9.3).

Molecular and Geologic Estimates Of Divergence Times

Numerous pairs of related fishes in the Atlantic and Pacific oceans have been classified at several taxonomic levels ranging from conspecific populations to congeneric species. In northern temperate areas most families of fishes are thought to have evolved in the North Pacific Ocean because it is geologically older and much larger than the Atlantic Ocean. Nonetheless, some families of fishes, such as the Gadidae and Clupidae, diversified in the Arctic–North Atlantic Basin at a time when northern seas were warmer (Svetovidov 1948, 1952). Skeletal fossils of gadids are common in European deposits from the Paleocene about 60 million years ago but are lacking in comparable Pacific Ocean deposits (Svetovidov 1948). Reciprocal faunal exchange between the North Atlantic and the North Pacific Ocean for cold-adapted species was not possible until the

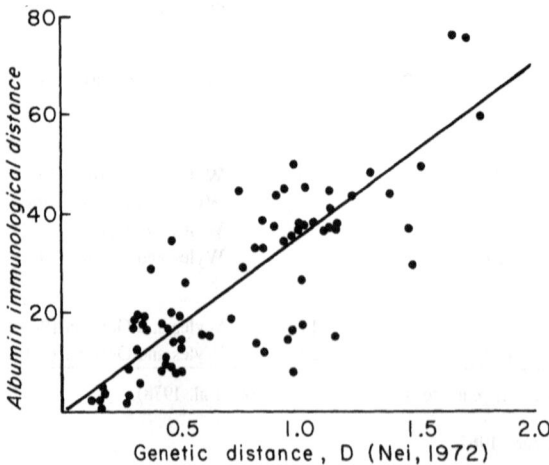

Fig. 9.3 Least-squares regression of albumin immunological distance (AID) with Nei's (1972) *D* for species pairs of the rodents *Dipodomys, Peromyscus* and *Thomnomys,* and for the reptiles *Hylad* and *Anolis.* The correlation coefficient, *r,* was 0.82 for 76 pairs. Taken from Sarich (1977).

mid-Pliocene about 3 million years ago. Therefore, estimates of divergence time for related North Atlantic and North Pacific fishes may not precede these dates.

The possibility that northern fishes dispersed from one ocean to the other through the Panama Strait before its closure must also be considered. However, it seems unlikely that boreal and cold temperate fishes passed through tropical seas having warmer-than-present temperatures. None of the northern fishes considered in this paper have present-day distributions reaching into the tropics.

The evolutionary history of related pairs of Atlantic and Pacific ocean fishes differs from that of north temperate fishes in that ancestral populations were subdivided by the formation of the Isthmus of Panama, which severed a seaway between the oceans that had existed for about 150 million years. Molecular estimates of divergence times should not be more recent than geologic estimates of the closure of the Panama seaway. Care must also be taken to choose pairs of fish that were actually separated by the rising land barrier, because some pairs may have speciated prior to the formation of the barrier. Disjunct distributions on either side of the Isthmus may be indirect evidence of allopatric speciation.

Bering Strait. The North Pacific and North Atlantic oceans were isolated from each other from the Mesozoic Era, (200 million years ago) until the Bering Strait opened in the late Cenozoic Era, between 3.0 and 3.5 million years ago. The formation of the Arctic Ocean reflects the tectonic complexity of the surrounding land masses. The basin between Alaska and Siberia is the oldest portion of the Arctic Ocean and was formerly an arm of the Pacific Ocean that was enclosed by the collision of the northward-moving Kolyma plate with the Alaskan and Siberian plates about 65 million years ago (Churkin and Texler 1980, Fujita and Newberry 1982) (Fig. 9.4). The Euroasian Basin is an extension of the North Atlantic Ocean that was formed by spreading of the sea floor along the Lomonosov Ridge in the Eocene about 55–60 million years ago (Vogt et al. 1979, Kitchell and Clark 1982). However, this portion of the Arctic Ocean probably remained isolated from the North Atlantic Ocean until 30–35 million years ago, when connections formed, possibly through Baffin Bay (Gradstein and Srivastava 1980) or what is more likely, between Greenland and Europe (Eldholm and Thiede 1980).

Hopkins and Marincovich (in press) suggest that the time the Bering Strait opened may be inferred from three lines of biostratigraphic evidence. First, fossil remains of *Pusa,* an arctic seal of Atlantic origin, have been found well below strata on the Pacific coast of southern Alaska, which date back 1.8 million years (Repenning 1983). Second, similar molluscan faunas in deposits to the north and south of Bering Strait indicate that the Bering Strait was open by 2.2 million years ago (Hopkins 1967). Third, a massive invasion of boreal Pacific molluscs in Pliocene deposits of Iceland indicates that the Bering Strait was open between 3.0 and 3.5 million years ago (Einarsson et al. 1967, Gladenkov 1979). The Arctic Ocean has since served intermittently between pe-

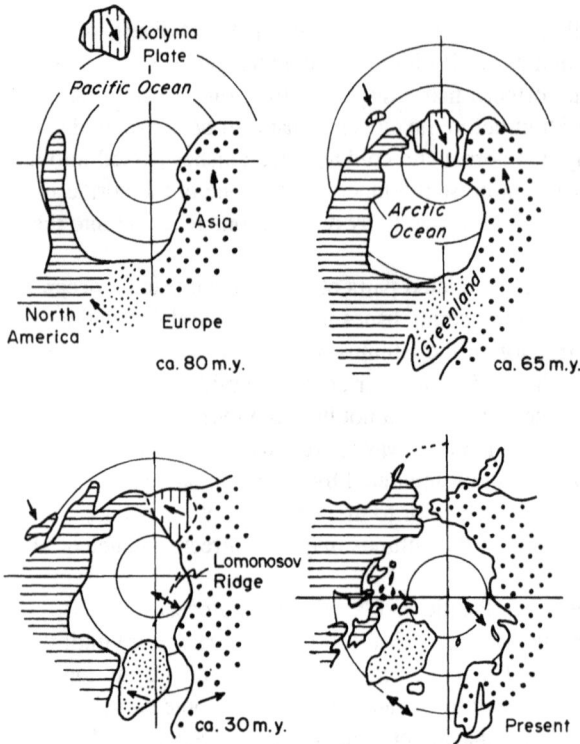

Fig. 9.4 Geologic evolution (in millions of years) of the Arctic Basin summarized from Churkin and Trexler (1980), Eldholm and Thiede (1980), and Gradstein and Srivastava (1980).

riods of glaciation as an avenue for marine dispersal between the Pacific and Atlantic oceans.

The average sea surface temperature in the Arctic Ocean dropped from 7°C in the mid-Miocene to 2°C in the Pliocene (Emiliani 1961), but it is not clear when perennial pack ice first formed or what its effects were on fish dispersal. Estimates of when the first permanent Arctic pack ice range appeared from 3 million years ago or earlier (Clark 1982, Gilbert and Clark 1983) to as recently as 0.7 million years ago (Herman 1970, Herman and Hopkins 1980). Hopkins and Marincovich (in press), however, suggest that these studies have been based on inadequate geochronologies of the Arctic Ocean sediment cores. According to their research, paleoclimates of the land surrounding the Arctic Ocean indicate that mild conditions existed until late Pliocene times. Pack ice probably formed on the Arctic Ocean in concert with Pleistocene glaciation on the surrounding land masses. Subsequently, several warm interglacial episodes may have permitted dispersal across the Arctic Ocean.

North temperate fishes. Genetic distances and the corresponding estimates of divergence times between four pairs of north temperate fishes are presented in Table 9.2. Between three of these pairs in the genera *Clupea* (herring), *Gadus* (cod), and *Hippoglossus* (halibut), D averaged 0.257 with an average standard error of 0.10. This represents an average divergence time of

Table 9.2 Average heterozygosity (H), average intraspecific genetic distance, (D), and interspecific D for related North Atlantic and North Pacific fishes.

	Number samples	Number loci	Average H	SE[1]	Average D	SE[1]	Divergence time interval (million years)	Reference
Clupea (herring)								
Atlantic *vs.* Pacific	26	40	—	—	0.264	0.086	3.4-6.7	Grant 1981
C. harengus (Atlantic)								
Between populations	3	24	0.070	—	0.002	—	—	Andersson et al. 1981
	6	40	0.069	0.023	0.001	0.000	—	Grant and Utter 1984
C. pallasi (Pacific)								
Between populations	21	40	0.086	0.026	0.003	0.002	—	Grant and Utter 1984
Between races	21	40	—	—	0.039	0.021	0.3-1.1	Grant and Utter 1984
Gadus (cod)								
Atlantic *vs.* Pacific	13	41	—	—	0.390	0.134	4.9-10.0	Grant and Ståhl, unpublished
G. morhua (Atlantic)								
Between populations	1	30	0.090	0.028	—	—	—	Mork et al. 1982
	2	41	0.108	0.031	0.033	0.020	—	Grant and Ståhl, unpublished
G. macrocephalus								
Between populations	11	40	0.026	0.034	0.008	0.007	—	Grant, Teel, Zhang, and Koybayashi, unpublished
Between races	11	40	—	—	0.023	0.019	—	Grant, Teel, Zhang, and Koybayashi, unpublished

Table 9.2 (continued) Average heterozygosity (H), average intraspecific genetic distance, (D), and interspecific D for related North Atlantic and North Pacific fishes.

Hippoglossus (halibut)								
Atlantic vs. Pacific	4	35	—	—	0.123	0.075	0.9-3.8	Grant et al. 1984
H. hippoglossus (Atlantic)								
Between populations	1	35	0.024	0.014	—	—	—	Grant et al. 1984
	3	25	0.004	0.004	—	—	—	Mork and Haug 1983
H. stenolepis (Pacific)	3	35	0.060	0.027	0.003	0.001	—	Grant et al. 1984
Reinhardtius hippoglossoides (Greenland halibut)								
Atlantic vs. Bering Sea	5	16	—	—	0.013	0.010	—	Fairbairn 1981[2]
Atlantic Ocean	4	14	0.048	0.033	0.000	0.000	—	Fairbairn 1981[2]
Bering Sea	1	16	0.079	0.053	—	—	—	Fairbairn 1981[2]

[1] Standard error (Nei and Roychoudhury 1974). For heterozygosities, SE was averaged over samples. For genetic distance, SE was averaged over all pairs for which D was calculated.

[2] Genetic distance between Bering Sea and Atlantic Ocean samples was recalculated using allelic frequency in that paper.

4.9 ($\pm$ 1.9) million years and is broadly consistent with a mid-Pliocene opening of Bering Strait. *D* between Atlantic and Pacific cod was somewhat larger than *D* between the other pairs of fishes. This may be due to a bottleneck effect in founding population size of Pacific cod. Corroborating evidence for this is that Pacific cod have a significantly lower average heterozygosity (H = 0.026, $\pm$ 0.034; 41 loci) than the presumably ancestral Atlantic cod (H = 0.108, $\pm$ 0.031; 41 loci) (Grant and Ståhl, unpublished).

In contrast to these three pairs, the genetic distance between Atlantic and Bering Sea populations of Greenland halibut (*Reinhardtuis hippoglossus* was only 0.013 (SE = 0.010) (Fairbairn 1981, but see Grant et al. 1984 for recalculation of *D*), which is typical of genetic distances between conspecific populations. This suggests either that there is ongoing migration across the Arctic Ocean or that there has been recent gene flow, perhaps during the postglacial hypsithermal period, when the Arctic Ocean was warmer than present (Stewart and England 1983).

Isthmus of Panama. The tectonic history of Central America has been summarized by Malfait and Dinkelman (1972) and is briefly outlined in Figure 9.5. During the late Mesozoic and early Cenozoic eras, a portion of the Pacific Ocean plate was thrust in a northeasterly direction between the converging North and South American plates. In the Oligocene the intruding plate detached from the Pacific Ocean plate and began to move eastward, pulling portions of North America with it. The Isthmus of Panama rose to its present posi-

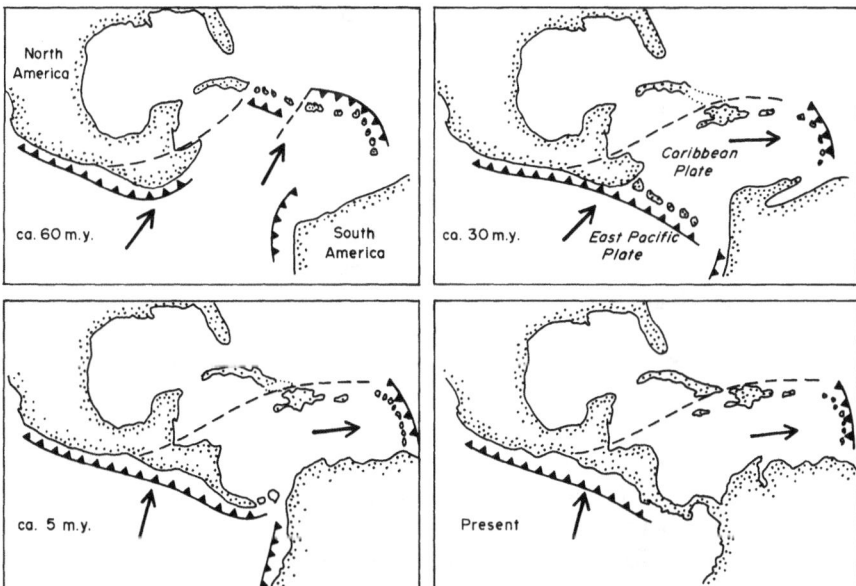

Fig. 9.5 Geologic evolution (in millions of years) of Central America summarized by Malfait and Dinkelman (1972). Triangles show areas of plate subduction, dashed lines show areas of plate shear, and arrows show movement of oceanic plates.

tion above sea level as a result of isostatic uplift following the cessation of subduction of the Pacific plate under the now independent Caribbean plate.

One estimate of the time of closure of the Panama seaway was made by Emiliani et al. (1972), who found carbonate-poor sediments in deep-sea cores from the western Atlantic Ocean that presumably resulted from the interruption of cold Pacific currents into the Atlantic Ocean. They estimated that the sediments were about 5.7 million years old using the biostratigraphic chronology of Berggren (1969).

However, Saito (1976) suggested a more recent date. He found that after the closure the planktonic foraminifera, *Pulleniatina*, disappeared from Atlantic Ocean sediments. When it reappeared in more recent sediments, the direction of shell coiling had reversed relative to Pacific Ocean *Pulleniatina* at the same strata. Since the direction of coiling appears to reflect temperature, the discordance between Atlantic and Pacific samples reflected a divergence in hydrographic conditions following the appearance of the isthmus. Using paleomagnetic stratigraphy to calibrate the age of the samples, he estimated the date of the closure to be about 3.5 million years ago.

Keigwin (1978) extended these results using additional and more complete cores taken from locations nearer to the Panama isthmus and revised the date to 3.1 million years, with the disappearance of *Pulleniatina* as the event marking the closure. Keigwin (1982) further analyzed the $^{12}C/^{13}C$ and $^{16}O/^{18}O$ ratios in foraminiferan shells and found an enrichment of ^{13}C in the Caribbean benthic foraminifera about 3.0 million years ago. There was a similar divergence in the oxygen ratios of the cores beginning about 4 million years ago. He interpreted these patterns as reflecting oceanographic divergence between the Atlantic and Pacific oceans after their separation.

Tropical fishes. Several pairs of fishes with representatives on either side of the Isthmus of Panama have been studied by protein electrophoresis (Gorman et al. 1977, Gorman and Kim 1976, Vawter et al. 1980), and these data are summarized in Table 9.3. D between seven pairs of related species separated by the Isthmus ranged from 0.131 to 0.361, and aeraged 0.234. Even for pairs of conspecific populations separated by the isthmus, D ranged between 0.142 and 0.320 and averaged 0.194. The corresponding estimates of divergence times for these pairs extended from 2.5 to 6.9 million years and generally conform to the expectation of the molecular clock hypothesis that divergence times should be greater than about 3.0–3.5 million years.

However, genetic distances between related pairs of sea urchins separated by the Panama isthmus showed less congruence with the molecular clock (Table 9.4). Lessios (1979, 1981) found that for three pairs D ranged from 0.329 to 0.638 and averaged 0.505, but for a fourth pair D was only 0.026, a value that was similar to that between populations within each species. These values of D correspond to divergence times ranging from 0.5 to 12.5 million years.

The small number (18) of loci used in this survey and the probabilistic

Table 9.3 Average heterozygosities (*H*) and genetic distances (*D*) between pairs of species or conspecific populations of Atlantic and Pacific Ocean fishes separated by the Isthmus of Parama.

SPECIES PAIRS	Number locations	Number loci	Average H	SE¹	Average D	SE¹	Divergence time interval (million years)	Reference
Abudefduf								
Atlantic vs. Pacific	4	28	—	—	0.320	0.114	3.9-8.2	Gorman and Kim 1977
A. saxatilis (Atlantic)	2	28	0.050	0.026	0.021	—	—	Gorman and Kim 1977; Vawter et al. 1980
A. troschelii (Pacific)	2	28	0.037	0.018	0.026	—	—	Gorman and Kim 1977; Vawter et al. 1980
A. taurus vs. A. concolor	2	27	—	—	0.194	0.174	3.7	Vawter et al. 1980
Bathygobius								
B. soporator vs. B. ramosus	2	26	—	—	0.402	0.142	5.3-10.7	Gorman et al. 1976
B. sporator vs. B. andrei	2	26	—	—	0.146	0.077	1.3-4.2	Gorman et al. 1976
B. ramosus vs. B. andrei	2	26	—	—	0.418	0.141	5.3-10.6	Gorman et al. 1976
B. andrei (Pacific)	1	26	0.027	0.014	—	—	—	Gorman et al. 1976
B. ramosus (Pacific)	1	26	0.056	0.021	—	—	—	Gorman et al. 1976
b. soporator (Atlantic)	1	26	0.069	0.030	—	—	—	Gorman et al. 1976
Rypticus								
R. bicolor vs. R. saponoceus	2	39	—	—	0.142	—	2.7	Vawter et al. 1980
Mulloidichthys								
M. dentatus vs. M. martinicus	2	23	—	—	0.168	—	3.2	Vawter et al. 1980

Table 9.3 (continued) Average heterozygosities (*H*) and genetic distances (*D*) between pairs of species or conspecific populations of Atlantic and Pacific Ocean fishes separated by the Isthmus of Panama.

CONSPECIFIC POPULATIONS

Diodon holocantus	2	41	—	0.192	—	3.6	Vawter et al. 1980
Diodon hystix	2	30	—	0.361	—	6.9	Vawter et al. 1980
Gerres cinereus	3	22	—	0.282	—	5.4	Vawter et al. 1980
Haemulon steindachneri	3	35	—	0.131	—	2.5	Vawter et al. 1980
Scorpaena plumeri							
Atlantic *vs.* Pacific	4	28	—	0.206	—	3.9	Vawter et al. 1980
Atlantic	2	25	—	0.031	—	—	Vawter et al. 1980
Pacific	2	26	—	0.008	—	—	Vawter et al. 1980

[1]Standard error after Nei and Roychoudhury (1974). For heterozygosities, SE was averaged over samples. For genetic distance, SE was averaged over all pairs for which *D* was calculated.

Table 9.4 Genetic distances (*D*) between pairs of invertebrates across barriers to migration.

	Number locations	Number loci	Average D	SE[1]	Divergence time interval (million years)	Reference
ECHINODERMATA						
Diadema						
Atlantic *vs.* Pacific	4	18	0.026	—	0.05	Lessios 1981
D. antillarium (Atlantic)	2	18	0.036	—	—	Lessios 1981
D. mexicanum (Pacific)	2	18	0.015	—	—	Lessios 1981
Echinometra						
E. lucunter vs. E. vanbrunti	4	15	0.549	—	10.4	Lessios 1981
E. lucunter vs. E. viridis	4	15	0.114	—	2.2	Lessios 1981
E. vanbruntis vs. E. viridis	4	15	0.638	—	12.5	Lessios 1981
E. lucunter (Atlantic)	2	15	0.009	—	—	Lessios 1981
E. vanbrunti (Pacific)	2	15	0.021	—	—	Lessios 1981
E. viridis (Atlantic)	2	15	0.007	—	—	Lessios 1981
Eucidaris						
E. tribuloides vs. E. thouarsi	4	15	0.329	—	6.2	Lessios 1981
E. thouarsi (Pacific)	2	15	0.024	—	—	Lessios 1981
E. tribuloides (Atlantic)	2	15	0.016	—	—	Lessios 1981
COELENTERATA						
Bunodosoma						
B. cavernata (Atlantic)	2	12	0.040	0.018	—	McCommas 1982
Atlantic *vs.* Gulf of Mexico	3	12	0.280	0.016	2.3-8.3	McCommas 1982
B. granulifera (Atlantic) *vs.*						
B. california (Pacific)	2	12	0.188	0.089	1.9-5.2	McCommas 1982

[1]Standard error after Nei and Roychoudhury (1974a).

nature of the molecular clock have been invoked to explain these results (Vawter et al. 1980). However, a third possibility has not been considered. This is that urchins may have been carried inadvertently across the isthmus by ships traveling through the Panama Canal (Menzies 1968, Rubinoff and Rubinoff 1968). Such a dispersal event appears to be highly improbable but not impossible.

There is another set of data for sea anemones (Coelenterata) (McCommas 1982) separated by the Isthmus of Panama and for sea anemones separated by the Florida Peninsula, which rose to its present level between two and five million years ago. In this study, related species across the Isthmus of Panama had a genetic distance of 0.185 and the disjunct populations across the Florida Peninsula had an average *D* of 0.280. Estimates of divergence times based on these results agree with geologic estimates of the formations of these two marine barriers.

DIVERGENCE OF LIFE HISTORY PATTERNS

As we said earlier, traditional systematic methods for tracing divergence are based largely on the analysis of morphological variation. This approach however, may not always reveal the true genetic relationships between taxa because of convergence or because of unequal rates of evolution in different lineages. Therefore, to clarify taxonomic relationships, additional gauges are needed which are more or less independent of morphology (King and Wilson 1975, Maxson and Wilson 1975). Biochemical genetic distance is one such gauge, but there must be a reasonable correspondence between the degree of molecular divergence and traditional taxonomic categories if genetic distance is to serve as a useful guide.

Several taxonomic surveys have shown that such a correspondence does exist in general, but also that molecular-taxonomic scales may vary from one group of organisms to the next (Avise 1974, Ayala 1975, Thorpe 1982, Avise and Aquadro 1982). For fishes, Shaklee et al. (1982) found that D averaged 0.05 (range 0.002–0.065) between conspecific populations, 0.30 (0.025–0.609) between congeneric species, and 0.90 (0.580–1.21) between confamilial genera. The boundaries between taxonomic categories are not sharp, because genetic divergence is a continuous process and because there are no absolute criteria for assigning an organism to a particular morphologically based category.

The species concept rests on the assumptions that populations of related species are reproductively isolated so from populations of other species and that sufficient genetic change has taken place since isolation so that sympatric reassociation will not lead to genetic assimilation (Lewontin 1974). After isolation, populations begin to diverge from one another because of genetic drift and natural selection. In addition to morphological and molecular divergence, changes in life history patterns may also be apparent. The objective of this section is to ask how much life history divergence has accompanied a given amount of molecular divergence between related Atlantic and Pacific ocean fishes.

Herring (Clupea)

Atlantic *(C. harengus)* and Pacific *(C. pallasi)* herring are morphologically very similar, showing only minor differences in body size and shape, average numbers of fin rays and vertebrae, and age at maturity (Leim and Scott 1966, Hart 1973). As a result of this similarity they are now considered to be subspecies (American Fisheries Society, 1980). However, Grant (1981) estimated a genetic distance between them of 0.264 (± 0.086), a distance that is more typical of a species level of divergence. Additionally, cross-fertilization experiments with cryopreserved Atlantic herring sperm showed that there may be postzygotic developmental barriers between these species (Rosenthal et al. 1978).

There are some life-history differences between the two species of herring. Pacific herring spawn in very shallow water only in spring (late winter in southern areas and early summer in northern areas), whereas groups of Atlantic herring spawn on offshore banks in autumn and inshore in spring. Kornfield et al. (1982a) reportedly found allele-frequency differences between these Atlantic herring spawning types on the North American side of the Atlantic Ocean, but the loci showing differences were not the same ones over three successive years. A gene diversity analysis (Nei 1973a) of these data shows that there was more genetic heterogeneity among years or among locations than there was between spawning types (Grant 1984). Other data suggest that fish may change from one spawning type to another in response to environmental changes (Messieh 1972).

The geographic structuring of populations also appears to differ between the two species of herring. Grant and Utter (1984) showed that there are two distinct geographic races of Pacific herring in the North Pacific having an average D between populations in each group of 0.039 (± 0.021), whereas in Atlantic herring there is virtually no transoceanic differentiation (Grant 1984). The average D between samples of Atlantic herring collected on both sides of the Atlantic Ocean was only 0.003 (± 0.002). The races of Pacific herring may have arisen as a result of repeated Pleistocene coastal glaciation along the coast of Alaska, which created a barrier between eastern and western Pacific Ocean populations. Pacific herring are particularly susceptible to coastal influences, because they spawn intertidally or in very shallow water. Grant (1984) has postulated that the lack of differentiation between widely separated populations of Atlantic herring may be due to extinction on one side of the Atlantic Ocean during Pleistocene glaciation followed by a postglacial radiation of a few genetically similar populations.

Cod (Gadus)

Atlantic *(G. morhua)* and Pacific *(G. macrocephalus)* cod are similar to one another but show sufficient morphological differences to be considered separate species (Svetovidov 1948). The genetic distance between these two species was estimated to be 0.390 (± 0.134) (Grant and Ståhl unpublished), and confirms their taxonomic treatment as full species.

The amounts of genetic divergence between populations within each species are similar. In a survey of genetic variation in populations of Pacific cod, Grant et al. (unpublished) found an Asian race and a North American race having an average D of 0.023 (± 0.019) between populations. However, the geographic boundary between these races was not the Alaska Peninsula, as it was for the races of Pacific herring, but somewhere in the western Bering Sea. There was little genetic differentiation between North American populations extending from the Bering Sea to the Washington coast, where the average D between populations was 0.008 (± 0.007). Similar transoceanic genetic differences between populations of Atlantic cod were reported by Grant and Ståhl (unpublished), who found a genetic distance of 0.033 (± 0.020) between a sample of European cod and a sample of North American cod. Other genetic studies of European populations show clinal variation with latitude for *Hemoglobin I*, but not for transferrin, *Gpi-1*, and *Ldh-3* over the same area (summarized in de Ligny 1969).

The most striking genetic difference between Atlantic and Pacific cod is their average levels of protein heterozygosity. Grant et al. (unpublished) estimated an average heterozygosity for Pacific cod to be 0.026 (±0.034), whereas that for Atlantic cod was estimated to be 0.108 (±0.031) by Grant and Ståhl (unpublished) and 0.090 (±0.028) by Mork et al. (1982). Apparently Pacific cod have also lost gene expression for *Me-3* (Grant and Ståhl, unpublished). These data suggest that founding populations of Pacific cod may have experienced an extreme bottleneck in size so that much of the genetic variation in the ancestral Atlantic populations was lost.

Halibut (Hippoglossus)

There are two morphologically similar species of halibut inhabiting the colder waters of the North Atlantic *(H. hippoglossus)* and North Pacific *(H. stenolepis)* oceans. These taxa appear to be at an early stage of evolutionary divergence with a genetic distance of 0.123 (±0.075) (Grant et al. 1984). This level of divergence is somewhat lower than that of other pairs of Atlantic and Pacific fishes and suggests either that ancestral stocks of Pacific halibut did not disperse into the North Atlantic until sometime after the Bering Strait opened or that there has been recent interocean migration.

Atlantic halibut also appear to have a lower average heterozygosity than do Pacific halibut. Grant et al. (1984) found an average heterozygosity of 0.060 (±0.027) for Pacific halibut, but only 0.020 (±0.010) for Atlantic halibut. Mork and Haug (1983) also found a low value of 0.004 (±0.004) for Atlantic halibut. Therefore, the appearance of Atlantic halibut may represent another instance of speciation by a very small founding population.

Both of these species spawn pelagic eggs at considerable depths near the edges of continental shelves (Leim and Scott 1966, Hart 1973), and tagging data show that both species are capable of migration over very long distances (McCracken 1958, Best 1981). This large potential for gene flow at both larval and adult stages is reflected in very low levels of genetic differentiation between populations within each species (Tsuyuki et al. 1969, Mork and Haug 1983, Grant et al. 1984).

Greenland Halibut (Reinhardtius)

In a study of morphometric and meristic characters, Hubbs and Wilimovsky (1964) concluded that Atlantic and Pacific populations of Greenland halibut *(R. hippoglossoides)* belong to a single species. More recently, Fairbairn (1981) reported a genetic distance of 0.098 between populations from eastern Canada and the Bering Sea. However, this estimate contained data for polymorphic loci only and cannot be used to make taxonomic inferences. When this genetic distance is recomputed with all loci examined in that study (D =

0.013), it is more typical of genetic distances between conspecific populations. The geographic distributions of this fish reach far into Arctic waters (Leim and Scott 1966), and the lack of interocean divergence suggests recent or continuing gene flow between Atlantic and Pacific populations.

Goby (Bathygobius)

There are three morphologically similar species of goby around Panama. *Bathygobius ramosus* and *B. andrei* are found on the Pacific side and *B. soporator* is found on the Atlantic side. Gorman et al. (1976) found a somewhat larger genetic distance between *B. ramosus* and the other two species (average *D* = 0.42) than the genetic distance between the allopatric pair *B. andrei* and *B. soporator* (*D* = 0.146) and postulated that *B. ramosus* may have speciated from ancestral stocks before the Panama seaway closed.

In breeding experiments in which sexes of different species were kept together, Rubinoff and Rubinoff (1971) found that there was occasional spawning between the closely related allopatric species but not between these species and the more distantly related *B. ramosus*. In fact, hybrids between the two former species could be raised to reproductive maturity.

CONCLUSIONS

This paper has examined the use of electrophoretic genetic distance for estimating divergence times between related species of fishes and as a taxonomic tool with specific reference to related Atlantic and Pacific ocean fishes. One of the major obstacles for testing the validity and the accuracy of the electrophoretic clock has been the lack of information on the times of specific speciation events. The geologic events that subdivided tropical Pliocene populations of fishes or that permitted reciprocal dispersal of boreal fishes from one ocean to the other represent independent, albeit indirect, markers of cladogenetic events. The results presented here show that genetic distance between specific pairs of taxa cannot be used with any degree of accuracy to estimate time since their divergence because of errors from several sources. The most important sources of error appear to be

- the inclusion of enzymes having different rates of amino-acid substitution,
- the calibration of genetic distance with time,
- errors resulting from unknowable population genetic and speciation effects on genetic distance, and
- stochastic errors inherent in the molecular clock itself.

In practice, therefore, it is not possible to distinguish between cladogenetic events in the late Miocene and Pliocene, for example, that are separated by as much as 5 million years.

The requirements for inferring phylogenetic relationships from genetic distances are more relaxed than those for estimating absolute time because only relative measures of divergence are needed. Even though rates of molecular evolution appear to vary among some higher taxa, e.g., reptiles and birds (Avise and Aquadro 1982), evolutionary rates among closely related taxa appear to be more similar because of similarities in genetic construction and life-history patterns. Electrophoretic data are most reliable for inferring systematic relationships below the level of genus, because the nonlinearity of D at values greater than about 1.5 compresses the apparent phylogenies at higher taxonomic levels.

The genetic data that were reviewed here for nominal conspecific populations of tropical fishes separated by the Isthmus of Panama show that several taxonomic revisions are needed. Although these pairs of fishes show little morphological divergence, their genetic distances accurately reflect that they have been isolated from one another since the closure of the Isthmus of Panama about 3 million years ago. Genetic distances are ideally suited to aid in the reclassification of these fishes at this level of evolutionary divergence.

Although new molecular methods such as DNA sequencing (Maxam and Gilbert 1977) or DNA restriction fragment analysis (Lansman et al. 1981) have been developed for studying natural populations, protein electrophoresis will continue to be   valuable tool in such areas as population genetics and taxonomy and in the applied field of fishery management. No other method exists which can produce large amounts of population data in short periods of time at a reasonable cost.

However, the limitations of this method must be recognized. As demonstrated here, one limitation is the estimation of divergence times on an absolute scale, and conclusions based on genetic distance commonly found in fishery literature must be treated cautiously.

10.
Use of Genetic Marks in Stock Composition Analysis

Jerome J. Pella and George B. Milner

It is essential to the effective management of mixed stock fisheries that the stocks which compose the mixture be identified and the extent of their contribution assessed (Larkin 1981). Without the benefit of such information, fisheries agencies may base their regulations only on the presence of numerically strong populations and be unaware that weaker stocks are being depleted; or they may focus on protecting weaker populations, to the end that the stronger populations are underharvested. However, in spite of its importance, stock-composition strategy has been practiced only rarely in fishery management because of the difficulty in adequately identifying stock groups that contribute to the fishery.

One thinks of species as being typically distinguishable from one another by obvious morphological differences, but exceptions to this general rule abound in fishes. Subadult forms such as eggs and larvae are often indistinguishable between species although differences among the adults are obvious (Mork et al. 1983). Even as adults, some sibling species lack readily discernible morphological differences (e.g., Shaklee and Tamaru 1981). Within species, morphological distinction between individuals belonging to different populations (stocks) is generally very difficult or impossible.

The general absence of natural and readily visible marks within species has resulted in a variety of more or less successful artificial marking procedures. Traditional methods such as coded-wire tagging and fin clipping have given managers valuable tools for identifying specific groups of hatchery fish, but these methods are difficult to use on wild populations. They also require considerable effort and cost (Ihssen et al. 1981).

Natural marks requiring more complicated procedures for their detection have also been tested for usefulness. They include scale analyses (Messinger and Bilton 1974), parasite infestations (Margolis 1963), morphometric and meristic characters (Amos et al. 1963, Fukuhara et al. 1962), elemental chemical composition (Mulligan et al. 1983), and genetic characters (Ridgway et al. 1962). All of these characters except the genetic ones can be strongly affected by environmental conditions, thus typically requiring yearly examination and revision of the standards.

The effective use of genetic characters has only recently become possible through the development of procedures for identifying allelic variants at a large number of loci (Chapters 1 and 2). The genotypic composition of an individual or a population can be regarded as a mark, a genetic mark that in many situations distinguishes that individual or population from other ones. Because of thier differentiating ability, genetic marks are becoming an important part of fishery management and research.

In this chapter, we discuss the use of allelic variants as genetic marks, with particular emphasis on estimating the contribution of different stocks to mixed stock fisheries. The first section of the chapter outlines the general principles and applications of genetic marks in fish and gives an example of how the composition of mixed stock is estimated. The second section contains a formal discussion of the mathematical basis for analyzing the contributing components of stock mixtures. It is intended as a reference for developing urgently needed procedures in fishery management; it may be of limited interest to the reader who is not primarily interested in the details of the statistical approach to such estimation procedures.

GENERAL PRINCIPLES AND APPLICATIONS

Genetic Marks

Allelic variants have attributes that make them uniquely valuable as marks for various applications in fish biology and fishery management:

- They occur naturally and so are equally useful for the identifying both wild and hatchery stocks.
- They are inherited in a Mendelian manner which transmits the marks from the parents of one generation to the offspring of the next.
- They are usually expressed throughout the life cycle of an individual; thus, juveniles and adults are equally identifiable.
- Allele frequencies at protein-coding loci tend to be fairly constant over time (generations), which reduces the need to revise the standards for characterizing a population.
- Imposing the mark does not require the handling of individual fish, and so no trauma results from mark application and retention.
- The mark can be read (and the genotype determined) at a reasonable effort and cost.

Genetic marks have two major applications: delineation of genetic population structures and—when the population structure is known—examination of population mixtures on the basis of this information. Most of the other chapters of this book deal with the first of these two applications. This chapter deals with the second.

Examination of Population Mixtures

Fixation for alternate alleles. The simplest case for examination of population mixtures occur when each contributing population is fixed (i.e., homozygous) for one or more alleles that do not occur in the other populations contributing to the mixture. In this case, every individual can be unambiguously classified with respect to the populations to which it belongs.

The most prevalent examples of alternate fixation pertain to situations in which species identification is required in the absence of obvious morphological species characteristics, as mentioned above. An obvious application in fishery management refers to species identification of fish eggs and larvae. Unfortunately, such applications have been largely neglected to date (but see Johnson et al. 1975 and Mork et al. 1983). Likewise, sibling species may be easily identified through fixed allelic differences, but have received greater attention. First-generation hybrids of different species are also unambiguously identifiable if the parental species have fixed allelic differences at one or more loci (Chapter 7).

Fixation for alternate alleles among naturally occurring conspecific populations is not common, although a few cases having apparent management implications have been reported (e.g., Ryman et al. 1979). However, fixation for alternate alleles can be readily achieved in cultured populations by intentionally breeding individuals with defined genotypes (e.g. Moav et al. 1976, Allendorf and Utter 1979). In a similar manner, progeny of experimental crosses may be unambiguously identified. For instance, Schroder (1982) performed paternity analyses for determination of spawning success using chum salmon males homozygous for different alleles.

Varying frequencies of the same alleles. In the above instances, the relative contribution of the populations to the mixture can be determined directly because the origin of every individual is known when the populations are fixed for different alleles. However, in the absence of fixed allelic differences, determining the composition of population mixtures becomes more complicated, because then every individual in a mixture cannot be unambiguously identified with the population to which it belongs. Although populations may be uniquely characterized by their allele and genotype frequencies, many individual genotypes are common to all populations that segregate for the same alleles. These genotypes, then, cannot be classified as belonging to a particular population with absolute certainty. The only genotypes that can be classified as belonging to a particular population are those which carry an allele that is unique to that particular population.

For most management purposes, however, it is not essential that every individual in a mixture is unequivocally identified with respect to the population to which it belongs. For example, if only a fraction of a particular group contributing to a mixture possess a mark, it is still possible to estimate the relative contribution of that group from the information provided by the marked fraction. Information provided by traditional tagging procedures is, in fact, typically based on such incomplete marking.

Similarly, if the same allele occurs at different frequencies in two populations, it is possible to estimate the proportion of individuals from each of the two populations when they occur in a mixture. For instance, consider a situation in which two (and only two) populations A and B intermingle. Assume that the frequency of an allele in population A is 0.20 (p_A) and in population B, 0.60 (p_B). Assume that in a sample from the mixture, the frequency of that particular allele is 0.50 (p_m). The composition of this mixture is then estimated to comprise 25% from population A and 75% from population B. This estimate (p_m) is obtained from the expression giving the weighted average of two allele frequencies,

$$p_m = f p_A + (1 - f) p_B \; ,$$

where f is the relative contribution from population A and $(1-f)$ is the contribution from population B. This procedure has been used to estimate compositions of simple mixtures of fish populations (e.g., Allendorf and Utter 1979, Murphy et al. 1983, Lane 1985).

In reality, most population mixtures are more complex than that described above; they commonly involve more than two populations, and the information from more than a single locus may have to be considered. In such cases, the information provided by the joint genotypic distribution at all the loci examined must also be considered, and the analysis is based on genotypic rather than allele frequency distributions of samples from the mixture and the contributing populations, respectively. If the number of genotypes equals or exceeds the number of populations, the composition of mixtures can be resolved. However, more complicated statistical procedures are then required to estimate the composition of such mixtures (Fig. 10.1).

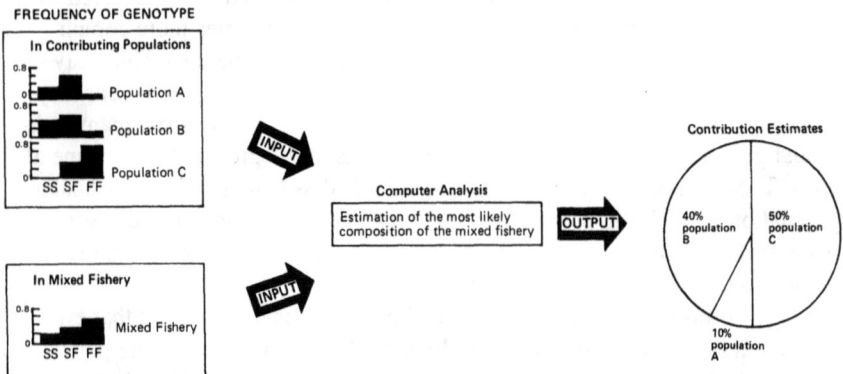

Fig. 10.1 Outline of procedure for estimating the composition of a stock mixture on the basis of genetic data. The mixture is composed of three populations, and the genotypic data refer to a single locus segregating for two alleles (S and F). From Milner et al. (1985).

The first estimates of complex mixed stock compositions with multiple loci data from fish populations were obtained for runs of sockeye salmon *(Oncorhynchus nerka)* returning to the Kenai Peninsula region of Alaska (Grant et al. 1980). Those estimates, derived by a maximum likelihood procedure, the EM algorithm, (Dempster et al. 1977), though consistent with other sets of biological data relating to these runs, could not be tested directly for accuracy.

A test for accuracy was subsequently performed by Milner et al. (1981) using data from populations of hatchery-reared chinook salmon *(0. tshawytscha)* of the Columbia River, an immense watershed of the northwestern United States and southwestern Canada. The numerous chinook salmon populations of the Columbia River intermingle with one another and with populations from other drainages, and are exploited by fisheries both before and after they return to the river. Here we use one example from that accuracy test to demonstrate the application of genetic data for estimating stock compositions of population mixtures. In the next section we take the same example to illustrate the properties of some different procedures for obtaining estimates and evaluating their precision.

The test of accuracy was constructed to mimic, as realistically as possible, a situation in which a number of populations potentially contribute to a mixed stock fishery and the manager wants to estimate what their relative contribution is. The populations potentially contributing to the mixtures were represented by 14 hatchery stocks of chinook salmon (Fig. 10.2), the progeny of parents returning to the river in spring months

Allele frequencies for each of these stocks were estimated for nine pro-

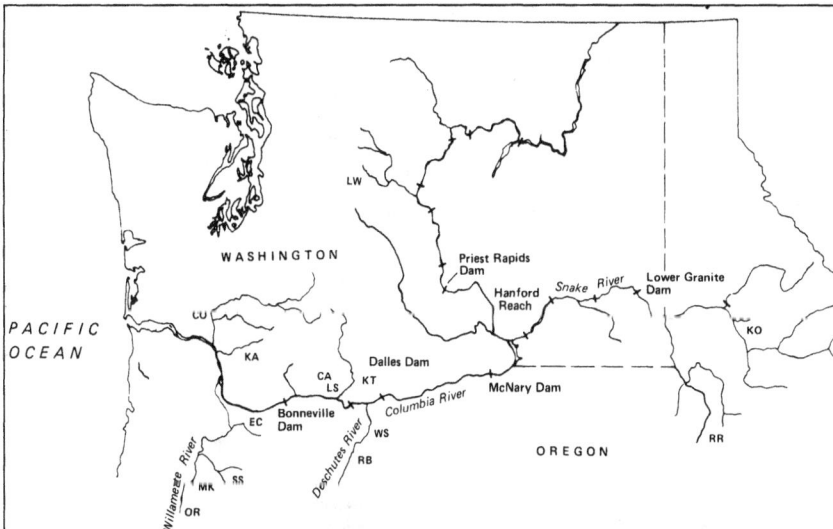

Fig. 10.2 Origin of 14 hatchery stocks of chinook salmon from the Columbia River used in blind test of mixed stock composition based on electrophoretically detectable allele frequency differences among stocks. (See Table 10.1 for abbreviations).

tein-coding loci (sample size was 200 fish per stock). The amount of genetic divergence among these stocks was typical of that commonly observed among local stock of anadromous salmonids (e.g., Chapter 9). All common alleles occurred in all populations at varying frequencies (no alternate fixations), but some alleles occurred uniquely at low frequencies in some populations. The range of allele frequency differences among samples of the most common allele at these nine loci—abbreviations for enzymes follow Allendorf and Utter 1979—were as follows: *ADH*, 0.04; *GL-1*, 0.03; *IDH-3,4*, 0.17; *LDH-4*, 0.02; *LDH-5*, 0.02; *LGG*, 0.16; *MDH-3,4*, 0.11; *PMI*, 0.57; and *SOD*, 0.47.) The 14 potentially contributing populations are frequently referred to as baseline populations, and the allele frequency estimates for these populations are referred to as baseline data.

The genetic relationships among the 14 potentially contributing populations are illustrated in the form of a dendrogram in Fig. 10.3. The two major groupings represent populations above and below Bonneville Dam (i.e., clusters A, B, and C from above the dam and cluster D from below.

Mixtures of fish from the 14 potentially contributing (baseline) populations were made in varying proportions (different samples were used for generating baseline data and for constructing populations mixtures). The stock mixtures were produced as blind samples, assembled by persons other than those involved in the collection and analyses of the individual and mixed stock data (Milner et al. 1981).

Maximum likelihood estimates (EM algorithm) of the relative contributions from each of the potentially contributing (baseline) populations were ob-

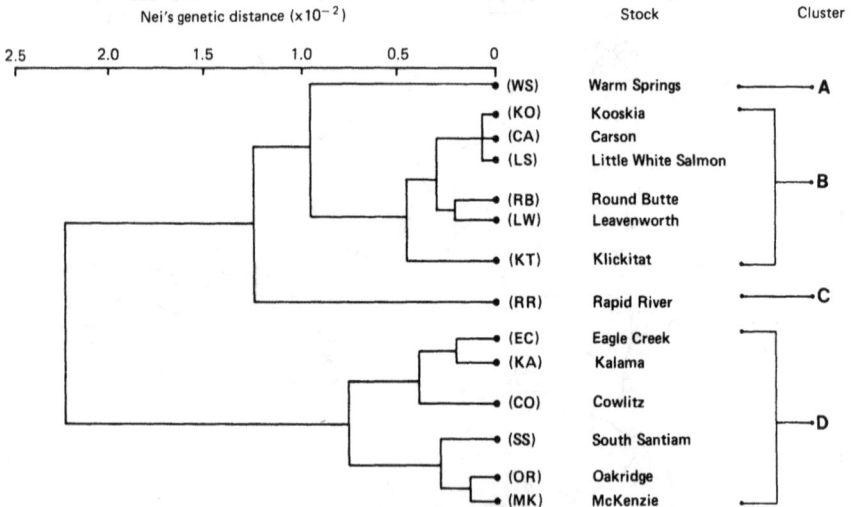

Fig. 10.3 Dendrogram (UPGMA) method (Sneath and Sokal 1973) summarizing genetic relationships (estimated from nine polymorphic loci) among 14 hatchery stocks of chinook salmon used in blind test of mixed stock composition (see Figure 10.2). The clusters A–D designate the four major branches of the dendrogram.

tained by comparing the genotypic distributions of each population with those of the mixture. In this particular case, the genotypic distributions of each of the baseline populations were estimated from the allele frequencies assuming random combination of alleles and loci (Hardy-Weinberg and gametic phase equilibria; Milner et al. 1981). The estimates obtained in one such stock mixture of 1,504 fish are summarized in Table 10.1. (These data are also contained in Tables 10.2 and 10.3 in comparisons evaluating bias and precision based on numerical studies of the next section.)

Although the baseline populations are not very different genetically, there is generally a very good correspondence between the actual and the estimated contribution to the mixture (Table 10.1). Not surprisingly, accuracy of estimates does not deteriorate when combining groups of populations which are genetically similar, such as those above and below Bonneville Dam (the bottom lines of Table 10.3). Even without combining populations into groups, all the major contributing populations are identified as making substantial contributions. Likewise, most stocks absent are indicated as providing small (<1%) or no contributions. It should be noticed, however, that apparent discrepancies occur between actual and estimated contributions. Particularly evident are the indicated presence of stocks that are actually absent (over 18% of the total mixture) and, consequently, the general underestimation of contributions of stocks present in the mixture.

Table 10.1 Actual and estimated contributions (%) to blind test mixture of 14 hatchery stocks of chinook salmon (cf. Figs. 10.2 and 10.3). The estimates were computed using the EM algorithm. SE = standard error.

Stock/ Group of Stocks	Genetic Cluster	Contribution to Mixture		
		Actual	Estimated	SE
Warm Springs (WS)	A	5.8	1.7	1.5
Kooskia (KO)	B	29.2	21.6	7.5
Carson (CA)	B	10.3	17.7	4.1
Little White Salmon (LS)	B	0	0.4	5.5
Round Butte (RB)	B	0	0.4	2.9
Leavenworth (LW)	B	0	0.8	4.2
Klickitat (KT)	B	0	12.3	4.8
Rapid River (RR)	C	30.3	27.6	0.9
Eagle Creek (EC)	D	0	0.0	1.6
Kalama (KA)	D	0	0.3	1.9
Cowlitz (CO)	D	12.2	4.9	2.2
South Santiam (SS)	D	0	4.0	2.4
Oakridge (OR)	D	12.2	8.3	1.9
McKenzie (MK)	D	0	0.0	3.3
Total		100.0	100.0	
Total for stocks in cluster B		39.5	53.2	3.0
Total for stocks in clusters A and B		45.3	54.9	2.5
Total for stocks in clusters A, B, and C (above Bonneville Dam)		75.6	82.5	2.6
Total for stocks in cluster D (below Bonneville Dam)		24.4	17.5	2.6

Table 10.2 Actual stock contribution (proportion) of Columbia River test sample chinook salmon and corresponding estimates[1,2] (including measures of variability[3]) for three analyses: (1) observed stock and mixture samples (OBSERVED); (2) bootstrapped stock samples and simulated resampling of the mixture (SIMULATE); and (3) bootstrapped stock and mixture samples (BOOTSTRAP).

Stock	Actual Contribution	OBSERVED Estimate	SE	SIMULATE Average	SD	BOOTSTRAP Average	SD
Warm Springs (WS)	0.058	0.017	0.0153	0.052	0.0216	0.018	0.0158
Kooskia (KO)	0.292	.216	.0747	.191	.0724	.200	.0814
Carson (CA)	0.103	.177	.0411	.135	.0617	.175	.0761
Little White Salmon (LS)	0	.004	.0555	.057	.0414	.013	.0158
Round Butte (RB)	0	.004	.0285	.015	.0163	.009	.0128
Leavenworth (LW)	0	.008	.0423	.005	.0076	.017	.0186
Klickitat ( )	0	.123	.0480	.030	.0262	.105	.0511
Rapid River (RR)	0.303	.276	.0092	.279	.0316	.275	.0356
Eagle Creek (EC)	0	.000	.0164	.008	.0105	.000	.0000
Kalama (KA)	0	.003	.0189	.010	.0131	.011	.0134
Cowlitz (CO)	0.122	.049	.0218	.091	.0247	.052	.0276
South Santiam (SS)	0	.040	.0243	.013	.0133	.046	.0254
Oakridge (OR)	0.122	.083	.0189	.107	.0211	.076	.0253
McKenzie (MK)	0	0.000	0.0331	0.008	0.0099	0.001	0.0036

[1]The EM algorithm searched for the maximum likelihood solutions until 100 iterations had occurred or until some iteration was reached at which none of the estimates changed by more than 10^{-6}. The initial point of search was that at which all stocks contributed equally.

[2]Average estimates across 100 resamplings are reported except the point estimate under OBSERVED.

[3]Conditional asymptotic standard error (SE) is reported for OBSERVED; otherwise, standard deviation (SD) of estimates across resamplings is reported.

Table 10.3 Actual stock contribution (proportion) of aggregates of Columbia River stocks (see Fig. 10.3) and corresponding estimates[1] from three analyses: (1) observed stock and mixture samples (OBSERVED); (2) bootstrapped stock samples and simulated resampling of the mixture (SIMULATE); (3) bootstrapped stock and mixture samples (BOOTSTRAP).

Aggregates of stocks and clusters	Actual contribution estimate	OBSERVED Estimate	SD	SIMULATE Estimate	SD	BOOTSTRAP Estimate	SD
A	0.058	0.017	0.0153	0.052	0.0216	0.018	0.0158
B	0.395	0.532	0.0301	0.432	0.0511	0.519	0.0449
A, B	0.453	0.549	0.0247	0.484	0.0403	0.537	0.0408
C	0.303	0.276	0.0092	0.279	0.0316	0.275	0.0356
A, B, C	0.756	0.825	0.0261	0.763	0.0241	0.813	0.0290
D	0.244	0.175	0.0261	0.237	0.0241	0.187	0.0290

[1]Average estimates across 100 resamplings and standard deviation (SD) of estimates are reported except for point estimate under OBSERVED where conditional asymptotic standard error (SE) is provided.

What, then, is required for obtaining accurate estimates of mixed stock compositions from genetic data? A single example, of course, is insufficient to evaluate a statistical estimation procedure. The numerous factors that must be considered in each particular situation preclude a simple answer to the question. First, the accuracy of the estimates is highly dependent on the number of potentially contributing stocks, the amount of genetic divergence among them, and whether all the potentially contributing populations are known. Second, the precision is dependent on the actual composition of the mixture. Third, the precision of the final composition estimates is itself dependent on the precision and the reliability of the underlying estimate of the genotypic distributions in the mixture as well as in the contributing populations.

In spite of the numerous factors affecting the estimates of mixed stock compositions based on genetic data, the procedures applied to date appear sufficiently robust to provide estimates that are reliable enough for practical use in many situations. Such analyses have, in fact, been evaluated and used in a number of recent studies (Grant et al. 1980, Fournier et al. 1984, Beacham et al. 1985a, 1985b, Milner et al. 1985). The information can also be obtained at a modest cost, particularly after the baseline data have been collected (see Milner et al. 1985).

STATISTICAL THEORY AND APPLICATION

In using genotypic frequencies to assess composition of a mixture, the approach we emphasize presumes that a complete list of potential contributing stocks is available. If the only useful information available is the genotypic composition of a sample from the mixture, assessment of composition and stock genotypic frequencies must be done by latent structure analysis (e.g., Goodman 1974, Everitt 1984). In simple situations of few contributory stocks, such an analysis may be successful (e.g., Makela and Richardson, 1977). However, precision improves greatly when information is available on the separate stocks which may contribute as well as baseline samples for genotypic frequencies from each. In practice, a truly complete stock list may be a rarity; checks for adequacy of the sampling models used may then be of some consolation. Nevertheless, research information external to baseline sampling is advisable in determining the list; for example, mixture individuals might be marked and monitored later when their origin can be identified.

Applications of methods described require common genotypic frequencies for members of a stock whether they are discrete or in the mixture. The individuals of a stock in both the mixture sample and the baseline sample are presumed to be drawn from a common frequency distribution of genotypes. Any cause or event that alters underlying stock genotypic frequencies between mixture and baseline samplings would make the composition estimates by the following approaches of questionable value.

General Approaches

Two general approaches have been used to resolve composition of a mixture of stocks from genotypic frequencies: classification and direct maximum likelihood. Actually, both approaches are based on the maximum likelihood method (Millar, in press), but the classification approach applies the estimation criterion after individuals in the mixture have been assigned to the contributing stocks by a rule developed from the baseline samples. The direct maximum likelihood approach does not identify stocks of individuals in the mixture sample, but finds the stock composition for which the observed genotypic frequencies would be most probable.

Classification. The classification approach is appealingly simple and computationally frugal. The frequency of occurrence within each stock of each genotype in the mixture sample can be estimated from the stock samples. Assuming we do not have prior estimates of contributions of the stocks to the mixture, we would assign fish of each genotype in the mixture to the stock in which that genotype is most frequent. The relatively complex observation representing any genotype in the mixture sample is reduced to a simple index variable representing the stocks. Such a classification is the Bayes minimum error rule (e.g., Hand 1981) if cost of misclassification is equal among stocks. The apparent composition resulting from this classification of the mixture sample becomes biased by misclassification. Corrections for such bias were described by Odell and Basu (1976) and independently by Cook and Lord (1978). Pella and Robertson (1979) evaluate statistical bias and precision of adjusted estimates. Millar (in press) amends the composition estimates, ensuring that only feasible (non-negative) values occur.

Our experience with the classification approach is limited, but we found it unsatisfactory because some genotypes are frequent relative to most other genotypes in most contributing stocks. The classification rule, which assigns individuals in the mixture sample to stocks, does so by partitioning all possible genotypes into disjoint subsets, each of which is classified to a particular stock. The genotypes frequent in most contributing stocks will be common in the mixture, and each will be classified to the stock associated with its partition regardless of the true composition of the mixture. Serious bias of apparent composition can result. Still, if classification is reasonably accurate in any application, reduction of bias from misclassification and evaluation of precision of composition estimates are straightforward (see Pella and Robertson 1979 and Millar, in press).

Fournier et al. (1984) report that their use of the classification approach achieved a rough agreement with maximum likelihood estimates (next discussed) from the same set of electrophoretic samples. The observed genotypes in samples from the stocks and mixture were coded according to Smouse et al. (1982), so the observations on each individual formed a discrete-valued vector. Multivariate discriminant analysis, with classification based on minimum Mahalanobis distance (BMDP7M) (Dixon et al. 1983), was used to assign mixture individuals to stocks. This process of analysis approximates the simpler rule we gave earlier in that both rules classify a fish in the mixture to that stock in which its genotype is estimated to be most frequent.

The bias of apparent composition from classification would be reduced or eliminated directly if, instead of classifying all mixture individuals of a genotype to a single stock, they were allocated to each of the stocks in proportion to the probability that they originated from these stocks. The second approach to estimating mixture composition utilizes the maximum likelihood criterion applied directly to the likelihood function (Edwards 1972) for the genotypic frequencies in the baseline and mixture samples. We shall see that one numerical algorithm utilized to find the maximum likelihood composition estimates from genotypic frequencies includes such an allocation.

Maximum likelihood. The maximum likelihood estimate of stock composition is that which maximizes (together with the associated maximum likelihood estimates of genotypic frequencies of the contributing stocks) the joint probability of the observed genotypic frequencies in the mixture sample and baseline samples. Maximization is accomplished by operations on the likelihood function for the genotypic frequencies in the samples, which is proportional to their joint probability; and which views them as fixed and the unknown parameters—composition and underlying relative genotypic frequencies in the stocks—as variable.

The multinomial distribution is commonly used to describe the sampling of genotypes from the stocks and mixture and will be used to clarify terms above. In the abstract situation, the multinomial generalizes the binomial distribution for the case of n repeated independent trials, but where each trial can have one of several (two in the case of the binomial) outcomes. Let the possible outcomes of each trial be $E_1, E_2, ..., E_r$ and suppose the probability of the occurrence of E_i at each trial is γ_i ($i = 1, 2, ..., r$). The probability that in n trials E_1 occurs k_1 times, E_2 occurs k_2 times, and so on, is

$$P = \frac{n!}{k_1!k_2! \ ... \ k_r!} \ \gamma_1^{k_1} \ \gamma_2^{k_2} \ ... \ \gamma_r^{k_r} \qquad 0 \le \gamma_i \le 1$$

$$\sum_{i=1}^{r} \gamma_i = 1 , \qquad (1)$$

where

$$n = \sum_{i=1}^{r} k_i .$$

The probability function P at (1) provides the probability of any set of outcomes $k_1, k_2, ..., k_r$ when $\gamma_1, \gamma_2, ...,\gamma_r$ are known. In this sense, the k_i are variable and γ_i are fixed. On the other hand, if a sample $k_1, k_2, ..., k_r$ is available and the γ_1, γ_2, ...,γ_r are unknown, the roles become reversed when estimating γ_1, γ_2, ..., γ_r. The maximum likelihood estimates of γ_1, γ_2, ..., γ_r *are those values* $\hat{\gamma}_1$, $\hat{\gamma}_2$, ..., $\hat{\gamma}_r$ at which P is maximal given k_1, k_2, ..., k_r. Equivalently, the maximum likelihood estimates maximize the likelihood function,

$$L = \gamma_1^{k_1}\, \gamma_2^{k_2}\, \ldots\, \gamma_r^{k_r}\,,$$

which differs from P by only a constant coefficient (viewing γ_1, γ_2, ..., γ_r as the only variables). The maximum likelihood estimates can be derived from the necessary conditions for an extremum,

$$\frac{\partial L}{\partial \gamma_i} = 0\,,$$

and can be explicitly expressed as

$$\hat{\gamma}_i = k_i/n \qquad (i = 1, \ldots, r)\,.$$

Consider the situation in which simple random samples of n_1, n_2, ..., n_c fish are taken from c separate stocks contributing proportions p_1, p_2, ..., p_c of a mixture. A simple random sample of m fish is also taken from the mixture. The same loci are assayed from each fish and so its multiloci genotype is determined. Let $\mathbf{X}$ be a discrete-valued vector representing the genotype of an individual. For example, the ith coordinate of $\mathbf{X}$ could correspond to the ith locus and the corresponding single locus genotype could be represented by an integer at this coordinate. Let the value $g\,(\mathbf{x};\boldsymbol{\theta}_i)$ *be the relative frequency in the ith stock of the genotype represented by* $\mathbf{X} = \mathbf{x}$. (The function g is called the genetic model hereafter, and $\boldsymbol{\theta}_i$ is the vector of genetic parameters for the ith stock.) The probabilities of the observed genotypic frequencies in each sample can be approximated well by the multinomial probability function. Multiloci genotypes replace the abstract possible outcomes, and sampled fish correspond to trials. Probabilities of multiloci genotypes in baseline samples from the stocks are given by the genetic model, and in the mixture sample by sums of these probabilities weighted by corresponding stock proportions p_1, p_2, ..., p_c.

The joint probability of the genotypic frequencies in the baseline and mixture samples is just the product of the probabilites of each sample because the samples are drawn independently. Therefore, the corresponding likelihood function for the genotypic frequencies in the baseline and mixture samples is:

$$
\begin{aligned}
L &= \left[\prod_{i=1}^{c}(\gamma_{i1}^{k_{i1}}\,\gamma_{i2}^{k_{i2}}\,\cdots\,\gamma_{ir(i)}^{k_{ir(i)}})\right][\lambda_1^{u_1}\lambda_2^{u_2}\,\ldots,\,\lambda_s^{u_s}] \\
&= \left[\prod_{i=1}^{c}\prod_{j=1}^{n_i} g(\mathbf{x}_{ij};\boldsymbol{\theta}_i)\right]\left[\prod_{h=1}^{m}\sum_{i=1}^{c} p_i g(\mathbf{x}_h;\boldsymbol{\theta}_i)\right]\,,
\end{aligned}
\tag{2}
$$

where $k_{i1}, k_{i2}, \ldots, k_{ir(i)}$ are the numbers of fish (greater than zero) of each of the $r(i)$ genotypes actually observed in the sample from the ith stock; $u_1, u_2, \ldots, u_s$ are the numbers of fish (greater than zero) of each of the s genotypes actually observed in the sample from the mixture; $\gamma_{i1}, \gamma_{i2}, \ldots, \gamma_{ir(i)}$ are the relative frequencies in the ith stock of each of the $r(i)$ genotypes observed in its sample; $\lambda_1, \lambda_2, \ldots, \lambda_s$ are the relative frequencies in the mixture of the s genotypes observed in its sample; and $\mathbf{x}_{ij}$ and $\mathbf{x}_h$ are specific values of X for the jth fish in the sample from stock i, and hth fish from the mixture sample, respectively.

At Eq. (2), $g(\mathbf{x}_{ij};\boldsymbol{\theta}_i)$ represents the relative frequency in the ith stock of the genotype of the jth fish in the sample from that stock. The relative frequencies in the mixture of the s genotypes observed in its sample are weighted relative frequencies of the corresponding genotypes in the stocks, with weights equal to the stock proportions of the mixture. Specifically, the relative frequency in the mixture of the genotype of the hth fish in the mixture sample is the weighted relative frequency,

$$\sum_{i=1}^{c} p_i\, g(\mathbf{x}_h;\boldsymbol{\theta}_i) \ .$$

Commonly, many possible genotypes will not be observed in some or all of the samples. Such genotypes appear to be omitted from the likelihood because their associated relative frequencies raised to the zero power equal one. However, their existence is accounted for in that the relative frequencies of the observed genotypes will sum to less than one, i.e.,

$$\sum_{j=1}^{r(i)} \gamma_{ij} \leq 1 \qquad (i = 1, 2, \ldots, c)$$

and

$$\sum_{i=1}^{s} \gamma_i \leq 1 \ .$$

The likelihood function for situations in which subsets of loci are assayed on some fish (e.g., missing data) is only more complex notationally. Then the likelihood function can be written as a product of factors like that above, each corresponding to fish on which the same subset of loci was observed.

Let us consider a specific genetic model g. Suppose that genetic variation is assayed at L loci, each with two alleles; that Hardy-Weinberg conditions apply at each locus; and that the loci are statistically independent (i.e., linkage disequilibrium is absent). Let $\boldsymbol{\theta}_{il}$ be the relative frequency for stock i of one of the alleles at the lth locus. Define x_{ijl} and x_{hl} to be the observed number of this allele present at the lth locus of the jth fish of the samples from stock i and the hth fish in the mixture sample, respectively. Then, for example, the hth fish of the mixture sample is represented by $\mathbf{x}_h = (x_{h1}, x_{h2}, ..., x_{hL})$. Define for the mixture sample the variable

$$\delta_{hl} = \begin{cases} 1, & \text{if } x_{hl} = 0, 2 \ ; \\ 0, & \text{if } x_{hl} = 1 \ . \end{cases}$$

Then the genetic model for the hth fish in the mixture sample if it originated in stock i is

$$g(\mathbf{x}_h; \boldsymbol{\theta}_i) = \prod_{l=1}^{L} 2^{1-\delta_{hl}} \theta_{il}^{x_{hl}} (1 - \theta_{il})^{2-x_{hl}} \ .$$

An analogous expression for the jth fish in the sample from stock i requires only replacing $\mathbf{x}_h$ by $\mathbf{x}_{ij}$ δ_{h1} by δ_{ij1}, and x_{h1} by x_{ij1} in this expression.

The researcher assessing composition of a mixture must provide the genetic model g describing genotypic frequencies of the stocks, keeping in mind the advice of Fournier et al. (1984) to introduce no more parameters than needed. In the above example, L parameters, $\theta_{i1}, \theta_{i2}, ..., \theta_{iL}$, describe the relative genotypic frequencies in each stock. Total possible number of genotypes (distinct $\mathbf{X}$ vectors) are 3^L. If no structure were provided, a total of $3^L - 1$ parameters would be needed to describe the relative genotypic frequencies. With L large and stock or mixed sample size fixed, most genotypes will be unobserved; the corresponding relative frequencies under the structureless model would be estimated as zero. As a result, reduced precision of estimates of relative genotypic frequencies would degrade precision of composition estimates. Clearly, the structureless nonparametric model for relative genotypic frequencies is usually unsatisfactory and a genetic model that is simpler (in terms of number of parameters) must be found.

Genetic models become cumbersome when loci are not independent (e.g., Crow and Kimura 1970, Roughgarden 1979). The number of parameters increases as well as complexity of expressions for relative genotypic frequencies. Fortunately, the biochemical genetic loci employed usually seem to be independent and Hardy-Weinberg models to hold; if so, the genetic functions are simpler. Nonetheless, researchers applying this technique must be alert to deviations from simple functions. Complexity of genetic functions presents no special difficulties in calculations of composition estimates, as high-speed computers and flexible programs are required in any event.

Composition Estimation by Maximum Likelihood

An early demonstration of feasibility of estimating composition of a two-population mixture from genotypic frequencies by the maximum likelihood method considered only information from a mixture sample (Makela and Richardson 1977). The situation is like that underlying latent class analysis (e.g., Everitt 1984). The likelihood function is obtained from (2) by omitting the baseline samples; i.e.,

$$L' = \prod_{h=1}^{m} \sum_{i=1}^{c} p_i \, g(\mathbf{x}_h; \boldsymbol{\theta}_i) \, . \tag{3}$$

Maximization of the likelihood function was accomplished with regard to both composition and parameters of the genetic model. Explicit expressions for the maximizing values of the parameters could not be obtained. Instead, a multidimensional grid search over the parameter space was used with the likelihood function evaluated and compared at points in the search path. Although local maxima of the likelihood function at the boundary of the feasible parameter space ($p_j = 0$) can occur (Goodman 1974), multiple maxima were not reported.

Milner et al. (1981) expanded the scope of composition estimation by considering an arbitrary number of stocks in the mixture. Baseline samples were available and were considered sufficiently large that relative genotypic frequencies in the stocks were assumed known; these were estimated by maximizing the likelihood functions for the baseline samples; i.e.,

$$L_i'' = \prod_{j=1}^{n_i} g(\mathbf{x}_{ij}; \boldsymbol{\theta}_i) \qquad (i = 1, \, 2, \, \ldots, \, c) \, .$$

The genetic model g chosen to describe relative genotypic frequencies was essentially the example described earlier based on independence of loci and Hardy-Weinberg equilibrium, but with arbitrary numbers of alleles at the loci. Maximum likelihood estimates of relative allelic frequencies (parameters of the genetic model) are the observed relative allelic frequencies in the baseline samples. Relative genotypic frequencies are estimated by substituting estimates of relative allelic frequencies into the genetic model.

Milner et al. (1981) chose the conditional maximum likelihood estimate for composition with parameters of the genetic model fixed. The conditional likelihood function is the one at (3) with the $\boldsymbol{\theta}_i$ set equal to estimates obtained from the baseline samples. Fournier et al. (1984) and Millar (1985) show that the conditional likelihood function in ordinary applications will have a unique maximum. In particular, Millar develops a condition related to identifiability (discussed later) which, if met by the observed genotypes in the mixture sample, is sufficient to ensure this uniqueness.

Numerical algorithms. Procedures to compute conditional maximum likelihood estimates are available. Each method begins a search from an initial guess of composition of the mixture and finds a sequence of improved guesses with increasing likelihood which converge to the maximum likelihood estimate. The first two procedures we discuss in detail, the EM algorithm and parameter transformation, have been in use for several years, so their suitability is proven through experience. A third approach, fitting expectations by iteratively re-weighted least squares, has only recently been developed (Pella, in preparation); judgment of its practicability awaits further use. However, the new method provides further insight into the estimation problem.

The EM algorithm is developed by setting to zero the first derivatives of the conditional likelihood function with respect to the composition parameters (the necessary condition for the maximizing point). One finds that the following equations result:

$$\hat{p}_i = \frac{1}{m} \sum_{h=1}^{m} \hat{P}(i|\mathbf{x}_h) \qquad (i = 1, 2, \ldots, c), \qquad (4)$$

where

$$\hat{P}(i|\mathbf{x}_h) = \frac{\hat{p}_i g(\mathbf{x}_h; \boldsymbol{\theta}_i)}{\sum_{s=1}^{c} \hat{p}_s g(\mathbf{x}_h; \boldsymbol{\theta}_s)} . \qquad (5)$$

Here $\hat{P}(i|\mathbf{x}_h)$ is the estimated posterior probability from Bayes's formula that, given the genotype of the hth fish of the mixture is $\mathbf{x}_h$, the fish came from the ith stock. Equation (4) allocates individuals of a genotype to the stocks in proportion to an estimated probability they arose from the stocks.

Although an explicit solution for $\hat{p}_i$ ($i = 1, 2, \ldots, c$) is not possible, Eqs. (4) and (5) are used iteratively to obtain the solution. Initial guesses of $\hat{p}_i$ are used in (5) to evaluate $\hat{P}(i|\mathbf{x}_h)$; the latter are substituted into (4) to obtain new values for $\hat{p}_i$. These improved estimates of p_i are inserted into (5), and the process is repeated until some arbitrary criterion of convergence is achieved. The procedure is a particular application of the EM algorithm (Dempster et al. 1977) developed by Milner et al. (1981).

The EM algorithm produces a sequence of feasible estimates (positive estimates summing to one) such that the likelihood function is nondecreasing during the search (Dempster et al. 1977). Therefore, EM estimates are constrained in the sense that only feasible points occur during the search provided one begins at a feasible point. The algorithm can be slow to converge to the desired estimates. Furthermore, final estimates and amount of computing time to find them may be sensitive to the convergence criterion. If they are, trials using increasingly severe tests of convergence are advisable, with independence of final estimates on the criterion being the goal. Fortunately, the algorithm is simple to implement on a computer and easily handles varying numbers of loci assayed among fish.

The search for conditional maximum likelihood estimates of composition also can be performed using computer programs based on one of the many optimization methods that have been developed (e.g., Rao 1984). In contrast to the EM algorithm, these optimization methods may have no guarantee that the likelihood function will not decrease between points of the search for the maximizing estimates. As a result, unless the procedure is modified, the search could produce feasible estimates that are farther from the maximum likelihood estimate than the initial guess. Further, algorithms which allow stock contributions to equal zero (exactly) during the search may have to be modified to prevent observed genotypes in the mixture sample from becoming impossible at points in the search. The EM algorithm avoids this pitfall because the search includes a boundary ($p_j = 0$) if and only if the search begins at the boundary. Otherwise, composition estimates can approach arbitrarily near zero. Finally, many algorithms will not constrain the search to feasible points. A parameter transformation can be used to constrain the search both to feasible points and away from boundaries ($p_j = 0$).

Maximization of the conditional likelihood function (3) subject to the constraints

$$\sum_{j=1}^{c} p_j = 1$$

and

$$p_j > 0$$

can be accomplished by methods for unconstrained optimization. Fournier et al. (1984) eliminate the constraints by parameter transformation, substituting

$$p_j = \exp(\mu_j) / \sum_{k=1}^{c} \exp(\mu_k) \qquad j = 1, \ldots, c$$

into the conditional likelihood function (3). However, the μ_j are not uniquely determined by the reparameterization because substituting $\mu_j + b$ for μ_j using any constant b leaves p_j unchanged. To remove this degeneracy, they maximize

$$L' - \left[\log_e \left(\sum_{k=1}^{c} \exp(\mu_k) \right) \right]^2$$

$$= \prod_{h=1}^{m} \sum_{i=1}^{c} \left[\exp(\mu_i) / \sum_{k=1}^{c} \exp(\mu_k) \right] g(\mathbf{x}_h; \boldsymbol{\theta}_i) - \left[\log_e \left(\sum_{k=1}^{c} \exp(\mu_k) \right) \right]^2$$

The μ_j ($j = 1, \ldots, c$) which maximize this expression also maximize L' at (3) and satisfy the arbitrary condition

$$\sum_{k=1}^{c} \exp(\mu_k) = 1.$$

The maximum likelihood estimates of p_j ($j=1, ..., c$) are computed from these μ_j ($j=1, ..., c$) using the defining equation of the reparameterization above. In their application, Fournier et al. (1984) used a computer optimization routine based on a quasi-Newton method (Rao 1984) to maximize the conditional likelihood function by this parameter tansformation technique.

Finally, the new approach to finding the conditional maximum likelihood estimates is by use of the general procedure of fitting expectations by iteratively reweighted least squares (Charnes et al. 1976, Jennrich and Rahlston 1979, Green 1984). Details of this application of the general method as well as a computer program to perform the calculations are provided by Pella (in preparation). A brief outline of the procedure is included both for completeness as well as for the insight the method provides of the estimation problem.

Consider a complete list of the observed genotypes (say H genotypes) in the mixture sample. The number of individuals in the mixture sample corresponding to genotype h is m_h, $h=1, 2, ..., H$. The remaining unobserved genotypes (if any) will be pooled to form an $(H+1)$st category with $m_{H+1}=0$. The probability of occurrence of the genotype h ($h=1, 2, ..., H$) when a fish is randomly taken from the mixture is

$$\lambda_h = \sum_{i=1}^{c} p_i g(\mathbf{z}_h; \boldsymbol{\theta}_i) , \tag{6}$$

where $\mathbf{z}_h$ is the discrete valued vector representing genotype h. The probability of occurrence of one of the genotypes in the unobserved category when a fish is randomly taken from the mixture is

$$\lambda_{H+1} = 1 - \sum_{h=1}^{H} \lambda_h .$$

The equation (6) can be rewritten to eliminate redundancy due to the constraint that the p_i sum to one:

$$\lambda_h - g(\mathbf{z}_h; \boldsymbol{\theta}_c) = \sum_{i=1}^{c-1} d_{hi} p_i . \tag{7}$$

where

$$d_{hi} = g(\mathbf{z}_h; \boldsymbol{\theta}_i) - g(\mathbf{z}_h; \boldsymbol{\theta}_c) \qquad i = 1, 2, ..., c-1 .$$

The method of fitting expectations is applied to the sample analogue of the linear model (7),

$$y_h = \sum_{i=1}^{c-1} d_{hi}\, p_i + e_h \qquad h = 1, 2, ..., H$$

$$y_{H+1} = \sum_{i=1}^{c-1} \left(\sum_{h=1}^{H} d_{hi} \right) p_i + e_{H+1} , \tag{8}$$

where

$$y_h = \frac{m_h}{m} - g(\mathbf{z}_h; \boldsymbol{\theta}_c) , \quad h = 1, 2, \cdots, H$$

$$y_{H+1} = 1 - \sum_{h=1}^{H} g(\mathbf{z}_h; \boldsymbol{\theta}_c) ,$$

e_h is the residual between the observed and expected values of y_h, and m, d_{hi}, and p_i were defined earlier. The estimates of parameter $\mathbf{p} = (p_1, p_2, \ldots, p_c)$ are iteratively computed by constrained ($p_i \geq 0$) weighted least squares. The estimates can be constrained to the feasible space by utilizing Wolfe's (1959) modification of the simplex algorithm for linear programing as applied to quadratic programing problems (see Rao 1984).

The weights for the y_h are given by

$$r_h = \frac{m}{\lambda_h} \qquad h = 1, 2, \ldots, H + 1 . \tag{9}$$

At each step the weights are computed anew from the present estimates of the parameters $\mathbf{p}$ inserted into λ_h. Revised constrained ($p_i \geq 0$) weighted least squares estimates of $\mathbf{p}$ are computed using the new weights. The process is continued to convergence, and the result, $\hat{\mathbf{p}}$, is the maximum likelihood estimate.

If the mixture sample contains fish for which varying numbers of loci were examined, the sample must be partitioned into classes of fish such that the same loci and only those loci were examined on each fish of a class. A set of equations (8) and weights (9) corresponding to each class is developed. The iterative fitting process goes on as described earlier, but now it is applied simultaneously to the sets of equations (8) of the classes.

The method of fitting expectations is both an iteratively reweighted Gauss-Newton algorithm and the classical Fisher scoring algorithm. As such, the procedure is not guaranteed to converge to the conditional maximum likelihood estimates. Wild oscillations can occur from one iteration to another because of overshooting (projected maximizing point at an iteration is in a direction for which likelihood increases, but too far from present point) when the value of the conditional likelihood decreases between iterations. The procedure is easily modified to prevent overshooting, however.

We applied the method of fitting expectations to the Columbia River test sample that had been analyzed earlier using the EM algorithm. The new method was more efficient as judged by number of iterations required to find composition estimates with likelihood values as great as those found by the EM algorithm. Both procedures were provided an initial guess of equal contributions by the 14 stocks. The likelihood value of the composition estimate at the seventh iteration by the new method exceeded that from 100 iterations by the EM algorithm. The EM algorithm required more than 500 additional iterations to increase the likelihood value by approximately 0.01%; at that point, not one of the composition estimates for a stock changed in the sixth decimal place between successive iterations. The method of fitting expectations required four more iterations for this increase in likelihood. Composition estimates for all stocks by the two algorithms agreed to at least three decimal places at this stage. The method of fitting expectations requires more computation per iteration, so computer time and cost are factors that must be considered.

Precision of estimates. Evaluation of precision of composition estimates is essential in assessing the mixture composition. The probability distribution of the nonredundant elements of the conditional maximum likelihood estimator, say

$$(\hat{p}_1, \hat{p}_2, \ldots, \hat{p}_{c-1}),$$

is well approximated by the multivariate normal distribution provided the following requirements are met:
- Genotypic frequencies in the stocks are known without error;
- Actual composition **p** is in the interior of the feasible parameter space, so additional estimates from potential mixture samples of the size observed would not encounter boundaries of the feasible sample space; and
- Sample size from the mixture is large.

The mean of the distribution of such potential estimates equals the actual composition, and the covariance matrix V is

$$V = T^{-1} : \ (c-1) \times (c-1),$$

where

$$t_{ij} = -E\frac{\partial^2 \log L}{\partial p_i \partial p_j} = M \sum_{\mathbf{x}_s \in S} \frac{[g(\mathbf{x}_s; \boldsymbol{\theta}_i) - g(\mathbf{x}_s; \boldsymbol{\theta}_c)][g(\mathbf{x}_s; \boldsymbol{\theta}_j) - g(\mathbf{x}_s; \boldsymbol{\theta}_c)]}{\sum_{k=1}^{c} p_k \ g(\mathbf{x}_s; \boldsymbol{\theta}_k)}$$

and the sum is over the entire set S of possible genotypes represented by values $\mathbf{x}_s$ taken on by $\mathbf{X}$.

Milner et al. (1981) develop the estimate of the covariance matrix, $\hat{V}$, from the inverse of the observed information matrix, $\hat{T}$, i.e.,

$$\hat{V} = \hat{T}^{-1} : \ (c-1) \times (c-1),$$

where

$$t_{ij} = -\frac{\partial^2 \log L}{\partial p_i \partial p_j} = \sum_{h=1}^{m} = \frac{[g(\mathbf{x}_h; \boldsymbol{\theta}_i) - g(\mathbf{x}_h; \boldsymbol{\theta}_c)][g(\mathbf{x}_h; \boldsymbol{\theta}_j) - g(\mathbf{x}_h; \boldsymbol{\theta}_c)]}{[\sum_{k=1}^{c} \hat{p}_k g(\mathbf{x}_h; \boldsymbol{\theta}_k)]^2};$$

and now the sum is over the fish of the mixture sample. Confidence statements for composition can be developed from normal theory using the estimates of composition and of the covariance matrix (e.g., see Pella and Robertson 1979). Naturally, neither of the first two conditions is ordinarily completely met, and so such confidence statements must be treated with circumspection.

A resampling (Efron 1981, 1982) approach has been recommended by Fournier et al. (1984) to account for effects of inequality constraints as well as baseline and mixture sampling variation. Composition estimates are computed repeatedly from resampled observations. Samples equal in size to those available are drawn at random with replacement from the original stock and mixture samples. Each resampled composition estimate maximizes the likelihood function conditional on the genotypic frequencies of the stocks equaling those from resampling. Nonparametric confidence intervals can be constructed using the percentile method, for example (see Efron 1982). This procedure provides a realistic evaluation of precision of composition estimates, but the amount of computing is considerable. Nonetheless, computing power continues to increase while cost decreases, making the resampling approach continually more feasible and attractive.

If the resampling approach is prohibitive in computing cost for some application, the conditional covariance matrix of composition estimates provides a rough indication of precision via normal theory. We have already noted the estimate of this matrix from asymptotic theory provided by Milner et al. (1981). However, Fournier et al. (1984) note that this estimate of V is inappropriate when the conditional maximum likelihood estimate lies on or near the constraints, because the formulas do not apply to the estimator if it is influenced by inequality constraints (also see Bard 1974, pp. 180–181). If inequalities come into play and sampling errors of genotypic frequencies in the stocks are adequately contained, actual variances of composition estimates will be overestimated by the asymptotic formulas. For example, Millar (in press) noted a general overstatement of variances for the Columbia River chinook example, especially for estimated stock contributions near zero. Although this asymptotic approach may understate precision in some situations, precision may be overstated in other cases. Fournier et al. (1984) noted that the conditional asymptotic covariance matrix of Milner et al. (1981) gave optimistic estimates of precision in their simulation studies, probably because it assumed a complete knowledge of genotypic frequencies in the stocks.

Millar (in press) examined alternative approaches to evaluating the covariance matrix of composition estimates. His results indicate that the matrix is better approximated by the infinitesimal jackknife procedure (Efron 1982) than by the conditional asymptotic covariance matrix of Milner et al. (1981). Expressions for evaluating the infinitesimal jackknife estimate of the covariance matrix are provided by Millar (in press).

Bias. Some bias of composition estimates from constrained maximization of the conditional likelihood function can be expected when unconstrained estimates would fall in or out of the feasible parameter space on repeated samplings. For example, if a stock does not contribute, only by allowing estimates less than zero can the estimator on average estimate the true composition. Constrained estimation will overestimate contributions from absent and rare stocks of the mixture and underestimate contributions of abundant stocks with similar relative genotypic frequencies. Increase of size of mixture or stock samples will increase precision and so decrease bias by reducing the frequency with which unconstrained estimates would encounter constraints. Outcomes of simulation studies by Fournier et al. (1984) and Beacham et al. (1985b) illustrate this behavior.

Fournier et al. (1984) found that bias in composition estimates (absent stocks estimated as contributors and abundant stocks underestimated) is reduced when stock genotypic frequencies are known (equivalent to very large stock sample sizes) rather than estimated. Remaining bias (Fournier et al. 1984, Table 1) presumably occurs because of finite mixture samples: stock composition and genotypic frequencies in the mixture samples deviate from true values due to sampling variation. Therefore, composition estimates may again encounter constraints.

Beacham et al. (1985b, Fig. 4) summarized effect of size of mixture sample on bias in estimating the composition of stock aggregates (subsets of the contributing stocks). Bias of estimated contributions of rare or common stock aggregates is reduced when mixture sample size increases. Similarities between stocks from different aggregates or sampling errors of genotypic frequencies in stock samples apparently produce additional bias less or not affected by mixture sample size.

Biased composition estimates may be viewed as a transformation from the composition space

$$(0 \leq \theta_i \leq 1; \quad \theta_1 + \theta_2 + \cdots + \theta_c = 1)$$

into itself. Calibration experiments based on resampling known mixtures can be used to examine these transformations. Optimistically, a unique inverse transformation would exist by which the biased estimates can be corrected. Each biased composition estimate would correspond to a unique point in the composition space.

No attempt has been made to correct biased composition estimates for situations involving more than two stocks treated severally. However, when only two stocks or stock aggregates occur in the mixture, this approach is tractable and appealing because it removes all bias, whatever its cause. (If stock aggregates are considered, it is necessary to know the relative frequencies of stocks within stock aggregates occurring in the mixture.) Plots of average (over resampling experiments) estimated proportion versus actual proportion of either stock in simulated mixtures for a range of stock proportions provide the transformation and its inverse. Beacham et al. (1985b) were the first to use calibration experiments to correct for bias.

Instead of removing bias by calibration, potential for bias can be reduced if similar stocks are combined into aggregates, as was done earlier in the analysis of the Columbia River test sample. Millar (in press) noted the combining can be done before or after application of the maximum likelihood estimation procedure. If the combining is done beforehand, some measure of stock similarity (e.g., Nei 1972) is needed; if it is done after composition estimation, the estimate of the covariance matrix can be used to compute the correlation of estimates between stocks. Stocks with strong negative correlations would be combined. In either case, the estimates of composition in terms of such aggregates must have practical use.

If combining is done after application of the maximum likelihood procedure, composition estimates for aggregates are obtained simply by summing composition estimates for member stocks. If combining is done beforehand, composition estimates are provided directly from the maximum likelihood procedure. In that case, however, a special problem arises: the relative genotypic frequencies in the stock aggregates must be provided. Millar (in press) suggests that, when combining is done beforehand, these relative frequencies be computed as a weighted average of relative frequencies of member stocks, provided some prior or independent information is available on their relative abundance in the mixture.

The question of when stocks should be combined to reduce bias is an open one at present. Numerical studies of the type discussed next may help to make this decision for a particular application.

Columbia River test sample revisited. We have performed two resampling experiments on the Columbia River chinook sample to clarify and extend points made in the text: (1) bootstrap and (2) simulation studies. In both, random samples of alleles at each locus of each stock, equal in size to the original, were generated by sampling the original samples with replacement. In bootstrap studies, multiloci genotypes observed in the mixture sample were similarly resampled to obtain a random mixture sample.

In the simulation studies, resampling of the original mixture was imitated. First, multinomial variation in stock composition of the mixture sample was generated with expected composition equal to the actual composition reported earlier (Table 10.1, column 1). Second, multiloci genotypes of the resulting members of the mixture sample were generated by sampling (with replacement) pairs of alleles at each locus from the appropriate stock sample of each individual; underlying this process is implicit independence of loci and Hardy-Weinberg equilibrium.

Composition estimates were calculated via the EM algorithm after each resampling of stocks and mixture. Genotypes occurring in the mixture, which were theoretically impossible based on allele frequencies in resampled stock samples, were rare and were omitted from EM computations. The effect of such individuals will be negligible because of their small numbers relative to the size of the mixture sample.

We shall refer to estimates generated under the two forms of resampling as BOOTSTRAP and SIMULATE estimates. SIMULATE estimates represent our best approximation to the behavior of conditional maximum likelihood estimates upon repetition of complete sampling of stocks and mixture; underlying a mixture analysis, one such possible outcome would be postulated to be OBSERVED, the composition estimates from the actual stock and mixture samples. In practice, simulation studies cannot be used to analyze a mixture directly because the unknown composition is required for the computations; on the other hand, the bootstrap procedure can so be used.

Our simulation studies reflect on the distributions of the maximum likelihood estimates of composition, their bias and precision, and on the observed mixture sample itself. First, distributions of SIMULATE estimates of individual contributions of most stocks (over resamplings) were asymmetrical and apparently multimodal. Estimates for stocks absent had their mode at or near zero and were skewed to the right (right tail drawn out). Multimodal appearance did not disappear when stocks were combined.

Estimates of contributions by stocks or their aggregates must concentrate at nodes in the parameter space and occur at reduced density away from such nodes. Sampling errors of allelic frequencies in genetically similar stocks probably cause shifts among nodes. These features of distributions of composition estimates have not been reported for other applications and so may be uncommon.

Second, like the OBSERVED estimates SIMULATE estimates for individual stocks are biased; that is, they generally underestimate stocks present and overestimate stocks absent. On the average, more than over 14% of the mixture were identified as originating from stocks that were in fact absent (Table 10.2). Bias is reduced for coarser aggregates (Table 10.3) and becomes negligible ($<1\%$) between coarsest aggregates above and below Bonneville Dam.

Third, most asymptotic estimates of variance from the observed sample provide crude approximations to those observed in simulation. Standard errors from the asymptotic equations should equal standard deviations of SIMULATE estimates (Table 10.2). Some asymptotic estimates of precision, e.g., those for the Rapid River and McKenzie stocks, were in serious error. The measure of variation based on asymptotic methods overestimates what is actually observed in the simulations for stocks absent (whose corresponding estimates would encounter the constraints). Undoubtedly, the jackknife procedure would have provided improved estimates of the covariance matrix had time been available for the computations. However, as we see next, the degree of concurrence is moot because the observed sample may not be representative of the sampling framework of the simulations.

Characteristics of the OBSERVED and SIMULATE estimates can be compared to judge whether the observed sample originated from sampling processes assumed in the simulations. The simplest comparisons are between distributions of SIMULATE estimates of individual stock contributions and the corresponding estimates from the observed sample. The observed Columbia

River test sample arouses suspicion that the sampling processes underlying the simulations are not fully satisfied.

Observed composition estimates for many stocks stand out from corresponding distributions from simulation; for example, observed estimates for Klickitat and Little White Salmon were beyond the range of corresponding estimates in the simulations. At the coarsest level of combining stocks (above versus below Bonneville Dam), the observed estimate of composition again was beyond the range of estimates computed in the simulation; estimates for stocks below Bonneville Dam ranged from 0.179 to 0.291 while the corresponding estimate from the observed sample was 0.175.

Our first thought in attempting to explain these inconsistencies between observed and simulated estimates was to examine records of the assembling of the blind mixture samples; however, none had been kept for this purpose. We do know that one fish in the mixture sample has a genotype that was estimated from stock samples as being absent in all stocks but Klickitat (Klickitat is purported to be absent from the sample), and that the other three test samples contained between 12.8% and 15.0% of the Klickitat stock (our estimate of the contribution for Klickitat was 13.4%). The possibility of some confusion among the stocks while the four blind mixture samples were being assembled cannot be dismissed.

It is not known what caused the inadequacy of the assumptions underlying the simulations. In actual applications, such violations must be tracked down and estimation equations amended. We will evade this responsibility for now because our primary purpose is to illustrate and compare methods rather than evaluate a particular mixture. The essential lesson of the simulation exercise is that we could check the plausibility of our understanding of stocks present in the mixture.

The use of bootstrap sampling has been recommended for estimating statistical bias as well as for realistically evaluating precision of estimation, particularly if estimation involves complex calculations (Efron 1982). Although the simulation sampling processes may not adequately represent those from which the mixture arose, the BOOTSTRAP estimates remain of interest for comparisons concerning bias and precision. Averages of BOOTSTRAP estimates lie very near those of OBSERVED (Tables 10.2 and 10.3); any bias correction is small. Results for estimating the precision of composition estimates are mixed between asymptotic methods and bootstrap approach (criterion for comparison is SIMULATE estimates) if individual stocks are examined (Table 10.2). Neither approach was better than the other in providing close estimates (stock by stock) of variation of composition estimates, but average absolute difference from variation of SIMULATE estimates is smaller for BOOTSTRAP estimates. These observations hold for stocks present or absent. However, in contrast to bootstrap results, asymptotic calculations consistently overestimate variation observed in simulations for stocks absent. At the coarser levels of aggregation, variability of (BOOTSTRAP estimates matches that of SIMULATE estimates closely (Table 10.3), although BOOTSTRAP estimates are slightly more variable.

The utility of standard measures of variability is compromised by the skewness and probable multimodal form of distributions of composition estimates. While such measures remain useful for indicating the precision of estimations as well as for indices for comparison, confidence statements for compositions are probably more accurately determined (if extra computing time is available) by the bootstrap approach provided bias can be reduced to negligible levels. Our resampling experiments have shown that if the conditions of the simulation are met, the conditional maximum likelihood approach generally will produce composition estimates with small bias, at least for aggregates of stocks.

Unconditional maximization. Full maximization of the likelihood function with respect to genetic parameters as well as composition can be valuable. If stock sample sizes are small relative to the mixture sample, considerable loss of information occurs when conditional maximization is used. In the limiting case (e.g., Makela and Richardson 1977), stock samples may not be available, yet latent structure analysis (e.g., Goodman 1974) could succeed in estimating composition and genetic parameters. If full maximization is wanted, the equations for the EM algorithm are usually fairly straightforward (see Everitt and Hand 1981, pp. 9–10). In our earlier example of a c stock mixture with L independent loci, each with two alleles, Eqs. (4) and (5) need but one change; the θ_i becomes $\hat{\theta}_i$. In addition, a third set of equations for estimating these θ_i is needed:

$$\hat{\theta}_{il} = \frac{\sum_{j=1}^{n_i} x_{ijl} + \sum_{j=1}^{m} \hat{P}(i|\mathbf{x}_j) x_{jl}}{2\left[n_i + m\sum_{j=1}^{m} \hat{P}(i|\mathbf{x}_j)\right]}$$

$$i = 1, 2, \ldots, c$$

$$l = 1, 2, \ldots, L. \tag{10}$$

The alleles observed at the lth locus in the mixture sample are allocated among the stocks in proportion to the estimated posterior probabilities. The estimate of the frequency of a chosen allele is the simple proportion observed in the stock and allocated from the mixture. The three equations are solved by iteration. Guesses of the $\hat{p}_i$ and $\hat{\theta}_{il}$ are inserted into (5) to compute $\hat{P}(i|\mathbf{x}_j)$. The estimates of these posterior probabilities are used to compute improved values of $\hat{p}_i$ and $\hat{\theta}_{il}$ in Eqs. (4) and (10), respectively. The process is repeated until an arbitrary convergence criterion is met.

To our knowledge full maximization has never been applied in stock identification when stock samples were available. The number of parameters in practical problems becomes large rapidly as the numbers of stocks and loci increase; this, combined with the nonlinear nature of the algorithm, has probably discouraged its use. Nevertheless, the associated asymptotic covariance matrix would be useful in program design: selecting loci to be sampled, and choosing stock and mixture sample sizes. However, the asymptotic covariance matrix involves some heavy algebra, even with simple genetic models.

Identifiability and its Relation to Uniqueness of Composition Estimates

A basic concern in assessing composition of the mixture is whether knowledge of the genotypic frequencies in the stocks and mixture determines a unique stock composition; if such is the case, the stocks are said to be *identifiable*. Identifiability is a statistical concept concerning the feasibility of distinguishing between two or more explanations of the same observed phenomenon. In the present instance, stock composition of the mixture would not be identifiable if two or more stock mixtures (different **p**) were indistinguishable even if relative genotypic frequencies in the stocks and the mixture were known without error. Such would be the situation if the distributions of genotypes in the mixture sample were identical for the different mixtures, and this occurrence is assured if expected relative frequencies for all possible genotypes in the mixture sample are identical for the different mixtures.

Evidently the number of distinct possible genotypes (say G) in the mixture must equal or exceed the number of stocks (c) for stocks to be identifiable. The $(G-1)$ independent relative genotypic frequencies in the mixture, λ_h, are weighted linear sums of the corresponding relative genotypic frequencies in the stocks, with weights being the stock proportions [see e.g. (6), but now applied to all possible genotypes]. Unless the number of such independent equations, $G-1$, equals or exceeds the $(c-1)$ independent p_i, identical expected genotypic frequencies in the mixture can be produced by different stock mixtures. However, this condition alone does not guarantee identifiability of stock composition; the relative genotypic frequencies in the stocks must satisfy a condition of distinctness.

We know from standard results on inverse transformations in calculus [viewing the λ_h as functions which define implicitly the $(c-1)$ independent p_i] that a unique composition is determined by the relative genotypic frequencies in the mixture provided the rank of an associated matrix of partial derivatives, D*, equals $c-1$. One form for these matrix elements is

$$d_{hi}^* = \frac{\partial \lambda_h}{\partial p_i} = g(\mathbf{x}_h; \boldsymbol{\theta}_i) - g(\mathbf{x}_h : \boldsymbol{\theta}_c)$$

$$i \neq c$$

$$h = 1, 2, \ldots, G. \tag{11}$$

The elements are differences in relative genotypic frequencies between pairs of stocks. If two or more stocks have identical relative genotypic frequencies, for example, the matrix, D^*, has rank less than $c - 1$; and the composition is obviously not unique. Other less evident situations could occur. If the relative genotypic frequencies of some stock can be expressed as a linear combination of the remaining stocks, linear dependence among the genotypic frequencies of stocks is said to occur; and again the rank of D^* is less than $c - 1$.

The condition on the matrix D^* ensuring identifiability includes all possible genotypes. A corresponding condition when only genotypes observed in the mixture sample are included in a reduced version of matrix D^*, say D, guarantees the existence of an unique global maximum of the likelihood function. The condition that rank of D equals $c - 1$ is sufficient but not necessary to assure this uniqueness (Millar 1985).

The matrix D can be seen to be the matrix of regressors (excluding the last row) in the linear model (8) of the method of fitting expectations. If rank of D is less than $c - 1$, the method of fitting expectations will fail. Linear dependencies for observed genotypes in the mixture sample exist between relative genotypic frequencies of the stocks. The method of fitting expectations is based on regression of observed relative genotypic frequencies in the mixture on relative genotypic frequencies in the stocks. Classical linear regression analysis requires that no exact linear relations hold among the observed values of the regressors; if there are such relations, composition cannot be estimated by this approach.

In practice, we may never find an exact linear relation between true or estimated relative genotypic frequencies in the stocks (i.e., rank of D will equal $c - 1$), but such relations may be closely approximated. In regression analysis, the corresponding problem is called multicollinearity. Estimates become subject to extreme sampling variations under this condition. If a near-linear relationship holds, the individual contributions of each stock within the relationship cannot be determined. When the relative genotypic frequencies of a group of stocks from among all the stocks of the mixture are nearly identical, the individual contributions (i.e., regression coefficients) may take on arbitrary values, and only their sum is meaningful. When presented with multicollinearity of relative genotypic frequencies in the stocks, other than pooling of stocks, search for new loci with stock differences is an obvious direction of research. Ridge regression (e.g., Hoerl and Kennard 1970a, 1970b, Draper and Smith 1981) was developed to cope with multicollinearity, but no one has examined its utility in the present context.

If the rank of D is less than $c - 1$, those attempting to determine the mixture composition should be alert to the problem so that a check can be made on identifiability of the stocks. If stocks are not identifiable, combining of stocks will be necessary until aggregations are identifiable. If stocks are identifiable but rank of D is less than $c - 1$, another method must be used to assess composition by the maximum likelihood method (see Millar, in press, for situations of estimating composition from a mixture sample consisting of a single fish). Such situations seem of little importance for realistic applications, however.

CONCLUDING REMARKS

The importance of assessing stock composition will increase as the competition for fishery resources intensifies. The use of genetic data presents unique opportunities for obtaining efficiently and inexpensively such composition estimates of mixed stock fisheries of many species.

The major problem in identifying the composition of population mixtures from genetic data relates to the difficulty in identifying and characterizing each of the contributing populations genetically. Allele frequency differences at protein-coding loci have been demonstrated to provide adequate discrimination in a number of cases. A research team using genetic information to identify contributions of stocks to a mixture must first determine whether or not genetic differences among the contributing populations are sufficient to permit such genetically based estimates. The number of informative loci needed will depend upon the amount of variation found among stocks as well as on the number of stocks which must be identified in the mixture. Identifiability conditions reported in this chapter provide minima for the number of loci and the degree of variation among stocks.

If the preliminary survey for genetic variation has satisfied the minimal conditions, numerical studies must be undertaken to determine whether the information will provide composition estimates with satisfactory precision and with either negligible or correctable bias. Simulated samplings from the contributing stocks as well as from mixtures of known composition are used to approximate the distributions of composition estimates. These observed distributions of simulated composition estimates are used when judging the appropriate conditions under which adequate information can be obtained for a particular set of contributing populations and loci.

If the amount of genetic variation observed appears insufficient for meaningful composition estimates, additional genetic variation among populations must be found, either from assaying additional loci or from intentional breeding (applicable only to cultured stocks). The original evaluation procedure must then be repeated to determine adequacy of the new conditions for obtaining meaningful composition estimates. When the information base from the several stocks is judged adequate, sampling of the mixture and estimation of composition may begin. At this stage it is advisable to continue checking the observed samples from the mixture for plausibility of representing the presumed contributing populations.

The limited amount of genetic divergence usually observed among conspecific populations requires use of efficient statistical methods of estimation. A general approach based on the direct application of the maximum likelihood criterion to genotypic frequencies in the mixture sample meets this condition of efficiency. Numerical techniques used so far to evaluate the estimates of composition include the EM algorithm and direct maximization of the likelihood function by nonlinear optimization algorithms. A new algorithm based on the method of fitting expectations is described.

Evaluation of composition estimates and their precision is hampered by the boundary conditions of the likelihood function. Constrained estimation to accommodate boundary conditions produces bias in composition estimates. Such bias can be reduced by increasing the sample size from the mixture, sample sizes from the stock, or the number of discriminatory loci. Some combination of these increases will reduce bias to acceptable levels. Further, the distribution of composition estimates can be distinctly skewed and perhaps multimodal, lessening utility of variance estimates as measures of precision. Resampling from stock and mixture samples to generate empirically the distribution of composition estimates appears to resolve the problem of determining precision, although at the expense of heavy computation.

11.
The Utility of Mitochondrial DNA In Fish Genetics and Fishery Management

Stephen D. Ferris and William J. Berg

A primary objective of fishery management is to obtain detailed descriptions of fish stocks and to understand how these stocks interact with each other and with their environment. Inherent in the term "stock" is the genetic basis of a contiguous biological unit (Ihssen et al. 1981) with a shared informational source—DNA—capable of producing a limited range of phenotypes given certain environmental constraints (Eden 1967). Sound management decisions require a thorough knowledge of the genetic and spatial discreteness of stocks as well as an understanding of the impact of particular fishery practices on the genetic structure of a stock (Larkin 1981). Ideally, the genetic delineation of fish stocks would produce information sets reflecting a stock's genetic structure, without reflecting any environmental factors. This ideal has only recently been approached (Chapter 2).

Phenotypic data sets have generally been accompanied by considerable uncertainty with regard to the accuracy of delineation of putative fish stocks. The problem of extrapolating from phenotypic data to the underlying genotypic information has been discussed by Frelin and Vuilleumier (1979, see particularly Fig. 1). As the distance grows between the characters studied and their informational source (i.e., DNA), genetic information decreases and the probability of error in precise stock identification increases.

One of the traditional methods of distinguishing fish stocks has been the comparative examination of morphological characters such as number of scales in a lateral series or relative body depth. However, it is well documented that such characters are highly sensitive to environmental variation (see Lindsey 1981 for coregonid fishes and Nordeng 1983 for Artic char; also Cherry et al. 1978 and Clayton 1981).

Protein electrophoresis, by providing rapidly collected and purely genetic data sets, has opened new ways of examining fishery management problems. Allelic protein data in many ways approach the ideal data set described above for stock delineation (Allendorf and Utter 1979). However, there are many instances in which electrophoresis fails to identify genetically discrete stocks (e.g., Utter 1981, chap. 2, fn. 5). This failure is at least partially a reflection of the relative insensitivity of a procedure which directly reflects differences at the DNA level but is nevertheless two steps removed from the gene itself. It is clear that although protein electrophoresis will continue to be a useful tool in fishery management, it simply cannot detect all the genetic variation that may be of value in understanding stock dynamics. We believe that, of all existing procedures, restriction endonuclease analysis of mitochondrial DNA most closely approaches the ideal method for quantifying genetic differences among populations.

MITOCHONDRIAL DNA

Approximately 99% of the DNA in eukaryotic cells resides within the nucleus, but of the remaining 1%, the mitochondrial DNA (mtDNA) may prove to be of major importance to fishery biologists. This distinctive bit of the cell's genetic machinery may be examined directly with restriction endonucleases to yield exact genotypic data containing more genetic information than has been previously obtained. Restriction endonucleases are enzymes which cleave DNA, producing fragments at specific recognition sites which consist of short, palindromic sequences of 4, 5, or 6 nucleotides (Boyer 1971, 1974). Although published accounts of studies of teleost mtDNA are limited (Avise et al. 1984, Avise and Saunders 1984, Berg and Ferris 1984, Graves et al. 1984, Gyllensten and Wilson, Chapter 12) work on mammalian systems predict great strides in our understanding of fish stock discreteness, migration patterns, hybridization among fish stocks, systematics of stocks and higher taxa, and the potential for genetic marking of stocks.

Properties of Mitochondrial DNA

Vertebrate mtDNA is a covalently closed circular molecule of between 15,000 and 18,000 base pairs (bp). Recently, determination of the complete nucleotide sequence of human, cow, and mouse mtDNA has revealed a highly conserved gene order with some unique characteristics (Anderson et al. 1981, 1982, Bibb et al. 1982). Each molecule codes for 2 ribosomal RNAs, 22 transfer RNAs, 13 polypeptides (some of unknown function), and the D loop region involved in replication (Borst and Grivell 1981). These genes are tightly packed with no introns and at least two cases of coding sequence overlap (Anderson et al. 1982, Montoya et al. 1983). The entire mitochondrial genome is transcribed as a unit with translation effected by a distinctive genetic code, different from the "universal genetic" code (Borst and Grivell 1981, Barrell et al. 1979).

Comparative sequencing and restriction enzyme mapping indicate that the greatest rate of mutational events occurs in the D loop and the least in the RNA genes (Brown and Simpson 1981, Ferris et al. 1981a, Cann and Wilson 1983; Greenberg et al. 1983; see especially Cann et al. 1984). Although mutations appear to occur somewhat randomly throughout the genome, fine-scale mapping reveals hot spots and other nonhomogeneities in their distribution

(Brown et al. 1982). Furthermore, small deletions and insertions of 1–10 bp appear to be more common in mtDNA than in nuclear DNA (Cann and Wilson 1983). A good review of mtDNA properties is given by Brown (1983).

Two features of mtDNA are most salient to stock identification: rapid evolution and maternal inheritance. Mitochondrial DNA evolves 5–10 times as rapidly as nuclear DNA (Brown et al. 1979), thereby providing a magnified view of the differences between populations and closely related species. The rate of base substitution in mammalian mtDNA has been estimated to be 1% replacement per million years (Brown 1980, Ferris et al. 1983c).

The second property, maternal inheritance (Hutchison et al. 1974, Giles et al. 1980), has already been exploited in tracing maternal lineages in a variety of organisms, including parthenogenetic lizards (Brown and Wright 1979), inbred mice (Ferris et al. 1982), and freshwater sunfish (Avise and Saunders 1984). It is as yet unresolved whether there is a rare paternal contribution. Upon entering an egg with its several thousand mitochondria, the few paternal mitochondria possibly brought in by the sperm (Gresson 1940) are presumably excluded. We should be mindful that rare events of this sort could complicate genealogical analysis, but so far this appears not to be a factor. Avise et al. (1984) found no evidence (at the 5% level) of paternal mtDNA transmission. Interspecific hybridization and backcrossing experiments with *Mus* suggest that mtDNA inheritance may be at least 99.9% maternal (Gyllensten et al. 1985). Lansman et al. (1983b) critically tested the possibility of a paternal contribution of mtDNA and estimated that the upper limit of such contribution must be on the order of 1 molecule per 25,000 per generation. Thus, maternal inheritance of mtDNA enhances its usefulness as a genealogical marker.

Considering segregational patterns of organelles (Birky et al. 1983) and the fact that mtDNA apparently does not undergo recombination (Brown 1983 and references therein), one would expect intra-individual mtDNA variation (heteroplasmy) to be quite rare. However, recent studies indicate that, at least in certain species, heteroplasmy does occur and may be common (Potter et al. 1975, Solignac et al. 1983, Harrison et al. 1985, Densmore et al. 1985).

A final property of mtDNA that needs to be examined is *invasiveness*. Invasiveness may be defined as the limited distribution of mtDNA across a hybrid zone without concomitant nuclear gene flow. This phenomenon has only been seen a few times, in semispecies of house mice and *Drosophila* (Ferris et al. 1983b, Powell 1983). Ferris et al. (1983b) clearly demonstrated that although the mtDNA of one species may have invaded and displaced the resident mtDNA of the second species, their nuclear genomes have remained completely distinct. One implication of their findings is that mtDNA may have its own selective distribution independent of nuclear genes. Takahata and Slatkin (1984) have shown that, under the assumption of selective neutrality for mtDNAs, invasiveness is largely a function of the maternal immigration rate and not of selection against the nuclear genome. Furthermore, in developing appropriate population genetic theory for mitochondrial genes, Birky et al. (1983) noted that since only migrating females may distribute organelle genes between populations while nuclear genes are carried by both sexes, under differential rates of migration, a population may be subdivided for mitochondrial genes and yet appear panmictic for nuclear genes. Thus, situations will arise concerning migration and hybridization between taxa that may be resolved only by examining both mitochondrial and nuclear genomes.

Correspondence of MtDNA And Protein Electrophoresis Data

The properties discussed above indicate that analysis of mtDNA may yield valuable data to the fishery biologist, but what evidence is there that population structures or genealogies revealed by mtDNA correspond to those obtained by other methods? In perhaps the largest study to date, Ferris et al. (1983b) examined mtDNA diversity among 208 animals, representing populations from throughout the world belonging to 10 nominal species and subspecies of the mouse subgenus *Mus*. Using 11 restriction endonucleases they identified approximately 200 nucleotide substitutions within 300 cleavage sites and an insertion of 12 bp located in or near the D loop. Systematic analysis indicated seven distinct lineages within this group and supports synonymy of three nominal subspecies of *Mus (M.) domesticus*. These results corroborate the findings of Marshall and Sage (1981) based on protein electrophoresis and comparative morphological analysis.

Avise et al. (1979a) used both restriction endonuclease analysis of mtDNA and protein electrophoresis to study the genetic relationships of populations of pocket gophers *(Geomys pinetis)*. Comparison of data obtained from each method is of interest. They screened 171 animals for 25 proteins encoded by nuclear genes. Although there was little allozymic variation, they were able to distinquish eastern and western forms of pocket gophers. Analysis of the mtDNA of 87 animals with only six restriction enzymes revealed 23 distinct mtDNA clones (matriarchal lineages). Construction of a phylogenetic network not only confirmed the reality of the eastern and western forms but also yielded an order of magnitude greater information concerning the genetic relationships of populations within each form than was obtained by protein electrophoresis.

Avise and Smith (1974) used protein electrophoresis to characterize the geographic distribution of two subspecies of bluegill sunfish *(Lepomis macrochirus)*. In this study they found an area of secondary contact and presumptive hybridization. Detection of very distinctive, essentially subspecies specific, mtDNAs (Avise et al. 1984) confirmed and extended their earlier work. Data from one site in the area of hybridization, Lake Oglethorpe, Georgia, indicated complete introgression of both nuclear and mitochondrial genomes.

In studies of closely related groups such as stocks of fish or individual families, the ability to resolve small differences precisely is very important. Humans have been studied in more detail with (protein electrophoresis) than any other organism; and in spite of the fact that hundreds of electromorphs have been identified, no fixed allele differences have been found between races of man. Brown (1980) estimated that the pairwise nucleotide diversity in human mtDNA was only 0.36% and yet was able to characterize each of the 21 individuals in the study on the basis of their mtDNA fragment patterns.

A quantitative comparison of the information content of data sets obtained by protein electrophoresis and restriction endonuclease analysis of mtDNA is summarized in Table 11.1 Protein electrophoresis of 50 genetic loci detected seven which were diagnostic for the mouse semispecies, while analysis of their mtDNA with 11 restriction endonucleases detected 54 diagnostic fragments. Quantifying this difference suggests that restriction endonuclease analysis of mtDNA may be more sensitive than protein electrophoresis in the detection of nucleotide substitutions. Although it is now possible to electrophoretically survey more than 100 loci (e.g., Morizot and Siciliano 1984), perhaps doubling the number of diagnostic electromorphs, there are over 10 times the number of restriction endonucleases used in this study which are available, each of which may be used to independently examine the mtDNA for specific sequences.

The ability to achieve greater resolution of genetic differences by restriction endonuclease analysis of mtDNA than by protein electrophoresis is more than simply a function of comparing direct genotypic versus phenotypic data, although this is a major factor. Greater resolution is also a function of the apparently higher evolutionary rate of mtDNA (Brown et al. 1979). Comparing sequence data from the nuclear genome (Kreitman 1983) to sequence data from the mitochondrial genome (Johnson et al. 1983) indicates that the nucleotide substitution rate is approximately six times higher for mtDNA than for nuclear DNA (Berg, unpublished). This estimate is in complete agreement with that made by Miyata et al. (1982) concerning the differential rate of silent substitutions in mitochondrial genes and nuclear genes. Thus, it is clear that analysis of mtDNA may yield data sets with hundreds of discrete genetic characters useful for fishery management.

Table 11.1 Comparative information yield of protein electrophoresis and restriction endonuclease analysis of mitochondrial DNA for discrimination of mouse semispecies.

Protein	mtDNA
50 loci	150 restriction sites
7 diagnostic loci	54 diagnostic fragments
7 nucleotide substitutions	18 nucleotide substitutions
1,000 nucleotides per locus	5 nucleotides per site
50,000 nucleotides surveyed	750 nucleotides surveyed
1.4×10^{-4} detected nucleotide substitution rate	240×10^{-4} detected nucleotide substitution rate

Note: Protein data from Sage (1981); mtDNA data from Ferris et al. (1983c).

RESTRICTION ENDONUCLEASE ANALYSIS
OF MITOCHONDRIAL DNA

The great appeal of mtDNA in genetic studies lies primarily in its ease of purification relative to other forms of DNA (Brown 1983). In this section we describe procedures for collecting specimens, purification, and data analysis. Before doing so, however, it is useful to digress somewhat and consider the place of mtDNA analysis with respect to a technique employing nuclear DNA. Table 11.2 summarizes salient points for four methods: hybridization, fragment studies, restriction endonuclease cleavage mapping, and sequencing. Time and cost generally increase in this order, but increasingly detailed information is obtained.

When double-stranded DNA is heated in an appropriate buffer solution, it will "melt" (denature) into the single-strand form. Slow cooling allows these single strands to reanneal, forming the original double-stranded helix. The temperature at which melting is 50% complete, T_{50}, is considered the reference temperature. Single-stranded DNA from different sources may be reannealed to form hybrid duplexes. These hybrid duplexes are then remelted, noting their T_{50}. The less similar the nucleotide sequences of the single strands are, the lower the thermal stability will be in the hybrid duplex, as indicated by a lower T_{50}. It has been estimated that a 1°C decrease in T_{50} indicates 1% sequence divergence (Britten et al. 1974). This method has been used in a few teleost studies (Gharret et al. 1977, Hanham and Smith 1980, Schmidtke and Kandt 1981). Inasmuch as this method examines the overall sequence similarities and thus yields an average similarity value, it lacks the detailed resolving power of restriction methods (but see Sibley and Ahlquist 1984). Furthermore, insertions, deletions, and transpositions of sequences within the compared strands will greatly affect this technique's accuracy. This method is probably not sensitive enough to be used successfully in intraspecific population surveys.

The remaining three methods provide discrete characters in the form of DNA fragments, restriction sites, or nucleotide sequences. These may be obtained in large numbers and summarized as to degree of relatedness using methods which will be described later. Fragment sharing is the quickest

Table 11.2 DNA methods used in population genetics and phylogenetic studies and appropriate taxonomic levels of application.

Method	Limit of resolution		Cost	Number of characters
	Nuclear DNA	mtDNA		
Hybridization	Genus & above	——	+	+
Fragment sharing	No limit	Intrafamilial	+ +	+ +
Cleavage mapping	No limit	Intrafamilial	+ + +	+ + +
Sequencing	No limit	Intrafamilial	+ + + +	+ + + +

method. Simply put, restriction enzymes are used to digest purified DNA at specific recognition sites (a site is an enzyme specific sequence of 4, 5, or 6 bases). As a result of this digestion the DNA molecule is cleaved into a number of fragments of various lengths, which may be separated by electrophoresis, forming a characteristic pattern. Fragment lengths are estimated by comparing their electrophoretic mobility (which is largely a function of length) rather than their charge, as in protein electrophoresis, to the mobilities of known size standards (Schaffer 1983).

Approximate locations of the restriction sites in the genome may be determined by the more time consuming mapping process, usually employing double digestions. Mapping allows more rigorous explanation of the mutations which produce the various fragment patterns seen in a population. Mapping of restriction sites often allows detection of deletions and other rearrangements. Sequencing provides the most data, but it is also the most time consuming and usually involves cloning procedures. Nevertheless, several comparative studies have examined DNA sequences at the population and species level. Partial mtDNA sequences have been determined in primates (Brown et al. 1982), rodents (Brown and Simpson 1982), and humans (Aquadro and Greenberg 1983), and numerous studies have examined nuclear DNA (e.g., Kreitman 1983).

Sampling Procedures

First we consider the geographic sampling strategy. Where allozyme projects ideally require hundreds of individual samples, those using mtDNA may require fewer because of its greater resolving power. Further, since mtDNA is clonally inherited, individuals from one locality are likely to be very similar in sequence. Indeed, such tendency to homogeneity has been observed in pocket gophers (Avise et al. 1979a), mice (Ferris et al. 1983b), and bluegill sunfish (Avise and Saunders 1984). Findings of large differences among mtDNAs from individuals sampled at the same locality would suggest an admixture of previously isolated populations (Avise et al. 1984, chap. 7). This expectation may not hold for highly migratory fish such as tuna, but should hold for those fish which home to natal streams (e.g., salmon). Thus, initial sampling should collect a few individuals from many different locales, rather than the reverse. In fact, statistically, this is the preferred strategy (Nei and Li 1979, Tajima 1983).

The ideal procedure is to keep the collected fish alive until purification of DNA. Freezing may reduce yield by 50% due to membrane rupture and nuclear DNA contamination. We have obtained good yields from salmon kept on wet ice 6 hours (5 μg mtDNA/200 g). Dry ice should be used if samples must be frozen. Quick freezing is apparently more important for certain marine fishes, like tunas, to avoid nuclease degradation of DNA. Another way to decrease nuclease activity is by a fivefold increase (to 5 mM) in the amount of EDTA in the homogenization solution. The experimenter should try a variety of procedures to find one which optimizes yield and is convenient to collectors.

Isolation and Purification
Of Mitochondrial DNA

Figure 11.1 describes the basic procedures used in restriction endonuclease analysis of mtDNA. In theory mitochondrial DNA may be isolated from any tissue, but best results are obtained from fresh "soft" tissues such as liver, spleen, and kidney. If one has 50-100 g of tissue available, the procedure is essentially that of Brown (1980). Maternal inheritance of mtDNA indicates that a female and all of her offspring will be identical for their mtDNA, so if individual fish are too small to yield adequate tissue samples, tissue from maternal half-sibs may be combined. Mitochondrial DNA has a greater buoyant density than linear nuclear DNA and thus may be separated into two distinct bands by cesium chloride ultracentrifugation. The lower band formed by mtDNA may be removed for subsequent analysis. Berg and Ferris (1984) found that in salmonids the lower mtDNA band may be obscured by a wide nuclear DNA band. Collecting the bottom 8-10 drops of this wide band and rebanding by cesium chloride ultracentrifugation resulted in a distinct mtDNA band. In some situations the simplified procedure of Ferris et al. (1983a, 1983b), which requires only 1-5 g of tissue, may be used. It has worked successfully with salmon (Chapter 12) and tuna (Ferris et al., unpublished).

An alternative procedure based on the Southern DNA hybridization technique (Southern 1975) is available for old, frozen, small, or otherwise stubborn samples. Total cellular DNA (mtDNA + nuclear DNA) is isolated, digested with restriction endonucleases, subjected to electrophoresis, denatured into single-stranded DNA, and transferred to a nitrocellulose filter (see also Lansman et al. 1981). Radioactively labeled mtDNA prepared from other sources may be used as a hybridizing probe to reveal the location of mtDNA fragments. In preparing the probe, mtDNA may be pooled from several individuals, or even from individuals of another species. In fact, one of us (SDF) has obtained adequate hybridization between the mtDNAs of salmon and tuna, species which diverged millions of years ago (Graves et al. 1984).

Restriction Endonucleases

Restriction enzymes are the principal tools in the analysis. These enzymes are derived from various bacteria, and more than a hundred of them are available commercially at a reasonable cost. These enzymes recognize 4-base, 5-base, or 6-base sequences. For instance, *Xba* I, isolated from *Xanthomonas badrii* (Zain and Roberts 1977) recognizes and cleaves the nucleotide sequence T^CTAGA (position of hatch,^, indicates point of cleavage). The enzyme cuts the DNA wherever this recognition site occurs in the molecule. The *Xba* I sequence occurs six times in *Salmo gairdneri*, resulting in 6 fragments with lengths of approximately 5,690, 3,150, 3,080, 2,310, 1,480, and 690 bp (Berg and Ferris 1984). Another enzyme, *Mbo* I, recognize ^GATC. This sequence occurs more than 25 times in *Salmo*. In general, 4-base cutters generate 3-6 times as many fragments as 6-base cutters. More detail concerning restriction

A. Preparation of mt DNA

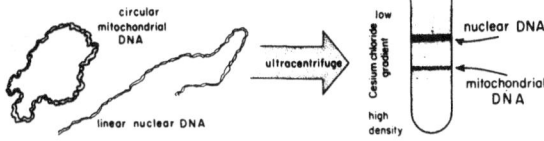

Recognition sequence of restriction enzyme *Xba* I: T↓CTAGA

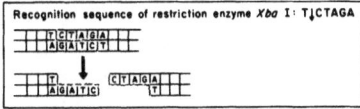

B. Restriction Enzyme Digest

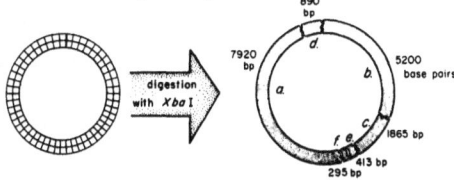

C. Separation by Gel Electrophoresis

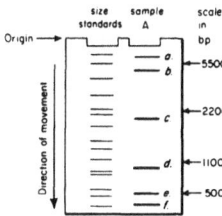

D. Interpretation of Results

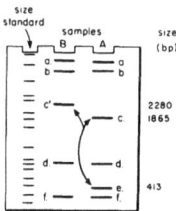

QUANTITATIVE RESULTS:

1. Number of mutations $\approx \dfrac{\text{band differences}}{3}$ $\left[\dfrac{3}{3} = 1\right]$

2. Percent sequence difference $\approx \dfrac{\text{number site changes}}{\text{total bp}(f \times l)}$ $\left[\dfrac{1}{6 \times 6} = 3\%\right]$

where f = total number of fragments
l = recognition length of enzyme

Fig. 11.1 Steps in the isolation and analysis of mitochondrial DNA.

A. Mitochondria, with circular DNA, are enriched by low-speed centrifugation. Mitochondrial pellets are lysed, releasing mtDNA and some contaminating nuclear DNA. Mitochondrial DNA and nuclear DNA are separated in CsCl density gradient, mtDNA appearing as a lower (higher buoyant density) band. Digestion with restriction endonclease *Xbal* (recognition site TCTAGA).

B. Six TCTAGA recognition sites are cleaved (example given is mtDNA).

C. Fragments are electrophoresed on agarose gel. Fragment is visualized by ethidium bromide staining (under UV light) or ^{32}P end-labeling. Fragment sizes are determined by reference to size standards.

D. Variant pattern B differs from common pattern A by three fragments (bands). Fragment length sum (A,c + A,e) equals length of fragment B,c. Hypothesize a single base change within TCTAGA recognition site, resulting in site loss. Cleavage mapping confirms that fragments A,c and A,e are adjacent. Fragments or sites are characters for further analysis.

enzymes may be found in Nathans and Smith (1975) and Roberts (1982). The fragments generated are then electrophoresed on 1% agarose for 5-base and 6-base cutters or 3%–5% acrylamide for 4-base cutters.

Visualization

Visualization is easy if one is fortunate enough to work with larger fish from which 5-20 μg of mtDNA may be purified. After electrophoresis, the gel is stained with ethidium bromide and viewed under ultraviolet light. The mobility of the fragments are then measured carefully and compared with size standards of known length, such as the fragments obtained from lambda phage DNA or mouse mtDNA which has been cut with a restriction enzyme. This comparative mobility data may be used to estimate fragment lengths with great precision (Schaffer 1983).

The above procedure requires about 200 ng of DNA per digest. With smaller yields, the more time consuming endlabeling procedure described by Brown (1980) is used. This procedure has the virtues, however, of requiring only 10 ng per digest, and allowing for the detection of very small fragments, as low as 30 bp (Ferris et al. 1983b, Brown 1980). In this method, ^{32}P-labeled dNTPs are used to label digests with the enzyme DNA polymerase Large Fragment (Klenow Fragment).

The second method of visualization, the Southern technique, has been mentioned earlier. Its main advantage is that only very small amounts of material are needed. It is an extremely sensitive technique in that one can detect mtDNA in the cellular preparation although mtDNA makes up less than 1% of the total cellular DNA. There are two ways to prepare the probe, endlabeling or the more sensitive "nick translation." Radioactive probe with tenfold higher specific activity may be achieved by nick translating mtDNA with the enzyme *Polymerase* I. Kits are available from Bethesda Research Laboratories at modest cost.

A drawback of the Southern technique is that it requires large amounts of restriction enzymes to digest the 2–4 μg of cellular DNA used. On the other hand, it is relatively easy to purify the DNA from a large number of individuals in a short time, perhaps as many as 100 in a week. Another disadvantage has been the loss of binding efficiency for fragments shorter than 300 bp. Thus, use of 4-base enzymes such as *Mbo* I, which produce 30–40 fragments in salmon mtDNA (Berg and Ferris 1984), tend to be impractical as many smaller fragments may not be resolved. There are several 4-base cutters, such as *Fnu* DII, *Hpa* II, *Hae* II, which yield 5–15 fragments, and these may be used with the Southern method. Recent technical advancements (e.g., Frossard et al. 1983) may improve fragment resolution to below 100 bp. If one is lucky, several of the enzymes employed will be diagnostic for the stock in question. In later screening for identification, only those enzymes which produce diagnostic fragments need be employed, thus reducing time and expense.

DATA ANALYSIS

Fragment Patterns

The simplest level of analysis begins by identifying each unique fragment pattern for a given restriction digest. A pattern consists of the array of fragments generated by a restriction enzyme. There are typically 1–10 fragments produced by 6-base cutters and 10–50 fragments for 4-base cutters. Two patterns differ if they have at least one fragment of different mobility. The most common outcome for intraspecific pattern variation is to differ by three bands. This is illustrated in Fig. 11.1, patterns A and B. Subsequent mapping of the fragments on the circular mitochondrial genome indicates that the two small unique fragments of B lie adjacent to each other on the map. The large unique fragment in A can be explained as the sum of the sizes of the two smaller bands. Thus, the restriction site at the junction of the two fragments may be hypothesized to have been lost due to a single nucleotide substitution within the recognition site and so the enzyme no longer cleaves there. Conversely, the large fragment in A could have gained a site in the interior, thus giving the smaller fragments. Only phylogenetic analysis can reveal which is more likely to have occurred.

There are several reasons why patterns may differ by fewer than three bands. One is that a small fragment has migrated off the gel. Another is that two nonhomologous fragments coincide in mobility. A third reason is that the two molecules are homologous, except that one has a small deletion within the bounds of one of the fragments; the shorter fragment would migrate faster than its counterpart.

The patterns are labeled with a letter. There is no strong convention so far, except that the reference pattern is usually called A. This may be the most common pattern or the one which has been sequenced, as in the case of mouse (Ferris et al. 1983c). The others may be labeled in order of discovery. Thus, the data may be summarized as a table with restriction enzymes across the top and taxa vertically. Each element is a pattern. This table contains considerable information, and at a glance it is possible to see which taxa are most similar based on the number of identical patterns. In the table, each row having at least one different pattern is a morph. With 5 enzymes and 4 taxa, we might have three morphs, 1 = aaabc, 2 = aaaba, and 3 = aaaab; thus the first three enzymes reveal conservative sequences, the fourth more variable ones, and the fifth the most variable. The next step is to relate the patterns and morphs to one another quantitatively.

Quantitative Analysis

What appears to be a growth industry has sprung up around developing estimators of within and between sample diversity and methods of phylogenetic inference from comparative mtDNA data (Upholt 1977, Gotoh et al. 1979, Aoki et al. 1981, Engels 1981, Ewens et al. 1981, Kaplan and Risko 1981, Hudson 1982, Felsenstein 1983, Lansman et al. 1983a, Templeton 1983b). Although those methods strictly concerned with producing a similarity index may be based on very different statistical grounds or use different data bases, the estimates obtained are often quite similar. Methods used to make phylogenetic inferences are also variable and may not yield the same results.

When estimating sample diversity, several assumptions are usually made concerning the underlying evolutionary processes. The most commonly made assumptions are that the nucleotides are randomly arranged throughout the genome, that only base pair substitutions occur (no insertions, deletions, or duplications), and that the substitution rate among the four nucleotides is equal. While each assumption may not always hold (Adams and Rothman 1982, Brown and Simpson 1982, Brown et al. 1982, Aquadro and Greenberg 1983, Cann and Wilson 1983, Harrison et al. 1985, Densmore et al. 1985), reliable estimates are obtainable.

It was recently found that, in mtDNA, transitions occur more frequently than transversions; that is, a purine nucleotide is much more likely to be substituted for by a purine than by a pyrimidine, and vice versa [e.g., $P_{A-G} > > P_{A-C/T}$] (Brown and Simpson 1982, Brown et al. 1982, Aquadro and Greenberg 1983, Cann et al. 1984). One of the implications of this is that back mutations and parallel mutations may accumulate very quickly between two lineages and any sequence estimate will likely underestimate the true level of sequence divergence. Holmquist (1983) has presented an interesting model to deal with this problem, but before applying it we need to obtain more precise estimates of the base pair substitution rates for various parts of the genome.

Perhaps the most commonly used similarity index is that of Nei and Li (1979) and Nei and Tajima (1981a). Use of their algorithm is quite straightforward, and calculations are easily done on a hand calculator. Although it is clear that some of the assumptions upon which their method is based are not realized, the correspondence between estimated and actual values is generally good (Brown et al. 1982). It should be noted that this method will give underestimates when the compared sequences have diverged more than 5% (George 1982). Templeton (1983a), by examining the components of this problem, confirmed that use of Nei and Li's algorithm at the intergeneric level results in loss of genetic information, and so its application may best be restricted to analysis of intraspecific groups, a level at which accurate estimates may be obtained (see Tables 11.3 and 11.4)..

Data from fragment patterns or mapped cleavage sites may be used in the Nei and Li (1979) equations to estimate sequence diversity $\hat{p}$. Comparative fragment data may be used in their Eq. (20). This method assumes that fragments of coincident mobility are homologous. This assumption may be erroneous; the comparative fragment method may not detect small length differences, and similar length fragments may be produced in different parts of the genome, especially if the restriction enzyme used produces many fragments.

Table 11.3 Within species, percent sequence divergence estimates based on data derived from restriction endonuclease analysis of mitochondrial DNA.

Species	Mean (percent)	Animals	Sites	Reference
Homo sapiens	0.40	21	217	W. Brown 1980
Homo sapiens	0.74	112	300	Cann and Wilson 1983
Homo sapiens	1.80	7	37	Aquadro and Greenberg 1983
Gorilla gorilla	0.55	4	50	Ferris et al. 1981a
Pan paniscus	1.00	3	50	Ferris et al. 1981a
Pan troglodytes	1.30	10	50	Ferris et al. 1981a
Pongo pygmaeus	3.50	5	50	Ferris et al. 1981a
Ovis aries	1.50	2	30	Upholt and Dawid 1977
Capra hircus	0.75	3	27	Upholt and Dawid 1977
Mus domesticus	0.77	30	166	Ferris et al. 1983c
M. musculus	0.92	4	165	Ferris et al. 1983c
M. molossinus	0.00	46	22	Yonekawa et al. 1981
Rattus norvegicus	1.00	21	25	G. Brown and Simpson 1981
R. rattus	4.00	26	25	G. Brown and Simpson 1981
Geomys pinetus	1.80	87	30	Avise et al. 1979a
Peromyscus polionotus	1.11	68	48	Avise et al. 1983
P. maniculatus	4.13	5	80	Avise et al. 1983; Lansman et al. 1983a
P. leucopus	1.42	14	48	Avise et al. 1983
Lepomis machrochirus	0.0-8.5	10	62	Avise and Saunders 1984; Avise et al. 1984
L. cyanellus	0.0-1.4	7	66	Avise and Saunders 1984
L. spp. (6 species)	0.00	2-5	n.a.	Avise and Saunders 1984
Katswonus pelamis	<1.00	16	~40	Graves et al. 1984
Salmo clarki	2.00	n.a.	n.a.	Gyllensten and Wilson, Chapter 12
S. trutta	0.50	29	n.a.	Gyllensten and Wilson, Chapter 12

Table 11.4 Within genus, mean percent interspecific sequence divergence estimates based on data derived from restriction endonuclease analysis of mitochondrial DNA.

Genus	Number species examined	Mean (percent)	Reference
Pan	2	3.7	Ferris et al. 1981a
Mus	6	7.6	Ferris et al. 1983c; Yonekawa et al. 1981
Rattus	2	16.0	G. Brown and Simpson 1981
Peromyscus	3	2.4	Avise et al. 1983
Lepomis	9	20.6	Avise and Saunders 1984
Salmo	4	4.0-9.0	Berg and Ferris 1984 Gyllensten and Wilson, Chapter 12

Although uncommon, intraspecific length mutations have been detected in primates (Ferris et al. 1981b, Cann and Wilson 1983, Aquadro and Greenberg 1983), lizards (Densmore et al. 1985), *Drosophila* (Solignac et al. 1983), and crickets (Harrison et al. 1985).

To avoid bias in diversity estimates it is necessary to establish the genome size. Size estimates may be obtained by summing the estimated fragment sizes after digestion. For example, the enzyme *Sac* II produced two cuts in the ribosomal genes of human and mouse mtDNA (Anderson et al. 1981, Bibb et al. 1981). *Sac* II also produced two fragments of equal mobility in four salmonid species, constituting strong evidence that their mitochondrial genomes are of the same size (Berg and Ferris 1984).

Both fragments of rainbow trout mtDNA produced by *Sac* II digestion were slightly larger than those of human mtDNA; therefore, we estimated that the salmonid mitochondrial genome was about 100 base pairs (bp) larger than human mtDNA, or about 16,670 bp. This estimate was corroborated by summing the fragment sizes for 12 additional restriction enzymes (Berg and Ferris 1984). Avise et al. (1984) estimated the mtDNA of bluegill sunfish to be approximately 16,200 bp, while Graves et al. (1984) determined skipjack tuna mtDNA to be about 16,900 bp.

Restriction site data may be used in Nei and Li's (1979) Eq. (16) to estimate sequence divergence. Estimates derived from restriction site data are considered more accurate than those obtained by the comparative fragment method. Estimates of $\hat{p}$ between sequenced DNAs are straightforward and may be considered as a reference for the other methods. Table 11.5 compares estimates of percent divergence derived from these three methods. Although the data are incomplete, there is general agreement between methods.

Phylogenetics

A branching diagram or tree (dendrogram) which portrays the genealogical relationships and relative order of genetic isolation events (e.g., dispersal, local extinction of intermediate populations, speciation) occurring between taxa is a phylogenetic tree (Wiley 1981). A phylogenetic tree hypothesizes the evolutionary process.

There are two basic methods of manipulating restriction data in preparation for tree construction. In the first, all available data, without bias, may be summed by a method such as Nei and Li's (1979) into an overall distance index. A matrix of all pairwise values serves as input data. These indices are then grouped by a phenetic method, linking taxa by their similarity or lack of it. Some methods such as the unweighted pair-group method using arithmetic averages (UPGMA, Sneath and Sokal 1973) assume an equal rate of evolutionary divergence (= constant nucleotide substitution rate) and sequentially averages the indices across taxa, resulting in a neat, orthogonal diagram. One may find many arguments against the use of this method in phylogenetics (Farris 1981), but it is sufficient to point out that the evolutionary rate of mtDNA as

Table 11.5 Comparison of sequence divergence estimates derived from three mitochondrial DNA methods.

	Percent sequence divergence			
Taxa compared	**Sequence (number bp)**	**Mapping (number sites)**	**Fragment (number fragments)**	**References**
Homo sapiens	0.20 (896)	——	——	W. Brown et al. 1982
Homo sapiens	——	0.40 (217)	——	W. Brown 1980
Homo sapiens	——	0.74 (300)	——	Cann and Wilson 1983
Homo sapiens	1.70 (899)	1.90 (37)	——	Aquadro and Greenberg 1983
Pan paniscus vs. *P. troglodytes*	——	4.00 (50)	3.00 (140)	Ferris et al. 1981a Ferris and Wilson, unpublished
H. sapiens vs. *P. troglodytes*	9.00 (896)	13.00 (50)	8.00 (140)	W. Brown et al. 1982 Ferris et al. 1981a Ferris and Wilson, unpublished
H. sapiens vs. *Gorilla gorilla*	10.00 (896)	15.00 (50)	8.00 (140)	W. Brown et al. 1982 Ferris et al. 1981a Ferris and Wilson, unpublished
H. sapiens vs. *Hylobates lar*	18.00 (896)	18.00 (50)	14.00 (140)	W. Brown et al. 1982 Ferris et al. 1981a Ferris and Wilson, unpublished
Rattus rattus vs. *Mus domesticus*	32.00[a]	35.00 (35)	——	Ferris and Wilson, unpublished Borst and Grivell 1981
M. domesticus vs. *M. spretus*	——	16.70 (31)	10.70 (165)	Ferris et al. 1983c
M. domesticus mtDNA A vs. MtDNA B	——	1.02 (165)	1.20 (165)	Ferris et al. 1983c
M. domesticus mt DNA A vs. mtDNA C	——	0.79 (165)	0.81 (165)	Ferris et al. 1983c
R. rattus mtDNA A vs. mtDNA B	8.00 (829)	15.00 (25)	——	G. Brown and Simpson 1982

[a]Data from part of ribosomal gene; number bp not available.

measured by nucleotide substitutions is far from constant. It varies not only from region to region within the genome (Brown and Simpson 1981, Cann and Wilson 1983) but also between individual nucleotides (Aquadro and Greenberg 1983).

To avoid invoking the restraint of a constant evolutionary rate, Farris (1972) developed the distance Wagner method based on fitting data via a Manhattan distance estimator. This method has appropriate applications at the intraspecific level but is not without limitations, primarily because of the potential loss of information when particulate data are subsumed into a single index (Penny 1982). It is also important to note that with the inherent underestimation characteristic of all indices when sequence divergence exceeds about 5%, confounding mutations may occur and simply never be detected. Additional problems with using indices as estimators of phylogeny have been discussed elsewhere (Farris 1981).

An alternative to using indices as representative of interrelationship is to allow each taxon, be it a single individual or an entire species, to be defined by its individual particulate data; thus, distinct fragments, mapped restriction sites, or sequence data may be used in constructing a phylogenetic hypothesis. Each datum is considered a character which may be coded simply by its presence or absence or by the frequency with which it appears. In general, those characters which are common to all the groups in question are considered to lack phylogenetic information concerning the relationships between the groups. Surely we can imply that the ancestor common to all the groups probably possessed that character, say an *Xba* I fragment, 955 bp in length, but that may be the extent of that character's value. Similarly, a character which is unique to a particular group bears little additional information. Usually the input data consist of those characters which are shared by at least two taxa but not all. Characters of this type have been called phylogenetically informative (Ferris et al. 1981a).

Once the input data set has been obtained, there are three general methods which may be followed. Two of them, which will not be discussed further, are compatibility analysis (Templeton 1983a, 1983b) and maximum likelihood methods (Felsenstein 1983). The most commonly used method of analysis, though not necessarily the best, is based on the principle of maximum parsimony. As reviewed by Felsenstein (1983), the maximum parsimony principle was first used in molecular studies by Eck and Dayhoff (1966). The primary objective of maximum parsimony is to construct a minimal spanning network between the taxa under study such that the branch lengths connecting taxa would represent the minimum number of mutations required to explain the observed data. Piecing together the minimal network by inspection may be virtually impossible as the number of alternative networks soon becomes astronomical with increasing numbers of taxa (Felsenstein 1978). Algorithms have been developed to aid in construction of these networks and enhance the probability of approximating the "best" network (Farris 1970, Fitch 1971, see Felsenstein 1983), and maximum parsimony networks have become a standard method of reporting restriction enzyme data (e.g., Avise et al. 1979a, 1979b, 1983, Ferris et al. 1981b, 1983c, Aquadro and Greenberg 1983, Lansman et al. 1983a, Berg and Ferris 1984).

These networks are often "unrooted"; that is, they connect the taxa in a $n-1$ (n = number of taxa) dimensional hyperspace without defining the most ancestral position. Without a root, one can state that evolutionary changes, mutations, have occurred but one cannot state the order or direction of that change. The most reasonable method of rooting a network is to add a taxon to the study which is accepted, by data other than those obtained in the study, as being related to but distinct from the rest of the taxa as a whole. Thus, Berg and Ferris (1984) used brook trout, *Salvelinus fontinalis,* a member of the Salmoninae tribe Salvelini, to represent the outgroup in their study of three species belonging to the tribe Salmonini. Addition of an outgroup establishes the direction of evolutionary change and orders the isolating events, allowing interpretation of phylogenetic relationships. Phylogenetic trees so constructed are not based on assumptions of evolutionary rate or direction, and interpretation of them will often lead the investigator to identify previously undetected back and parallel mutations.

GENETIC VARIATION IN STOCKS

Interpopulation Versus Intrapopulation Divergence

There are two ways to measure variation in a population. One considers divergence between mtDNA types in a population, the other weights the types according to the number of individuals in the population having a particular type of sequence. Both may have as their starting point the estimated percent sequence difference, or this number expressed as a fraction of 1, namely p. In stock analysis, we want to know both heterogeneity within populations and genetic divergence between populations. Perhaps the most appropriate estimate of intrapopulation mtDNA heterogeneity is given by the nucleon diversity of the population (Nei and Tajima 1981a). An application of this method is given in Chapter 12 and will not be dealt with here.

An answer to the question of how much genetic divergence between conspecific populations we might expect to find may be suggested by the data in Table 11.3. In general, the average mtDNA sequence divergence estimates between conspecific mammals and fish are on the order of only a few percent. In only a few cases do the intraspecific sequence divergence levels exceed 3%–4% with the highest intraspecific mtDNA differentiation being found between two subspecies of *Lepomis macrochirus* (Avise and Saunders 1984). This relatively uniform low level of intraspecific divergence is markedly different from the wide ranging estimates obtained at the intrageneric level (Table 11.4).

Mitochondrial DNA diversity within a species is generally described as the mean sequence divergence among the unique mtDNA types identified within all surveyed populations. This statistic may also be used to understand the relative ages of the populations. The assumption is that a population (or species) with many divergent mtDNA types is older than one with fewer, more closely related types. If the rate of nucleotide substitutions is approximately constant, pairwise comparisons of mean $\hat{p}$ values yields an average relative age of the molecular lineage. The most divergent mtDNAs in a population may have an estimated divergence time that actually predates the formation of the species to which the population belongs. This is possible especially if population sizes historically have been large. Let us assume that a large population of fish with many mtDNA types is split in half by a geologic event, resulting in an absolute barrier to reciprocal gene flow. Each daughter population would evolve independently, and while some of the original mtDNA types might persist, new mtDNA variants would appear. Detection of the same mtDNA types in both daughter populations would suggest that those mtDNA types predated the formation of the barrier to gene flow. When comparing populations polymorphic for their mtDNAs, Nei and Li (1979) recommend subtracting the average intrapopulation divergence from the average interpopulation mtDNA sequence divergences to arrive at a true picture of the interpopulation divergence.

Experimentally, large populations such as house mice (Ferris et al. 1983b) and deer mice (Lansman et al. 1983a) have been found to have mtDNAs of considerable intrapopulation sequence divergence, on the order of 1%. Both are thought to have existed for over a million years and thus to have experienced anagenic mtDNA evolution. To date no studies have been published of intrapopulation mtDNA diversity of documented recent populations or ones with very small population sizes to check the expectation of low genetic variation.

The Effect of Bottlenecks on Variability

Variability of mtDNA is expected to be more profoundly affected by bottlenecks than is variability of nuclear DNA. The effective population size for mtDNA is immediately reduced by half because only the female mtDNA types are inherited, and intra-individual mtDNA variation is extremely rare. Consider a hypothetical case of an extreme bottleneck, a single mated pair dispersing to a new habitat. While the female carries only one type of mtDNA, they are both probably heterozygous at several nuclear DNA loci. Thus, their descendants will be homogeneous for one mtDNA type for perhaps thousands of generations, until a new variant arises and yet heterozygous (admittedly at a reduced level) at some nuclear loci. The exciting possibility for the future is that mtDNA may be able to give clues to historical bottlenecks by comparing mtDNA heterogeneity with nuclear DNA heterogeneity. The bottleneck issue has been discussed by Ferris et al. (1981a, 1983c), Brown (1980), and Yonekawa et al. (1981).

APPLICATIONS OF MITOCHONDRIAL DNA VARIABILITY

Genetic Branding

Like the cattle rancher, the ocean rancher needs to brand his stock to distinquish it from those of other ranchers and from wild stocks. Fish may be branded either mechanically or genetically. Mechnical branding includes such methods as fin clipping, cold branding, and coded wire tags. All these methods are labor intensive (and thus expensive) and need to be repeated each year. Fin clips are perhaps the least desirable of these three methods because regeneration of fins and the normal abrasive action on unclipped fins may confound data retrieval. Cold brands, administered by "burning" a brand upon a fish with a liquid nitrogen cooled marker, may disappear over time due to scar tissue formation or scale regeneration. Implanting coded wire tags, the most labor-intensive method, requires individual placement of internal snout tags, clipping the adipose fin for later identification, and, upon return, tag decoding. Mechanical branding must be done every generation.

Identification of fish stocks with genetic brands rather than by any available mechanical methods has some very attractive features. Perhaps the most attractive feature is that a genetic brand is, by definition, a heritable trait. It therefore does not need to be manually applied to each fish in each generation; rather, it is a "built-in" characteristic of that fish stock. Corollary to this is that genetic brands are stable through time; they do not wear off. Another desirable feature is that there are hundreds of protein-encoding loci and nucleotide sequences, any one of which is potentially useful in identifying specific fish stocks.

The feasibility of using electrophoretically detected allozymes to distinquish fish stocks has been shown in many studies (e.g., Ryman et al. 1979, Grant and Utter 1984). A potential drawback to using allozymes is that it is unusual to find fixed allele differences between fish stocks, and it may therefore be necessary to screen relatively large samples to obtain statistical confidence in stock identification. This requirement for large sample sizes may be abrogated by using a mtDNA-based genetic brand.

To develop mtDNA-branded fish, a preliminary survey of potential source populations, those currently under hatchery culture and nearby wild populations, would be required to yield needed baseline data. The survey probably would not require exhaustive application of many different restriction endonucleases. Selection of those enzymes which are known to cleave mtDNA many times, thus producing many fragments (e.g., *Fnu* DII, *Ava* II, *Mbo* I, or *Hin* fI), may yield distinctive profiles, fragment patterns, of the source populations. Maternal inheritance of mtDNA assures that, regardless of the paternal source, prudent selection of females for broodstock would produce an entire generation identical for the chosen mtDNA brand.

It should be noted that use of a single female to produce the first generation of mtDNA branding would result in all half-sibs regardless of the number
of male parents (Chapter 3). This potential for inbreeding may be minimized by
prudent selection of several female parents from the same matriarchal lineage,
thereby reducing the impact on the overall genetic variability of the stock. Furthermore, selection for desirable phenotypic characteristics such as
efficaciousness of return, return weight, and meat quality—characters presumably under nuclear DNA control—could go on as before.

Stability of mtDNA through generations would assure the hatchery manager that returning fish could be unambiguously identified as released stock or
strays. Screening for new genetic variants might be required only every few
generations (Hauswirth and Laipis 1982).

Mixed Fishery Analysis

Recently, maximum likelihood methods have been developed to estimate
component contributions of various stocks to a mixed fishery (Milner et al.
1981; Fournier et al. 1984, chap. 10). These methods may use gene frequency
data obtained by starch gel electrophoresis as baseline data and have produced
very resonable estimates in several applications (Grant et al. 1980, Miller et al.
1983, Fournier et al. 1984). These methods of analysis may also be applicable
to mtDNA data.

Several studies have demonstrated extensive intraspecific mtDNA variation. Avise et al. (1979a,b) found absolutely distinctive mtDNA fragment patterns for several endonucleases in local populations of field mice *(Peromyscus)*
and pocket gophers *(Geomys pinetus)*. Brown (1980) was able to individually
characterize 21 humans on the basis of their mtDNA fragment patterns. Other
studies which have a bearing on this point include Brown and Simpson's (1981)
work on *Rattus,* the examination by Denaro et al. (1981) of ethnic variation in
humans, and perhaps most especially the recent papers by Johnson et al. (1983)
and Lansman et al. 1983a). Johnson et al. (1983) used only five endonucleases
but were able to quantitatively and qualitatively characterize 200 humans to
their major ethnic group and to estimate the evolutionary relationships of those
ethnic groups. Lansman et al. (1983a) examined 135 individuals of *Peromyscus
maniculatus* collected from throughout North America. Using only eight endonucleases they found 61 distinctive mtDNA morphs which were assignable to
five geographic assemblages. Current data are limited concerning the amount of
variability in teleost mtDNA, but in a recently completed study of *Salmo trutta*
and *S. salar* by Gyllensten and Wilson (Chapter 12) levels of variation are being
found which were comparable to that of mammals.

If some intraspecific teleost groups (stocks) are found to have diagnostic
mtDNA fragment patterns, a powerful, direct-count method of estimating proportional stock contributions to mixed fisheries will be available. Accepting
that all stocks may not be fixed for diagnostic fragment patterns, the frequency

of any distinctive patterns may still lead to relatively simple estimates of stock contribution.

As an aside, we point out that if a particular stock does not possess at least distinctive fragment patterns, either the stock as originally delineated may not be a stock—that is, it may be a subunit of a larger group—or it may be a very recently formed unit.

Migration and Hybridization

Once we have identified genetically branded domestic stocks and native stocks, we may address the issues of migration and hybridization. Ihssen et al. (1981) pointed out that tagging data have been inappropriately used to estimate gene flow between stocks. It is impossible to state with certainty that tagging data measure anything other than movement. Obviously, to speak of gene flow we need diagnostic tags which determine whether migrants are breeding with other stocks. While both protein data and mtDNA may serve as valuable tags (or brands), mtDNA may reveal previously unavailable information.

Figure 11.2 illustrates how one may determine whether a hybridization event has taken place. Given two populations which possess very similar morphology—either intraspecific groups such as fish stocks of the northern and southern Pacific skipjack tuna *(Katsuwonus pelamis)* or interspecific sibling species as discussed above—when a hybridization event takes place we may find the previously diagnostic mtDNA of one form occurring in both forms. Detailed phylogenetic analysis of morphological or nuclear gene traits may still distinquish the forms, but they obviously have exchanged some genetic material, their mtDNAs. We may feel confident that mtDNA provides a genetic marker which acts as a migratory tag and as an indicator of hybridization.

It is clear that the evolutionary pressures upon the mitochondrial genome are quite different from those affecting the nuclear genome. As pointed out by Lansman et al. (1982), the level of diversity among mtDNAs appears to be uncoupled from that of morphology or nuclear gene diversity. In fact, it has

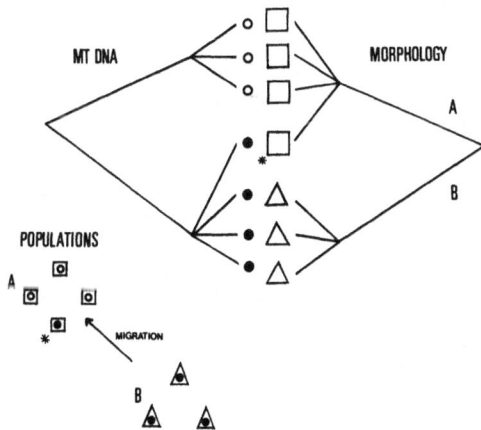

Fig. 11.2 Detecting a migration event through mtDNA genealogical analysis. Morphologically distinct and geographically isolated populations, A and B, are shown at lower left. Sometime in the past, there was a migration/hybridization event. As a result of backcrossing, present-day descendant is virtually indistinguishable from A. However, its mtDNA type is unchanged (or only slightly diverged from original). Event is detected by parsimony analysis of the mtDNA sequences, as shown in center. The descendant of the migrant has type B mtDNA and type A morphology.

been shown that stocks may be discrete for their mtDNA and yet possess uniform nuclear gene complements (Birky et al. 1983). It is possible that when intraspecific hybridization events take place there may be undetected exchange of nuclear gene material. Distribution of characteristic mtDNAs across stocks may be the only indication of gene flow between what were previously considered discrete stocks.

Current Studies

Berg and Ferris (1984) examined the systematic relationships of the mtDNAs of four salmonid species: rainbow trout, chinook salmon, brown trout, and brook trout. On the basis of 272 mtDNA fragments produced by 15 restriction endonucleases, they were able to establish that the salmonid mitochondrial genome is about 100 bp larger than human mtDNA, or about 16,670 bp. Estimates of sequence divergence obtained by the comparative fragment method (Nei and Li 1979, Nei and Tajima 1981a) indicate that rainbow trout mtDNA differ from chinook salmon mtDNA by only 3.4%, brown trout mtDNA differ from the Pacific salmonids by 7.2%, and brook trout mtDNA have diverged by approximately 10%. Qualitatively, each species' mtDNA had distinctive fragment patterns for 11 of the 15 endonucleases employed. This study confirms the close phylogenetic relationship between rainbow trout *(Salmo (Parasalmo) gairdneri)* and chinook salmon *(Oncorhynchus tshawytscha)* and argues for a more appropriate nomenclatural treatment of these taxa (see Kendall and Behnke 1984).

How much variation exists in highly pelagic marine fishes such as tuna? Are skipjack tuna from the Atlantic genetically isolated from those of the Pacific? These questions were recently addressed by Graves et al. (1984) in a study of skipjack tunas from Hawaii, Puerto Rico, and Brazil. They found that these populations showed very little mtDNA differentiation in spite of the presumptive 3-million-year barrier between them, the Isthmus of Panama. With only three nucleotide substitutions detected, the average sequence divergence was estimated to be less than 0.5%, a value to be expected for a single wide ranging panmictic unit. This very low value, coupled with a lack of protein variation (Sharpe, personal communication), will force fishery biologists to rethink whether one or two stocks of skipjack tuna exist and to reexamine the question of the amount of migration between the two oceans, perhaps by way of South Africa. Studies like these may eventually allow us to tackle the important question of how speciation occurs in the ocean, including when and where differentiation of the DNA occurs.

Only recently has hybridization in fishes been studied using restriction endonuclease analysis of mtDNA. Avise et al. (1984) and Avise and Saunders (1984) examined natural inter- and intraspecific hybridization events in the freshwater sunfish genus *Lepomis*. They found that in most interspecific hybridization events the female parent belongs to the less common species. This conclusion was based on the characteristic of strict maternal inheritance of mtDNA.

Current research by U. Gyllensten and A. C. Wilson on mtDNA hetero- geneity in various Salmonidae is summarized elsewhere in this volume. As pre- liminary results they state that (1) natural populations possess moderate levels of mtDNA variation, (2) a reduction in genetic diversity may have occurred in hatchery stocks, and (3) phylogenetic relationships based on mtDNA data are largely in accord with previous studies.

CONCLUDING REMARKS

The use of restriction endonucleases in the analysis of mitochondrial DNA allows direct examination of perhaps the best understood piece of the ver- tebrate genome. From the modest beginnings of Brown and Vinograd (1974) to the extensive surveys of today, the ease of analysis, rapid rate of nucleotide di- vergence, and maternal inheritance of mitochondrial DNA have yielded a wealth of information. Many questions which have perplexed industry and sci- ence alike will soon be addressed in new ways.

12.
Mitochondrial DNA of Salmonids
Inter- and Intraspecific Variability
Detected with Restriction Enzymes

Ulf Gyllensten and Allan C. Wilson

Without a doubt, our efforts to exploit or manage a fish species can alter the genetic structure of that species. We have already observed the alterations produced by intensive harvesting of natural populations and by large-scale planting of populations that have been propagated in the hatcheries for many generations. As such activities increase in scope and intensity, the need arises for powerful analytical tools that allow us to describe the genetic characteristics of a species and thereby to assess the magnitude of the alterations. Specifically, there is a need for

- detailed descriptions of the genetic structure of natural and hatchery-propagated populations,
- genetic markers with which single fish can be assigned unambiguously to a specific hatchery strain or natural population,
- systems for monitoring quantitative and qualitative changes in the genetic resources resulting from various management activities (for instance, sensitive methods for estimating the loss of genetic variation in both natural and hatchery populations), and
- genetic identification systems for patenting strains.

Protein electrophoresis has been used successfully to address some of these needs. For instance, the distribution of allelic variants of enzymes (allozymes) has been used both to describe the genetic population structure of a number of fish species (Ryman et al. 1979, Kornfield et al. 1982a, Grant and Utter 1980, 1984, Gyllensten, 1985) and to measure the genetic changes during artificial propagation (Allendorf and Phelps 1980, Ryman and Ståhl 1980). However, protein electrophoresis has only a limited ability to resolve existing genetic variability. It detects only a fraction of the amino acid substitutions and the investigator is restricted to the part of the genome that codes for soluble enzymes.

More genetic variation can be found by working at the DNA level. This allows for studies of both the variation in other parts of the genome and nucleotide variation that does not affect the gene product (synonymous substitutions). Variation at the DNA level is detectable with the aid of restriction enzymes, i.e., enzymes that recognize specific nucleotide sequences of from four to six bases and cleave the DNA wherever such a sequence occurs.

There is one class of DNA that accumulates substitutions very rapidly, namely mitochondrial DNA or mtDNA (W. M. Brown et al. 1979, 1982, Wilson et al. 1985). Restriction enzyme analysis of mtDNA can give a magnified view of the genetic distances between closely related taxa. The mtDNA work of Ferris et al. (1981a,b, 1983c) on closely related mammals permits us to surmise that the resolving power provided by restriction enzymes can greatly exceed that of conventional protein electrophoresis.

Here we report the use of restriction enzymes for detecting variation in salmonid mtDNA and discuss our findings in relation to management and gene resource conservation. We also review some management aspects of two other studies. The first of these (by Gyllensten, Wilson, and Ryman, unpublished) concerns variation within and among hatchery and natural populations of brown trout and Atlantic salmon, and the second (Gyllensten et al. 1985a) is a study of the hybridization of two subspecies of cutthroat trout.

THE MITOCHONDRIAL GENOME

The mtDNA of land vertebrates is the most intensively studied piece of eukaryotic DNA (for reviews see Brown 1983, Sederoff 1984, Wilson et al. 1985, Ferris and Berg Chapter 11). This small circular genome is 16–18 kilobases long and is tightly packed with 37 genes that code for 13 polypeptides, 2 ribosomal RNAs, and 22 transfer RNAs. Most of these macromolecules function as components of the energy-yielding and protein-synthesizing machines inside the mitochondria. In addition, one of the mtDNA genes appears to influence the structure of an antigen on the cell surface (Ferris et al. 1983a. Fischer-Lindahl et al., unpublished).

The complete base sequence of mtDNA is known for a single individual in the case of four species of tetrapod vertebrates: human (Anderson et al. 1981), house mouse (Bibb et al. 1981), domestic cow (Anderson et al. 1982), and clawed frog (Roe et al. 1985). Although these four sequences differ greatly from one another, owing mainly to base substitutions, all have the same gene order.

The mtDNA of fish is much less well known. Nevertheless, the evidence obtained by Berg and Ferris (1984), Avise and Saunders (1984), and our studies is consistent with the possibility that the mtDNA of salmonids resembles that of land vertebrates in size and substitution rate.

Dynamics

A notable feature of mtDNA is its mode of inheritance, which seems to be strictly maternal in all animals tested (Gyllensten et al. 1985b and references therein). This feature makes mtDNA a valuable genealogical tool for tracing the history of female lineages within and among populations (Ferris et al. 1982, Avise and Lansman 1983). In contrast, the evolution of the paternal genome cannot be traced using this molecule.

In most cases, each individual appears from restriction analysis to have only one type of mtDNA, even though there are thousands of mtDNA molecules in a typical somatic cell and from 10^5 to 10^8 mtDNA molecules in an egg. (For an exception in which heteroplasmy has been found, see Solignac et al. 1983.) Since individuals are effectively haploid as regards mtDNA, the effective population size for mtDNA is probably smaller than for nuclear DNA (Nei and Tajima 1981a). Therefore, mtDNA variation will be lost more easily than nuclear DNA variation when a population goes through a bottleneck in population size.

Analysis of Mitochondrial DNA

Although mtDNA variability can be studied directly by base sequencing (Brown and Simpson 1982, Brown et al. 1982, Greenberg et al. 1983, Higuchi et al. 1984), it is more practical at present for fisheries biologists to use restriction enzymes for comparing mtDNAs. We draw attention to four methods of restriction enzyme analysis in Table 12.1.

The method of highest resolving power (I) employs a set of four-base restriction enzymes. Each enzyme recognizes a specific sequence of four bases and cuts mtDNA wherever that recognition site occurs. The resulting fragments of mtDNA are made detectable by attaching radioactive phosphate to their ends. A typical four-base enzyme cuts vertebrate mtDNA into about 30 fragments, most of which can be separated from one another by electrophoresis through a long (40 cm) polyacrylamide gel (Brown 1980, Ferris et al. 1982, 1983a,b,c, Cann and Wilson 1983, Cann et al. 1984). The fragment patterns, made visible by the use of X-ray film, are then compared for different mtDNAs that have been treated with the same enzyme. Differences between fragment

Table 12.1 Four restriction methods for comparing mtDNAs.

Method[1]	Way of detecting DNA fragments	Type of restriction enzyme	Minimum mtDNA divergence[2] detectable (%)	Resolving power
I	Radioactive labeling	4-base	0.05	High
II	Radioactive labeling	6-base	0.2	Intermediate
III	Southern blotting	6-base	0.5	Low
IV	Ethidium staining	6-base	0.5	Low

[1]For illustrations of these methods, see: for I, W. Brown (1980); for II, Ferris et al. (1981a,b); for III, Denaro et al. (1981); for IV, W. Brown and Wright (1979), W. Brown et al. (1979).
[2]Assuming the use of about ten enzymes.

patterns are usually due to substitutions that cause restriction sites to be gained or lost (Brown and Simpson 1982, Brown et al. 1982, Greenberg et al. 1983). The four-base, end-labeling method (I) is capable of distinguishing between mtDNAs that differ by less than 0.05% in base sequence (see, for example, Ferris et al. 1983c).

Six-base enzymes are also valuable for studies of mtDNA variation (see II–IV, Table 12.1). Six-base methods require less technical expertise and have less resolving power because the number of six-base restriction sites in mtDNA is usually only about three per enzyme. Moreover, ethidium staining and Southern blotting (III and IV) ignore fragments less than 250 bases long. With those methods it is difficult to distinguish between mtDNAs that have diverged at fewer than 0.5% of their base pairs. Since mtDNA divergence within salmonid species does not often exceed 0.5% (see below), our work necessarily relies heavily on the end-labeling methods (I and II).

MITOCHONDRIAL DNA VARIATION AMONG FIVE SALMONID SPECIES

We used restriction analysis of mtDNA to study nine hatchery populations and one natural population representing five salmonid species. Included were two hatchery stocks each of Atlantic salmon *(Salmo salar)* and brown trout *(Salmo trutta)* and three stocks of rainbow trout *(Salmo gairdneri)*. One of the Atlantic salmon stocks came from rivers draining into the North Sea (Ätran) on the Swedish west coast and the other came from the Baltic Sea (Lule River). The brown trout were from Gullspång and Åvaån; both were maintained at the Älvkarleby Hatchery, Sweden. The three rainbow trout strains examined were (1) Arlee rainbow trout, maintained by the Montana Department of Fish, Wildlife and Parks, (2) Eagle Lake rainbow trout, a California strain maintained by the Montana Department of Fish, Wildlife and Parks, and (3) European rainbow trout (Sweden), originally derived from a California strain.

Only one stock was examined in each of the other taxa, namely two subspecies of cutthroat trout *(Salmo clarki bouvieri,* the Yellowstone subspecies, and *Salmo clarki lewisi,* the westslope subspecies) and the brook trout *(Salvelinus fontinalis)*. Westslope cutthroat trout were obtained from the Creston National Fish Hatchery in Creston, Montana, and Yellowstone cutthroat trout were from the Yellowstone River State Trout Hatchery in Big Timber, Montana. The brook trout were caught by gillnets in Wings Pond on the east coast of Newfoundland.

In each case we examined from three to five individuals per hatchery stock using 13 restriction enzymes (*Ava*I, *Ava*II, *Bam*HI, *Bgl*I, *Bgl*II, *Fnu*DII, *Hind*III, *Hinc*II, *Hpa*I, *Pst*I, *Pvu*II, *Sma*I, and *Xba*I) that recognize four-, five-, and six-base sequences.

Methods

The mtDNA from fish specimens was extracted routinely from 1–2 g of frozen tissue by a modification of the procedure of Brown et al. (1979). Liver and heart tissue was homogenized in cold 0.25 M sucrose, 5 mM EDTA, and 10 mM Tris using a chilled, hand-driven, Dounce blender. Intact nuclei were removed by centrifugation for 5 minutes at 1000 × g, and the mitochondria pelleted at 12,000 × g and lysed in 1% sodium dodecyl sulfate. MtDNA of high purity was obtained by repeated isopycnic centrifugations in a solution containing cesium chloride and propidium diiodide.

Nanogram amounts of mtDNA were digested with 1 unit of each restriction enzyme according to the supplier's recommendation. Fragments were labeled by filling the 3′ ends with the appropriate alpha[32]PdNTP, in the presence of the large (Klenow) fragment of DNA polymerase I, and separated electrophoretically in 1.2% horizontal agarose gels or native 3.5% polyacrylamide gels (Brown 1980). Fragments of known size from the phages PM2, phiX174, and lambda were used as size standards. The fragments were visualized by autoradiography of dried gels at room temperature. An example of an autoradiogram of brown trout mtDNA is shown in Fig. 12.1.

The nomenclature used follows the system proposed by Ferris et al. (1982, 1983a,b). Each fragment pattern resulting from cleavage with a restriction enzyme is denoted by a capital letter. Note that the alphabetical proximity of two patterns does not necessarily reflect genetic similarity.

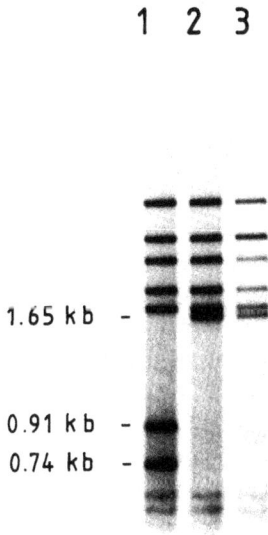

Fig. 12.1 Autoradiograph of brown trout mtDNA from three individuals cleaved with *Hinc*II and separated by electrophoresis in 1.2% agarose. DNA fragments have been labeled with alpha[32]PdNTP prior to separation. Two mtDNA fragment patterns are shown: *X* with 10 fragments (lane 1) and *Y* with 9 fragments (lanes 2,3). A single nucleotide substitution may explain the difference between the two types; it causes the loss of a site in type *Y* and the joining of the 0.91 kb (kilobase) and 0.74 kb fragments into a 1.65 kb fragment. By similar analyses of the fragment patterns resulting from digestion with different enzymes, the average number of nucleotide substitutions between mtDNAs can be estimated (Nei and Li 1979).

Pairwise estimates of percent sequence divergence between mtDNAs were made from the fraction of fragments having identical electrophoretic mobilities with Eq. (20) of Nei and Li (1979). Separate estimates were calculated for enzymes recognizing four-, five-, and six-nucleotide sequences, and these values were weighted according to the total number of nucleotides examined in the different groups. This method is satisfactory for mtDNAs that have diverged by less than 5%. For more divergent sequences the fragment method usually underestimates the true divergence (George 1982, Ferris et al. 1983a,c).

The diversity of mtDNA lineages within a population *(h)* was estimated with Nei and Tajima's (1981) nucleon diversity:

$$h = \frac{1}{n-1}\left[n\left(1 - \sum_{i=1}^{l} x_i^2\right)\right] ,$$

where x_i is the frequency of the *i*th type of mtDNA in a population sample of n specimens and l is the number of mtDNA types. This measure is usually used to estimate the heterozygosity at nuclear loci, but it may also be employed to estimate the diversity of maternal lineages within and among populations.

Results

The fragment patterns observed by cleavage of mtDNA from the 10 taxa with each of the 13 restriction enzymes are given in Table 12.2, while the number of fragment differences and estimated percent sequence divergence among the taxa appear in Table 12.3. The fragments used to calculate these estimates are listed by enzyme in the *Appendix* 12.1.

A total of 221 distinct DNA fragments were detected, and the number of fragment differences between the 10 stocks ranged from 1 to 104 in pairwise comparisons (Table 12.3). No variation was found within any of the hatchery stocks, a result which contrasts with that of our survey of natural populations (see below). This contrast implies that each hatchery-propagated population may have originated from a very small number of females.

The mean extent of sequence divergence between hatchery populations within a species ranged from 0.1% to 2.0%. The only case in which we observed an intraspecific divergence value above 1% was provided by the cutthroat trout, whose two subspecies diverged by 2%. These estimates of salmonid mtDNA variability are within the range of values observed between conspecific populations of land vertebrates (Wilson et al. 1985, Avise and Lansman 1983) and *Lepomis* fish (Avise and Saunders 1984). The percent sequence divergence between species was much larger, ranging from 3% to 14% within the genus *Salmo*. Between the two genera *(Salmo* and *Salvelinus),* the divergence was greater still, roughly 13%, in satisfactory agreement with the estimate made by Berg and Ferris (1984) for another (partially overlapping) set of restriction enzymes.

Table 12.2 Mitochondrial DNA fragment patterns in five salmonid species.

Species	Sample size	*Ava* I	*Ava* II	*BamH* I	*Bgl* I	*Bgl* II	*FnuD* II	*Hinc* II	*Hind* III	*Hpa* I	*Pst* I	*Pvu* II	*Sma* I	*Xba* I
Brown trout														
Gullspång	5	A	A	A	A	A	A	A	A	A	A	A	A	A
Åvaån	4	A	B	A	A	A	A	A	A	A	A	A	A	A
Atlantic salmon														
Åtran	4	B	C	B	B	A	B	B	B	B	B	B	A	B
Lule	5	B	D	B	B	A	B	B	B	B	B	B	A	B
Westslope cutthroat trout														
Creston	3	C	E	C	C	B	C	C	C	C	C	C	A	C
Yellowstone cutthroat trout														
Big Timber	3	C	F	C	D	C	D	D	D	C	C	C	A	C
Rainbow trout														
Swedish hatchery	4	D	G	D	E	D	E	E	E	C	D	C	B	C
Arlee Lake	5	E	H	D	E	E	E	E	F	C	D	C	B	C
Eagle Lake	4	F	I	D	E	F	F	F	E	C	D	C	B	C
Brook Trout														
Wings Pond	3	G	J	E	F	G	G	G	G	D	E	D	C	D

Note: A capital letter (A, B, C, etc.) denotes a particular fragment pattern obtained with a given restriction enzyme. Fragment patterns with same letter for different enzymes are not related.

Phylogenetic Tree

A phylogenetic tree relating the 10 types of mtDNA appears in Fig.12.2. The mtDNA in the four species of *Salmo* falls into two groups. On the one hand, there is a fairly close mtDNA relationship between the European brown trout and the Atlantic salmon. On the other hand, the two species whose native range is in western North America—the rainbow and cutthroat trouts—are grouped together. The brook trout *(Salvelinus)* appears to be distantly related to the *Salmo* species.

The mtDNA tree fits with views based on protein electrophoresis (Utter et al. 1973, Ferguson and Fleming 1983). There is also concordance between our mtDNA tree and that presented by Berg and Ferris (1984) for three of the same species (brook, rainbow, and brown trouts). Furthermore, on the basis of the small divergence (3%–4%) that Berg and Ferris (1984) found between the mtDNA of rainbow trout and one of the Pacific salmon species *(Oncorhynchus tshawytscha),* it seems likely that this *Oncorhynchus* mtDNA lineage descended from the lineage that also gave rise to the trout mtDNAs of western North America (represented by rainbow trout and cutthroat trout in Fig. 12.2). A close nuclear relationship between western trout and Pacific salmon is also evident from the fact that they share certain protein polymorphisms (Bailey et al. 1970, Ferguson and Fleming 1983).

Table 12.3 Number of fragment differences (above diagonal) and percent sequence divergence (below diagonal), estimated according to Nei and Li (1979; Eq. 20), among the 10 salmonid taxa.

| | Brown trout | | Atlantic salmon | | Cutthroat trout | | Rainbow trout | | | Brook trout |
| | Gullspång | Åvaån | Ätran | Lule | Westslope | Yellowstone | Swedish hatchery | Arlee | Eagle Lake | Wings Pond |
No.	1	2	3	4	5	6	7	8	9	10
1	—	6	69	70	82	83	85	90	84	92
2	0.36	—	70	70	83	86	86	92	86	93
3	6.18	6.52	—	1	94	94	101	104	99	99
4	6.45	6.45	0.12	—	94	93	100	102	98	98
5	9.20	9.20	13.20	13.11	—	29	54	59	52	88
6	9.44	9.88	11.45	11.38	1.98	—	49	58	51	91
7	9.70	9.05	13.21	13.14	4.09	3.39	—	15	14	90
8	9.27	10.16	14.19	14.13	4.82	4.27	0.82	—	19	93
9	8.28	8.96	13.14	13.06	4.21	3.72	0.75	1.08	—	92
10	13.38	13.32	18.05	18.00	12.00	12.43	11.43	11.48	11.36	—

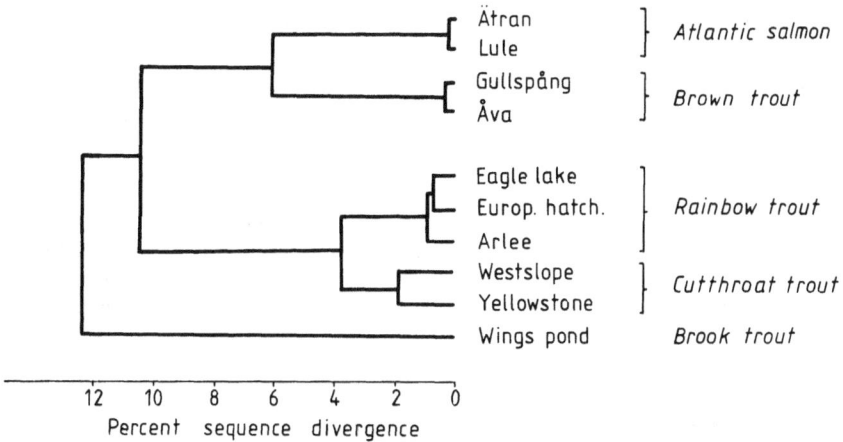

Fig. 12.2 Phylogenetic tree relating mtDNA from nine hatchery populations representing four species of *Salmo* and one natural population of a species of *Salvelinus* (brook trout). Thirteen restriction enzymes were used to cleave the mtDNA. The fragments produced were labeled with alpha^{32}P and compared electrophoretically. The extent of sequence divergence between pairs of lineages was estimated from the fraction of fragments shared with Eq. (20) of Nei and Li (1979); see Table 12.3. The order of branching of the lineages leading to the ten types of mtDNA was determined by the UPGMA method (Sneath and Sokal 1973). The parsimony method, using fragments as characters (cf. Ferris et al. 1981a,b), was also used to study the relationships among lineages. Of a total of 219 DNA fragments, 141 (64%) were phylogenetically informative; that is, they occurred in more than one but not in all of the taxa. The branching order of the most parsimonious network (requiring 176 steps) is identical to that derived by the UPGMA method. The study by Berg and Ferris (1984) indicates that *Oncorhynchus* may tentatively be expected to branch off from the lineage leading to rainbow trout and cutthroat trout.

Differential Introgression

Contrasting views of the genetic relatedness between the two subspecies of cutthroat trout and rainbow trout emerge from comparative studies of mtDNA and of proteins encoded by nuclear DNA. Whereas the two cutthroat subspecies belong to a mtDNA lineage separated from that leading to the three rainbow trout strains (Fig. 12.2), protein electrophoretic analysis (Leary et al. 1984) indicates that the westslope cutthroat trout is more similar to the rainbow trout than to other cutthroat subspecies (Fig. 12.3). This difference is further

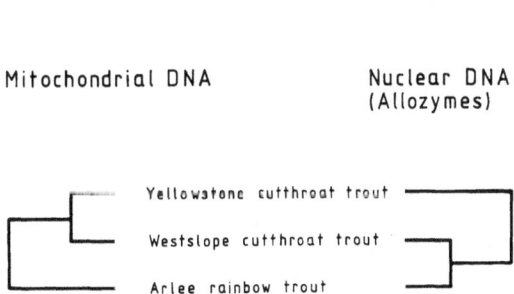

Fig. 12.3 Relationship between two forms of cutthroat trout and the Arlee strain (Montana, USA) of rainbow trout with respect to mitochondrial DNA (mtDNA) and nuclear DNA (nDNA) (estimated from protein electrophoresis). Results of mtDNA comparisons are from Fig. 12.2; electrophoretic comparisons made by Leary et al. 1984. The order of branching of the lineages was determined by the UPGMA method.

supported by more extensive interspecific comparisons which indicate a closer relationship of coastal, Lahontan, and westslope subspecies of cutthroat trout than to Yellowstone cutthroat trout based on nuclear genes (Leary et al., in press). A possible explanation for this contrast is a recent introgression of nuclear genes between taxa that did not affect the mitochondrial gene pools. Such introgression could be mediated by males. Although this explanation is speculative at present, it heightens our awareness that this type of gene flow may be much easier to detect when independent sources of information regarding the phylogenetic relationships are available. Parallel studies of mitochondrial and nuclear genes may thus increase the potential for detecting gene flow between partially reproductively isolated units, in nature as well as in hatchery populations.

VARIATION WITHIN NATURAL POPULATIONS OF BROWN TROUT

In a different study by Gyllensten, Wilson, and Ryman, intraspecific variation among eight natural populations of brown trout from Sweden and Ireland was assessed with a set of seven restriction enzymes *(Ava*II, *Fnu*DII, *Hae*III, *Hinc*II, *Hinf*I, *Hpa*II, and *Taq*I). This survey revealed a wealth of variation, with 15 maternal lineages among the 29 fish tested (2–8 per population) and a mean intrapopulation diversity *(h)* of 0.72.

The heterogeneity among mtDNAs from the same natural population may reflect variation pre-existing in the colonizing populations; it could also be the result of recent migration. A third alternative is that this diversity arose by accumulation of substitutions *in situ*. This last alternative is unlikely to account for much of the diversity because there is a tendency for a given mtDNA type to be found in more than one population. Furthermore, the magnitude of sequence divergence between representative lineages from the same locality is about 0.2%–0.5%, which indicates that most of these lineages diverged long before these lakes formed (maximum 8,000 years ago; Kerr 1983). This conclusion is based on the assumption that the base substitution rate of salmonid mtDNA is not substantially different from that of mammalian mtDNA.

REDUCED DIVERSITY IN HATCHERY STOCKS OF BROWN TROUT

With the same set of seven restriction enzymes, Gyllensten, Wilson, and Ryman (unpublished) looked for variation within six hatchery populations of brown trout. Using the diversity *(h)* of maternal lineages, they evaluated the effect of using a limited number of parents in the propagation of fish stocks. Fig. 12.4 shows the distribution of h in natural populations and hatchery stocks of brown trout (the fragment patterns of the lineages in some of these populations are given in Table 12.4). A statistically significant reduction of h is seen in hatchery stocks (mean $h = 0.2$) relative to that found in natural populations

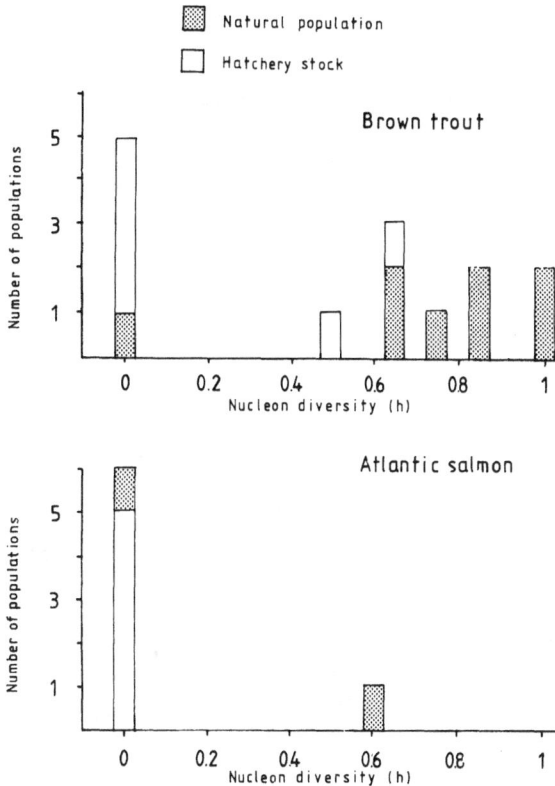

Fig. 12.4 Diversity of mitochondrial DNA lineages estimated by nucleon diversity (*h;* Nei and Tajima 1981) in natural populations and hatchery strains of brown trout and Atlantic salmon. Data from Gyllensten, Wilson, and Ryman (unpublished).

(mean *h* = 0.7) (Mann-Whitney test; P < 0.05). On average, the Swedish hatchery stocks have retained only 25% of the mtDNA variability of the natural populations.

The high fecundity of salmonid species has often tempted managers of hatcheries to limit the number of fish used in propagation of the stocks. Also, the deterioration of spawning grounds has resulted in lower returns of wild fish and has forced hatcheries to use a small number of parents. Studies of nuclear markers have shown that genetic variation has been lost in hatchery strains by using a small number of parents (Allendorf and Phelps 1980, Ryman and Ståhl 1980, Ståhl 1983, Allendorf, Ryman, and Utter, Chapter 1). Although we lack information at present on the relative loss of nuclear and mitochondrial gene variation from the same populations, it appears that the loss of variability in general has been more severe for the mitochondrial genome than for the nuclear genome. For example, the study by Allendorf and Phelps (1980) indicates that the hatchery population of cutthroat trout they examined has retained about 80% of the nuclear genetic variation (average heterozygosity) in the natural population used to found the stock. Our data on brown trout indicate a reduction of the diversity of mtDNA lineages by a factor of 4 in hatchery popula-

Table 12.4 Variation of mtDNA within and between five Swedish hatchery stocks and one mixed stock of brown trout (*Salmo trutta*).

Hatchery stock	Sample size	Fragment patterns produced by enzymes							Diversity of lineages (*h*)
		*Ava*II	*Fnu*DII	*Hae*III	*Hinc*II	*Hinf*I	*Hpa*II	*Taq*I	
Åvaån	4	B	A	C	A	A	C	A	0
Fituna	3	A	A	A	A	B	A	C	0
Weichsel	3	A	A	B	A	C	D	A	0
Gullspång	3	A	A	A	A	B	A	C	0
Dalälven	4								0.5
1		B	A	A	A	C	A	A	
2-4		A	A	C	A	A	C	A	
Lule[1]	9								0.72
1-5		A	A	A	A	B	A	C	
6		B	A	A	A	A	C	A	
7		E	A	B	A	C	A	B	
8		A	A	B	A	B	A	B	
9		B	A	C	A	C	A	A	

[1]Putative mixed stock (see text).

tions. The greater loss of variation for the mitochondrial genes is probably due to a smaller effective population size for mitochondrial DNA than for nuclear genes. Consequently, the effect of genetic drift is expected to be stronger on the mtDNA variation.

This suggests that the occurrence of bottlenecks in the history of a strain may be traced more effectively by combining mtDNA and nuclear DNA analyses than by nuclear DNA analysis alone. We note that analysis of mtDNA may indicate only population bottlenecks affecting the females; variation of nuclear genes may be brought into a population by migration of males without affecting the mitochondrial gene pool.

ADMIXTURE AND HYBRIDIZATION

Analyses of admixture or hybridization between populations are contingent upon the availability of genetic characters diagnostic for each taxon. Protein markers or single genes with a major morphological effect traditionally have been employed to detect hybridization. The usefulness of such characters relies on the occurrence of alleles that are unique to each group or occur at sufficiently distinctive frequencies in different taxa. Diagnostic biochemical genetic characters are usually available for different species but not for conspecific populations. Certain of the characteristics of mtDNA make it more appropriate than nuclear DNA for analysis of hybrid populations. The high substitution rate gives a potentially greater power to identify genetic characters unique to each taxon. Also, the uniparental inheritance allows for an examination of the direction of introgression.

Hybridization Between Cutthroat Trout Subspecies

The utility of mtDNA for studies of admixture will depend, however, on whether the proportion of different mtDNA types can be taken to indicate the proportion of nuclear genes of the different taxa, e.g., if mtDNA is a reliable marker for the nuclear genome. To test this hypothesis we, in collaboration with Fred W. Allendorf and Robb F. Leary at the University of Montana, have studied the hybridization between two subspecies of cutthroat trout in a small lake in Montana (Gyllensten et al. 1985a). This lake originally contained a natural population of westslope cutthroat trout, but repeated (1937, 1952, and 1954) attempts by fishery managers to stock the lake with Yellowstone cutthroat trout produced a hybridization with the remnants of the westslope population.

Starch gel electrophoresis was used to score 33 individuals at 45 nuclear loci, 11 of which were diagnostic. The mtDNA in 15 of these fish was studied using two restriction enzymes (*Hind*III and *Hinc*II) which previously had been shown to differentiate between the mtDNAs of the pure subspecies. Identical estimates of the proportion of Yellowstone genes in the hybrid swarm were obtained from the mtDNA and allozyme analyses (Gyllensten et al. 1985a). Concerning the genetic composition of individual fish, different proportions of nuclear genes from the two subspecies were found in the 15 fish, but there was no connection between the mtDNA type and a specific set of allozymes. The mtDNA results showed that the introgression was successful in both directions.

Composition of a Mixed Commercial Stock Of Brown Trout

The use of single individuals as operational taxonomic units in mtDNA studies opens up new ways to identify potential hybrid populations and to determine the relative magnitude of the contribution of a source population in a mixed population. An example of a data set (from Gyllensten, Wilson, and Ryman, unpublished) that can be used in such an analysis appears in Table 12.4. This table gives the fragment patterns (denoted by letters) for mtDNA from five pure hatchery stocks of brown trout and one putative mixed stock. The detailed history of the latter is not known, but there is substantial evidence that several different source populations were used during various phases of the breeding program. The mtDNA of fish from each stock was digested with seven restriction enzymes; each maternal lineage is therefore characterized by a combination of seven letters.

Five maternal lineages were found among nine fish analyzed from the hybrid population; the diversity of lineages expressed by h is the highest for any hatchery population examined so far (Table 12.4). This is a strong indication of mixing of strains.

The commonest of the maternal lineages in the mixed stock appears identical to that of Gullspång and Fituna hatchery stocks. Gyllensten, Wilson, and Ryman (unpublished) also found this same mtDNA line in a natural population. By contrast, some of the other mtDNA lineages in the mixed stock appear unique to this population; an extended survey of mtDNA variation in natural and hatchery-propagated populations also failed to detect these particular lineages elsewhere. This contrast suggests an unusually heterogeneous and unique genetic background for this population.

PATENTING OF GENOTYPES

As the use of hatchery strains in breeding programs intensifies the need will arise, as it did in plant breeding, for genetic identification systems by which commercial breeders can put a proprietary stamp on their product (strain). This problem can best be approached with a combination of nuclear and mtDNA markers.

The security of such an approach is clear. The combination of two independent marker systems would increase the probability of finding a unique combination of genetic characteristics for each strain. To duplicate the genetic constitution of a strain that is distinct with respect to allozyme frequencies and mtDNA type, a breeder would not only have to select for a specific set of allozymes but would also have to change the type of mitochondria of the material. These procedures are so laborious that few breeders would attempt them.

Since submitting this article, we have learned of four new publications which extend knowledge of the types of mtDNA present in salmonid populations (Wilson et al. 1985, 1986, Thomas et al. 1986, and Birt et al. 1986). Altogether, more than 60 populations belonging to nine species have now been surveyed.

Appendix 12.1 MtDNA fragments in five salmonid species; ● indicates an observed fragment. The species are brown trout (1, 2), Atlantic salmon (3, 4), cutthroat trout (5, 6), rainbow trout (7, 8, 9), and brook trout (10). The *Ava*I 3400 fragment is a possible duplet, and the *Bgl*I 4500 fragment is a possible duplet in taxa 2-6.

Fragment (size)	1	2	3	4	5	6	7	8	9	10
AvaI										
6700			●	●						
6400	●	●								
6200					●	●	●			
5030			●	●						
3800										●
3400			●	●						
3260										●
3100					●	●				
3030	●	●								
2750	●	●			●	●	●	●	●	●
2280					●	●				
2060	●	●			●	●	●	●	●	●
1690							●	●	●	
1580			●	●						
1310										●
1240					●	●	●	●	●	
1210							●	●	●	
1140			●	●						
1120							●			
955										●
910										●
890	●	●								
770										●
760							●			
680	●	●								
500	●	●								
AvaII										
6100										●
6000							●	●	●	
4800					●					
4500				●						
4100			●	●						
4000	●	●								
3800							●			
2800	●					●	●		●	
2500	●				●					
2350			●	●						
2300					●	●	●		●	
2100			●	●						
1950			●	●						
1945										●
1710										●
AvaII (continued)										
1700									●	
1655		●								
1650									●	
1610	●						●	●		●
1500								●		
1450	●	●	●	●						
1400	●	●			●	●	●	●	●	●
1340				●	●					
1320							●	●		
1230							●	●		
1200								●		
1100							●	●	●	
1000								●		
980										●
950	●	●	●							
940					●	●				
840										●
790					●					
760			●							
700					●					
610	●	●	●	●		●				
BamHI										
14200										●
13400					●	●				
12100								●	●	●
10100			●	●						
6400			●	●						
6200	●	●								
5900	●	●								
4400	●	●					●	●	●	
3100					●	●				
2300										●
BglI										
10800										●
5700										●
5500								●	●	●
5400					●	●				
5250				●						
4500	●	●	●	●	●	●	●	●	●	
4260	●	●								
4200							●			

Appendix 12.1 (continued) MtDNA fragments in five salmonid species; • indicates an observed fragment. The species are brown trout (1, 2), Atlantic salmon (3, 4), cutthroat trout (5, 6), rainbow trout (7, 8, 9), and brook trout (10). The PstI 3900 fragment is a possible duplet in taxa 2 and 3, and the FnuDII 1180 fragment is a possible duplet in taxa 7, 8, and 10.

Fragment (size)	1	2	3	4	5	6	7	8	9	10
BglI (continued)										
3800							•	•	•	
2800	•	•								
2630	•	•								
2250					•	•	•	•	•	
1200			•	•						
950			•	•						
700							•	•	•	
PstI										
16500										•
12600	•	•								
10500					•	•	•	•	•	
10350			•	•						
6150					•	•				
4300							•	•	•	
3900	•	•	•	•						
1700							•	•	•	
PvuII										
12400	•	•								
10350										•
6500					•	•				
6400						•	•	•	•	•
6150										•
5100					•	•	•	•	•	
4100	•	•								
3300			•	•						
2600					•	•	•	•	•	
2500			•	•						
2400			•	•						
2300					•	•	•	•	•	
1400			•	•						
SmaI										
16500	•	•	•	•	•	•				
11700							•	•	•	
11100										•
5400										•
4800							•	•	•	
XbaI										
6900	•	•								
5900										•
5300			•	•						
5200					•	•	•	•	•	•
3800	•	•	•	•	•	•	•	•	•	

Fragment (size)	1	2	3	4	5	6	7	8	9	10
XbaI (continued)										
3300	•	•	•	•	•	•	•	•	•	•
2400					•	•	•	•	•	•
2200				•	•					
2150	•	•								
2000				•	•					
1500						•	•	•	•	•
BglII										
15600	•	•	•	•						
15200										•
11800									•	
11600							•			
7000								•		
6800					•	•				
4700				•						
4500								•		
3600						•	•	•	•	
2500					•	•				
2150					•	•				
1250						•	•			•
1100								•	•	
900	•	•	•	•						
FnuDII										
5550					•					
5030				•	•					
4880				•	•					
4420	•	•								
4260										•
3600						•	•	•	•	•
3100									•	
3000									•	
2900	•	•	•	•		•	•	•	•	•
2800					•					
2750						•				
2700										•
2520						•				
2300				•	•					
2200								•	•	
1560										•
1420	•	•								
1360							•	•		
1270							•	•	•	
1250										•
1180	•	•	•	•	•	•	•	•	•	•

Appendix 12.1 (continued) MtDNA fragments in five salmonid species; ● indicates an observed fragment. The species are brown trout (1, 2), Atlantic salmon (3, 4), cutthroat trout (5, 6), rainbow trout (7, 8, 9), and brook trout (10). The 3290 and 2300 *Hinc*II fragments are possible duplets in taxa 10.

Fragment (size)	1	2	3	4	5	6	7	8	9	10
*Fnu*DII (continued)										
1150					●					
1110	●	●								●
1100					●					
980					●	●	●			
950										●
790	●	●	●	●	●	●	●	●	●	●
700										●
560						●	●	●		
530	●	●	●	●	●	●	●	●	●	●
500	●	●								
*Hinc*II										
5600						●	●	●	●	
5150			●							
4540			●	●						
4300			●							
4200				●						
3800	●	●		●		●	●	●		
3700			●	●						
3450						●	●	●		
3290										●
3000										●
2850	●	●			●	●	●	●	●	
2400	●	●			●					
2300			●	●						●
1900	●	●								
1720	●	●	●	●						
1700										●
1690									●	
1680				●	●					
910	●	●	●	●						
900						●	●	●		
740	●	●	●	●						
680	●	●	●	●						
620	●	●	●	●						

Fragment (size)	1	2	3	4	5	6	7	8	9	10
*Hind*III										
8500	●	●	●	●						
6100					●	●	●	●	●	
5100										●
4100					●	●				
4000										●
3800									●	
3500	●	●	●	●	●	●	●		●	
3030										●
2300	●	●								
2150					●	●	●	●	●	
1750	●	●			●		●	●	●	●
1350						●				
1260										●
1220					●	●	●	●	●	
1110					●	●	●	●	●	
450						●				
*Hpa*I										
11900					●	●	●	●	●	
11400	●	●								
8300										●
6700					●	●				
5000					●	●				
4600					●	●	●	●	●	
3900										●
3400					●	●				
3200										●
2400	●	●								
1700	●	●	●	●						
1050										●
1000	●	●								

13.
Chromosome Manipulation and Markers in Fishery Management

Gary H. Thorgaard and Standish K. Allen, Jr.

Induced polyploidy and gynogenesis are chromosome manipulation techniques which have been investigated in fish. Induced polyploidy, which involves the production of individuals with extra sets of chromosomes, has been of interest because triploid fish appear to be sterile. Induced sterility could be useful in fish management and aquaculture as a method of preventing over-population and improving growth and survival in fish after the age of sexual maturity. Gynogenesis involves the production of diploid individuals with both chromosome sets from the female parent. It may be valuable as a method for rapid inbreeding of fish and in the production of all-female populations.

The majority of research in chromosome manipulation to date has concentrated on methods of efficiently inducing polyploidy and gynogenesis. Research has concentrated on species for which effective means of population control are needed, such as the grass carp *(Ctenopharyngodon idella),* and on economically important species, such as the common carp *(Cyprinus carpio),* rainbow trout *(Salmo gairdneri),* and channel catfish *(Ictalurus punctatus).* Only recently have data begun to accumulate on the performance of these animals in laboratory and natural situations. We intend to describe the methods which have emerged, review the current data on characteristics of triploids, and identify what we feel to be the most likely management roles of chromosome manipulations.

Natural chromosome variations within a species, like protein variants, can be exploited as genetic markers to differentiate populations. The analysis of these chromosome markers is time consuming, but it can demonstrate differences among populations that are not apparent with other methods. We will describe some potential chromosome markers and situations in which they might prove useful.

INDUCED POLYPLOIDY

Inducing triploidy can be remarkably simple, requiring only a rudimentary understanding of fertilization events and the time and space for trial and error. Figure 13.1 shows a generalized scheme for inducing triploidy in externally fertilizing animals. Time of fertilization is carefully controlled so that treatments may be administered uniformly. Shortly after fertilization (e.g., an inter-

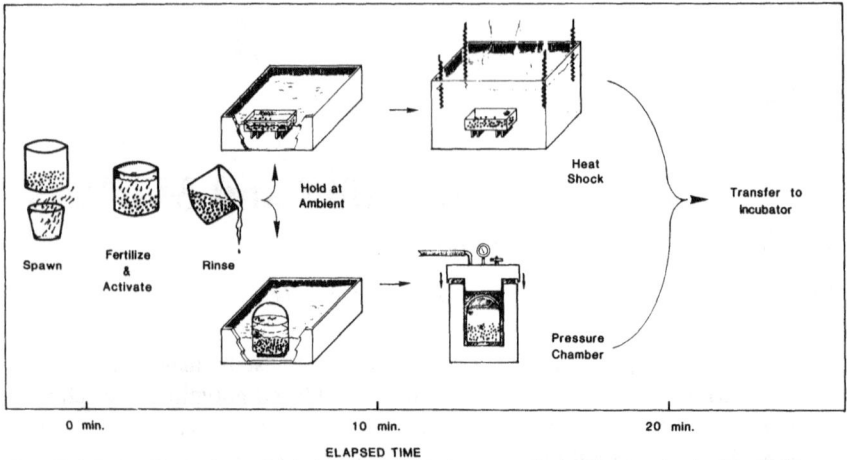

Fig. 13.1 Generalized scheme for inducing triploidy in externally fertilizing animals. The timing and duration of the heat shock or pressure treatments may vary between species.

val of 10 minutes in rainbow trout) the eggs are treated to inhibit the extrusion of the second polar body (early treatment, Fig. 13.2). In experiments this treatment has consisted of chemicals (cytochalasin B, Refstie et al. 1977, Allen and Stanley 1979, Allen et al. 1982), heat shocks (Chourrout 1980, Thorgaard et al. 1981, Utter et al. 1983, Refstie 1983), cold shocks (Chourrout 1980, Wolters et al. 1981), and hydrostatic pressure (Streisinger et al. 1981, Yamazaki 1983, Benfey and Sutterlin 1984, Chourrout 1984, Lou and Purdom 1984, Allen and Myers 1985).

Thermal shocks are the most readily amenable procedure for present-day culture operations. As shown in Fig. 13.1 thermal shocks require inserting only a single step into the production scheme. The type, intensity, and duration of the treatment depend on the species being cultured. Trials to identify the optimal intensity of thermal shock begin by finding the upper (or lower) lethal limit and then "backing off" to allow acceptable survival. The lethal level, and also the most effective level of treatment, varies with species: generally, cold water species respond to heat shock, while warm water species are affected by cold shocks (Table 13.1). In either case, uniform thermal shocks can be assured by using large volumes of water to treat thin layers of eggs.

An induction technique currently receiving some attention is the use of hydrostatic pressure. Allen and Myers (1985) found that pressures of 10,000 psi produced at least 90% triploidy in five full-sib chinook salmon *(Oncorhynchus tshawytscha)* families; heat shocks in these same families produced 30%-100% triploidy. Benfey and Sutterlin (1984a) produced 100% triploidy in landlocked Atlantic salmon *(Salmo salar)* using 10,150 psi. In other reports, pressure was shown to be highly effective in inhibiting polar body extrusion (Dasgupta 1962, Tompkins 1978, Muller et al. 1978, Chourrout 1984). Pressure treatments could also be incorporated into production schemes, although an initial investment in

Table 13.1 Species and biotypes of some fishes in which triploidy has been induced by thermal shock.

Species	Biotype	Shock	Reference
Loach	warm	cold	Romashov and Belyaeva 1965
Plaice	cool	cold	Purdom 1969, 1972
Tilapia	warm	cold	Valenti 1975
Carp	warm	cold	Ojima and Makino 1978; Nagy et al. 1978
Catfish	warm	cold	Wolters et al. 1981
Brook trout	cold	cold	Lemoine and Smith 1980
Rainbow trout	cold	cold, heat better	Chourrout 1980
		heat	Thorgaard et al. 1981
Stickleback	warm	heat, cold better	Swarup 1959
Sturgeon	cool	heat	Vasetskii 1967
Chinook salmon	cold	heat	Utter et al. 1983
Pink salmon	cold	heat	Utter et al. 1983
Coho salmon	cold	heat	Utter et al. 1983

an apparatus is necessary. A pressure cylinder for production might hold a liter or more and be capable of generating 10,000 psi for a maximum of 15 minutes. There are some species on which pressure shock would be logistically easier to use, e.g., fish having gelatinous egg masses, shellfish with microscopic eggs, or even animals which control fertilization such as prawns, because the entire brooding female can be pressurized.

The possibility of creating tetraploid broodstock to cross with diploids for production of sterile triploids has been raised a number of times, (e.g., Refstie et al. 1977, Thorgaard et al. 1981, Chourrout 1982a, 1984), but progress in producing and rearing tetraploids has been slow. Tetraploidy can be induced by treatments which block the first cleavage division (late treatment, Fig. 13.2). Chourrout (1984) produced viable tetraploid rainbow trout using hydrostatic pressure treatments. More recently, Chourrout (personal communication) has successfully crossed tetraploid male rainbow trout to diploid females to produce triploids. If the viability and fertility of tetraploids are high, crossing tetraploids to diploids might eventually be the preferred method for generating triploid fish.

Interest in triploids arises from two characters triploids might display: increased growth due to larger cell size, and sterility. However, of all the species in which triploids have been produced, increased juvenile size has been demonstrated in only one (*Tilapia,* Valenti 1975). The benefits of polyploidy thus appear to lie primarily in sterility and its attendant side effects.

Sterility, Growth, and Survival

Stocks of fish often are planted into ecosystems that do not support a natural spawning habitat; rainbow trout, landlocked Atlantic salmon, and striped bass *(Morone saxatilis)* are three such examples. Sterile triploids could be useful in these situations. By circumventing gonadal maturation, triploids of

CHROMOSOME SET MANIPULATION

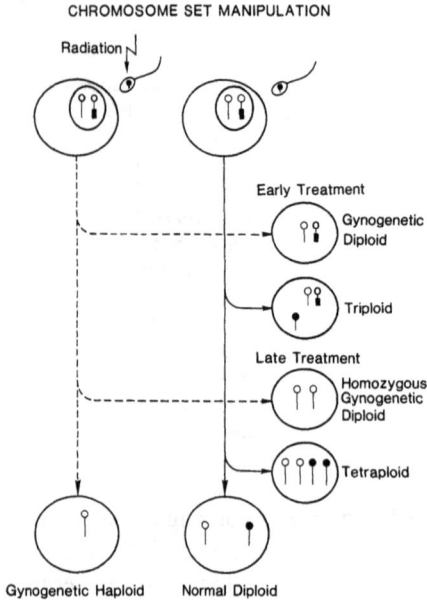

Fig. 13.2 Alternative chromosome set manipulation procedures for externally fertilizing animals. Before fertilization, the egg contains two nonidentical sets of chromosomes; the figures in the nucleus represent these sets from the egg and the second polar body. Fertilization with radiation-inactivated sperm induces gynogenesis (dashed line). An early pressure or temperature shock treatment causes retention of the second polar body and results in partially homozygous gynogenetic diploids or in triploids. A late treatment to block the first cleavage results in totally homozygous gynogenetic diploids or in tetraploids.

several species can grow faster than diploids (see Thorgaard 1983a for review). However, triploids outperform diploids only when they have the opportunity to reach the age of sexual maturity (Lincoln 1981a, Wolters et al. 1982). That is, triploids are not larger because they are triploid per se but because they are sterile and do not shunt somatic growth to gonadal development. Thus, it is not likely that sterile fish will be a significant management tool in "put and take" fisheries with high fishing pressure. On the other hand, triploids may outperform diploids in fisheries where fish have the opportunity to grow older. Substantial benefits might be realized if sterility extends longevity, as in Pacific salmon (Robertson 1961, Hunter and Donaldson 1983), or prevents the deterioration of flesh quality inherent to maturation. Hormonal sterilization (reviewed by Donaldson and Hunter 1982) would result in similar long-term benefits.

While there is little doubt that triploids of any species will be sterile, i.e., unable to produce viable offspring (Lincoln 1981b, Lincoln and Scott 1984), gonads may appear and differentiate to some degree. Substantial testis development was demonstrated in male triploid rainbow trout (Thorgaard and Gall 1979, Lincoln and Scott 1984), Atlantic salmon (Benfey and Sutterlin 1984b), plaice *(Pleuronectes platessa)* and plaice × flounder *(Platichthys flesus)* hybrids (Lincoln 1981b), and channel catfish (Wolters et al. 1982); female triploids of these species showed substantially less development than diploids (see also Lincoln 1981c). It is not known if this residual gonad development in triploids affects longevity.

Triploid Hybrids

One of the more exciting and untapped potentials of induced triploidy resides in the creation of triploid interspecific or intergeneric hybrids (Allen and Stanley 1981, Chevassus 1983, Thorgaard 1983a). Spontaneous polyploids observed following intergeneric crosses—e.g., grass carp × common carp (Vasilev et al. 1975), rainbow trout × brook trout, *Salvelinus fontinalis* (Capanna et al. 1974, Ueda et al. 1984), and grass carp × bighead carp, *Aristichthys nobilis* (Marian and Kraznai 1978)—probably reflect increased survival of triploid hybrids over diploid hybrids. Triploid amphibian hybrids also frequently appear to survive better than diploid hybrids (Bogart 1980, Elinson and Briedis 1981). Incompatibilities between the genomes of the two species apparently are somehow moderated by the doubled maternal gene dose. Studies with induced triploid salmonids (Chevassus et al. 1983, Scheerer and Thorgaard 1983, Arai 1984) and triploid tilapia (Chourrout and Itskovich, in press) have provided direct evidence that triploid hybrid fish survive better than diploid hybrids.

Hybridization followed by retention of two maternal chromosome sets could be an approach to the creation of new "species" for sport and commercial fisheries. Unlike diploid hybrids, triploids are likely to carry a majority of maternal traits, possibly satisfying market demands in commercial fisheries. Triploidy in conjunction with hybridization would almost certainly assure sterility (Allen and Stanley 1981).

Some examples of possible uses of triploid hybrids follow.
- Refstie et al. (1982) found that coho *(Oncorhynchus kisutch)* × chinook salmon hybrids grew much faster than pure coho, but with higher mortality and increased incidence of deformities. Triploid coho × chinook hybrids might retain the hybrid vigor of the diploids while showing lower mortality and fewer deformities.
- The Yellowstone cutthroat trout *(Salmo clarki bouvieri)* × rainbow trout cross is recognized as a vigorous hybrid, but managers fear introgression of rainbow genes into cutthroat trout populations. Triploid hybrids would allow a robust sport fishery while preventing unwanted interbreeding with native cutthroat trout (Rohrer 1982).
- Utter et al. (1983) produced reciprocal crosses of pink *(Oncorhynchus gorbuscha)* and chinook salmon, hypothesizing that a valuable new product could emerge by blending the desirable qualities of the parents, early seawater adaptation of pinks and the more desirable flesh quality of chinook.

Control of Reproduction

Several factors have limited the use of pure (intraspecific) triploids for population control. First, studies of induced triploids have just recently emerged from the "biotechnical" era of trial and error with a wide variety of species. Experimental lots of eggs treated for polyploidy were often not large enough to commence field trials or were terminated during their evaluation. Second, in all species examined thus far, pure triploids are morphologically indistinguishable from diploids. Although methods for identifying triploids have become increasingly sophisticated (e.g., Thorgaard et al. 1982, Allen 1983), it is not always practicable to separate many hundreds of triploid individuals from diploids.

The grass carp provides an example of a species whose triploid possibilities are being evaluated by fisheries managers. These herbivorous fish were introduced in the United States in 1963 and have since been advocated as a biological weed control agent. A possible solution to concerns about the establishment of undesired reproducing populations of this exotic was found when Marian and Kraznai (1978) reported that the F_1 hybrid progeny of female grass carp and male bighead carp were triploid. Because the hybrid receives two doses of the maternal genome, it retains many of the characters of the grass carp and is far more vigorous than the diploid hybrid. Cassani (1981) reported that the feeding behavior of under-yearling triploid hybrids was similar to that of the grass carp. The possibility was raised that the triploid could supplant the pure diploid grass carp as a weed control agent. Kraznai et al. (1982) reported that the gonads of hybrids "were less developed in each case than the parental species of the same age." However, there is no other clear documentation to date concerning the fertility of this fish. Recent feeding studies indicate that surgically sterilized and diploid grass carp consume significantly more vegetation than did triploid hybrids of similar size (Beatty et al. 1983, Shireman et al. 1983).

More recently (fall 1983) the production of "pure" (i.e., not hybridized) triploid grass carp was announced by J. M. Malone and Son, a commercial fish breeder in Lonoke, Arkansas. By a combination of proprietary techniques of egg manipulation and careful sorting of fingerlings, they obtained 98% triploids (Allen, unpublished data, and Anon. 1983). Triploid grass carp were shown to consume as much hydrilla *(Hydrilla verticillata)* as diploids and 3–4 times as much as triploid hybrids (Wattendorf and Anderson, in press). Because of its greater weed-eating ability and the prospects for equivalent sterility, the pure triploid has supplanted the hybrid as the preferred herbivore. It appears that most states will require 100% triploidy before they allow stocking in major drainages. Systematic screening of diploids and triploids is performed by sizing erythrocyte nuclei with a Coulter counter and channelizer (R. J. Wattendorf, personal communication, and J. M. Malone, personal communication).

If the gonads of male triploids "mature," will they spawn? If they spawn, they may interact with diploids in natural populations; sperm from triploid male plaice (Lincoln 1981b) and triploid male rainbow trout (Lincoln and Scott 1984) were used to fertilize normal eggs, and in both cases all the resulting embryos died, probably from aneuploidy.

The possibility of spawning triploids raises an intriguing possibility and a potential problem. Intriguing is the possibility that triploid males of "undesirable" species, e.g., some spiny ray or cyprinid species, might be used in population control if they mated with diploids to produce inviable offspring (Thorgaard 1983a). Similar approaches have been widely used in insect population control (Smith and von Borstel 1972). Gamete production in triploid males would suggest, however, that releasing triploids of more desirable species into natural populations might interfere with natural reproduction. The latter threat could be reduced by creating all female (XXX) triploid populations which do not sexually mature (Lincoln and Scott 1983).

Reproductive behavior of triploids is still largely an unanswered question. Will the hormones produced by meiotically aberrant testes or rudimentary ovaries elicit normal sexual behavior? What effect will sterility have on the migratory behavior of triploid salmonids? Will anadromous species ever return from ocean waters? These questions await the second stage of investigations into the efficacy of induced polyploidy as a management tool. Because of the progress realized thus far on techniques to induce polyploidy in fish, many of the substantial potential benefits can now be examined in field trials from a fisheries management perspective.

INDUCED GYNOGENESIS

Gynogenesis involves producing diploid individuals with both chromosome sets from the female. In species with homogametic (XX) females, gynogenesis is expected to result in all-female offspring. Gynogenesis has two major potential applications: sex control by production of monosex populations, and rapid inbreeding for the generation of inbred lines (Stanley and Sneed 1974, Purdom 1976, 1983, Cherfas 1981, Chourrout 1982b).

Techniques for inducing gynogenesis are similar to those used for inducing polyploidy, except that radiation-treated or chemically inactivated sperm is used in the place of normal sperm (Fig. 13.2). Relatively few studies to date have used chemically inactivated sperm for gynogenesis (e.g., Uwa 1965, Tsoi 1969, Mantelman 1980). Both gamma and ultraviolet radiation have been used frequently to inactivate sperm. Gamma radiation (often using a Co^{60} source) has the advantage of excellent penetration, which facilitates treatment of large quantities of sperm. However, residual chromosome fragments are frequently found in gynogenetic offspring after fertilization with gamma-irradiated sperm (Ijuri 1980, Chourrout and Quillet 1982, Onozato 1982, Chourrout 1984, Thorgaard et al., 1985). Such fragments may reduce survival in the gynogenetic offspring; consequently, ultraviolet radiation may be the preferred method for sperm chromosome inactivation.

Ultraviolet (UV) treatment of sperm has been used successfully in a number of gynogenesis studies (e.g., Stanley 1976, Streisinger et al. 1981, Chourrout 1982c, Thorgaard et al. 1983). The procedure is relatively simple: sperm is diluted in an appropriate extender, spread in a thin film in a dish, and exposed to UV light from a germicidal lamp. Large numbers of eggs can be fertilized by adding the eggs directly to the dish in which sperm inactivation takes place. Normally at high UV exposures, no residual paternal characters (Ijiri and Egami 1980) or chromosome fragments (Chourrout 1982c) are passed on by the sperm. To confirm that paternal genes are not transmitted, sperm donors of the same species homozygous for dominant genetic markers (Nagy et al. 1978, Chourrout 1980, Streisinger et al. 1981) or of a foreign species (Stanley and Jones 1976) should be used.

After fertilization with inactivated sperm, the eggs can be treated as previously described (Fig. 13.1) to produce gynogenetic diploids. Treatments applied soon after fertilization (early treatment, Fig. 13.2) will produce partially homozygous gynogenetic diploids. If totally homozygous gynogenetic diploids are desired, as for studies aimed at producing homozygous inbred lines, similar treatments can be used to block the first cleavage division (late treatment, Fig. 13.2; Streisinger et al. 1981; Thorgaard et al. 1981; Chourrout 1982a, 1984).

Sex Control

Much of the interest in gynogenesis in fish has focused on the generation of all-female populations. The production of such populations after gynogenesis could be one means of preventing reproduction. Gynogenesis has several disadvantages for control of reproduction, however.

One problem is that gynogenetic fish are inbred. This is an advantage in the generation of inbred lines, but a disadvantage if fish are to be released in nature. The survival of gynogenetic fish of many species has been shown to be substantially lower than that of normal fish (reviewed in Thorgaard 1983a), presumably a reflection of inbreeding.

A second problem with the use of gynogenesis for control of reproduction is that although the offspring may all be of one sex, they remain fertile. Accidental introduction of males or occasional production of males after gynogenesis (e.g., Chourrout and Quillet 1982) could then lead to the establishment of an unwanted population. The induction of sterility through induced triploidy or hormone treatments seems preferable to control of reproduction by gynogenesis.

However, gynogenesis may prove to be a valuable step in the indirect production of all-female populations. In this approach, a gynogenetic (all-female) population is hormonally sex-reversed into functional males (Nagy et al. 1981, Donaldson and Hunter 1982, Jensen et al. 1983). The resulting fish are genetically female but functionally male, and will produce all-female offspring when mated to normal females. The method has several advantages:

- The fish being treated are all females, so that all functional males recovered after the hormone treatment are known to be genetically XX.
- If sex-reversed gynogenetic fish are mated to unrelated females, the all-female offspring in the succeeding generation will be outbred.
- Sperm from the XX males can be used to produce large numbers of females in the next generation.

Manipulating the sex ratio using the gynogenesis-sex reversal approach has a number of possible applications (Donaldson and Hunter 1982). Jensen et al. (1983) propose this method for generating all-female grass carp for population control. In Pacific salmon, females are more valuable because fewer individuals mature sexually at a young age and because the eggs are highly valued for caviar. Female rainbow trout, unlike males, normally do not mature sexually before the time of commercial harvest and consequently are more desirable. In general, the manipulation of sex ratios in favor of females in hatchery populations could allow a greater egg production from the same number of broodstock. Shifting the sex ratio, however, requires that hatchery managers take special care to avoid inbreeding problems, because a skewed sex ratio may result in a drastically reduced *effective* number of parents (see Chapter 3, this volume).

Inbred Lines

The application of gynogenesis for the production of inbred lines will certainly be important in research, but the likelihood of practical application in fish culture and management remains uncertain. Inbred lines have been important tools in such areas as mouse research (Green 1981) and maize breeding (Allard 1960). The major obstacle to the use of inbred lines in fish research and breeding has been the considerable length of time needed to generate them. Gynogenesis could dramatically shorten the time required to produce inbred lines in fish.

The approach used by Streisinger et al. (1981) in zebra fish *(Brachydanio rerio)* provides a model for the development of inbred lines using gynogenesis. They first produced homozygous diploid zebra fish by gynogenesis after blocking the first cleavage division. Another cycle of gynogenesis then resulted in fully homozygous offspring that were genetically identical to each other, a fully inbred line in two generations. Streisinger et al. (1981) also observed that hybrids between inbred lines showed increased vigor and that more vigorous lines could be produced by further selection. These results suggest that inbreeding by gynogenesis may have merit in practical breeding programs.

There are many obstacles to the use of gynogenesis in breeding programs, however. The techniques involved are more complicated than mass or family selection schemes for improvement. Gynogenesis by blocking the first cleavage, while maximizing inbreeding (100% in a single generation), also maximizes mortality and abnormalities in the offspring. More gradual approaches to inbreeding using gynogenesis by retention of the second polar body combined with sib-mating may prove more feasible in practice (Nagy and Csanyi 1984). Gynogenesis by retention of the second polar body results in approximately 45%-65% inbreeding in a single generation, depending on the amount of crossing-over in meiosis (Thorgaard et al. 1983). Sib-sib mating results in 25% inbreeding, while self-fertilization results in 50% inbreeding.

The use of inbred lines in breeding programs also raises the danger of generating a monoculture that might prove more susceptible to a particular disease or to environmental problems (Streisinger et al. 1981). The use of gynogenetic inbred lines will probably remain at the research stage until clear evidence of a practical potential is demonstrated.

Two other applications of gynogenesis for research, while not likely to be immediately useful in fishery management, are nevertheless worth discussing here. Gynogenesis after second polar body retention is a useful method for mapping genes in relation to their centromeres on chromosomes (Thompson et al. 1981, Nagy and Csanyi 1982, Thorgaard et al. 1983, Guyomard 1984, Thompson and Scott 1984). This is useful for studying chromosome rearrangements in evolution and in estimating the rate of inbreeding after this method of gynogenesis. Because it allows rapid inbreeding, gynogenesis could also facilitate the isolation of new strains of fish that are homozygous for recessive alleles and could be used for genetic studies (Stanley and Sneed 1974, Streisinger et al. 1981).

Androgenesis, which involves the inactivation of the egg chromosomes followed by the doubling of the chromosomes contributed by the sperm, is an alternate approach to gynogenesis for rapid inbreeding in fish. Relatively little work has been done to date on this technique in fish (Romashov and Belyaeva 1964, Purdom 1969, Arai et al. 1979, Parsons and Thorgaard 1984, 1985), but encouraging results in amphibians (Gillespie and Armstrong 1980, 1981) and the fact that males may frequently mature earlier than females in fish suggest that this method for rapid inbreeding has merit.

CHROMOSOME MARKERS

Although a karyotype consisting of 48 acrocentric chromosomes is quite common among fish species, most closely related species of interest have distinct karyotypes (Sola et al. 1981). Chromosome polymorphisms within fish species are also well documented. Greatly improved chromosome preparation techniques have been developed recently (e.g., Kligerman and Bloom 1977, Yamazaki et al. 1981, Blaxhall 1983a), but chromosome analysis is still much more time consuming than alternative identification techniques such as protein electrophoresis. Consequently, chromosome markers are likely to be most useful in management situations where protein electrophoresis does not provide a clear distinction between groups, as in some comparisons within species.

A common feature of the various chromosome polymorphisms we will describe is that they appear to have little obvious phenotypic effect because they do not involve changes in the amount of functional genetic material. One type of chromosome variation that seems to be relatively common is variation in chromosome number. A common type of chromosome rearrangement in fish and other animals are centric fusions and fissions, sometimes termed Robertsonian rearrangements, in which one-armed chromosomes fuse to form two-armed chromosomes, or two armed chromosomes split to form one-armed chromosomes (White 1973). This results in a change in chromosome number without changing the chromosome arm number or substantially changing the amount of genetic material. Many of the chromosomal differences between fish species involve this type of rearrangement (Gold 1979, Sola et al. 1981).

Robertsonian variation has been found within a number of fish species, including rainbow trout (e.g., Ohno et al. 1965, Thorgaard 1976, 1983b, Hartley and Horne 1984), Atlantic salmon (reviewed in Gold 1979, Hartley and Horne 1984), cutthroat trout (Gold et al. 1977, Loudenslager and Thorgaard 1979) and black sea bass, *Spicara flexulosa* (Vasilev et al. 1980). A surprising aspect of this variation has been the many reports, beginning with one by Ohno et al. (1965) on rainbow trout, of Robertsonian variation within individual fish (reviewed by Gold 1979, Kirpichnikov 1981). An unknown but possibly substantial fraction of this variation reported within individual fish may actually reflect errors in chromosome counting (Thorgaard 1976). Much of the Robertsonian variation within species, such as that found among rainbow trout native to the Pacific coast of North America (Thorgaard 1983b), probably represents stable, inherited variation among populations. More examples are likely to be found as more species are examined in detail.

Variation in the number of short second arms of chromosomes has been found in rainbow trout (Thorgaard 1976, 1983b) and cutthroat trout (Gold et al. 1977, Loudenslager and Thorgaard 1979). This type of variation has been found in rodents (e.g., Murray and Kitchin 1976) and is a common type of chromosomal difference between fish species (Sola et al. 1981).

Variation in the nucleolus organizer regions (NORs) of the chromosomes may also be common. The NORs are the achromatic stalks seen on one or several pairs of chromosomes in most species. These regions are specifically stained using a "silver-staining" technique (Goodpasture and Bloom 1975, Howell and Black 1980, see Fig. 13.3). Variations in these regions include elongated stalks, enlarged satellites on the end of the stalks, and double satellites which are observed when the region has been duplicated. These types of varia-

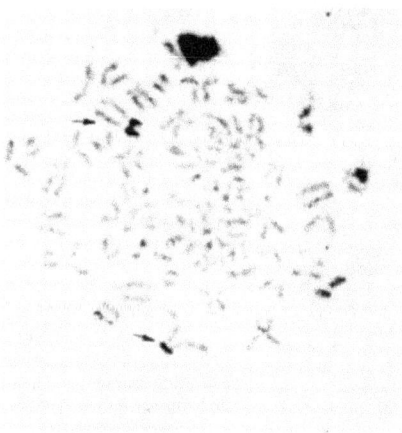

Fig. 13.3 Metaphase spread from brown trout *(Salmo trutta)* showing silver staining of nucleolus organizer regions (NORs), indicated by arrows. The two NORs differ in size.

tions are found in humans and other animals (Rocchi et al. 1971, Ward 1977, Markoviac et al. 1978, Ruiz et al. 1981). Several studies have also found variations in the NORs in fish species (Thorgaard 1976, Foresti et al. 1981, Phillips 1983, Gold 1984).

Variations in the heterochromatin in the centromere and telomere regions of chromosomes are well documented in humans and other animals (John and Miklos 1979, Therman 1980). These can be detected using both C-banding for heterochromatin and Q-banding using the fluorescent dye quinacrine. Phillips and Zajicek (1982) demonstrated Q and C band variation in individual metacentric chromosomes in lake trout, *Salvelinus namaycush* (see Fig. 13.4). This type of heterochromatic variation is likely to be found in other fish species after careful study. However, to identify such variations, the individual chromosome pairs need to be distinguishable. This is sometimes possible on the basis of chromosome size and centromere position, but it will be increasingly feasible as improved methods for producing bands along the chromosome arms are developed (e.g., Ojima and Ueda 1982, Blaxhall 1983b, Delany and Bloom 1984).

The time required for chromosome analysis means that chromosome markers will only be practical for stock identification where few protein differences between populations are available. Chromosome markers show potential application in lake trout, a species in which Q-band variation but little protein variation has been found (Phillips and Zajicek 1982), and among rainbow trout populations along the west coast of North America, which show Robertsonian variation but little protein variation among some populations (Thorgaard 1983b). Other situations where chromosome markers could prove useful in stock identification are likely to arise in the future.

Until recently, cytogenetic studies in fish have been of considerable

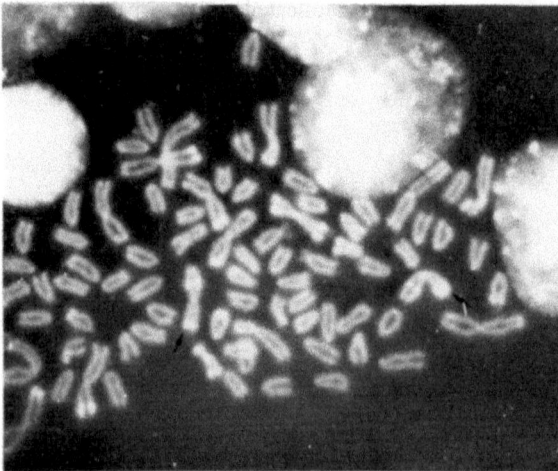

Fig. 13.4 Metaphase spread from lake trout *(Salvelinus namaycush)* stained using the Q-banding technique. The Q bands on chromosome 7 (indicated by arrows) differ in size.

basic interest but of limited value to management programs. The developments in the manipulation and detailed analysis of fish chromosomes that we have described suggest that cytogenetics is likely to play a larger role in fish management in the future.

14.

Stock Transfer Relative to Natural Organization, Management, and Conservation of Fish Populations

Yuri P. Altukhov and Elena A. Salmenkova

From its inception at the beginning of this century, population genetics has focused on gaining a better understanding of the population dynamics of evolutionary processes (e.g., Chetverikov 1926, Fisher 1930, Wright 1931, Haldane 1932, Dubinin and Romashov 1932, Dobzhansky 1955, Lewontin 1974). An extensive theoretical and empirical body of knowledge has accumulated and resulted in a commonly held belief that man has almost infinite power to willfully transform the biosphere in whatever manner he chooses because of the genetic variability existing within and among its component organisms. However, today it is more and more evident that man's carelessness is unintentionally but, unless checked, inevitably guiding the biosphere toward drastic and potentially disastrous and irretrievable losses of genetic resources (see Chapter 15). Thus, it is impossible to consider natural variability a sufficient buffer to compensate for any or all changes caused by human activity.

It is therefore necessary to understand that the stability and evolutionary potential of populations, species, and ecosystems are not infinite. Further, it is evident now that successful management of natural populations is associated mainly with the factors and conditions of their genetic stability rather than with their evolutionary potential. Thus, the adaptive genetic structure of fish populations has generally remained unknown or ignored. The long-term consequences of such actions are losses of genetic diversity and, ultimately, productivity (e.g., Chapter 15, Altukhov 1983, Altukhov and Varnavskaya 1983). Stock losses due to harvest beyond the point of recovery are the most obvious examples of what can happen when one ignores the genetic diversity within a species, i.e., the typological concept that one individual of a species has the same genetic potential as another.

A less obvious but equally harmful example of typological activity is the transfer of stocks among areas without consideration of the genetic attributes of the donor stocks and indigenous populations in the areas of introduction. This problem has received considerable attention recently because of growing concern about negative genetic and ecological effects of such introductions (see Chapter 15 and associated references).

This chapter addresses that concern by presenting data on transplanted chum salmon *(Oncorhynchus keta)* populations of the Soviet Union. Since such data have been reported in several previous publications (Altukhov 1974, 1981, Altukhov et al. 1980, Altukhov and Salmenkova 1981), we shall focus on the results of our most recent studies. The effects of these transplantations are identified and discussed, and followed by a more general discussion relating to the management and conservation of fish populations.

STABILITY OF ALLELIC FREQUENCIES IN CHUM SALMON POPULATIONS

Populations of many fish species, particularly the salmonids, are characterized by a complex structure of subpopulations representing historically developed *population systems*. Their most important features include common spawning area and time, integrity of morphological and behavioral characters, and a high degree of isolation as compared with subpopulations (for details, see Altukhov 1974, 1975, 1977, 1981). Such population systems as a whole are characterized by long-term genetic stability due to reciprocal balance between such dynamic factors as random genetic drift, gene migration, and selection.

If allelic frequencies for a particular population remain stable over generations and year classes (Altukhov 1973, Utter et al. 1974, Ryman 1983), the possibility arises for the effective application of biochemical genetic data to studies of the population structure. With such stability even a single estimate may be useful to characterize a population for extended periods, and data collected at various times for a particular population and locus may be pooled to increase the precision of the overall estimate. Contrary to earlier expectations (e.g., Williams et al. 1973), reports are accumulating that demonstrate the temporal stability of allelic frequencies in unperturbed populations of diverse organisms. Such data have been accumulated on rainbow trout *(Salmo gairdneri)* (Allendorf and Utter 1979, Utter et al. 1980), chinook salmon *(Oncorhynchus tshawytscha)* (Chapter 10), sockeye salmon *(Oncorhynchus nerka)* (Grant et al. 1980, Altukhov et al. 1975 a,b, Altukhov 1981, Ryman 1983), pink salmon *(Oncorhynchus gorbuscha)* within-year classes (Utter et al. 1980, Salmenkova et al. 1981) chum salmon (Altukhov 1974, 1981, Okazaki 1978), and brown trout *(Salmo trutta)* (Ryman 1983). Further stability of allelic frequencies in chum salmon populations is documented here as a basis for subsequent comparisons of data from transplanted stocks.

Allelic frequency data for several protein loci were collected from the Kalininka River on Sakhalin Island and the Kurilka River on Iturup Island (Fig. 14.1) for a number of consecutive years (Table 14.1). Each of the Kalininka and Kurilka populations consists of several (at least three) subpopulations; i.e., they represent two population systems. If spawners are examined over all of the spawning run, allele frequency differences are revealed between separate samples taken within a year, as repeatedly demonstrated for the Kalininka River chum salmon (Altukhov et al. 1980, Altukhov 1981). Nevertheless, for this pop-

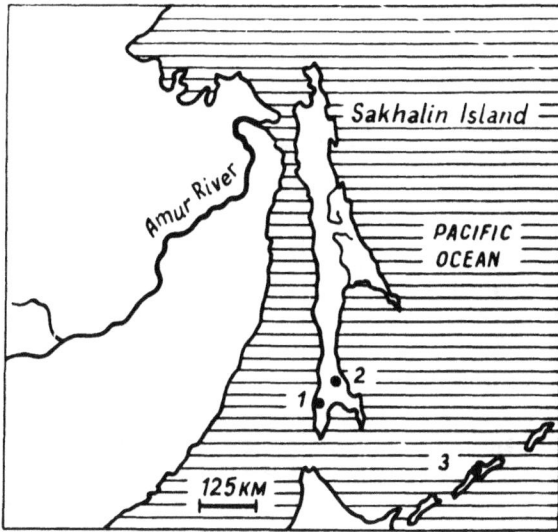

Fig 14.1 Location of chum salmon populations concerned in egg transfer: (1) Kalininka River; (2) Naiba River; (3) Kurilka River.

ulation as a whole, allelic frequencies at different loci are characterized by considerable stability over the period of observation (Table 14.1). The Kurilka River population has been investigated in less detail (no more than two samples within a spawning season), but a similar stability was observed.

The allelic frequencies at a number of loci differ (without overlapping) in samples taken from the Kalininka and Kurilka rivers. Such clear-cut differences constitute the genetic basis for identification of fish transplanted from one river to another.

TRANSPLANTATION DATA

Minimal stock transfer of chum salmon to the Kalininka River occurred prior to 1976, when a deficiency of mature fish resulted in the transfer of about 34 million eggs from the Kurilka hatchery. These transplanted fish were expected to return as four-year-olds in 1980 and five-year-olds in 1981. Therefore, a detailed examination of allelic frequencies was made of the returns to the Kalininka River in those years in an attempt to detect the presence of Kurilka River fish and to estimate their contribution.

An examination of these data (Table 14.2) in comparison with those of Table 14.1 gives no indication of anything but Kalininka River fish in the first six collections of 1980 (September 3 through October 3). However, the frequencies in the last two samples (October 13 and 20) are indistinguishable from those of the Kurilka River (Table 14.1); late-run data of the Kalininka River of previous years coincide with those of earlier runs within the population. The conclusion that these late-run fish of 1980 are those transplanted from the Kurilka River is further supported by the generally later mean return time of chum salmon to the Kurilka River; specifically, the eggs transplanted in 1976

Table 14.1 Frequencies of the most common alleles of *MDH, AAT, LDH-1, ME-2, IDH-2, IDH-3* and *PGD* loci in the Kalininka (1969-1981) and Kurilka (1970-1980) chum populations.

River, year	Number of samples	Number of fish	Allele frequencies						
			MDH[a]	*AAT*[a]	*LDH-1*	*ME-2*	*IDH-3*	*IDH-2*	*PGD*
Kalininka River									
1969	3	191	0.919						
1970	4	349	0.920						
1971	4	187	0.924		0.985				
1974	4	392	0.902		0.996				
1975	2	99	0.909		1.0				
1976	3	125	0.908	0.918	1.0				
1977	3	150	0.923	0.908	0.983			0.650	0.997
1978	3	148	0.910	0.935	0.993			0.685	0.997
1980	6[b]	585	0.910	0.937	0.980	0.905	0.908	0.677	0.996
1981	5[b]	492	0.910	0.927	0.989	0.883	0.898	0.662	0.990
Total		2,718	0.912	0.928	0.989	0.895	0.904	0.669	0.994
χ^2			3.08	3.81	17.59[c]	2.74	0.62	1.38	4.1
d.f.			9	4	7	1	1	3	3
Kurilka River									
1970	1	95	0.979		0.845				
1971	2	81	0.985		0.810				
1972	2	94	0.987		0.830				
1977	1	31	0.968		0.871			0.484	1.0
1978	1	97	0.995	0.954	0.845		0.768	0.569	1.0
1979	1	95	0.992	0.958	0.898		0.801	0.506	1.0
1980	1	97	0.977	0.952	0.874	0.745	0.737	0.505	1.0
Total		590	0.984	0.955	0.852	0.745	0.768	0.522	1.0
χ^2			4.48	0.08	7.26		2.23	2.50	
d.f.			6	2	6		2	3	

[a] Allele frequencies for duplicated loci *MDH* and *AAT* are calculated on the basis of two loci for *AAT* and two loci for *MDH* per individual. The frequency of each allele is assumed to be the same at each *AAT* and each *MDH* locus (May et al. 1975).

[b] Average allele frequencies are given for the samples from September 3 to October 3, 1980, and from September 3 to October 6, 1981, in which admixture of Kurilka River fish is absent (see details in Table 14.2 and in the text).

[c] $P < 0.01$. In this case the significant value of χ^2 can be explained by the low expected frequencies of the less frequent allele.

were collected during mid-to-late October. These findings parallel those of Okazaki (1978), who reported persistence of allele frequencies and timing of areas of origin in chum salmon transplanted in the Tokachi River (Hokkaido) in Japan.

The results of the 1981 returns to the Kalininka River are less definitive regarding the proportion of five-year-old fish transplanted from the Kurilka River. This difference from the 1980 returns was expected because only a small proportion of Kurilka River fish return as five-year-olds (with a range of

Table 14.2 Frequencies of the most common alleles of *MDH, AAT, LDH-1, ME-2, IDH-2, IDH-3* and *PGD* loci in samples from the Kalininka chum population, 1980-1981.

Date	Number of fish	MDH	AAT	LDH-1	ME-2	IDH-2	IDH-3	PGD
1980								
Sept. 3	98	0.888	0.940	0.970	0.914	0.697	0.890	1.0
Sept. 13	97	0.917	0.955	0.977	0.914	0.655	0.910	1.0
Sept. 19	98	0.920	0.940	0.995	0.885	0.711	0.905	0.980
Sept. 25	96	0.913	0.895	0.975	0.880	0.662	0.919	1.0
Sept. 30	96	0.903	0.951	0.978	0.944	0.655	0.909	0.995
Oct. 3	100	0.917	0.943	0.985	0.895	0.680	0.900	1.0
Oct. 13	98	0.947	0.942	0.861	0.804	0.545	0.830	1.0
Oct. 20	100	0.967	0.958	0.876	0.795	0.475	0.795	1.0
1981								
Sept. 3	96	0.920	0.967	0.979	0.872	0.655	0.870	0.985
Sept. 10	98	0.913	0.929	0.995	0.885	0.640	0.900	1.0
Sept. 20	100	0.905	0.890	0.990	0.891	0.660	0.920	0.985
Sept. 29	98	0.897	0.933	0.980	0.914	0.657	0.890	0.985
Oct. 6	100	0.910	0.905	0.990	0.865	0.697	0.910	0.995
Oct. 12	99	0.925	0.927	0.950	0.853	0.568	0.830	0.995
Oct. 20	100	0.903	0.936	0.936	0.865	0.604	0.899	1.0

Note: Allele frequencies for duplicated loci *MDH* and *ATT* are calculated on the basis of two loci for *ATT* and two loci for *MDH* per individual. The frequency of each allele is assumed to be the same at each *AAT* and each *MDH* locus (May et al. 1975).

2%–19%). Age determinations from scale readings indicated that 26% (12 of 46) of the fish collected on October 6 were five years old, while 43% (22 of 51) in the October 12 collection were five years old. It is notable that in this collection the allelic frequencies of each of the loci are intermediate between the mean values for the Kalininka and Kurilka rivers given in Table 14.1, in contrast with values close to the means for the Kalininka River in the other collections.

The above data are interpreted to reflect a mixed origin of the October 12, 1981, sample. The apparent absence of Kurilka River chum in the October 20, 1981, sample is somewhat anomalous, perhaps because this sample came from the very last fish returning in that year.

An estimation of the relative contributions of the two populations to the 1980 and 1981 collections by the proportionate contributions of alleles in the subsamples (Altukhov 1974; see also Allendorf and Utter 1979) support the conclusion that a late segment of the 1980 Kalininka returns predominantly represented transplanted Kurilka fish and a reduced proportion of such fish returned in 1981. The estimates of Kurilka River fish were 85% in the October 12 and October 20 samples of 1980, and only 39% in the October 12, 1981, sample. There were no indications of Kurilka chum in the other 1980 and 1981 samples.

The hatchery statistics on the size of runs coupled with the results of this study indicate that 0.03% of the total Kurilka River chum releases returned to the Kalininka River. This figure is less than one-tenth the estimated proportions for the Kalininka (0.4%) and Kurilka (0.7%) fish returning to their native areas, and clearly reflects the inadaptive nature of the transplantations.

INADAPTIVE NATURE OF TRANSPLANTED POPULATIONS

From the viewpoint of genetics, acclimatization is adaptation to a new environment. The efficiency of acclimatization can be assessed only after the formation of a self-reproducing population with a stable, integrated gene pool which is able to persist indefinitely through many generations. Unfortunately, there is little evidence at present in favor of this phenomenon.

However, there are many contrary examples. Ricker (1972), who summarized a vast body of data on transfer of North American salmonid populations, showed that the return of the first generation to a "foreign" river takes place rather often but significantly less often than to a "native" river; in subsequent generations the return to a foreign river is sharply reduced or does not occur at all. Pink salmon introduced almost annually since 1956 from Sakhalin Island into rivers of Kola Peninsula have returned only in small numbers and as a rule only from odd-year lineages (Dyagilev and Markevich 1979). It is obvious, however, that these introduced fishes do not reproduce in the natural way.

According to data of Okazaki (1978, 1982) from numerous chum transplantations in Japan, only a few transplantations within Hokkaido were successful (i.e., were observed for three generations after transplanting). The phenomenon of acclimatization in salmon has been reviewed recently by Withler (1982), who concluded from an extensive review of the literature that transplants of Pacific salmon within their natural range have been singularly unsuccessful in producing new anadromous stocks.

These and many other facts suggest the unique and conservative character of local adaptations formed by natural selection throughout thousands of generations in a particular environment with which the entire life history of a population is related. When we disturb the relationships between components of the ecosystem, by removing the population from its own historically formed environment and transferring it into a new environment, the genetic stability acquired in one environment is usually insufficient for the new one. The facts that the transferred population returns for reproduction at the same time as it did in the native river and that the measurable genetic structure of the transplants remains constant despite the very low coefficient of return suggest only that selection under the new conditions plays a catastrophic role representing nonselective elimination of the protein genotypes examined.

It is not quite clear what stages of ontogenesis are the most sensitive in transplanted chum. It has been demonstrated in pink salmon that the survival rate of transplanted fish during incubation and the level of return of adult fish to the new coastal region conform to the survival rate and the corresponding level of return for the native population (Bams 1976). However, returns of transplanted pink salmon to the particular river where they were hatched constituted only a small part of the returning native stock, but the marked fish did not return to the donor river either. In short, homing of the transplants appeared to be disturbed. It should be noted, however, that from the viewpoint of population genetics, pink salmon is an "unspecialized" species; it does not display the pronounced spatial genetic differentiation that chum salmon does, and so there are greater possibilities for "interchangeability" of some populations (Salmenkova et al. 1981, Altukhov et al. 1983).

As for chum salmon, according to the data of the Kalininka hatchery, the loss among the transplanted Kurilka chum eggs during incubation only slightly exceeded the loss of the native chum eggs; thus, the earliest stages of ontogenesis in both chum and pink salmon appear to be adaptable to new hatchery environments. In natural conditions the situation may be different; one cannot rule out increased mortality of transplants during migration and juvenile growth in coastal regions. Whatever the cause, unfavorable effects of transplantation of salmon eggs are substantial and, as indicated above, related to the conservative character of adaptation of each particular population.

Is it possible to increase the efficiency of transplantation, particularly in instances where it is necessary to restore depleted stocks or to create new ones? What we have learned about the genetics of natural populations permits the following working hypothesis to be proposed for experimental checking by future investigations: Efficiency of acclimatization must be related to the degree of genetic similarity of populations having common origin.

The first step toward testing this hypothesis can be taken now by the comparing genetic distances with the return rate for spawners for three pairs of populations investigated by us in this study and in previous work (Table 14.3; see Altukhov et al. 1980). An inverse proportionality between genetic distance and rate of return is apparent from these data. The largest genetic distance in the Kurilka-Kalininka pair corresponds to the smallest return rate. The genetic

Table 14.3 Genetic distances D (by Nei 1972) between populations (rivers) and rates of return of transplanted fish.

Rivers			Rate of return
Donor	Recipient	$D \times 10^{-5}$	(percent)
Kalininka	Naiba	233	0.10
Kurilka	Naiba	279	0.11-0.17
Kurilka	Kalininka	398	0.03

distances in the two other population pairs are smaller and similar to each other, and the corresponding rates of return are larger and also similar.

It should also be considered that when eggs are transplanted, usually only part of the genetic variability of a population system is transferred. Therefore, the possibilities for adaptation to the new environment are already limited to the extent that adaptation is associated with genetic diversity.

This phenomenon has been considered already by us in detail (Altukhov 1981) and is most clearly seen in the course of artificial reproduction of fish populations in hatcheries (see also Chapter 6). For instance, a sharp reduction of genetic diversity in a hatchery population of cutthroat trout *(Salmo clarki)* as compared with a native donor population in Montana has been described by Allendorf and Phelps (1980), who studied 35 electrophoretically identified protein loci. After 14 years of artificial maintenance of the hatchery population the proportion of polymorphic loci had decreased by 57%, the average number of alleles per locus had decreased by 29% and the individual heterozygosity had decreased by 21%.

A similar situation has been reported for Atlantic salmon *(Salmo salar)* in Sweden. A comparison of 37 enzyme loci between native and artificially maintained populations revealed that the amount of inter- and intrapopulation variability of hatchery populations was considerably lower than that of natural populations (Ståhl 1983). Swedish hatchery stocks of brown trout have also been shown to exhibit substantially lower levels of genetic variation and to have allelic and genotypic distributions very different from the natural populations they were assumed to represent (Ryman 1981, Ryman and Ståhl 1980, 1981).

Investigation of the relationship between genetic diversity over several generations for five polymorphic enzymes and coefficients of return for chum salmon spawners at two hatcheries in Sakhalin has revealed a positive relationship (Altukhov 1983). Lower levels of polymorphism and heterozygosity are correlated with lower numbers of fish returning.

All these data indicate that an optimal strategy for the conservation and management of fish populations must be adopted and carried out. What should this strategy be?

MANAGEMENT AND CONSERVATION OF FISH POPULATIONS

It follows from the above considerations that the most important task in developing a strategy of gene pool conservation is to coordinate knowledge of population structure with management practices. Considering our understanding of the structures of natural populations, we can apply our practical activities of exploitation, artificial reproduction, and acclimatization in ways that could also open the possibility of intensive economic management.

But sound coordination of genetics and management is possible only if the features of systemic organization of populations are taken into account, their historically formed hereditary heterogeneity is preserved, and the auto-regulation mechanisms are maintained to enable efficient adaptation of populations to normally varying environmental conditions. Only this approach permits the practical realization of an optimal strategy which allows for not only economic benefits but also continuous maintenance of biological resources.

To organize a regulated fishery, it is necessary to know the population structure of the species concerned, the degree of isolation of subpopulations, and the tempo of their reproduction and size dynamics. Finally, on the basis of that structure, it is necessary to provide the fishing industry with well-founded recommendations so that the intensity of fishing does not exceed the tempo of natural (or artificial) reproduction.

The subpopulation structure should also be taken into consideration in artificial reproduction of populations. It should be remembered that each population has its own evolutionally formed biological optimum, determined by the preceding evolution and the current position of a population in the ecosystem. Regarding the size of populations, the optimal limits are determined by their minimal and maximal levels, and steady estimates of these levels can be obtained only through averaging the results of long-term observations. It is clear that controlling the artificial reproduction of a system is very difficult unless one knows its stability range.

We recognize that these concepts contradict the view that some genotypes of fish (e.g., those relating to rapid growth, high fertility, and high maturation tempo) are important to man while other genotypes are less valuable and therefore undesirable, or at least expendable, a view that has resulted in the advocacy of directional selection in natural populations (Larkin 1981, Malinovsky and Mina 1976, Mina 1978). As is well known, our entire agricultural practice is based on this principle.

However, it should be remembered that extreme reinforcement of single characters and properties in the course of directional selection is inevitably due to the existence of negative correlations in the system of ontogenesis (Shmalhausen 1938, Belayev 1980) related to weakening and deterioration of other characters. It results in the reduction of genetic diversity of a breed or variety—a usual "cost of selection."

There are enough negative examples of this cost in agriculture, with its specific potentialities to maintain environmental stability (Chang 1979). As for fisheries, the situation is no better. Nevertheless, the negative experiences are seldom discussed. Usually the same, more or less successful, examples are cited again and again although they are few in number. Moreover, results documented in the literature are of rather short term in view of those time intervals which characterize the lifetime of historically established native populations. From the viewpoint of genetics, the time measure in living nature is generations rather than years. Although our approach has been criticized (Malinovsky and Mina 1976, Mina 1978; see also Altukhov 1977, 1983), its validity in the rational use of biological resources is recognized by many workers (Starobogatov

1975, Aronshtam et al. 1977, Dubinin 1976, Kirpichnikov 1981, Konovalov 1972, 1980, Konovalov et al. 1975, Thorpe et al. 1981) and has been implemented at fish hatcheries in Sakhalin, where the data indicate a positive effect.

In the more general sense, the material presented to this point has focused on a singular approach to prevent the disastrous loss of genetic variation; that is, our strategy of interaction with nature must be reorganized in a manner that will reverse the accelerating destruction of gene pools and promote the stable development and permanent coexistence of the "Man and Biosphere" system.

Such a model can be called a model of socio-ecological optimum. Its realization in practice requires great coordinated efforts by representatives of different sciences, and population genetics can and must make its contribution. From this point of view, it seems urgent to adopt an approach for the protection and rational management of natural resources which is based not only on evolutionary principles but also on the factors and conditions affecting genetic stability of populations and species.

CONCLUSIONS

The principles of population genetics are common to all biological species. Much attention has been paid to fish populations because of their practical importance, and the genetic structure of their populations has been intensively studied for the last 15–20 years. These studies, in turn, have made it possible to identify the systemic organization of isolated populations of other species in diverse ecosystems, to obtain new data on the mechanisms of their genetic stability, and to determine causes of disturbance of this stability in the course of our economic activity when natural laws are ignored. The natural laws emphasize an urgent necessity to stop polluting the environment, removing forests and farm lands, exhausting the soil and killing wild animals to an extent that exceeds their natural self-reproductive abilities. The absolutely vital and irreplaceable value of the genetic diversity of life should serve as a constant reminder that the remaining unique genetic pools must be preserved.

But how can this be done? At present the problem of gene pool protection is solved mainly by the creation of biosphere reserves or national parks; i.e., by creating slightly changed or, ideally, intact life reserves under conditions of sharply changing environments (Sokolov 1981). Such an approach is very important, but it is also necessary to control those channels of anthropogenic pressure on the biosphere that are associated with exploitation and artificial reproduction of biological resources.

Conservation of genetic diversity of the remaining population systems, restoration of those with disturbed structures, and creation of new population systems are the main problems whose solution will help to eliminate or at least diminish the observed conflict between mankind and the biosphere. The same principle of conservation of the evolutionarily formed genetic diversity must be true of any level of biological organization in the sense of resistance to external influences. It can be demonstrated that in all these cases the stability conditions of ecosystems remain unchanged due to self regulation through interaction of relatively independent structural components exchanging information on their own state and the state of the environment.

The long-term existence of a protected or a newly formed community in a stable environment and its ability to respond adequately to external influences which do not exceed the adaptation optimum are possible only on the basis of preservation, restoration, and imitation of the historically determined directions and intensity of the informational and energetic flows. We must implement such measures:

- We must turn from qualitative formulations to the creation of mathematical and experimental models that permit long-term forecasting of the behavior of natural populations under different types of industrial and other influences on their gene pools.
- Population genetic studies must be carried out on those animal and plant species which are the most important for man.
- The contacts must be strengthened between scientific and industrial organizations directly engaged in exploitation and reproduction of biological resources.

15.
Genetical Conservation
Of Exploited Fishes

Keith Nelson and Michael Soulé

Recent reviews have adequately described the current state of fish conservation throughout the world (Berst and Simon 1981, FAO/UNEP 1981, Ryman 1981a). The threats are manifold and serious. They include pollution, siltation and other habitat destruction, introduction of exotics, and overfishing. Rather than elaborating once again this litany of horrors, we have instead focused upon a more restricted subject of utmost concern to fisheries management: what are the effects of the fishing and hatchery practices themselves upon the gene pools of the harvested species?

In a larger philosophical context, our posture here must be self-consciously anthropocentric. To address the subject at all we must assume as axiomatic the following points:

- Fishes exist for the welfare of humans, or at least no moral laws forbid our killing and eating them.
- It is in our interest as a species to maintain and increase the productivity of fish populations in order to provide more and better food for people, and to keep open the exploitative options of future generations.

But we question other assumptions that are implicitly or explicitly maintained by large segments of humanity, no doubt including each of us at one time or another:

- We, individually and collectively, have a right to fish where and when we please, so long as we do not violate the property or treaty rights of other humans.
- It is always in the best interest of society for a fishery to maximize individual and collective wealth by maximizing current and future economic returns.

These last two assumptions are literally counterproductive. They have been instrumental, we believe, in the progressive decline in fisheries production worldwide, and in the collapse of scores of individual fisheries in almost all rivers, lakes, and oceans. The "scramble frenzy" mentality and the eye on the profit margin may permanently compromise the future welfare of the communities that depend on the affected fisheries. Humanity has yet to overcome its primitive greed. It continues to maximize short-term returns and to discount the future.

What would a non-anthropocentric essay on fish conservation propose? Perhaps such an attempt is logically impossible, for our thinking is necessarily human. We cannot exchange our feet for fins or our hair for scales. But we can proclaim unashamedly our love of nature and of all biological diversity. It may not be anthropomorphic to project onto other species our own "racial" desire to survive, and accordingly to grant them the space and time to exist and evolve. Again, we do not imply that it is evil to enjoy lox, calamari, and caviar, for carnivory is not inherently immoral. It is only excess that is offensive to nature. And when we destroy the ability of species to survive and to maintain their ecological position, when we destroy their habitats and their capacity to evolve, that is excess.

Awareness has grown in recent years that genetics is of central concern in problems of biological conservation, whatever one's philosophical persuasion (Soulé and Wilcox 1980, Frankel and Soulé 1981, Schonwald-Cox et al. 1983). Several recent conferences (Berst and Simon 1981, FAO/UNEP 1981, Ryman 1981a) have stressed the importance of genetic considerations to rational fisheries management. In this chapter we examine the evidence for genetic changes in fish populations consequent to man's exploitative activities, and outline preventive and remedial measures suggested by quantitative and population genetics. Necessarily, we will also assess the likely impact of various management policies upon the genetic structure of fish populations. We assume throughout that the preservation of genetic variation, the avoidance of artificial selection, and the maintenance of subpopulation structure are desirable goals for fisheries management.

All the changes we discuss can be characterized as "loss of genetic diversity." Genetic variation may be partitioned hierarchically, from the diversity among higher taxa down to the level of the individual genotype (Chapter 4). Man's activities can cause losses within a taxon or losses resulting from breakdown between phylads, as with interspecific hybridization of a resident population with an introduced species.

Loss of diversity is most evident in the progressive changes in species composition in intensive and selective fisheries, although these changes often may be confounded with losses brought about by eutrophication and other forms of pollution (Smith 1968, Regier 1973). Study of such species succession in fisheries may offer clues to the nature of more subtle fishery-caused genetic erosions that may be occurring below the species level. Of these, selective elimination of subpopulations or stocks has attracted the most attention (Berst and Simon 1981).

Within a stock, we must distinguish between undirected loss of genetic variation (inbreeding, genetic drift) and directed change (selection), for the problems and solutions they represent are distinct. Both will be obscured by the large nongenetic (and often reversible) responses to exploitation shown by most fish populations (e.g., Nikolskii 1969). The theoretical background necessary to sort out these various responses is beginning to emerge at the junction of fish population dynamics, life-history theory, and population and quantitative genetics (Dingle and Hegmann 1982). The question of directional selection in fisheries was last addressed by Miller (1957), who could find little evidence to support the idea. In the last 25 years evidence has accumulated, and it is time once again to review the subject.

Hybridization can also lead to genetic attrition. It results from the breakdown of barriers, geographic and otherwise, to reproduction between stocks or species. Because hybridization is so often the result of introducing one stock into the territory of another, we discuss hybridization and introduction together. We have emphasized conservation of natural populations and de-emphasized hatchery strategies, which are discussed elsewhere in this book. Similarly, we have stressed the ounce of prevention over the pound of cure.

PROCESSES CAUSING LOSS OF GENETIC DIVERSITY

Succession in Intensive Fisheries

Regier and Loftus (1972), Regier (1973), and Spangler et al. (1977) have described the often-observed succession of species during the "fishing up" sequence in freshwater communities: initial exploitation of the larger, later maturing species higher on the food chain and, as these decline, successive concentration on smaller species, perhaps at lower trophic levels. Size selection occurs because the largest species usually have the greatest value and higher catchability, at least initially; it also results from the usually size-selective nature of the fishing gear itself.

A classic example concerns the bottom-trawl fishery for the cichlid species-flock of southern Lake Malawi, begun in 1968 (Turner 1977). By 1975 in one arm of the lake the larger species had disappeared from the catch entirely, and medium-sized species were greatly reduced in abundance. Turner attributes the changes to the size-selective mesh in the cod end of the trawl: the larger, fast-growing species were caught and therefore could not reproduce.

Similar species successions occur in intensive marine fisheries, but apparently the process is more often modified by competitive interactions and changes in utilization, current, and climate patterns (Garrod 1973a, Regier and McCracken 1975, Daan 1980, Merriman 1982). Fishery collapses are prevalent, especially in the great pelagic fisheries (Cushing 1968, Saville 1980), but local extinction may be followed by repopulation from refugia, as in the case of the recovery of the Japanese sardine (*Sardinops melanosticta;* Kondo 1980).

Stock selection within species. Successive elimination of stocks, those more or less reproductively isolated subpopulations differentiated by time or location of spawning, has characterized man's impact upon both marine and freshwater species (Berst and Simon 1981). Often the sequential loss was not noted in the catch statistics (Regier and Loftus 1972, Selgeby 1982), and the pattern could only be reconstructed retrospectively through historical records and interviews.

Where large morphological differences existed, the most valued form typically disappeared first (for example, the fat "siskowet" lake trout of Lake Michigan; Brown et al. 1981). Probably more often, morphological differences between stocks were not pronounced. Loftus (1976) compiled the evidence available at that time showing that when stocks differed in life history characteristics, the less productive and the less resilient to exploitation were the first to disappear. Had current techniques of stock identification been available (Casselman et al. 1981; Ihssen et al. 1981; Pella and Milner, Chapter 10; Ferris and Berg, Chapter 11), the richness of stock structure might have been appreciated before the components were irretrievably lost.

Erosion of stock structure has consequences that are easily grasped, and fisheries biologists have become increasingly aware of the need to prevent it (Berst and Simon 1981, FAO/UNEP 1981, Ryman 1981a). Without special precautions, selective loss of stocks from an intensive multistock fishery may be inevitable, especially if there are differences in life-history features and fishing mortality is considerable (Ricker 1958, 1973, Larkin 1963, Paulik et al. 1967). It is likely that the first stocks to disappear would be those with properties most desirable to the fishery and to future enhancement or aquaculture efforts, e.g., rapid growth and high catchability (Thorpe and Koonce 1981).

If the fishery itself is unselective, life-history differences can still result in the selective disappearance of stocks, particularly the less resilient and less productive ones (Loftus 1976, Larkin 1977). Hatchery-maintained stocks, for example, possess an artificial advantage in early survival. Exposure to a common fishery may then result in the selective elimination of native stocks even in the absence of "swamping" through hybridization, and all in spite of the goal of rehabilitation (Paulik et al. 1967, Helle 1981, Larkin 1981).

Other apparently sound management policies may have contrary results. For example, fishing the maximum sustained yield was supposed to provide for the preservation of the fishery. But Ricker (1973) and Larkin (1977) have shown how the policy could only result in disaster for the less productive members of a stock complex. Again, the provision of adequate escapement of fish to breed is the rationale behind closure of inshore areas to Pacific salmon fishermen (Loftus 1976). But these are the areas in which the stocks diverge to their natal streams. The opposite policy of confining fishing to these inshore areas facilitates separate management of the stocks. Judicious use of temporal closure and controlled access may still provide for adequate escapement.

The likelihood that a population is composed of spatially and/or temporally isolated stocks is a function of topography, life history, behavior, and historical circumstances (Spangler et al. 1981). Without adequate stock identification (Ihssen et al. 1981; Pella and Milner, Chapter 10) accurate estimates of the composition of stock mixtures (Fournier et al. 1984; Pella and Milner, Chapter 10), and knowledge of life history differences, rational management of multistock fisheries may be impossible.

Fishing pressure is not the only agent of man in the selective erosion of stock structure. Such activities as the construction of hydroelectric power plants often interfere directly with spawning migrations (Ros 1981). Often, tolerances to water conditions are quite narrow, especially in stocks adapted to oligotrophic conditions (Loftus and Regier 1972). As a result, fish stocks are differentially endangered by human activities that increase water temperature, increase the load of suspended particles, change the rate of flow, or affect salinity or pH.

Even less dramatic changes such as progressive siltation of spawning sites, milder industrial pollution, eutrophication, and introduction of exotic species may render a stock more vulnerable to the stress of overexploitation by man (Berst and Simon 1981, FAO/UNEP 1981, Ryman 1981a). For example, the progressive disappearance from Lake Michigan of the larger species of chub (or cisco, *Leucichthys,* a subgenus of *Coregonus*) was thought to result from size-selective predation by the introduced sea lamprey *(Petromyzon marinus),* but the estimated "catches" by the sea lamprey and by the size-selective human fishery during the 1950s were probably equivalent (Smith 1968).

Undirected Genetic Erosion Within Stocks

Effective population size and loss of genetic variation. Survival of a stock may depend upon maintenance of adequate numbers in many ways besides the tautologically obvious one. Reduction of spawning stock below some critical point may have a *depensatory* effect—that is, the self-reinforcing or positive feedback of stock size upon the numbers recruited in the following generation (Neave 1954)—upon the ability of the stock to recover. For example, predators may consume a certain number of juveniles, and recruitment failure may occur if the number of juveniles is not greater than this (Clark 1976). Similarly, schooling may not protect a stock if the protection from predation offered by schooling is a function of school size (Clark 1974). Fishing pressure may be similarly depensatory, as when a constant-yield harvesting strategy by positive feedback paradoxically increases the variance in numbers of the catch and, incidentally, of the stock (Beddington and May 1977). Catchability may actually increase as the stock shrinks in numbers or area. Even with fishery closure, its incidental catch in other fisheries may exceed recruitment (Garrod 1973a, Gulland 1977) and drive the stock to extinction.

In such a stock (and in those already lost) the additional question of genetic drift or inbreeding-induced loss of genetic variation may be moot, but in other cases and in hatchery stocks, inbreeding depression may accelerate the processes leading to extinction. This is because "fitness characters" related to reproduction, with low heritabilities but great dominance variation, are most sensitive to inbreeding (Robertson 1955). In normally outbred species, a 10% increase in the inbreeding coefficient ΔF ($\Delta F = 1/2N_e$, where N_e is the effective population size) may result in a 5%–10% decrease in fitness for a particular reproductive trait and a total decrease in reproductive performance of 25% or more (Frankel and Soulé 1981).

It therefore behooves the manager to keep inbreeding to a minimum where the possibility exists; Frankel and Soulé suggest a maximum ΔF of approximately 1%. This translates to a minimum N_e of about 50. If it is further desired that an isolated native population maintain its historical level of genetic variation, and thus retain its long-term adaptive potential, minimum N_e appears to be more on the order of 500 (Franklin 1980, FAO/UNEP 1981).

Although inbreeding is a major concern with hatchery stocks (Chapter 1 and Chapter 3), and although loss of heterozygosity may characterize small relict or otherwise isolated populations (Soulé 1980), there seems to have been little effort expended in documentation of inbreeding or genetic drift resulting from human influence upon natural aquatic populations. Northern elephant seals *(Mirounga angustirostris)* display no electrophoretically detectable polymorphisms, a fact that possibly reflects loss of genetic variation as a result of reduction to extremely small population sizes by hunting during the last century (Bonnell and Selander 1974). But elephant seals are quite different from most populations in growth parameters, breeding structure, and fecundity.

Factors affecting N_e and its estimation. There are as yet very few trustworthy estimates of N_e in natural populations of any sort (Ryman et al. 1981), and it may or may not be that the dangers of inbreeding, especially in heavily fished continental waters, have been seriously underestimated. In one important study, average N_e in a series of lake- and river-spawning subpopulations of sockeye salmon *(Oncorhynchus nerka)* was estimated to be approximately 200 (Altukhov 1981). Although N_e appears low in this example, interchange between subpopulations occurs, and they seem in little danger of inbreeding (Allendorf and Phelps 1981b).

N_e is extremely difficult to measure. Except in heavily managed hatchery populations, it must be very much smaller than the actual population size. It is important to realize that juveniles and nonbreeding adults do not enter into its calculation. High fecundity and density dependence of recruitment may mask a serious depletion in the stock. There may seem to be plenty of fish still around when in fact they are the progeny of relatively few adults (Cushing 1973, Garrod 1973b).

N_e is decreased by an unbalanced sex ratio. It is closer to the number of the less frequent sex, because it is a function of the harmonic mean of the sexes (Crow and Kimura 1970). More importantly, it is an inverse function of the variance in lifetime family size, which is apt to be especially high in iteroparous fishes. This is, first, because of enormous fecundity and the great importance of density-independent mortality factors in recruitment, as reflected for example in orders-of-magnitude variation in year-class strength and survival (Garrod 1973b, Cushing 1973).

Second, variance in lifetime family size is high because of the peculiar age distribution of that fecundity in iteroparous fishes. In populations with overlapping generations, the variance in lifetime family size is related to the reproductive value (expected future offspring discounted by future mortality and population increase) remaining at death. Even in humans, in which age-specific reproductive value, v_x, declines immediately following sexual maturity, the effect of variance in lifetime family size is to reduce N_e by two-thirds (Felsenstein 1971). Fish are unusual in that v_x is not inversely related to mortality-at-age, q_x, and does not decline at or soon after sexual maturity. Instead, as fecundity increases with weight (Bagenal 1973), v_x continues to rise with increasing size far beyond the age of maturity, whereas q_x remains approximately constant (Michod 1978). Except in heavily managed hatchery populations, the combination of the effects of skewed sex ratio and variance in lifetime family size may reduce N_e by an order of magnitude or more.

We are surprised by the result that a reproductively isolated population of as many as 5,000 adults (i.e., 10 times Franklin's minimum N_e of 500) might be in danger of ultimate loss of genetic variation. Many, perhaps most, smaller lentic, lotic, or migratory stocks are isolated or nearly so. Genes introduced by even a single migrant per generation may swamp the effects of inbreeding, however, regardless of the size of N_e (Wright 1969).

This may be used as a management tool to counteract allelic loss without destroying local adaptation. In cases in which it can be assumed that little selection is occurring and that equilibrium between migration and genetic drift is approached, the number of migrants may be estimated using the same electrophoretic allele frequency data obtained to determine stock structure (Allendorf and Phelps 1981b, Allendorf 1983). When recent isolation of stocks has occurred as a result of human activity, the estimate is rather of the former number of migrants, which might be used as a guide to the maintenance of an artificial level of migration.

Detection of stock reduction and inbreeding effects by the same allozyme electrophoresis technology is relatively straightforward and easier than attempting an accurate estimate of N_e. Reduction in heterozygosity in longitudinal studies or relative to undisturbed populations should be detectable by means of isozyme analysis (FAO/UNEP 1981; Allendorf et al., Chapter 1; Utter et al., Chapter 2). If the stocks differ morphometrically, osseometrically, or meristically, these characters may be used in combination with electrophoretic differences (Casselman et al. 1981, Ihssen et al. 1981). Incidence of deviant phenotypes or fluctuating asymmetry of morphometric or meristic variation (Valentine et al. 1973) could also be used to warn of loss of heterozygosity as well as presence of pollution and other forms of stress (Soulé 1982, Leary et al. 1985c).

The effects of management practices. When the parameters of the stock-recruitment relationship are "not uncertain," as in some Pacific salmon stocks, the optimum fishery management policy is to allow a fixed escapement of fish to breed each year (Walters and Hilborn 1976). We suggest that this would almost always be the optimum policy genetically, especially in fisheries on isolated stocks as discussed above. In the large marine fisheries, the dangers of loss of genetic variation may not be so apparent, and investment tends to fix fishing effort at a fairly constant level that cannot respond easily to a changing surplus above some escapement (Garrod 1973a). But especially when such fisheries are composed of multiple stocks, an escapement "floor" might best be established that is related to the numbers in the smaller stocks (Paulik et al. 1967, Larkin 1977). As the relationships between escapement and N_e and between stock and recruitment are so uncertain, the minimum allowable escapement should be chosen conservatively.

Walters and Hilborn (1976, 1978) advocate experimental "active adaptive management" programs which may at times call for deliberate overfishing to clarify the relationship of recruitment numbers to stock numbers. Part of the attractiveness of such programs lies in their apparent objectivity, part in the opportunity they seem to offer for learning more about the dynamics of exploited populations. But the population biology of a stock establishes a minimum size for it, and the exact course of the stock-recruitment relationship near the origin of the graph may be better left unknown.

Otherwise, the goal of utilizing management options to increase knowledge of the fishery is laudable, especially when it is suspected that a hitherto constant escapement level has been set too low. Certainly, "active adaptive management" is more reasonable than "pulse fishing," where, in the usual absence of any information about a stock other than that it is abundant, the strategy is to raid the stock and reduce it to a very low level, (Clark 1976, Regier and McCracken 1975). With some simplified models, pulse fishing may even appear to be an "optimal" economic solution (Botsford 1981; but see also Botsford and Wainwright 1985); that it is dangerous, especially to fisheries with uneven or unknown stock composition, cannot be overemphasized.

While there exist breeding schemes designed to maximize N_e (Frankel and Soulé 1981, Templeton and Reed 1983), most hatchery practices seem unconsciously designed, in the interests of short-term economy, to reduce N_e as much as possible (see Chapter 1 and Chapter 3). Technology is constantly suggesting new ways to do this. As an example, Helle (1981) points out that often one male salmon is used to fertilize several females, whereas in the natural situation as many as 25 males may fertilize one female's eggs. Production of monosex female groups of salmon by means of heterologous steroids (Hunter et al. 1983) is designed to distort the sex ratio even further. A popular account enthusiastically relates this scenario: "The manager ... decides to produce an all-female run.... In a few years the returning fish will be mostly females, with a few normal and sex-reversed males for spawning" (Biggs-Chisum 1984). We expect N_e then to be closely related to the numbers of these few males, and the long-term result to be irreversible loss of genetic variation.

Such conflicts between short-term gain and long-range benefit seem always to be resolved in favor of the former; "future value" is discounted in more ways than one. Fisheries exploit a commonly held resource base, and the destructive practices we encounter there are understandable economically if not justifiable in terms of the future of the resource (Christie and Scott 1965, Clark

1976). But the first concern of hatcheries should be preservation of the gene pool within which they must operate (FAO/UNEP 1981, Ryman 1981a).

Changes Resulting From Selection

Evidence of changes caused by fishing. Directional changes in gene frequency in natural populations under stress, perhaps surprisingly, are more difficult to detect than undirected changes, and even less well understood. It might appear obvious that if a fishery consistently takes more of the fast-growing animals, or those more easily caught, or those in the larger schools, the fish left to breed will be characterized in general by slower growth, lower catchability, or a reduced tendency to school. To the extent that variation in such traits is genetic, eventually the population characteristics should change.

Wohlfarth et al. (1975) have demonstrated just such genetic differences between Chinese and European carp stocks. The Chinese stocks are supposed to have been "selectively" fished for centuries whereas the European stocks were "husbanded," that is, the fish with the more desirable traits were saved for breeding. The Chinese carp are more difficult to catch than the European carp, have a faster juvenile growth rate but then grow more slowly, are younger and smaller when they reach sexual maturity, and have higher viability and fertility.

One must remember, however, that the Chinese carp was pond-cultured, not exploited by a fishery. Furthermore, to be an appropriate model of the results of selective fishing, it should be compared with carp populations that do not have a history of exploitation. Good examples of intrapopulation (as opposed to stock) selection are very few, notwithstanding an apparently widespread belief in the genetic dangers of fisheries selection (e.g., Moav et al. 1978, Kirpichnikov 1981, Dourojeanni 1982).

In the following discussion, we will restrict ourselves to consideration of the effects of size selection upon the age at sexual maturity, α, and size at sexual maturity, l_α), which together mediate the effect of growth rate upon fecundity.

Ricker's (1981 and earlier) thoughtful studies of the effects of fisheries selection on the simplified life histories of five British Columbia salmon species *(Oncorhynchus)* display in detail the kinds of complexity that can exist. All the species are semelparous; that is, they breed only at the end of life. Two have uniform or nearly uniform α. Three are exposed to inshore fishing mortality only as they approach maturity. By comparing the average weights of fish caught by the relatively unselective seine fishery and by the highly selective gill netting and trolling methods, Ricker et al. (1978) and Ricker and Wickett (1980) were able to establish a strong case for selection as the cause of marked declines in growth rate in the two species with uniform α (pinks, *O. gorbuscha,* and cohos, *O. kisutch);* however, their data do not completely rule out stock selection as opposed to intrastock selection.

In the species without uniform α, this simple picture of directional selective change is complicated by multiple ages at maturity and in the fishery. In the sockeye *(O. nerka),* Ricker (1982) found evidence for a kind of disruptive selection by gill netting: a gradual divergence in the size of 3- and 4-year age-at-maturity or α-groups, which he attributed to greater escape of *small* 3-year and *large* 4-year fish. If true, this would be a remarkable case of life-history-contingent selection and induced negative covariance between traits. Todd and Larkin (1971) had earlier pointed out that the direction of selection might be in-constant, because the relative selection profile of the gillnets shifted from year to year with changes in average size of members of the exposed cohorts of sockeye.

In chum salmon *(O. keta),* gill netting aimed primarily at other smaller species apparently preferentially removed the *smaller* (hence younger, and therefore more rapidly growing) individuals. Ricker (1980a) suggested that this accounts for an observed increase in α and a small but inconsistent decrease in l_α, but he acknowledges that such an effect would be opposed by selection for the larger (hence more rapidly growing) animals within an α-group. Finally, in the absence of data on age and size at maturity, Ricker (1980b) could only speculate on the causes for the decline in average weight of the catch of chinook *(O. tshawytscha),* the species with the most variable α and time exposed to the fishery.

For iteroparous species (in which at least some members breed more than once), even fewer examples of fishery-induced genetic change have been documented. Declines in growth rate and condition in the gillnet-fished popula-tion of lake whitefish *(Coregonus clupeaformis)* in Lesser Slave Lake, Alberta, led Handford et al. (1977) to suggest selection for slower-growing fish maturing later but at a smaller size. They appear to have ruled out alternative explana-tions which could not account for a concomitant significant decline in the co-efficient of variation in condition: increased interspecific competition, deterio-rating environmental conditions and the progression of large year-classes of slower growing fish through the fishery. They recognize that in order for the ge-netically slower growing fish to survive the winnowing, it is critically important that maturity be a function of age as well as of size and condition. Handford, Bell, and Dietz (unpublished) have described similar changes in another Lesser Slave Lake coregonid, *C. artedii.*

Other examples are less convincing. Sometimes the claim is based merely upon a shift in the catch to smaller sizes or ages, which happens in the "fishing up" process as older age classes are removed (Baranov 1918). Favro et al. (1979) suggested a genetic basis for such changes in the brown trout *(S. trutta)* of the Au Sable River, Michigan. R.J. Behnke (personal communica-tion) comments that if growth rate has indeed declined in these trout, it is most likely due to the recent oligotrophication of their environment. Schaffer and Elson (1975) similarly suggest that selective removal of the "larger (i.e., older) individuals" in the weakly iteroparous Atlantic salmon *(S. salar)* should result in genetically earlier breeding. Again, the evidence is more easily explained by the fishing-up effect. Rasmuson (1968) stated that because the Baltic salmon fishery takes the larger, older individuals, the early-maturing males (grilse) and the slow-growing fish have the best chances of returning to the spawning ground; she gave no evidence of the results of such selection. But in this species as with *Oncorhynchus* the genetic correlation between growth rate and developmental rate appears to be positive (Thorpe et al. 1983), and thus selection for early maturity and for slow growth would be opposed (as Ricker pointed out for chum and sockeye salmon).

With other (perhaps all other) cited examples, the difficulty is that changes in such traits as α, l_α, and growth rate do in fact occur in response to fishing and other factors, but without any necessary mediation of genetic change. Bell (1976) seems to have been the first to note this problem. Fish species with high fecundity and low juvenile survival show extraordinarily plastic responses to imposed shifts in demography. Increased mortality is *compensated* in a density-dependent way by increased juvenile growth and survival, sometimes higher later growth rate, and earlier α (Nikolskii 1969, Cushing 1973). Early fecundity also may increase, depending upon whether l_α increases, decreases, or remains the same (see Alm 1959) and upon whether growth ceases at maturity. These are precisely the *genetic* changes predicted by life history theory to occur following increased mortality (Stearns, 1976, reviews the earlier literature). Increased juvenile survival is supposed to lead to the evolution of increased reproductive effort, that is, to earlier maturation and increased early fecundity. Increased adult mortality imposed by a size-selective fishery should only reinforce the process by shifting the age structure further toward the younger classes.

But if fish populations have the phenotypic plasticity to make these adjustments without resort to changes in gene frequency (Garrod and Knights 1979, Stearns 1980, Caswell 1983), in particular cases we should not attribute such adjustments to genetic causes without further evidence. Thus we must set aside as not proven all claims in which only a reduction in α (Borisov 1978, Beacham 1983c) or l_α (Gwahaba 1973, Beacham 1982, 1983a) or both (Charnov 1981, Beacham 1983b,d) are claimed without accompanying evidence such as a declining growth rate to demonstrate that the changes are not part of the typical compensatory response to exploitation.

Detection of fisheries selection. In contrast to the overwhelming evidence for species and stock succession by fisheries selection, there is little hard evidence that fisheries selection occurs within stocks. A lack of evidence does not mean that such selection is occurring only rarely; rather, it underscores the difficulty of detection and suggests that perhaps the few well-established cases represent but the tip of the iceberg.

Proof of genetic change due to fisheries selection requires the establishment not only of the change but of a commensurate selection differential. By measuring the difference in size between seined, gillnetted, and trolled pink and coho salmon, Ricker et al. (1978) and Ricker and Wickett (1980) estimated plausible selection differentials for gillnetting and trolling, although they made

no attempt to incorporate fecundities correlated with growth rate in their estimates. With heritabilities that were perhaps overestimated they could account
for the decline in growth over time.

When more than one age enters the fishery, an accurate estimate of
growth-rate selection differentials is much more complex, though perhaps possible in ideal circumstances. It is instructive to study what would be involved.
We might begin by calculating size-specific selectivity curves or ogives
(Hamley 1975) and then translate these into age- and size-specific mortality
functions. This would require a knowledge of demographic structure not usually obtainable. One might use Lee's phenomenon (Lee 1912) as a shortcut.
This effect, growth-rate-selective mortality as displayed in length-at-age estimates from growth rings on scales or other hard parts, is observed even in un-
fished populations (cf. Ricker's 1969 discussion of the effect in the data of Hile
1936). These estimates are obtained by "back-calculation." In essence, the ratio
of the size of a growth ring to total size of the structure at capture is multiplied
by the length of the fish at capture to obtain an estimate of length at the time
the growth ring was deposited.

Calculation of Lee's phenomenon can be made to yield direct estimates
of growth-rate-specific mortality at age or at size. However obtained, the mortality schedules must then be weighted by (growth-rate-specific?) fecundity
schedules to yield selection differential as a function of growth rate (perhaps by
an approach similar to that of Doyle 1983 or Caswell 1982, 1983).

If change in α is suspected and reproductive maturity has left a mark on
the hard part used for back-calculation of length-at-age, a two or more dimen-
sioned *selective surface* or fitness surface may be estimated from phenotypic
means, variances, and covariances of α and age-specific growth rates (more
likely of their principal components) before and after selection, after the manner of Lande and Arnold (1983). But the selection is occurring *over time* within
a year-class, and fecundity schedules specific to growth rate and α should again
be incorporated into the fitness measure. If size-selective mortality began before all were mature, estimating the α of those that died before spawning (from
those with the same growth rate that did not die) would be necessary but not
without risk.

Rougher measures may still give useful estimates of selection that are
sufficient to determine whether a selection hypothesis is a realistic explanation
of the demographic changes in a given fishery. Perhaps all highly selective
modes of fishing should be suspect, but the response to selection must vary
with the relation of the selection function to the *distributions* of growth and fecundity functions and of sizes and ages at maturity, and with their inter-
correlations.

Handford et al. (1977) discuss several possibilities. At one extreme, if
maturity is entirely a function of age, the slower growing individuals may breed
before capture; if it is entirely a function of size, they will be disadvantaged by
exposure to the fishery for a longer period and by a longer mean generation

time. Handford et al. point out that "if age makes a substantial contribution ... selection may act upon the age-specific schedules as well as on the size-specific schedules; moreover, it will do so in fisheries ... where selection for size may be negligible," acting "in concert with density-dependent compensation" (as we discussed above).

Favro et al. (1979, 1980, 1982) were able to simulate the decline in size observed in Au Sable River trout with genetic size-selection models involving several loci affecting growth rate. They specifically removed from their models any size-mediated genetic effects upon fecundity. Inclusion of such effects causes their models rapidly to "run away" to fixation of the fastest-growing genotype, regardless of the distribution and amount of fishing and other mortality (K. Nelson and V. Nereo, unpublished results). Thus, the approximate concordance of their simulation results with the observed changes in the size distribution must be coincidental.

A particularly promising approach to detecting size selection in natural populations is occasioned by association of electrophoretically detectable loci with effects on growth and developmental rates. In a population of cod *(Gadus morhua)* in Trondheimsfjord, Norway, Mork et al. (1983, 1984a) detected differences in growth rate and age at first and subsequent gonadal maturities between the bearers of the two common alleles at a locus encoding hemoglobin *(HbI)*. They also deduced size selection at that locus from study of captured, genotyped, marked, and released fish from which the tags were retrieved at recapture in the fishery (Mork et al. 1984a). Specifically, the proportion of fish bearing the HbI[1] allele, associated with slower growth and maturation, increased with size (which they interpreted as age) in the original sample and with date and size at recapture (i.e., age again).

Unfortunately, age itself could not be measured, and there are difficulties in fitting certain of their other observations to their hypothesis of size selection. For example, frequency of HbI[1] decreased with distance of recapture, suggesting the possibility that selection was for decreased activity and catchability rather than for slower growth *per se;* that allele is apparently less well adapted to the temperatures found in Trondheimsfjord (Karpov and Novikov 1980).

Obviously, more study of the situation is needed, but the techniques employed by Mork's group may be extremely useful in verifying the existence of selection, not only in this and other cod populations (e.g., Borisov 1978, Beacham 1983b) but in other fisheries as well (such as rainbow trout; Allendorf et al. 1983). If the search for electrophoretically detectable loci associated with effects upon growth and development proves fruitless with a particular species, it may still be possible to incorporate allozyme markers (Chapter 13) into artificially reared slow- and fast-growing cohorts and then follow their progress through a fishery.

Detecting response to selection. We have already pointed out the difficulty encountered when the changes predicted by life-history theory are parallel to those of the purely phenotypic compensatory response. However, the predictions of life-history theory may not be an adequate guide even in cases of real genetic response to fisheries selection. First, until very recently (Caswell 1983) such theory has been cast in terms of the age distribution of mortality and fecundity (e.g., Michod 1978), whereas in fished populations they are structured by size. Thus, an exploited population achieves its nongenetic response in α and in early fecundity *via* increased growth rate; but intuitively at least, genetic size selection should usually be for reduced growth rate.

Second, the population may be under constraints which prevent it from responding in the "optimal" manner (Lewontin 1979, Stearns 1980). One such constraint might be the lack of additive genetic variation; another might be the presence of nonadditive variation, e.g., a correlation of fitness or growth rate with heterozygosity (Beardmore and Ward 1977, Zouros et al. 1980); another might be negative genetic correlation between the relevant life history traits. In general, quantitative genetic theory predicts that additive genetic variance in fitness-related traits will be low unless the traits are negatively correlated by pleiotropy and linkage (Robertson 1955, Lande 1982). This has been shown to be the case in several recent experimental studies (see Rose 1982). Thus, whether growth rate, maturation rate, and fecundity can respond "appropriately" in terms of life history theory may depend upon both the direct and the indirect effects of size selection upon each, as mediated by their genetic variances and covariances.

These latter quantities may never be determinable amid the "noise" of natural environments. Furthermore, we may distinguish an appropriate genetic change from a compensatory response only if life-history parameters fail to revert back upon release from exploitation, and sometimes not even then (in cases of multiple stock-recruitment equilibria; May 1977, Peterman 1977). Thus, in the short run we may recognize the genetic effects of selection only when they are at variance with the compensatory response and, ironically, with the predictions of age-structured life-history theory.

There remain several possible direct approaches. First, association of an electrophoretic locus with an effect upon growth rate or other character may allow the monitoring of selection-related differences between populations or of secular changes in allele frequency within a population (which do not seem to be occurring at the cod *Hbl* locus in Trondheimsfjord, notwithstanding significant differences between year-classes; Mork et al. 1984b, Mork and Sundnes 1985). Gauldie (1984) has suggested such an association between fisheries-induced changes in allele frequency on the one hand and shifts in the distribution of growth rates on the other. At a phosphoglucomutase locus in the tarakihi *(Cheilodactylus macropterus)* from New Zealand waters, significant differences in Bertalanffy growth parameters were found between genotype-sex categories. Curiously, the slowest growing genotype in the males (PGM^{MF}) was the fastest growing among the females, and (approximately) vice versa.

Gauldie attempts to correlate these differences with changes in allele frequency between exploited and unexploited populations, but his logic is unconvincing and his demonstration incomplete. He does not show that there are, in fact, growth-rate differences between the exploited and unexploited populations; as he acknowledges, other and simpler explanations exist. Krueger and Menzel (1979) found a correlation between Ldh-B^2 allele frequencies in natural populations of brook trout *(Salvelinus fontinalis)* and stocking intensity of hatchery brook and brown trout. They favored a selection explanation, as hybridization with the hatchery stockings seemed unlikely. However, evidence is slim for association of that locus with effects upon which selection could act.

Second, comparison of progeny from similar exploited and unexploited populations reared in the same environment may verify suspected genetic changes. Either method would be more powerful if a model complete with estimates of selection differential and of heritability or penetrance and expression were to make a predictive confirmation possible.

Management options and their effects. Hatcheries practice inadvertent selection upon sex ratio (Doyle 1983). Altukhov (1981) has shown that when there is an excess of males, hatchery selection of chum salmon from early in a run results in a preponderance of males in the whole run in later years. Fishery closures at the beginning or end of a run may have similar effects. Even measures to control the amount of fishing, especially if unallocated, often concentrate fishing into a short period at the beginning of the quota year (Gulland 1982). Such management policies also may result in artificial selective pressures for an earlier or later breeding season, upsetting the delicate adjustment of larval production to the natural production cycle of the plankton. As we have seen, in natural populations of iteroparous species the results of such fisheries selection are very difficult to verify. But fishery and management practices, again, seem at times almost designed to bring them about.

A case in point is the imposition of a legal minimum size to ensure that most fish have the opportunity to spawn at least once, or to avoid "growth overfishing" (the uneconomical harvesting of juveniles; see Cushing 1968). The effect depends upon the relationship between size and cumulative frequency of capture on the one hand (the "selection ogive") and the relationship between size and cumulative percentage of mature individuals on the other (the "size at maturity ogive"). If the mandated selection ogive is at all close to the size-at-maturity ogive, it should be maximally advantageous for an undersized fish to breed.

As discussed above, the outcome will depend upon many things, but rather than wait for the results of selection to become manifest, it might be prudent to question whether the size minimum is having the desired effect. Its removal, with provision of a market for the undersized fish, might eliminate the wasteful practice of throwing them back to die (Larkin 1978). "Growth overfishing" (i.e., of juveniles), after all, holds fewer dangers than "recruitment overfishing" (the destabilizing elimination of adult age classes; Cushing 1973,

Gulland 1977, Larkin 1978). It goes without saying that fishing effort should be reduced to the point that "growth overfishing" is not a danger.

An alternative means of reducing the effects of size selection on iteroparous species is to increase the median point of the selection ogive, so that relatively more fish are taken late in their reproductive lives. This is in accord with the principle of the "prudent predator" (Slobodkin 1968, Frankel and Soulé 1981, but see Caswell 1982) and has been suggested by Healey (1975) as an appropriate way of managing lake whitefish populations. It may not be practical in some fisheries, including those in which older specimens exhibit undesirable changes in texture (Roff 1983). But in other cases it might eventually increase the value of the catch: as Gulland (1982) puts it, "The great fish-eating public ... insists on preferring the larger, mostly predatory species, and the larger individuals of those species."

Hybridization and Introduction

Hybridization and loss of genetic diversity. Perhaps very few erosions and disappearances of fish stocks are random or unselective losses. Thermal pollution or acid rain selects individuals according to their tolerance of the pollutant, the introduced pathogen selects its host, and the introduced species or hatchery stock selects its nearest available relatives to hybridize with, to replace, or both (Allendorf and Phelps 1981a). Hubbs (1955) described a case in which minnows introduced for bait *(Gila orcutti)* hybridized with and seemed to be supplanting an apparently less well adapted native chub *(Siphateles [Gila bicolor] mojavensis)*. Stocking of rainbow trout *(Salmo gairdneri)* in western North American streams that already contained the closely related cutthroat trout *(S. clarkii)* has resulted in all possible combinations of the two species and their hybrids, from complete replacement by the introduced species to morphologically undetectable introgression of rainbow genes into cutthroat populations (Behnke 1972, Busack and Gall 1981, Leary et al. 1984). Such introgression and the threatened loss of a native stock from "swamping" by a hatchery stock are detectable with the technology of allozyme electrophoresis; in the latter case the task is simplified if the hatchery stock bears a genetic marker, an allozyme rare in the native population (Allendorf and Utter 1979, Thorgaard and Allen, Chapter 13).

Inasmuch as hybridization is often used to increase the amount of genetic variation in a population, it may seem strange that we think of it as a mode of loss of genetic diversity. Consider the merging of two similarly sized subpopulations fixed for different codominant alleles for some quantitative character. Even if there is no loss of *genic* diversity (i.e. no change in allele frequency), when the merged population comes to equilibrium it will have but half the *genotypic* variance of the two subpopulations; the proportion of homozygotes has been cut by half.

More generally, the effect of gene exchange between subpopulations is to increase the variance within groups, decrease the variance between groups, and decrease the total variance (the existence of dominance modifies these relationships somewhat; Crow and Kimura 1970). Moreover, if there exist co-adapted groups of genes within each of the subpopulations, merger may result in the breakup of these and the loss of the particular adaptedness of each subpopulation to its environment.

Note that loss of genotypic variance is essentially as irreversible as the loss of alleles from a population. One can no more sort out the genes into groups of genotypes resembling the original subpopulations than one can obtain the works of Shakespeare from a collection of monkeys and typewriters, and for similar reasons.

Nevertheless, there is a sense in which inbreeding and hybridization are opposites. We saw in our discussion of the former that one way to counteract it was to permit a limited amount of migration, on the order of one migrant per generation. Evidently a quantitatively small amount of hybridization between small neighboring subpopulations may provide them with a beneficial stimulus of new alleles without destroying local adaptation (Allendorf 1983).

"Genetic improvement" of natural populations. Hybridization between stocks or species has occasionally been proposed as a means of increasing genetic variation and production characteristics of natural populations (Calaprice 1969, Ihssen 1976). In 1978, Moav et al. unveiled a more detailed proposal for "genetic improvement" of wild fish populations, in particular those that had suffered "genetic deterioration" at the hands of man. Among the goals they announced were the breeding of pollution-resistant genotypes and the improvement of the overall economic value of harvested fish in their natural environments; the means to accomplish these goals were to be "specific selection of compensatory traits in a domesticated breed followed by its mating to a wild indigenous population."

Response to their proposal has apparently been muted, although Ryder et al. (1981) paraphrased H. Harvey as characterizing the first goal as analogous to "the genetic selection of a Welsh coal miner's canary that could live in conditions that would kill miners." Dr. Moav was one of the true pioneers of fish and aquacultural genetics. Neverless, we must take exception to the Moav et al. (1978) proposal and point out the dangers inherent in it.

There is a distinction to be made between confined ecological systems such as ponds and raceways, where genetic, epizoötic, and other ecological damage can be contained, and unconfined systems such as streams, lakes, and oceans, where it cannot (Helle 1981). Although Moav et al. (1978) propose to start with the former, they desire to extend the improvement to the latter as technical means become available. They claim (apparently without evidence other than data on weights of hybrids and backcrosses of several carp strains) that there is little danger of introgressed genes damaging the wild stocks. "When single-sex D [domesticated or donor] breeders that produce only sterile W [wild] × D hybrids are released, the natural gene pool cannot be contaminated or harmed." However, even they notice that "the pest control method of releasing sterile individuals of a single sex . . . is essentially identical" to their proposed procedure.

Even should the introduced hybrid be fertile, Moav et al. argue that there is little risk in their proposal. In fact, the available evidence (from salmonid studies) of the results of hybridization of native with hatchery stocks indicates that there is indeed much risk in such schemes. Bams (1976) demonstrated the superior homing ability conferred upon hybrids (TK) between a transplanted donor (KK) and a native stock (TT) of pink salmon by paternal genes from the native stock. But although the TK hybrids were far better at homing than the KK stock and even reached the native rivers as accurately as did TT fish from previous (alternate cycle) years, only about half as many actually homed to the natal tributary stream. Bams notes the danger presented by the KK and TK strays to the gene pools of nearby streams.

Reisenbichler and McIntyre (1977) demonstrated decreased juvenile stream survival rates in hybrids between hatchery and native steelhead (anadromous rainbow trout, *Salmo gairdneri),* and predicted a long-term decrease in the stock-recruitment relationship in such a contaminated stock. Altukhov (1981) attributes a literal decimation in population size of chum salmon in the Naiba River to "disturbance of genetic structure" of the native population, resulting from several years' introduction of eggs from a neighboring river.

Introgressive hybridization with rainbow trout and other hatchery introductions is thought to have been a major contributor to the decline of all other western North American *Salmo* species (references in Busack and Gall 1981). Introduction of hatchery rainbows, which hybridize readily with other species, has broken down the reproductive isolating barriers between sympatric native populations of rainbow trout and westslope cutthroat trout (*S. clarkii lewisi;* Behnke 1972, Allendorf and Phelps 1981a). Allozyme studies show that few pure cutthroat populations remain (Leary et al. 1984). The results of such introgression and loss are irreversible (Busack and Gall 1981). Similar breakdown of reproductive isolation between local populations after introduction of hatchery stock may be occurring as well in European brown trout (*S. trutta;* Ryman 1981b).

Introductions in general carry many other risks ignored by Moav et al. (1978); they include loss of native species through competition and predation, the concomitant introduction of exotic parasites and diseases, and even drastic change of the ecosystem at several trophic levels (Hurlbert et al. 1972, Smith 1968, 1972, Zaret and Paine 1973, Mann 1979, Rosenfield and Kern 1979, FAO/UNEP 1981). That the introduced hatchery stock is closely related to the native stock is no guarantee that such changes will not occur. If it has been cultured elsewhere, it has become an exotic stock.

Turner (1949, cited in Dean 1979) lists five criteria to govern the introduction of a non-native stock:

● It should fill a need created by the absence of a similar desirable local species;

- It should not compete with valuable local species;
- It should not hybridize with local species;
- It should not be accompanied by enemies, parasites or diseases which might attack native species; and
- It should live and reproduce "in equilibrium with its new environment, i.e. within the confines of the supply of space, food and other factors that limit its range and abundance."

Although each case must be evaluated individually, hybrids of local and domesticated stocks clearly fail the first three of these criteria and, with other kinds of introduced stocks, carry the risks of failing on the others—especially in unconfined bodies of water from which their genes and introduced parasites may spread to neighboring ecosystems.

Rehabilitation and preservation. Often, however, remedial schemes are called for where indigenous species have become rare or extinct. Loftus (1976), for example, estimates that very few of Canada's fisheries are *not* in need of rehabilitation. Ryder et al. (1981) suggest that restoration of the former ecosystem should be the guidepost for rehabilitation, not the development of stocks that can survive under the current degraded conditions. Krueger et al. (1981) outline two alternative strategies for obtaining fish for stocking:

- "Perform all possible crosses within and between" sources representing the entire genetic diversity of the species, and then stock the progeny; or
- Stock with the progeny of fish from refugia within the same drainage, or at any rate from a source as closely similar ecologically to that of the remnant or extinct stock as possible.

They point out that the main difficulty with the second alternative strategy is "whether the perception of environmental similarity by the biologist will be the same as that of the fish." With the first strategy the hope is that some recombinant genotypes will find the new conditions amenable. They warn of the dangers inherent particularly in the first strategy, and they recommend its use only after considerable thought and only in cases of severely changed habitat with multiple extinctions. We agree.

Thus far we have been speaking mainly of genetic conservation *in situ*— that is, in the native environment including freshwater and marine reserves that may have to be established. The consensus (Berst and Simon 1981, FAO/UNEP 1981, Ryman 1981a) is that today this is by far the better approach. But in some cases it may not be possible. Hynes et al. (1981) ask that we then consider the establishment of reference populations of culturally important stocks in natural refugia, which may mean sacrificing their current tenants either by intent or by inadvertence.

Often even this option will be closed, as when a domesticated hatchery population is the sole surviving remnant of the former stock (Ryman 1981b). Then the technology of *ex situ* artificial preservation, including rotational line crossing and cryopreservation of gametes to maintain an adequate N_e (Frankel and Soulé 1981, Schonewald-Cox et al. 1983), may be the only remaining alternative to extinction (FAO/UNEP 1981, Ryman 1981a).

DISCUSSION
Results of Good Intentions

In surveying the causes of loss of genetic diversity we are struck by how often the conspirators are not the expected Ignorance and Greed but, rather, the equally dangerous Partial Knowledge and Good Intentions. For example, hatchery introductions made with the laudable goal of stream rehabilitation have resulted in unforeseen consequences that include the loss of native stocks.

Loss of stocks has also resulted from attempts to fish a maximum sustainable yield (or more recently, maximum sustained economic rent), which failed to take into account fluctuation in recruitment and differences between stock components. Clark (1976, p. 32) has pointed out that, far from representing the action of unalloyed economic forces

> ...the concept of maximum sustained economic rent amounts to taking a particular and quite extreme position regarding intertemporal tradeoffs. Indeed, ... it amounts to setting the discount rate equal to zero. Consequently it is based on the underlying assumption that current sacrifices are immaterial if they ultimately lead to permanent increase in economic benefits.

In this respect at least, more sophisticated economic models in which future value is discounted require a retreat by managers from a more conservationist position.[1]

In the past, size limitations were rationalized as ensuring that a sufficient fraction of the recruitment were given an opportunity to breed, and also to maximize the present and future biomass of the catch (as part of the maximum sustainable yield program, to eliminate "waste"; Larkin 1978). However, size limitations often encourage a reduction in the number of breeding year classes to a destabilizing one or two and produce powerful size-selective forces with as

---

[1]   We should note here, however, that sometimes conservationist and economic priorities coincide. From a genetic point of view, any policy which reduces fishing mortality is likely to be beneficial. Economists have long understood that when fishermen have open access to a fishery, economic rent (receipts less costs) is dissipated: more and more enter the fishery, and fishing effort and total costs increase, until at equilibrium (no net entry) there is no profit left for anyone (the "Gordon-Shaefer principle"; Clark 1976). But if, on the other hand, the fishery behaves as if it had a sole owner (whether a private operator or a management agency), economic rent will tend to be maximized, at usually much lower levels of fishing effort (or mortality) and total costs. Thus, the policy of limited access is both economically and genetically attractive.

Botsford and Wainwright (1985) point out that when fixed costs (of vessels and gear) are high, the difficult-to-apply and expensive policy of pulse fishing may not be optimal. The optimal fishery policy may be one of low and relatively constant fishing effort with a quite low age (or size) of first entry into the fishery. Again, such a policy might coincide well with the genetic objective of reducing size selection by the fishery, and again limitation of entry may be the means of choice.

yet unknown consequences. And well-intentioned but injudicious temporal closures may produce selective pressures on sex ratio or on the timing of the breeding season, again with unknown consequences.

The intent of the Moav et al. (1978) proposal was the genetic *improvement* of wild fish populations. The methods proposed, however, were genetically and ecologically naive and the results potentially disastrous. As Bams's (1976) study has shown, it is not the wild but the domesticated stock that stands to be improved by hybridization with native stock, and this is reason enough for native stocks to be kept genetically inviolate.

Priorities for Future Research

It is doubtful that much can be done soon to eliminate Greed, and we do not for a moment suggest the abolition of Good Intentions. It follows that our efforts should be directed to the reduction of Ignorance and the amelioration of Partial Knowledge, always conceding that knowledge can never be complete. Others have made recommendations regarding research priorities (Berst and Simon 1981, FAO/UNEP 1981, and Ryman 1981a). Here we add ours.

In discussing inbreeding effects we have already noted the near total absence of accurate estimates of effective population size N_e. We intimated that the need for such estimates can be circumvented when inbreeding effects can be estimated by the methods of allozyme analysis. But there is another context in which such estimates of N_e turn out to be of crucial importance.

We need a dynamic and eclectic theory of subpopulation structure and of its breakdown under conditions of exploitation. Population genetics has provided a theoretical background (Wright 1969, Crow and Kimura 1970) but has not yet been integrated with historical changes in distribution and density, contemporary studies of migration, larval dispersal, schooling, and life history patterns; it is not yet predictive. Vague ideas have emerged concerning the combinations of species, ecology, and geography likely to develop extensive subpopulation structure or to be especially sensitive to exploitation (Berst and Simon 1981). Although each month brings new studies of stock structure that utilize the latest techniques of stock identification, we are far from understanding why and how subpopulation structure has developed and what happens as it is lost. When management can call upon a predictive theory, perhaps stock structure will finally be safe under exploitation.

For this reason we have been deliberately vague and inconsistent in our own usage of the "stock concept" in this review. We do not even know that "stocks" in different species represent similar biological entities, nor do we know the roles, if any, of such entities as "schools" and "age cohorts" in the maintenance and commingling of stocks.

Such a theory of stock structure would be based on the relationship between two variables, effective population size, N_e, and the fraction, M, of the population which is replaced by migrants from other populations each generation. In Wright's model, if M is very much smaller than $1/4N_e$, local population differentiation may be expected. As both terms have N_e in the denominator, N_e may be canceled out, and it is often stressed that it is the absolute number of migrants exchanged that determines whether differentiation will occur, regardless of the size of N_e. The implication is that knowledge of N_e is not really necessary, and perhaps this is part of the reason there have been so few serious attempts to estimate it accurately.

But surely in any theory of dispersal (e.g., as a diffusion process) the number of migrants per generation from a population is an increasing function of population size. We need, therefore, to know the behavior of N_e as population size decreases under exploitation. Such variables as sex ratio and variance of lifetime fecundity, upon which N_e is known to depend, may themselves be dependent upon population density (as is fecundity itself). The stock-recruitment relationship makes density in turn a function of N_e. Clearly, we need to know more about the interrelationship of these factors in exploited populations if we expect to understand the effects of exploitation upon migration and upon stock structure.

The determination of N_e is especially important for the study of stream-breeding species in which small, linearly arranged populations are more or less effectively isolated by precise homing behavior or inhospitable habitat, species in which inbreeding and drift effects are likely to be most pronounced. Perhaps the study of the temporal, spatial, and between-locus variation in gene frequency under such conditions will lead to more accurate and practical means of indirect estimation of N_e or its equivalent from allozyme or other data. Obviously, some new approaches are needed.

A theory of stock formation might also help us to learn when to expect an introduction to be dangerous. Before proceeding in the mixing of gene pools, especially of natural and domesticated stocks, it would be wise to learn through controlled experiments in confined ecological situations what the likely effects are to be. But such a seat-of-the-pants approach is insufficient. Just as there are factors that determine which species are likely to develop subpopulation structure, there must also be genetic and other factors that make a population resistant to swamping or to introgression. Again, we need to know the relationship of these to such things as the prior history of the species.

Somehow such a theory of stock structure must eventually be integrated with a practicable theory of the selective effects of exploitation. We have seen that such selection may occur at two levels, reduction or disappearance of stocks and directional change of gene frequency within a stock. We have suggested that study of changes resulting from size selection within stocks presents problems which seem almost insurmountable. We shall not repeat them here. Suffice it to say that much fundamental research is needed in the quantitative genetics of fish growth and life history before we can estimate the danger a particular selective regime presents to a population.

One initial question demands an answer. If there exists additive genetic variation in growth rate, if growth rate and maturation rate have a positive genetic correlation (as they appear to in salmonids at least), and if fecundity is more or less proportional to weight, enormous genetic variance in fecundity would appear to be implied. What sort of advantage must the slower growing, less fecund individuals possess so that their genes can remain even in the un-fished population?

We would know that we had a real theory of size selection when it was sufficiently robust to encompass the haphazard influence of density-dependent infant mortality and growth upon recruitment and upon selective adult mortality. Such a theory would stand as a major contribution to our understanding of the population biology of exploitation.

We have mentioned only briefly the possible selective effects upon school size from technologically advanced exploitation of pelagic species. Escape of smaller fishes and escape of smaller schools to breed are both forms of catchability selection. Is behavior that leads to the fragmentation of schools selected for by search and capture methods aimed at the larger schools? Again, a theory that is able to weigh the advantages and disadvantages of schooling under different conditions of exploitation and other predation is clearly needed.

Even the best-managed fisheries may be expected to exert more subtle forms of life-history selection not so easily subsumed by the "catchability selection" model. For example, life-history theory might note that providing a constant escapement over a long time might reduce the density-dependent fraction of the natural and often extreme fluctuations in juvenile survival which a species has come to expect in the course of its history. How would such an effect interact with the other selective forces faced by the species?

Conservation Genetics and Management

We come to the conclusion that the areas in which research is most sorely needed are the formation of stock structure and the response of populations to the selective pressures generated by exploitation. But more research is *always* needed, and it is usually expensive. What is the urgency in this case? We have stated that preservation of the existing gene pools is the primary obligation of fisheries management. Within our anthropocentric framework we have intimated that somehow managers must find a way to place the welfare of the exploited species above the particular short-term interests of the exploiters, lest they find themselves with nothing left to manage.

But managers are in a difficult position. They are beset by many other desiderata—political, economic, social—and there may be more immediate threats to their jobs than the vague possibility of disappearance of the resource sometime in the future. They may be asked, what is so wrong with the elimination of unproductive stocks? Can you prove that fisheries selection represents a danger to the resource? Aren't hybrids altogether more vigorous? Management must be able to substantiate the answers to such questions with solid information, not vague theory. They must be able to convey unequivocally the importance of the preservation of stock structures with adequate numbers, the avoidance of selection, the maintenance of the purity of gene pools. At the moment, and for some time to come, we must concede that the hard facts are not always there to enable them to do this. Hence the urgency.

In the meantime, what *can* management do? Our answer is that in the meantime the stance of management must be a conservative one. Policy must also be oriented to education, the dispelling of Ignorance with emphasis upon that which we *do* know. We do know that the disappearance of stocks is a too common occurrence. We do know that many factors, including inbreeding depression, may prevent recovery of a stock from a state of reduction. We do know that, once hybridization has occurred, introgression is difficult or impossible to control. We can see in many cases the disappointing results of size limitations. In species with simplified life histories we can measure both the selective pressures of size selection and the response to it. We can state that in all cases the direction of change favorable to conservation is toward reduction in fishing mortality, to which any selective pressures will usually be directly proportional. One might call this the principle of least fishing effort.

Genetic arguments should support a manager's resistance to the usual pressures for increased fishing effort. The conservative position of keeping effort (on smaller stocks especially) as low as political reality will allow may often be supported by economic considerations as well, as with markets in which demand is inelastic (Clark 1976) or for which fixed costs of fishing are high (Botsford and Wainwright, 1985).

The common ground with economic theory must be sought out. Therefore, scientific input in the presentation of available management options should include estimation of the likely and also the worst-case impact of each upon the general integrity of the gene pool and upon the economy. It should also include an estimate of the impact on the economic and social structure in the event of the gene pool's severe degradation or loss.

In the meantime, as we have said, the stance must be a conservative one. Management must be wary of change. In our ignorance we cannot know that a particular, well-meaning, innovative, seemingly innocuous policy (such as the production of all-female runs of salmon, or "active adaptive management") will not result in disaster. The principle of least fishing effort is for the moment perhaps the best guide in the evaluation of innovative fishing methods and policy.

LITERATURE CITED

Adams, J. and E. D. Rothman. 1982. Estimation of phylogenetic relationships from DNA restriction patterns and selection of endonuclease cleavage sites. *Proceedings of the National Academy of Sciences U.S.A.* 79:3560–3564.

Adams, J. and R. H. Ward. 1973. Admixture studies and the detection of selection. *Science* 180:1137–1143.

Adest, G. A. 1977. Genetic relationships in the genus *Uma* (Iguanidae). *Copeia* 1977:47–52.

Alberts, B., D. Bray, J. Lewis, M. Raff, K. Roberts, and J. Watson. 1983. *Molecular Biology of the Cell*. Garland Publishing, Inc., New York.

Ali, M. Y. and C. C. Lindsey. 1974. Heritable and temperature induced meristic variation in the mendaka, *Oryzias latipes*. *Canadian Journal of Zoology* 52:959–976.

Allard, R. W. 1960. *Principles of Plant Breeding*. Wiley, New York.

Allen, S. K., Jr. 1983. Flow cytometry: Assaying experimental polyploid fish and shellfish. *Aquaculture* 33:317–328.

Allen, S. K., Jr. and J. M. Myers. 1985. Chromosome set manipulation in three species of salmonids using hydrostatic pressure. Abstract in *Salmonid Reproduction: An International Symposium*, ed. R. N. Iwamoto and S. Sower (Washington Sea Grant Program, Seattle), p. 65.

Allen, S. K., Jr. and J. G. Stanley. 1979. Polyploid mosaics induced by cytochalasin B in landlocked Atlantic salmon, *Salmo salar*. *Transactions of the American Fisheries Society* 108:462–466.

Allen, S. K., Jr. and J. G. Stanley. 1981. *Polyploidy and gynogenesis in the culture of fish and shellfish*. International Council for the Exploration of the Sea, Cooperative Research Report Series B, C. M. 1981/F:28, 18 pp.

Allen, S. K., Jr., P. S. Gagnon, and H. Hidu. 1982. Induced triploidy in the shoft-shell clam: Cytogenetic and allozymic confirmation. *Journal of Heredity* 73:421–428.

Allendorf, F. W. 1977. Electromorphs or alleles? *Genetics* 87:821–822.

Allendorf, F. W. 1983. Isolation, geneflow, and genetic differentiation among populations. In *Genetics and Conservation*, ed. C. M. Schonewald-Cox et al. (Benjamin-Cummings, Menlo Park, Calif.), pp. 51–65.

Allendorf, F. W. and S. R. Phelps. 1980. Loss of genetic variation in a hatchery stock of cutthroat trout. *Transactions of the American Fisheries Society* 109:537–543.

Allendorf, F. W. and S. R. Phelps. 1981a. Isozymes and the preservation of genetic variation in salmonid fishes. In *Fish Gene Pools*, ed. N. Ryman. Ecological Bulletins (Stockholm), vol. 34, pp. 37–52.

Allendorf, F. W. and S. R. Phelps. 1981b. Use of allelic frequencies to describe population structure. *Canadian Journal of Fisheries and Aquatic Sciences* 38:1507–1514.

Allendorf, F. W. and G. H. Thorgaard. 1984. Tetraploidy and the evolution of salmonid fishes. In *Evolutionary Genetics of Fishes*, ed. B. Turner (Plenum Press, New York), pp. 1–53.

Allendorf, F. W. and F. M. Utter. 1979. Population genetics. In *Fish Physiology*, vol. 8, ed. W. S. Hoar, D. J. Randall, and J. R. Brett (Academic Press, New York), pp. 407–454.

Allendorf, F. W., F. M. Utter, and B. P. May. 1975. Gene duplication within the family

Salmonidae. II. Detection and determination of the genetic control of duplicate loci through inheritance studies and the examination of populations. In *Isozymes IV: Genetics and Evolution,* ed. C. L. Markert (Academic Press, New York), pp. 415–432.

Allendorf, F. W., N. Ryman, A. Stennek, and G. Ståhl. 1976. Genetic variation in Scandinavian brown trout *(Salmo trutta* L.): Evidence of distinct sympatric population. *Hereditas* 83:73–82.

Allendorf, F. W., N. Mitchell, N. Ryman, and G. Ståhl. 1977. Isozyme loci in brown trout *(Salmo trutta* L.): Detection and interpretation from population data. *Hereditas* 86:179–190.

Allendorf, F. W., D. M. Espeland, D. T. Scow, and S. Phelps. 1980. Coexistence of native and introduced rainbow trout in the Kootenai River drainage. *Proceedings of the Montana Academy of Sciences* 39:28–36.

Allendorf, F. W., K. L. Knudsen, and R. F. Leary. 1983. Adaptive significance of differences in the tissue-specific expression of a phosophoglucomutase gene in rainbow trout. *Proceedings of the National Academy of Sciences U.S.A.* 80:1397–1400.

Allendorf, F. W., G. Ståhl, and N. Ryman. 1984. Silencing of duplicate genes: A null allele polymorphism for lactate dehydrogenase in brown trout *(Salmo trutta). Molecular Biology and Evolution* 1:238–248.

Alm, G. 1939. Investigations of growth by different forms of trout. (Swedish with English summary.) *Reports of the Institute for Freshwater Research, Drottningholm* 40:6–145.

Alm, G. 1946. Reasons for the occurrence of stunted fish populations. *Meddelanden fran Statens Undersoknings och Foroksanstalt for Sotvattensfisket Stockholm* 25:1–146.

Alm, G. 1949. Influence of heredity and environment on various forms of trout. *Reports of the Institute for Freshwater Research, Drottningholm* 29:29–34.

Alm, G. 1959. The connection between maturity, size, and age in fishes. *Reports of the Institute for Freshwater Research, Drottningholm* 40:6–145.

Altukhov, Yu. P. 1973. Local fish stocks as genetically stable population systems. In *Biochemical Genetics of Fish* (Institute of Cytology, USSR Academy of Sciences, Leningrad). In Russian.

Altukhov, Yu. P. 1974. *Population Genetics of Fish.* (Translated by Fisheries and Marine Service, Canada, Translation Series No. 3548, 1975.) 247 pp.

Altukhov, Yu. P. 1975. Genetics of natural populations and biosphere resources. *Bulletin of the USSR Academy of Sciences* 10:37–45. In Russian.

Altukhov, Yu. P. 1977. Problems of populational-genetic organization of species in fish. *Journal of General Biology* (USSR) 38:893–907. In Russian.

Altukhov, Yu. P. 1981. The stock concept from the viewpoint of population genetics. *Canadian Journal of Fisheries and Aquatic Sciences* 38:1523–1528.

Altukhov, Yu. P. 1983. *Genetic Process in Populations.* Nauka, Moscow, 280 pp. In Russian with English summary.

Altukhov, Yu. P. and E. A. Salmenkova. 1981. Applications of the stock concept to fish populations in the USSR. *Canadian Journal of Fisheries and Aquatic Sciences* 38:1591–1600.

Altukhov, Yu. P. and N. V. Varnavskaya. 1983. Adaptive genetic structure and its connection with intrapopulation differentiation for sex, age and growth rate in sockeye salmon, *Oncorhynchus nerka* (Walb.). *Genetica* 19:796–807. In Russian.

Altukhov, Yu. P., E. A. Salmenkova, S. M. Konovalov, and A. I. Pudovkin. 1975a. Stationary distribution of the frequencies of lactate dehydrogenase and phosphoglucomutase genes in a system of subpopulations of a local fish stock of *Oncorhynchus nerka.* I. Stability of a stock over generations with simultaneous variability of the component subpopulations. *Genetica* 11:44–53. In Russian.

Altukhov, Yu. P., A. I. Pudovkin, E. A. Salmenkova, and S. M. Konvalov. 1975b. Stationary distributions of the frequencies of lactate dehydrogenase and phosphoglucomutase genes in a system of subpopulations of a local fish stock of *Oncorhynchus nerka*. II. Random genetic drift, migration and selection as factors of stability. *Genetica* 11:54–62. In Russian.

Altukhov, Yu. P., E. A. Salmenkova, G. D. Ryabova, and N. I. Kulikova. 1980. Genetic differentiation of chum slamon *Oncorhynchus keta* populations and effectiveness of some acclimatization measures. *Marine Biology* (Vladivostok) 3:23–28. In Russian.

Altukhov, Yu. P., E. A. Salmenkova, V. T. Omel'chenko, and V. N. Efanov. 1983. Genetic differentiation and population structure in the pink salmon of Sakhalin-Kuril region. *Marine Biology* (Vladivostok) 2:46–51. In Russian.

American Fisheries Society. 1980. *A List of Common and Scientific Names of Fishes From the United States and Canada.* American Fisheries Society Special Publication No. 12. American Fisheries Society, Bethesda, Md.

Amos, M. H., R. E. Anas, and R. E. Pearson. 1963. Use of a discriminant function in the morphological separation of Asian and North American races of pink salmon, *Oncorhynchus gorbuscha* (Walbaum). *International North Pacific Fisheries Commission Bulletin* 11:73–100.

Anderson, S., A. T. Bankier, B. G. Barrell, M. H. L. de Bruijn, A. R. Coulson, J. Drouin, I. C. Eperon, D. P. Nierlich, B. A. Roe, F. Sanger, P. H. Schreier, A. J. H. Smith, R. Staden, and I. G. Young. 1981. Sequence and organization of the human mitochondrial genome. *Nature* 290:457–465 (Erratum 291:168).

Anderson, S., M. H. L. de Bruijn, A. R. Coulson, I. C. Eperon, F. Sanger, and I. G. Young. 1982. The complete nucleotide sequence of bovine mitchondrial DNA: Conserved features of the mammalian mitochondrial genome. *Journal of Molecular Biology* 156:683–717.

Andersson, L., N. Ryman, R. Rosenberg, and G. Ståhl. 1981. Genetic variability in Atlantic herring *(Clupea harengus harengus)*: Description of protein loci and population data. *Hereditas* 95:69–78.

Angerbjörn, A. 1986. Gigantism in island populations of wood mice (*Apodemus*) in Europe. *Oikos* 47:47–56.

Anon. 1983. Produce sterile white amur. *Aquaculture Magazine,* Nov.-Dec.: 44–46.

Aoki, K., Y. Tateno, and N. Takahata. 1981. Estimating evolutionary distance from restriction maps of mitochondrial DNA with arbitrary $G+C$ content. *Journal of Molecular Evolution* 18:1–8.

Aquadro, C. F. and B. D. Greenberg. 1983. Human mtDNA variation and evolution: Analysis of nucleotide sequences from seven individuals. *Genetics* 103:287–312.

Arai, K. 1984. Developmental genetic studies on salmonids: Morphogenesis, isozyme phenotypes and chromosomes in hybrid embryos. *Bulletin of the Faculty of Fisheries Hokkaido University* 31:1–94.

Arai, K., H. Onozato, and F. Yamazaki. 1979. Artificial androgenesis induced with gamma irradiation in masu salmon, *Oncorhynchus masou. Bulletin of the Faculty of Fisheries Hokkaido University* 30:181–186.

Aronshtam, A. A., L. Ya. Borkin, and A. I. Pudovkin. 1977. Isozymes in population and evolutionary genetics. In *Genetics of Isozymes* (Nauka, Moscow), pp. 199–249. In Russian.

Aspinwall, N. 1974. Genetic analysis of North American populations of the pink salmon *(Oncorhynchus gorbuscha)*. Possible evidence for the neutral mutation-random drift hypothesis. *Evolution* 28:295–305.

Aulstad, D. and A. Kittlesen. 1971. Abnormal body curvatures of rainbow trout *(Salmo gairdneri)* inbred fry. *Journal of the Fisheries Research Board of Canada* 28:1918–1920.

Avise, J. C. 1974. Systematic value of electrophoretic data. *Systematic Zoology* 23:465–481.

Avise, J. C. and C. F. Aquadro. 1982. A comparative summary of genetic distances in the vertebrates: Patterns and correlations. *Evolutionary Biology* 15:151–185.

Avise, J. C. and J. Felley. 1979. Population structure of freshwater fishes. I. Genetic variation of bluegill *(Lepomis macrochirus)* populations in man-made reservoirs. *Evolution* 33:15–26.

Avise, J. C. and R. A. Lansman. 1983. Polymorphism of mitochondrial DNA in populations of higher animals. In *Evolution of Genes and Proteins,* ed. M. Nei and R. K. Koehn (Sinauer Associates, Sunderland, Mass.), pp. 147–164.

Avise, J. C. and N. C. Saunders. 1984. Hybridization and introgression among species of sunfish *(Lepomis):* Analysis by mitochondrial DNA and allozyme markers. *Genetics* 108:237–255.

Avise, J. C. and R. K. Selander. 1972. Evolutionary genetics of cave-dwelling fishes of the genus *Astyanax. Evolution* 26:1–19.

Avise, J. C. and M. H. Smith. 1974. Biochemical genetics of sunfish. I. Geographic variation and subspecific intergradation in the bluegill, *Lepomis macrochirus. Evolution* 28:42–56.

Avise, J. C. and M. J. Van Den Avyle. 1984. Genetic analysis of reproduction of hybrid white bass × striped bass in the Savannah River. *Transactions of the American Fisheries Society* 113:563–570.

Avise, J. C., J. J. Smith, and F. J. Ayala. 1975. Adaptive differentiation with little genic change between two native California minnows. *Evolution* 29:411–426.

Avise, J. C., C. Giblin-Davidson, J. Laerm, J. C. Patton, and R. A. Lansman. 1979a. Mitochondrial DNA clones and matriarchal phylogeny within and among geographic populations of the pocket gopher, *Geomys pinetis. Proceedings of the National Academy of Sciences U.S.A.* 76:6694–6698.

Avise, J. C., R. A. Lansman, and R. O. Shade. 1979b. The use of restriction endonucleases to measure mitochondrial DNA sequence relatedness in natural populations. I. Population structure and evolution in the genus *Peromyscus. Genetics* 92:279–295.

Avise, J. C., J. F. Shapira, S. W. Daniel, C.F. Aquadro, and R. A. Lansman. 1983. Mitochondrial DNA differentiation during the speciation process in *Peromyscus. Molecular Biology and Evolution* 1:38–56.

Avise, J. C., E. Bermingham, L. G. Kessler, and N. C. Saunders. 1984. Characterization of mitochondrial DNA variability in a hybrid swarm between subspecies of bluegill sunfish *(Lepomis macrochirus). Evolution* 38:931–941.

Ayala, F. J. 1975. Genetic differentiation during the speciation process. *Evolutionary Biology* 8:1–78.

Ayala, F. J. 1984. Molecular polymorphism: How much is there and why is there so much? *Developmental Genetics* 4:379–391.

Ayala, F. J., M. L. Tracey, L. G. Barr, J. F. McDonald, and S. Pérez-Salas. 1974. Genetic variation in natural populations of five *Drosophila* species and the hypothesis of the selective neutrality of protein polymorphism. *Genetics* 77:343–384.

Baake, B. M. 1977. Tough redband trout. *Salmon Trout Steelheader* 11:42–44.

Bagenal, T. B. 1973. Fish fecundity and its relations with stock and recruitment. *Rapports et Procès-verbaux des Réunions Conseil International pour l'Exploration de la Mer* 164:186–198.

Bailey, G. S., A. C. Wilson, J. E. Halver, and C. L. Johnson. 1970. Multiple forms of supernatant malate dehydrogenase in salmonid fishes. *Journal of Biological Chemistry* 245:5927–5940.

Baker-Cohen, K. F. 1961. Visceral and vascular transposition in fishes, and a comparison with similar anomalies in man. *American Journal of Anatomy* 109:36–55.

Balakrishnan, V. and L. D. Sanghvi. 1968. Distance between populations on the basis of attribute data. *Biometrics* 24:859–865.

Bams, R. A. 1976. Survival and propensity for homing as affected by presence or absence of locally adapted paternal genes in two transplanted populations of pink salmon *(Oncorhynchus gorbusha). Journal of the Fisheries Research Board of Canada* 33:2716–2725.

Baranov, T. I. 1918. On the question of the biological basis of fisheries. *Nauch. issledov. iktiol. Inst. Izv.* I(1):81–128. In Russian.

Bard, Y. 1974. *Nonlinear Parameter Estimation.* Academic Press, New York, 341 pp.

Barlow, G. W. 1961. Causes and significance of morphological variation in fishes. *Systematic Zoology* 10:105–117.

Barrell, B. G., A. T. Bankier, and J. Drouin. 1979. A different code in human mitochondria. *Nature* 282:189–194.

Barton, N. H. and G. M. Hewitt. 1981. Hybrid zones and speciation. In *Evolution and Speciation*, ed. W. R. Atchley and D. S. Woodruff (Cambridge University Press, Cambridge), pp. 109–145.

Barton, N. H. and G. M. Hewitt. 1983. Hybrid zones and barriers to gene flow. In *Protein Polymorphism: Adaptive and Taxonomic Significance*, ed. G. S. Oxford and D. Rollinson (Academic Press, London and New York), pp. 341–359.

Beacham, T. D. 1982. Median length at sexual maturity of halibut, cusk, longhorn sculpin, ocean pout, and sea raven in the Maritimes Region of the Northwest Atlantic Ocean. *Canadian Journal of Zoology* 60:1326–1330.

Beacham, T. D. 1983a. Variability in size and age at sexual maturity of argentine, *Argentina silus*, on the Scotian Shelf in the Northwest Atlantic Ocean. *Environmental Biology of Fishes* 8:67–72.

Beacham, T. D. 1983b. Variability in median size and age at sexual maturity of Atlantic cod, *Gadus morhua*, on the Scotian Shelf in the Northwest Atlantic Ocean. *U.S. National Marine Fisheries Service Fishery Bulletin* 81:303–322.

Beacham, T. D. 1983c. *Variability in size and age at sexual maturity of haddock* (Melanogrammus aeglefinus) *on the Scotian Shelf in the Northwest Atlantic.* Canadian Technical Report of Fisheries and Aquatic Sciences 1168, 33 pp.

Beacham, T. D. 1983d. *Variability in size and age at sexual maturity of American plaice and yellowtail flounder in the Canadian Maritimes Region of the Northwest Atlantic Ocean.* Canadian Technical Report of Fisheries and Aquatic Sciences 1196, 75 pp.

Beacham, T. D., R. E. Withler, and A. P. Gould. 1985a. Biochemical genetic stock identification of chum salmon (*Oncorhynchus keta*) in Southern British Columbia. *Canadian Journal of Fisheries and Aquatic Sciences* 42:437–448.

Beacham, T. D., R. E. Withler, and A. P. Gould. 1985b. Biochemical genetic stock identification of pink salmon *(Oncorhynchus gorbuscha)* in southern British Columbia and Puget Sound. *Canadian Journal of Fisheries and Aquatic Sciences* 42:1474–1483.

Beardmore, J. A. and R. D. Ward. 1977. Polymorphism, selection, and multilocus heterozygosity in the plaice, *Pleuronectes platessa* L. In *Measuring Selection in Natural Populations*, ed. F. B. Christiansen and T. M. Fenchel. Lecture Notes in Biomathematics, vol. 19, pp. 207–222.

Beatty, P. R., R. G. Thiery, R. K. Fuller, J. E. Boutwell, J. S. Thullen, and F. L. Nibling, Jr. 1983. *Impact of hybrid amur in two California irrigation systems.* Progress report, 1982, Coachella Valley Water District, Coachella, Calif., 70 pp.

Beddington, J. R. and R. M. May. 1977. Harvesting natural populations in a randomly fluctuating environment. *Science* 197:463–465.

Behnke, R. J. 1972. The systematics of salmonid fishes of recently glaciated lakes. *Journal of the Fisheries Research Board of Canada* 29:639–671.

Behnke, R. J. 1979. *Monograph of the Native Trouts of the Genus* Salmo *of Western North America.* Published jointly by the U. S. Forest Service, U. S. Fish and Wildlife Services, and U. S. Bureau of Land Management, 173 pp. Available from Regional Forester, P. O. Box 25127, Lakewood, CO 80225.

Beland, K. F., F. L. Roberts, and R. L. Saunders. 1981. Evidence of *Salmo salar* × *Salmo trutta* hybridization in a North American river. *Canadian Journal of Fisheries and Aquatic Sciences* 38:552–554.

Bell, G. 1976. On breeding more than once. *American Naturalist* 110:57–77.

Belayev, D. K. 1980. Destabilizing selection as a factor of domestication. In *Well-being of Mankind and Genetics,* MIR Publ., Moscow, vol. 1, pp. 64–80.

Benfey, T. J. and A. M. Sutterlin. 1984a. Triploidy induced by heat shock and hydrostatic pressure in landlocked Atlantic salmon *(Salmo salar* L.). *Aquaculture* 36:359–367.

Benfey, T. J. and A. M. Sutterlin. 1984b. Growth and gonadal development in triploid landlocked Atlantic salmon *(Salmo salar). Canadian Journal of Fisheries and Aquatic Sciences* 41:1387–1392.

Bennet, J. H. 1954. On the theory of random mating. *Annals of Eugenics* 18:311–317.

Berg, W. J. and S. D. Ferris. 1984. Restriction endonuclease analysis of salmonid mitochondrial DNA. *Canadian Journal of Fisheries and Aquatic Sciences* 41:1041–1047.

Berggren, W. A. 1969. Rates of evolution in some Cenozoic planktonic foraminifera. *Micropaleontology* 15:351–365.

Bermingham, E. and J. Avise. 1984. Genetics of zoogeography of freshwater fishes in the southeastern United States: Restriction analysis of mitochondrial DNA. *Genetics* 107: s10–s11.

Bernstein, F. 1931. Die geographische Verteilung der Blutgruppen und ihre anthropologische Bedeutung. In *Comitato Italiana per lo Studio dei Problemi della Populazione* (Instituto Poligrafico dello Stato, Rome), pp. 227–243.

Berst, A. H. and R. C. Simon (eds.). 1981. Proceedings of the Stock Concept Symposium. *Canadian Journal of Fisheries and Aquatic Sciences* 38(12).

Best, E. A. 1981. Halibut ecology. In *The Eastern Bering Sea Shelf: Oceanography and Resources,* ed. D. W. Hood and J. A. Calder (U.S. Dept. of Commerce, NOAA, Washington, D.C.), vol. 1, pp. 495–508.

Bhattacharyya, A. 1946. On a measure of divergence between two multinomial populations. *Sankhya* 7:401–406.

Bibb, M. J., R. A. Van Etten, C. T. Wright, M. W. Walberg, and D. A. Clayton. 1981. Sequence and gene organization of mouse mitchondrial DNA. *Cell* 26:167–180.

Biggs-Chisum, E. D. 1984. "High-tech" fish of the future. *National Fisherman* Feb.:4.

Birkey, Jr., C. W., T. Maruyama, and P. Fuerst. 1983. An approach to population and evolutionary genetic theory for genes in mitochondria and chloroplasts, and some results. *Genetics* 103:513–527.

Birt, T. P., J. M. Green, and W. S. Davidson. 1986. Analysis of mitochondrial DNA in allopatric anadromous and nonanadromous Atlantic salmon, *Salmo salar. Canadian Journal of Zoology* 64:118–120.

Blanken, R. L., L. C. Klotz, and A. G. Hinnebusch 1982. Computer comparison of new and existing criteria for constructing evolutionary trees from sequence data. *Journal of Molecular Evolution* 19:9–19.

Blaxhall, P. C. 1983a. Lymphocyte culture for chromosome preparation. *Journal of Fish Biology* 22:279–282.

Blaxhall, P. C. 1983b. Chromosome karyotyping of fish using conventional and G-banding methods. *Journal of Fish Biology* 22:417–424.

Bodmer, W. F. and L. L. Cavalli-Sforza. 1968. A migration matrix model for the study of random genetic drift. *Genetics* 59:565–592.

Bodmer, W. F., E. A. Jones, C. J. Barnstable, and J. G. Bodmer. 1978. The major human histocompatibility system. *Proceedings of the Royal Society of London B* 22:93–116.

Bogart, J. P. 1980. Evolutionary implications of polyploidy in amphibians and reptiles. In *Polyploidy: Biological Relevance*, ed. W. H. Lewis (Plenum Press, New York), pp. 341–378.

Bondari, K. 1983. Caudal fin abnormality and growth and survival of channel catfish. *Growth* 47:361–370.

Bonnell, M. L. and R. K. Selander. 1974. Elephant seals: Genetic variation and near extinction. *Science* 184:908–909.

Borgeson, D. P. 1980. Changing management of Great Lakes fish stocks. *Canadian Journal of Fisheries and Aquatic Sciences* 38:1466–1468.

Borisov, V. M. 1978. The selective effect of fishing on the population structure of species with a long life cycle. *Journal of Ichthyology* 18:896–904.

Borst, P. and L. A. Grivell. 1981. Small is beautiful—portrait of a mitochondrial genome. *Nature* 290:443–444.

Botsford, L. W. 1981. Optimal fishery policy for size-specific, density-dependent population models. *Journal of Mathematical Biology* 12:265–293.

Botsford, L. W. and T. C. Wainwright. 1985. Optimal fishery policy: An equilibrium solution with irreversible investment. *Journal of Mathematical Biology* 21:317–328.

Boyd, W. 1966. *Fundamentals of Immunology*, 4th ed. John Wiley and Sons, New York, 773 pp.

Boyer, H. W. 1971. DNA restriction and modification mechanisms in bacteria. *Annual Review of Microbiology* 25:153–176.

Boyer, H. W. 1974. Restriction and modifications of DNA: Enzymes and substrates. *Federation Proceedings* 33:1125–1127.

Brewer, G. 1970. *An Introduction to Isozyme Techniques*. Academic Press, New York, 186 pp.

Briles, W., W. McGibbon, and M. Irwin. 1950. On multiple alleles affecting cellular antigens in the chicken. *Genetics* 35:633–652.

Britten, R. J., D. E. Graham, and B. R. Neufeld. 1974. Analysis of repeating DNA sequences by reassociation. In *Methods in Enzymology*, ed. L. Grossman and K. Moldave (Academic Press, New York), vol. 29, part E.

Brown, A. H. D. 1975. Sample sizes required to detect linkage disequilibrium between two or three loci. *Theoretical Population Biology* 8:184–201.

Brown, A. H. D., M. W. Feldman, and E. Nevo. 1980. Multilocus structure of natural populations of *Hordeum spontanum*. *Genetics* 96:523–536.

Brown, E. H. Jr., G. W. Eck, N. R. Foster, R. M. Horrall, and C. E. Coberly. 1981. Historical evidence for discrete stocks of lake trout *(Salvelinus namaycush)* in Lake Michigan. *Canadian Journal of Fisheries and Aquatic Sciences* 38:1747–1758.

Brown, G. G. and M. V. Simpson. 1981. Intra- and interspecific variation of the mitochondrial genome in *Rattus norvegicus* and *Rattus rattus:* Restriction enzyme analysis of variant mitochondrial DNA molecules and their evolutionary relationships. *Genetics* 97:125–143.

Brown, G. G. and M. V. Simpson. 1982. Novel features of animal mtDNA evolution as shown by sequences of two rat cytochrome oxidase subunit II genes. *Proceedings of the National Academy of Sciences U.S.A.* 79:3246–3250.

Brown, W. M. 1980. Polymorphism in mitochondrial DNA of humans as revealed by restriction endonuclease analysis. *Proceedings of the National Academy of Sciences U.S.A.* 77:3605–3609.

Brown, W. M. 1983. Evolution of animal mitochondrial DNA. In *Evolution of Genes and Proteins*, ed. M. Nei and R. K. Koehn (Sinauer Associates, Sunderland, Mass.), pp. 62–88.

Brown, W. M. and J. Vinograd. 1974. Restriction endonuclease cleavage maps of animal mitochondrial DNA. *Proceedings of the National Academy of Sciences U.S.A.* 71:4617–4621.

Brown, W. M. and J. W. Wright. 1979. Mitochondrial DNA analyses and the origin and relative age of parthenogenetic lizards (genus *Cnemidophorus). Science* 203:1247–1249.

Brown, W. M., M. George, Jr., and A. C. Wilson. 1979. Rapid evolution of animal mitochondrial DNA. *Proceedings of the National Academy of Sciences U.S.A.* 76:1967–1971.

Brown, W. M., E. M. Prager, A. Wang, and A. C. Wilson. 1982. Mitochondrial DNA sequences of primates: Tempo and mode of evolution. *Journal of Molecular Evolution* 18:225–239.

Bruce, E. J. 1977. A study of the molecular evolution of primates using the techniques of amino acid sequencing and electrophoreses. Ph.D. dissertation, University of California, Davis.

Bruce, E. J. and F. J. Ayala. 1979. Phylogenetic relationships between man and the apes: Electrophoretic evidence. *Evolution* 33:1040–1056.

Busack, C. 1983. Comment on use of allelic frequencies to describe population structure. *Canadian Journal of Fisheries and Aquatic Sciences* 40:1323–1324.

Busack, C. A. and G. A. E. Gall. 1981. Introgressive hybridization in populations of Paiute cutthroat trout *(Salmo clarki seleniris). Canadian Journal of Fisheries and Aquatic Sciences* 38:939–951.

Busack, C., R. Halliburton, and G. A. E. Gall. 1979. Electrophoretic variation and differentiation in four strains of domesticated rainbow trout *(Salmo gairdneri). Canadian Journal of Genetics and Cytology* 21:81–94.

Busack, C. A., G. H. Thorgaard, M. P. Bannon, and G. A. E. Gall. 1980. An electrophoretic, karyotypic and meristic characterization of the Eagle Lake trout, *Salmo gairdneri aquilarum. Copeia* 1980:418–424.

Calaprice, J. R. 1969. Production and genetic factors in managed salmonid populations. In *Symposium on Salmon and Trout in Streams 1968.* H. R. MacMillan Lectures in Fisheries (University of British Columbia, Vancouver), pp. 377–388.

California Gene Resources Program. 1982. *Anadromous salmonid genetic resources: An assessment and plan for California.* National Council of Gene Resources, Berkeley, prepared under contract No. 9146, California Department of Food and Agriculture, 168 pp.

Campton, D. E. and J. M. Johnston. 1985. Electrophoretic evidence for a genetic admixture of native and non-native trout in the Yakima River, Washington. *Transactions of the American Fisheries Society* 114:782–793.

Campton, D. E. and F. M. Utter. 1985. Natural hybridization between steelhead trout *(Salmo gairdneri)* and coastal cutthroat trout *(Salmo clarki clarki)* in two Puget Sound streams. *Canadian Journal of Fisheries and Aquatic Sciences* 42:110–119.

Cann, R. L. and A. C. Wilson. 1983. Length mutations in human mitochondrial DNA. *Genetics* 104:699–711.

Cann, R. L., W. M. Brown, and A. C. Wilson. 1984. Polymorphic sites and the mechanism of evolution in human mitochondrial DNA. *Genetics* 106:479–499.

Capanna, E., S. Cataudella, and R. Volpe. 1974. Un ibrido intergenerico tra trota iridea e salmerino di fonte *(Salmo gairdneri × Salvelinus fontinalis). Bollettino di Pesca Piscicoltura e Idrobiologia* 29:101–106.

Carlson, S. S., A. C. Wilson, and R. D. Maxson. 1978. Do albumin clocks run on time? A reply. *Science* 200:1183–1185.

Carmelli, D. and L. L. Cavalli-Sforza. 1976. Some models of population structure and evolution. *Theoretical Population Biology* 9:329–359.

Case, S. M. 1975. Evolutionary studies in selected North American frogs of the genus *Rana* (Amphibia, Anura). Ph.D. dissertation, University of California, Berkeley.

Case, S. M., P. G. Haneline, and M. F. Smith. 1975. Protein variation in several species of *Hyla. Systematic Zoology* 24:281–295.

Cassani, J. R. 1981. Feeding behavior of underyearling hybrids of grass carp *(Ctenopharyngodon idella)* and bighead *(Hypophthalmichthys nobilis)* on selected species of aquatic plants. *Journal of Fish Biology* 18:127–133.

Casselman, J. M., J. J. Collins, E. J. Crossman, P. E. Ihssen, and G. R. Spangler. 1981. Lake whitefish *(Coregonus clupeaformis)* stocks of the Ontario waters of Lake Huron. *Canadian Journal of Fisheries and Aquatic Sciences* 38:1772–1789.

Caswell, H. 1982. Stable population structure and reproductive value for populations with complex life cycles. *Ecology* 63:1223–1231.

Caswell, H. 1983. Phenotypic plasticity in life-history traits: Demographic effects and evolutionary consequences. *American Zoologist* 23:35–46.

Cavalli-Sforza, L. L. 1969. Human diversity. In *Proceedings of the 12th International Congress on Genetics,* Tokyo, vol. 3, pp. 405–416.

Cavalli-Sforza, L. L. and W. F. Bodmer. 1971. *The Genetics of Human Populations.* W. H. Freeman and Co., San Francisco, 965 pp.

Cavalli-Sforza, L. L. and A. W. F. Edwards. 1967. Phylogenetic analysis: Models and estimation procedures. *American Journal of Human Genetics* 19:233–257.

Chakraborty, R. 1974. A note on Nei's measure of gene diversity in a substructured population. *Humangenetik* 21:85–88.

Chakraborty, R. 1977. Estimation of time of divergence from phylogenetic studies. *Canadian Journal of Genetics and Cytology* 19:217–223.

Chakraborty, R. 1980. Gene-diversity analysis in nested subdivided populations. *Genetics* 96:721–726.

Chakraborty, R. 1981. The distribution of the number of heterozygous loci in an individual in natural populations. *Genetics* 98:461–466.

Chakraborty, R. 1984. Detection of nonrandom association of alleles from the distribution of the number of heterozygous loci in a sample. *Genetics* 108:719–731.

Chakraborty, R. and M. Nei. 1974. Dynamics of gene differentiation between incompletely isolated populations of unequal sizes. *Theoretical Population Biology* 5:460–469.

Chakraborty, R. and M. Nei. 1976. Hidden genetic variability within electromorphs in finite populations. *Genetics* 84:385–393.

Chakraborty, R. and M. Nei. 1977. Bottleneck effects on average heterozygosity and genetic distance with the stepwise mutation model. *Evolution* 31:347–356.

Chakraborty, R., P. A. Fuerst, and M. Nei. 1977. A comparative study of genetic variation within and between populations under the neutral mutation hypothesis and the model of sequentially advantageous mutation. *Genetics* 86:s10–s11.

Chakraborty, R., P. A. Fuerst, and M. Nei. 1978. Statistical studies on protein polymorphism in natural populations. II. Gene differentiation between populations. *Genetics* 88:367–390.

Chakraborty, R., M. Haag, N. Ryman, and G. Ståhl. 1982. Hierarchical gene diversity analysis and its application to brown trout population data. *Hereditas* 97:17–21.

Chang, Te-Tzu. 1979. Genetics and evolution of the green revolution. In *Replies From Biological Research*, ed. Roman de Vicente (Conselo Superior de Investigaciones Cientificas, Madrid), pp. 187–209.

Charlesworth, B. 1980. *Evolution in Age-Structured Populations*. Cambridge University Press, Cambridge.

Charnes, A., E. L. Frome, and P. L. Yu. 1976. The equivalence of generalized least squares and maximum likelihood estimates in the exponential family. *Journal of the American Statistical Association* 71:169–171.

Charnov, E. L. 1981. Sex reversal in *Pandalus borealis:* Effect of a shrimp fishery? *Marine Biology Letters* 2:53–57.

Cherfas, N. B. 1981. Gynogenesis in fishes. In *Genetic Bases of Fish Selection*, ed. V. S. Kirpichnikov (Springer-Verlag, Berlin and New York), pp. 255–273.

Cherry, L. M., S. M. Case, and A. C. Wilson. 1978. Frog perspective on the morphological difference between humans and chimpanzees. *Science* 200:209–211.

Chetverikov, S. S. 1926. On certain aspects of the evolutionary process from the standpoint of genetics. *Journal of Experimental Biology U.S.S.R.* 2:3–54. In Russian. (English translation, *Proceedings of the American Philosophical Society* 105:167–195.)

Chevassus, B. 1983. Hybridization in fish. *Aquaculture* 33:245–262.

Chevassus, B., R. Guyomard, D. Chourrout, and E. Quillet. 1983. Production of viable hybrids in salmonids by triploidization. *Génétique Sélection d'Évolution* 15:519–532.

Chilcote, M., S. Leider, and R. Jones. 1981. *Kalama River salmonid studies*. Washington State Game Department, Progress Report 1980, Fishery Research Report 81-11.

Chourrout, D. 1980. Thermal induction of diploid gynogenesis and triploidy in the eggs of the rainbow trout *(Salmo gairdneri* Richardson). *Reproduction Nutrition Development* 209:727–733.

Chourrout, D. 1982a. Tetraploidy induced by heat shocks in the rainbow trout *(Salmo gairdneri* R.). *Reproduction Nutrition Development* 20:727–733.

Chourrout, D. 1982b. La gynogenèse chez les vertèbres. *Reproduction Nutrition Development* 22:713–724.

Chourrout, D. 1982c. Gynogenesis caused by ultraviolet irradiation of salmonid sperm. *Journal of Experimental Zoology* 223:175–181.

Chourrout, D. 1984. Pressure-induced retention of second polar body and suppression of first cleavage in rainbow trout: Production of all-triploids, all-tetraploids and heterozygous and homozygous diploid gynogenetics. *Aquaculture* 36:111–126.

Chourrout, D. and E. Quillet. 1982. Induced gynogenesis in the rainbow trout: Sex and survival of progenies. Production of all-triploid populations. *Theoretical Applied Genetics* 63:201–205.

Chourrout, D. and J. Itskovich. In press. Three manipulations permitted by artificial insemination in Tilapia: Induced diploid gynogenesis, production of all-triploid population and intergeneric hybridization. In *Proceedings of the Tiberias Symposium on Tilapia in Aquaculture*.

Christiansen, F. B. and O. Frydenberg. 1974. Geographic patterns of four polymorphisms in *Zoarces viviparus* as evidence of selection. *Genetics* 77:765–770.

Christensen, O. and P. O. Larsson. 1979. *Review of the Baltic salmon research*. International Council for the Exploration of the Sea Cooperative Research Report Series B 1979(89).

Christy, F. T. Jr. and A. Scott. 1965. *The Commonwealth in Ocean Fisheries*. Johns Hopkins Press, Baltimore, Md.

Churkin, M., Jr. and J. H. Trexler, Jr. 1980. Circum-Arctic plate accretion-isolating part of a Pacific plate to form the nucleus of the Arctic Basin. *Earth Planetary Sciences Letters* 48:356–362.

Clark, C. W. 1974. Possible effects of schooling on the dynamics of exploited fish populations. *Journal du Conseil international pour l'Exploration de la Mer* 36:7–14.

Clark, C. W. 1976. *Mathematical Bioeconomics: The Optimal Management of Renewable Resources.* J. Wiley and Sons, New York.

Clark, D. L. 1982. Origin, nature, and world climate effect of Arctic Ocean ice-cover. *Nature* 300:321–325.

Clayton, J. W. 1981. The stock concept and the uncoupling of organismal and molecular evolution. *Canadian Journal of Fisheries and Aquatic Sciences* 38:1515–1522.

Cockerham, C. C. 1969. Variance of gene frequencies. *Evolution* 23:72–80.

Cockerham, C. C. 1973. Analyses of gene frequencies. *Genetics* 74:679–700.

Cockerham, C. C. and B. S. Weir. 1977. Digenic descent measures for finite populations. *Genetical Research* 30:121–127.

Cocks, G. T. and A. C. Wilson. 1969. Immunological detection of single amino acid substitutions in alkaline phosphatase. *Science* 164:188–189.

Cook, R. C. and G. E. Lord. 1978. Identification of stocks of Bristol Bay sockeye salmon, *Oncorhynchus nerka*, by evaluating scale patterns with a polynomial discriminant method. *U.S. Department of Commerce, Fishery Bulletin* 76:415–423.

Cook, R. D. and S. Weisberg. 1974. A note on the estimation of individual admixture. *Annals of Human Genetics* 37:355–358.

Cooper, D. W. 1968. The significance level in multiple tests made simultaneously. *Heredity* 23:614–617.

Cooper, E. L. 1961. Growth of wild and hatchery strains of brook trout. *Transactions of the American Fisheries Society* 90:424–438.

Coyne, J. 1982. Gel electrophoresis and cryptic protein variation. *Isozymes: Current Topics in Biological and Medical Research* 6:1–32.

Coyne, J., A. Felton and R. C. Lewontin. 1978. Extent of genetic variation at a highly polymorphic esterase locus in *Drosophila pseudoobscura*. *Proceedings of the National Academy of Sciences U.S.A.* 75:5090–5093.

Cramp, S. and K. Simmons. 1980. Hawks to Bustards. Handbook of the Birds of Europe, the Middle East and North Africa, vol. 2. Oxford University Press.

Croizat, L., G. Nelson, and D. E. Rosen. 1974. Centers of origin and related concepts. *Systematic Zoology* 23:265–287.

Cronin, J. E. 1975. Molecular systematics of the order primates. Ph.D. dissertation, University of California, Berkeley.

Cross, T. F. and J. King. 1983. Genetic effects of hatchery rearing in Atlantic salmon. *Aquaculture* 33:33–40.

Cross, T. F. and R. D. Ward. 1980. Protein variation and duplicate loci in the Atlantic salmon, *Salmo salar* L. *Genetical Research* 36:147–165.

Crow, J. 1976. *Genetics Notes.* Burgess Publishing Co., Minneapolis, Minn., 278 pp.

Crow, J. F. and K. Aoki. 1984. Group selection for a polygenic trait: Estimating the degree of population subdivision. *Proceedings of the National Academy of Sciences U.S.A.* 81:6073–6077.

Crow, J. F. and M. Kimura. 1970. *An Introduction to Population Genetics Theory.* Harper and Row, New York, 591 pp.

Crow, J. F., and N. E. Morton. 1955. Measurement of gene frequency drift in small populations. *Evolution* 9:202–214.

Cruden, D. 1949. The computation of inbreeding coefficients. *Journal of Heredity* 40:248–251.

Cushing, D. H. 1968. *Fisheries Biology.* University of Wisconsin Press, Madison.

Cushing, D. H. 1973. *Recruitment and parent stock in fishes.* Washington Sea Grant Publication WSG 73–1, University of Washington, Seattle, 197 pp.

Cushing, J. 1956. *Observations on the serology of tuna.* U.S. Fish and Wildlife Service, Special Scientific Report—Fisheries No. 183, 14 pp.

Daan, N. 1980. A review of replacement of depleted stocks by other species and the mechanisms underlying such replacement. *Rapports et Procès-Verbaux des Réunions Conseil International pour l'Exploration de la Mer* 177:405–421.

Dangel, J. R., P. T. Macy, and F. C. Withler. 1973. *Annotated bibliography of interspecific hybridization of fishes of the subfamily Salmonidae.* NOAA Technical Memorandum NMFS NWFC-1, U.S. Department of Commerce, Washington, D.C., 48 pp.

Darwin, C. 1859. *The Origin of Species by Means of Natural Selection; or, the Preservation of Favoured Races and the Struggle for Life.* Mentor Books, New York (1966).

Dayhoff, M. O. (ed.). 1972. *Atlas of Protein Sequence and Structure,* vol. 5. National Biomedical Research Foundation, Silver Spring, Md.

Deakin, M. A. B. 1966. Sufficient conditions for genetic polymorphism. *American Naturalist* 100:690–692.

Dean, D. 1979. Introduced species and the Maine situation. In *Exotic Species in Mariculture,* ed. R. H. Mann (MIT Press, Cambridge, Mass.), pp. 149–164.

Delany, M. E. and S. E. Bloom. 1984. Replication banding patterns in the chromosomes of the rainbow trout. *Journal of Heredity* 75:431–434.

Dempster, A. P., N. M. Laird, and D. B. Rubin. 1977. Maximum likelihood estimation from incomplete data via the EM algorithm. *Journal of the Royal Statistical Society B: Methodology* 39:1–38.

Denaro, M., H. Blanc, M. J. Johnson, K. H. Chen, E. Wilmsen, L. L. Cavalli-Sforza, and D. C. Wallace. 1981. Ethnic variation in *Hpa*I endonuclease cleavage patterns of human mitochondrial DNA. *Proceedings of the National Academy of Sciences U.S.A.* 78:5768–5772.

Denniston, C. D. 1978. Small population size and genetic diversity. In *Endangered Birds: Management Techniques for Preserving Endangered Species,* ed. S. A. Temple (University of Wisconsin Press, Madison).

Dickerson, R. E. 1971. The structure of cytochrome *c* and the rates of molecular evolution. *Journal of Molecular Evolution* 1:26–45.

Dingle, H. and J. P. Hegmann (eds.). 1982. *Evolution and Genetics of Life Histories.* Springer-Verlag, New York.

Dixon, W. J., M. B. Brown, L. Engleman, J. W. Frane, M. A. Hill, R. I. Jennrick, and J. D. Toropek. 1983. *BMDP Statistical Software.* 1983. Printing with additions. University of California Press, Berkeley, 733 pp.

Dobzhansky, T. 1955. *Genetics of the Evolutionary Process.* Columbia University Press, New York.

Donaldson, E. M. and G. A. Hunter. 1982. Sex control in fish with particular reference to salmonids. *Canadian Journal of Fisheries and Aquatic Sciences* 39:99–110.

Dourojeanni, M. J. 1982. *Renewable Natural Resources of Latin America and the Caribbean: Situations and Trends.* World Wildlife Fund, Washington, D.C., 495 pp.

Dowling, T. E. and W. S. Moore. 1984. Level of reproductive isolation between two cyprinid fishes, *Notropis cornutus* and *N. chrysocephalus. Copeia* 1984:617–628.

Dowling, T. E. and W. S. Moore. 1985. Evidence for selection against hybrids in the family Cyprinidae (genus *Notropis*). *Evolution* 39:152–158.

Doyle, R. W. 1983. An approach to the quantitative analysis of domestication selection in aquaculture. *Aquaculture* 33:167–186.

Draper, N. R. and H. Smith. 1981. *Applied Regression Analysis,* 2nd ed. John Wiley and Sons, New York, 709 pp.

Dubinin, N. P. 1976. *General Genetics.* Nauka, Moscow, 590 pp. In Russian.

Dubinin, N. P. and D. D. Romashov. 1932. Genetic structure of species and its evolution. *Biological Journal of the U.S.S.R.* 1:59–95. In Russian.

Dubinin, N., D. Romashov, M. Heptner, and Z. Demidova. 1937. Aberrant polymorphism in *Drosophila fasciata* Meig (Syn. *melanogaster* Meig). *Biologicheskii Zhurnal Armenii* 6:311–354.

Dyagilev, S. E. and N. B. Markevich. 1979. Different time of maturation of odd- and even-year pink salmon as a major factor responsible for different results of their acclimatization in the North European part of the USSR. *Journal of Ichthyology* (USSR) 19:230–245. In Russian.

Eanes, W. F. and R. K. Koehn. 1978. Relationship between subunit size and number of rare electrophoretic alleles in human enzymes. *Biochemical Genetics* 16:971–986.

Eck, R. V. and M. O. Dayhoff. 1966. *Atlas of Protein Sequence and Structure.* National Biomedical Research Foundation, Silver Spring, Md.

Eden, M. 1967. Inadequacies of neo-Darwinian evolution as a scientific theory. In *Mathematical Challenges to the Neo-Darwinian Interpretation of Evolution,* ed. P. S. Moorhead and M. M. Kaplan (Wistar Institute Press, Philadelphia, Pa.).

Edwards, A. W. F. 1972. *Likelihood: An Account of the Statistical Concept of Likelihood and Its Application to Scientific Inference.* Cambridge University Press, New York, 235 pp.

Edwards, R. J. 1979. A report of Guadelupe bass *(Micropterus treculi)* × smallmouth bass *(M. dolomieui)* hybrids from two localities in the Guadelupe River, Texas. *Texas Journal of Science* 31:231–238.

Efron, B. 1981. Nonparametric standard errors and confidence intervals (with discussion). *Canadian Journal of Statistics* 9:139–172.

Efron, B. 1982. *The jackknife, the bootstrap, and other resampling plans.* National Science Foundation Conference Board of the Mathematical Sciences Monograph 38. SIAM, Philadelphia, 92 pp.

Einarsson, T., D. M. Hopkins, and R. R. Doell. 1967. The stratigraphy of Tjornes, northern Iceland, and the history of the Bering land bridge. In *The Bering Land Bridge,* ed. D. M. Hopkins (Stanford University Press, Stanford), pp. 312–325.

Eldholm, O. and J. Thiede. 1980. Cenozoic continental separation between Europe and Greenland. *Palaeography, Palaeoclimatology, Palaeoecology* 30:243–259.

Elinson, R. P. and A. Briedis. 1981. Triploidy permits survival of an inviable amphibian hybrid. *Developmental Genetics* 2:357–367.

Elston, R. C. 1971. The estimation of admixture in racial hybrids. *Annals of Human Genetics* 35:9–17.

Emigh, T. H. 1980. A comparison of test for Hardy-Weinberg equilibrium. *Biometrics* 36:627–642.

Emik, L. O. and C. E. Terrill. 1949. Systematic procedures for calculating inbreeding coefficients. *Journal of Heredity* 40:51–55.

Emiliani, C. 1961. The temperature decrease of surface seawater in high latitudes and of abyssal-hadal water in open oceanic basins during the past 75 million years. *Deep-Sea Research* 8:144–147.

Emiliani, C., S. Gartner, and B. Lidz. 1972. Neogene sedimentation of the Blake Plateau and the emergence of Central American Isthmus. *Paleography, Paleoclimatology, Paleoecology.* 11:1–10.

Endler, J. A. 1977. *Geographic Variation, Speciation, and Clines.* Princeton University Press, Princeton, New Jersey.

Engles, W. R. 1981. Estimating genetic divergence and genetic variability with restriction endonucleases. *Proceedings of the National Academy of Sciences U.S.A.* 78:6329–6333.

Ehrlich, P. R. and P. H. Raven. 1969. Differentiation of populations. *Science* 165:1228–1232.

Everhart, W. H. and W. D. Youngs. 1981. *Principles of Fishery Science*, 2nd ed. Cornell University Press, Ithaca.

Everitt, B. S. 1978. *Graphical Techniques for Multivariate Data.* North-Holland, New York, 117 pp.

Everitt, B. S. 1984. *An Introduction to Latent Variable Models.* Monographs on Statistics and Applied Probability, Chapman and Hall, London, 107 pp.

Everitt, B. S. and D. J. Hand. 1981. *Finite Mixture Distributions.* Monographs on Statistics and Applied Probability, Chapman and Hall, London, 143 pp.

Ewens, W. J. 1983. The role of models in the analysis of molecular genetic data, with particular reference to restriction fragment data. In *Statistical Analysis of DNA Sequence Data*, ed. B. S. Weir (Marcel Dekker, Inc., New York).

Ewens, W. J., R. S. Spielman, and H. Harris. 1981. Estimation of genetic variation at the DNA level from restriction endonuclease data. *Proceedings of the National Academy of Sciences U.S.A.* 78:3748–3750.

Fairbairn, D. J. 1981. Biochemical genetic analysis of population differentiation in Greenland halibut *(Reinhardtius hippoglossoides)* from the Northwest Atlantic, Gulf of St. Lawrence, and Bering Sea. *Canadian Journal of Fisheries and Aquatic Sciences* 38:669–677.

Falconer, D. S. 1981. *Introduction to Quantitative Genetics*, 2nd ed. Longman, London.

Farris, J. S. 1970. Methods for computing Wagner trees. *Systematic Zoology* 19:83–92.

Farris, J. S. 1972. Estimating phylogenetic trees from distance matrices. *American Naturalist* 106:645–668.

Farris, J. S. 1981. Distance data in phylogenetic analysis. In *Advances in Cladistics*, ed. V. A. Funk and D. R. Brooks (New York Botanical Gardens, New York), pp. 1–23.

Favro, L. D., P. K. Kuo, and J. F. McDonald. 1979. Population-genetic study of the effects of selective fishing on the growth rate of trout. *Journal of the Fisheries Research Board of Canada* 36:552–561.

Favro, L. D., P. K. Kuo, and J. F. McDonald. 1980. Effects of unconventional size limits on the growth rate of trout. *Canadian Journal of Fisheries and Aquatic Sciences* 37:873–876.

Favro, L. D., P. K. Kuo, J. F. McDonald, D. D. Favro, and A. D. Kuo. 1982. A multilocus genetic model applied to the effects of selective fishing on the growth rate of trout. *Canadian Journal of Fisheries and Aquatic Sciences* 39:1540–1543.

Feldman, M. and R. C. Lewontin. 1975. The heritability hangup. *Science* 190:1163–1168.

Felsenstein, J. 1971. Inbreeding and variance effective numbers in populations with overlapping generations. *Genetics* 68:581–597.

Felsenstein, J. 1978. The number of evolutionary trees. *Systematic Zoology* 27:27–33.

Felsenstein, J. 1983. Inferring evolutionary trees from DNA sequences. In *Statistical Analysis of DNA Sequence Data*, ed. B. S. Weir (Marcel Dekker, Inc., New York).

Felley, J. 1980. Analysis of morphology and asymmetry in bluegill sunfish *(Lepomis machrochirus)* in the southeastern United States. *Copeia* 1980:18–29.

Ferguson, A. 1980. *Biochemical Systematics and Evolution.* Blackie and Son Limited, Glasgow, 194 pp.

Ferguson, A. and C. C. Fleming. 1983. Evolutionary and taxonomic significance of protein variation in the brown trout *(Salmo trutta L.)* and other salmonid fishes. In *Protein Polymorphism: Adaptive and Taxonomic Significance*, ed. G. S. Oxford and D. Rollinson (Academic Press, London and New York), pp. 85–99.

Ferguson, A. and F. M. Mason. 1981. Allozyme evidence for reproductively isolated sympatric populations of brown trout *Salmo trutta* L. in Lough Melvin, Ireland. *Journal of Fish Biology* (Dublin) 18:629–642.

Ferris, S. D. and G. S. Whitt. 1978. Genetic and molecular analysis of nonrandom dimer assembly of the creatine kinase isozymes of fishes. *Biochemical Genetics* 16:811–830.

Ferris, S. and G. Whitt. 1979. Evolution of the differential regulation of duplicate genes after polyploidization. *Journal of Molecular Evolution* 12:267–317.

Ferris, S. D., W. M. Brown, W. S. Davidson, and A. C. Wilson. 1981a. Extensive polymorphism in the mitochondrial DNA of apes. *Proceedings of the National Academy of Sciences U.S.A.* 78:6319–6323.

Ferris, S. D., A. C. Wilson, and W. M. Brown. 1981b. Evolutionary tree for apes and humans based on cleavage maps of mitochondrial DNA. *Proceedings of the National Academy of Sciences U.S.A.* 78:2432–2436.

Ferris, S. D., R. D. Sage, and A. C. Wilson. 1982. Evidence from mtDNA sequences that common laboratory strains of inbred mice are descended from a single female. *Nature* 295:163–165.

Ferris, S. D., U. Ritte, K. Fischer Lindahl, E. M. Prager, and A. C. Wilson. 1983a. Unusual type of mitochondrial DNA in mice lacking a maternally transmitted antigen. *Nucleic Acids Research* 11:2917–2926.

Ferris, S. D., R. D. Sage, C.-M. Huang, J. T. Nielsen, U. Ritte, and A. C. Wilson. 1983b. Flow of mitochondrial DNA across a species boundary. *Proceedings of the National Academy of Sciences U.S.A.* 80:2290–2294.

Ferris, S. D., R. D. Sage, E. M. Prager, U. Ritte, and A. C. Wilson. 1983c. Mitochondrial DNA evolution in mice. *Genetics* 105:681–721.

Fisher, R. A. 1930. *The Genetical Theory of Natural Selection.* Oxford University Press, Oxford.

Fitch, W. M. 1971. Toward defining the course of evolution: Minimum change for a specified tree topology. *Systematic Zoology* 20:406–416.

Fitch, W. M. 1976. Molecular evolutionary clocks. In *Molecular Evolution,* ed. F. J. Ayala (Sinauer Associates, Sunderland, Mass.), pp. 160–178.

Fitch, W. M. and C. H. Langley. 1976. Protein evolution and the molecular clock. *Federation Proceedings* 35:2092–2097.

Fitch, W. M. and E. Margoliash. 1967. Construction of phylogenetic trees. *Science* 155:279–284.

Food and Agriculture Organization of the United Nations. 1981. *Conservation of the genetic resources of fish: Problems and recommendations.* Report of the Expert Consulation on the Genetic Resources of Fish, Rome, 9–13 June 1980, FAO Fisheries Technical Paper 217, 43 pp.

Foresti, F., L. F. Almeida Toledo, and S. A. Toledo. 1981. Polymorphic nature of nucleolus organizer regions in fishes. *Cytogenetics and Cell Genetics* 31:137–144.

Fournier, D. A., T. D. Beacham, B. E. Riddell, and C. A. Busack. 1984. Estimating stock composition in mixed stock fisheries using morphometric, meristic, and electrophoretic characteristics. *Canadian Journal of Fisheries and Aquatic Sciences* 41:400–408.

Frankel, O. H. and M. E. Soulé. 1981. *Conservation and Evolution.* Cambridge University Press, Cambridge.

Franklin, I. R. 1980. Evolutionary change in small populations. In *Conservation Biology: An Evolutionary-Ecological Perspective,* ed. M. E. Soulé and B. A. Wilcox (Sinauer Associates, Sunderland, Mass.), pp. 135–150.

Franklin, I. R. 1977. The distribution of the proportion of the genome which is homozygous by descent in inbred individuals. *Theories of Population Biology* 11:60–80.

Frelin, Ch. and F. Vuilleumier. 1979. Biochemical methods and reasoning in systematics. *Zeitschrift für zoologische Systematik und Evolutionsforschung* 17:1–10.

Frossard, P., K. Sisco, and D. Rucknagel. 1983. Transfer of small plasmid DNA fragments from polyacrymlamide gels onto nitocellulose paper. *Analytical Biochemistry* 134:265–268.

Fuerst, P. A. and R. E. Ferrell. 1980. The stepwise mutation model: An experimental evaluation utilizing hemoglobin variants. *Genetics* 94:185–201.

Fuerst, P. A., R. Chakraborty, and M. Nei. 1977. Statistical studies on protein polymorphism in natural populations. I. Distribution of single locus heterozygosity. *Genetics* 86:455–483.

Fujita, K. and J. T. Newberry. 1982. Tectonic evolution of northeastern Siberia and adjacent regions. *Tectonophysics* 89:337–357.

Fukuhara, F. M., S. Murai, J. J. LaLanne, and A. Sribhibhadh. 1962. Continental origin of red salmon as determined from morphological characters. *International North Pacific Fisheries Commission Bulletin* 8:15–109.

Gall, G. A. E. 1983. Genetics of fish: A summary of discussion. *Aquaculture* 33:383–394.

Gall, G. A. E. and B. Bentley. 1981. Para-albumin polymorphism: An unlinked two locus system in rainbow trout. *Journal of Heredity* 72:22–26.

Garrod, D. J. 1973a. Management of multiple resources. *Journal of the Fisheries Research Board of Canada* 30:1977–1985.

Garrod, D. J. 1973b. The variation of replacement and survival in some fish stocks. *Rapports et Procès-verbaux des Réunions Conseil International pour l'Exploration de la Mer* 164:43–56.

Garrod, D. J. and B. J. Knights. 1979. Fish stocks: Their life-history characteristics and response to exploitation. *Symposia of the Zoological Society of London* 44:361–382.

George, M., Jr. 1982. Mitochondrial DNA evolution in Old World monkeys. Ph.D. dissertation, University of California, Berkeley.

Gharrett, A. and F. M. Utter. 1982. Scientists detect genetic differences. *Sea Grant Today* 12(2):3–4.

Gharrett, A. J., R. C. Simon, and J. D. McIntyre. 1977. Reassociation and hybridization properties of DNAs from several species of fish. *Comparative Biochemistry and Physiology* 56B:81–85.

Gilbert, C. R. 1978. The nominal North American cyprinid fish *Notropis henryi* interpreted as an intergeneric hybrid, *Clinostomus funduloides* × *Nocomis leptocephalus*. *Copeia* 1978:177–181.

Gilbert, M. W. and D. L. Clark. 1983. Central Arctic Ocean paleoceanographic interpretations based on late Cenozoic calcareous dinoflagellates. *Marine Micropaleontology* 7:385–401.

Giles, R. E., H. Blanc. H. M. Cann, and D. C. Wallace. 1980. Maternal inheritance of human mitochondrial DNA. *Proceedings of the National Academy of Sciences U.S.A.* 77:6715–6719.

Gill, A. E. 1976. Genetic divergence of insular populations of deer mice. *Biochemical Genetics* 14:835–848.

Gillespie, J. and K. Kojima. 1968. The degree of polymorphism in enzymes involved in energy production compared to that in nonspecific enzymes in two *D. ananassae* populations. *Proceedings of the National Academy of Sciences U.S.A.* 61:582–585.

Gillespie, L. L. and J. B. Armstrong. 1980. Production of androgenetic diploid exolotls by suppression of first cleavage. *Journal of Experimental Zoology* 213:423–425.

Gillespie, L. L. and J. B. Armstrong. 1981. Suppression of first cleavage in Mexican axolotl *(Ambystoma mexicanum)* by heat shock or hydrostatic pressure. *Journal of Experimental Zoology* 218:441–445.

Gjedrem, T. 1983. Genetic variation in quantitative traits and selective breeding in fish and shellfish. *Aquaculture* 33:51–72.

Gjerde, B., K. Gunnes, and T. Gjedrem. 1983. Effect of inbreeding on survival and growth in rainbow trout (as cited by Kincaid, 1983).

Gladenkov, Yu. B. 1979. Comparison of late Cenozoic molluscan assemblages in northern regions of the Atlantic and Pacific Oceans. *International Geology Review* 21:880–890.

Glass, B. 1955. On the unlikelihood of significant admixture of genes from the North American Indians in the present composition of the Negroes of the United States. *American Journal of Human Genetics* 7:368–385.

Glass, B. and C. C. Li. 1953. The dynamics of racial intermixture—an analysis based on the American Negro. *American Journal of Human Genetics* 5:1–20.

Gold, J. R. 1979. Cytogenetics. In *Fish Physiology,* vol. 8, ed. W. S. Hoar, D. J. Randall, and J. R. Brett (Academic Press, New York), pp. 353–405.

Gold, J. R. 1984. Silver-staining and heteromorphism of chromosomal nucleolus organizer regions in North American cyprinid fishes. *Copeia* 1984:113–139.

Gold, J. R., J. C. Avise, and G. A. E. Gall. 1977. Chromosome cytology in the cutthroat series, *Salmo clarki* (Salmonidae). *Cytologia* 42:377–382.

Gold, J. R., W. J. Karel, and M. R. Strand. 1980. Chromosome formulae of North American fishes. *Progressive Fish-Culturist* 42:10–23.

Goodfellow, W. L., Jr., C. H. Hocutt, R. P. Morgan II, and Jay R. Stauffer, Jr. 1984. Biochemical assessment of the taxonomic status of *Rhinichthyes bowersi* (Pisces: Cyprinidae). *Copeia* 1984:652–659.

Goodman, L. A. 1974. Exploratory latent structure analysis using both identifiable and unidentifiable models. *Biometrika* 61:215–231.

Goodman, M., M. L. Weiss, and J. Czelusniak. 1982. Molecular evolution above the species level: Branching patterns, rates, and mechanisms. *Systematic Zoology* 31:376–399.

Goodpasture, C. and S. E. Bloom. 1975. Visualization of nucleolar organizer regions in mammalian chromosomes using silver staining. *Chromosoma* 53:37–50.

Gorman, G. C. and Y. J. Kim. 1977. Genotypic evolution in the face of phenotypic conservativeness: *Abudefduf* (Pomacentridae) from the Atlantic and Pacific sides of Panama. *Copeia* 1977:694–697.

Gorman, G. C., A. C. Wilson, and M. Nakanishi. 1971. A biochemical approach towards the study of reptilian phylogeny: Evolution of serum albumin and lactic dehydrogenase. *Systematic Zoology* 20:167–185.

Gorman, G. C., Y. J. Kim, and R. Rubinoff. 1976. Genetic relationships of three species of *Bathygobius* from the Atlantic and Pacific sides of Panama. *Copeia* 1976:361–364.

Gotoh, O., J.-I. Hayashi, H. Yonekawa, and Y. Tagashira. 1979. An improved method for estimating sequence divergence between related DNAs from changes in restriction endonuclease cleavage sites. *Journal of Molecular Evolution* 14:301–310.

Gottlieb, L. D. and N. F. Weeden. 1981. Correlation between subcellular location and phosphoglucose isomerase variability. *Evolution* 35:1019–1022.

Gowe, R. S., A. Robertson, and B. D. H. Latter. 1959. Environment and poultry breeding problems. 5. The design of poultry control strains. *Poultry Science* 38:462–471.

Gradstein, F. M. and S. P. Srivastava. 1980. Aspects of Cenozoic stratigraphy and paleooceanography of the Labrador Sea and Baffin Bay. *Palaeogeography, Palaeoclimatology, Palaeoecology* 30:261–295.

Grant, W. S. 1981. Biochemical genetic variation, population structure, and evolution of Atlantic and Pacific herring. Ph.D. dissertation, University of Washington, Seattle.

Grant, W. S. 1984. Biochemical population genetics of Atlantic herring, *Clupea harengus*. *Copeia* 1984:357–364.

Grant, W. S. and F. M. Utter, 1980. Biochemical genetic variation in walleye pollock, *Theragra chalcogramma:* Population structure in the southeastern Bering Sea and the Gulf of Alaska. *Canadian Journal of Fisheries and Aquatic Sciences* 37:1093–1100.

Grant, W. S. and F. M. Utter. 1984. Biochemical populations genetics of Pacific herring *(Clupea pallasi)*. *Canadian Journal of Fisheries and Aquatic Sciences* 41:856–864.

Grant, W. S., G. B. Milner, P. Krasnowski, and F. M. Utter. 1980. Use of biochemical genetic variants for identification of sockeye salmon *(Oncorhynchus nerka)* stocks in Cook Inlet, Alaska. *Canadian Journal of Fisheries and Aquatic Sciences* 37:1236–1247.

Grant, W. S., D. J. Teel, T. Kobayashi, and C. Schmitt. 1984. Biochemical population genetics of Pacific halibut (*Hippoglossus stenolepis*) and comparison with Atlantic halibut (*H. hippoglossus*). *Canadian Journal of Fisheries and Aquatic Sciences* 41:1083–1088.

Graves, J. E., S. D. Ferris, and A. E. Dizon. 1984. Close genetic similarity of Atlantic and Pacific skipjack tuna *(Katsuwonus pelamis)* demonstrated with restriction endonuclease analysis of mitochondrial DNA. *Marine Biology* 79:315–319.

Green, M. C. (ed.). 1981. *Genetic Variants and Strains of the Laboratory Mouse*. Gustav Fischer Verlag, New York.

Green, P. J. 1984. Iteratively reweighted least squares for maximum likelihood estimation and some robust and resistant alternatives (with discussion). *Journal of the Royal Statistical Society B* 46:149–192.

Greenberg, B. D., J. E. Newbold, and A. Sugino. 1983. Intraspecific nucleotide sequence variability surrounding the origin of replication in human mitochondrial DNA. *Gene* 21:33–49.

Greene, C. W. 1952. Results from stocking brook trout of wild and hatchery strains at Stillwater Pond. *Transactions of the American Fisheries Society* 81:43–52.

Greenfield, D. W. and T. Greenfield. 1972. Introgressive hybridization between *Gila orcutti* and *Hesperoleucus symmetricus* (Pisces: Cyprinidae) in the Cuyama River Basin, California. I. Meristics, morphometrics and breeding. *Copeia* 1972:849–859.

Greenfield, D. W., F. Abdel-Hameed, G. D. Deckert, and R. R. Flinn. 1973. Hybridization between *Chrosomus erythrogaster* and *Notropis cornutus* (Pisces: Cyprinidae). *Copeia* 1973:54–60.

Greenwood, P. H., D. E. Rosen, S. H. Weitzman, and G. S. Myers. 1966. Phyletic studies of teleostean fishes, with a provisional classification of living forms. *Bulletin of the American Museum of Natural History* 131:339–456.

Gresson, R. A. R. 1940. Presence of the sperm middle-piece in the fertilized egg of the mouse *(Mus musculus)*. *Nature* 145:425.

Griffiths, R. C. 1980. Genetic identity between populations when mutation rates vary within and across loci. *Journal of Mathematical Biology* 10:195–204.

Gross, M. 1985. Disruptive selection for alternative life histories in salmon. *Nature* 313(5997):47–48.

Gudelis, V. and L. K. Königsson. 1979. *The Quarternary History of the Baltic*. Almqvist and Wiksell, Stockholm, 279 pp.

Gulland, J. A. 1977. The stability of fish stocks. *Journal du Conseil International pour l'Exploration de la Mer* 37:199–204.

Gulland, J. A. 1982. Long-term potential effects from management of the fish resources of the North Atlantic. *Journal du Conseil International pour l'Exploration de la Mer* 40:8–16.

Guyomard, R. 1984. High level of residual heterozygosity in gynogenetic rainbow trout, *Salmo gairdneri* Richardson. *Theoretical and Applied Genetics* 67:307–316.

Gwahaba, J. J. 1973. Effects of fishing on the *Tilapia nilotica* (Linne 1757) population in Lake George, Uganda, over the past 20 years. *East African Wildlife Journal* 11:317–328.

Gyllensten, U. 1985. The genetic structure of fish: Differences in the intraspecific distribution of biochemical genetic variation between marine, anadromous, and freshwater species. *Journal of Fish Biology* 26:691–699.

Gyllensten, U., R. F. Leary, F. W. Allendorf, and A. C. Wilson. 1985a. Introgression between two cutthroat trout subspecies with substantial karyotypic, nuclear and mitochondrial genomic divergence. *Genetics* 11:905–915.

Gyllensten, U., D. Wharton, and A. C. Wilson. 1985b. Maternal inheritance of mitochondrial DNA during backcrossing of two species of mice. *Journal of Heredity* 76:321–324.

Haldane, J. B. S. 1932. *The Causes of Evolution*. Harper, New York-London, 234 pp.

Hamley, J. M. 1975. Review of gillnet selectivity. *Journal of the Fisheries Research Board of Canada* 32:1943–1969.

Hand, D. J. 1981. *Discrimination and Classification*. John Wiley and Sons, New York, 218 pp.

Handford, P., G. Bell, and T. Reimchen. 1977. A gillnet fishery considered as an experiment in artificial selection. *Journal of the Fisheries Research Board of Canada* 34:954–961.

Handford, P., G. Bell, and K. Dietz. Secular trends in growth-related parameters of a population of tullibee, *Coregonus artedii*. Unpublished report, Fish and Wildlife Division, Edmonton, Alberta.

Hanham, A. F. and M. J. Smith. 1980. Sequence homology in the single-copy DNA of salmon. *Comparative Biochemistry and Physiology* 65B:333–338.

Harris, H. 1966. Enzyme polymorphisms in man. *Proceedings of the Royal Society B* 164:298.

Harris, H. and D. Hopkinson. 1976. *Handbook of Enzyme Electrophoresis in Human Genetics*. American Elsevier, New York.

Harris, H., D. A. Hopkinson, and Y. H. Edwards. 1977. Polymorphism and subunit structure of enzymes: A contribution to the neutralist-selectionist controversy. *Proceedings of the National Academy of Sciences U.S.A.* 74:698–701.

Hart, J. L. 1973. Pacific fishes of Canada. *Fisheries Research Board of Canada Bulletin* 180:1–740.

Hartl, D. 1980. *Principles of Population Genetics*. Sinauer Associates, Sunderland Mass., 488 pp.

Hartley, S. E. and M. T. Horne. 1984. Chromosome relationships in the genus *Salmo*. *Chromosoma* 90:229–237.

Hartman, W. and R. Raleigh. 1964. Tributary homing of sockcyc salmon at Brooks and Karluk lakes, Alaska. *Journal of the Fisheries Research Board of Canada* 21:485–504.

Hauswirth, W. W. and P. J. Laipis. 1982. Rapid variation in the mammalian mitochondrial genotypes: Implications for the mechanism of maternal inheritance. In *Mitochrondrial Genes*, ed. P. Slonimski, P. Borst, and G. Attardi (Cold Spring Harbor Lab., New York).

Healey, M. C. 1975. Dynamics of exploited whitefish populations and their management with special reference to the Northwest Territories. *Journal of the Fisheries Research Board of Canada* 32:427–448.

Hedrick, P. W. 1971. A new approach to measuring genetic similarity. *Evolution* 25:276–280.

Hedrick, P. W. 1983. *Genetics of Populations*. Science Books International, Boston, Mass., 629 pp.

Hedrick, P. W., S. Jain, and L. Holden. 1978. Multilocus systems in evolution. In *Evolutionary Biology*, vol. 11, ed. M. K. Hecht, W. C. Steere, and B. Wallace (Plenum Press, New York and London), pp. 101–184.

Heggberget, T. G. and B. O. Johnsen. 1982. Infestations by *Gyrodactylus* sp. of Atlantic salmon, *Salmo salar* L., in Norwegian rivers. *Journal of Fish Biology* 21:15–26.

Heincke, F. 1898. *Naturgeschichte des Herings. I. Die Lokalformen und die Wanderungen des Herings in den europäischen Meeren*, vol. 2, Abhandlungen des deutschen Seefischereivereins, Berlin.

Henning, W. 1966. *Phylogenetic Systematics*. University of Illinois Press, Urbana.

Herman, Y. 1970. Arctic paleo-oceanography in late Cenozoic time. *Science* 169:474–477.

Herman, Y. and D. M. Hopkins. 1980. Arctic oceanic climate in late Cenozoic time. *Science* 209:557–562.

Hershberger, W. and R. Iwamoto. 1981. *Genetics Manual and Guidelines for the Pacific Salmon Hatcheries of Washington*. University of Washington, College of Fisheries.

Highton, R. and A. Larson. 1979. The genetic relationships of the salamanders of the genus *Plethodon*. *Systematic Zoology* 28:579–599.

Higuchi, R., B. Bowman, M. Freiberger, O. A. Ryder, and A. C. Wilson. 1984. DNA sequences from the quagga, an extinct member of the horse family. *Nature* 312:282–284.

Hile, R. 1936. Age and growth of the cisco, *Leucichthys artedi* (LeSueur), in the lakes of the northeastern highlands, Wisconsin. *U.S. Fish and Wildlife Service Fishery Bulletin* 38:211–317.

Hill, W. G. 1972. Effective population size with overlapping generations. *Theoretical Population Biology* 3:278–289.

Hill, W. G. 1974. Estimation of linkage disequilibrium in randomly mating populations. *Heredity* 33:229–239.

Hill, W. G. 1979. A note on effective population size with overlapping generations. *Genetics* 92:317–322.

Hillis, D. 1984. Misuse and modification of Nei's genetic distance. *Systematic Zoology* 33:238–240.

Hodgins, H. 1972. Serological and biochemical studies in racial identification of fishes. In *The Stock Concept in Pacific Salmon*, ed. R. Simon and P. Larkin. H. R. MacMillan Lectures in Fisheries (University of British Columbia, Vancouver), pp. 199–208.

Hoerl, A. E. and R. W. Kennard. 1970a. Ridge regression: Biased estimation for nonorthogonal problems. *Technometrics* 12:55–67.

Hoerl, A. E. and R. W. Kennard. 1970b. Ridge regression: Applications to nonorthogonal problems. *Technometrics* 12:69–82; correction, 12:723.

Holmquist, R. 1983. Transitions and transversions in evolutionary descent: An approach to understanding. *Journal of Molecular Evolution* 19:134–144.

Hopkins, D. M. 1967. The Cenozoic history of Beringia—a synthesis. In *The Bering Land Bridge*, ed. D. M. Hopkins (Stanford University Press, Palo Alto, Calif.), pp. 451–484.

Hopkins, D. M. and L. Marincovich, Jr. In press. Whale biogeography and the history of the Arctic Basin. In *Whaling and Whales in the Arctic*. Arktisch Centrum, University of Groningen.

Howell, W. M. and D. A. Black. 1980. Controlled silver staining of nucleolus organizer regions with a protective colloidal developer: A one-step method. *Experientia* 36:1014–1015.

Hubbs, C. L. 1920. Notes on hybrid sunfishes. *Aquatic Life* 5:101–103.

Hubbs, C. L. 1955. Hybridization between fish species in nature. *Systematic Zoology* 4:1–20.

Hubbs, C. L. and L. C. Hubbs. 1931. Increased growth in hybrid sunfishes. *Papers of the Michigan Academy of Science, Arts and Letters* 13:291–301.

Hubbs, C. L. and L. C. Hubbs. 1932. Experimental verification of natural hybridization between distinct genera of sunfishes. *Papers of the Michigan Academy of Science, Arts and Letters* 15:427–437.

Hubbs, C. L. and L. C. Hubbs. 1933. The increased growth, predominant maleness, and apparent infertility of hybrid sunfishes. *Papers of the Michigan Academy of Science, Arts and Letters* 17:613–641.

Hubbs, C. L. and L. C. Hubbs. 1947. Natural hybrids between two species of Catostomid fishes. *Papers of the Michigan Academy of Science, Arts and Letters* 31:147–167.

Hubbs, C. L. and K. Kuronuma. 1942. Hybridization in nature between two genera of flounders in Japan. *Papers of the Michigan Academy of Science, Arts and Letters* 27:267–306.

Hubbs, C. L. and K. Lagler. 1970. *Fishes of the Great Lakes Region*. University of Michigan Press, Ann Arbor, Michigan, 213 pp.

Hubbs, C. L. and J. J. Wilimovsky. 1964. Distribution and synomomy in the Pacific Ocean, and variation of the Greenland halibut, *Reinhardtius hippoglossoides* (Walbaum). *Journal of the Fisheries Research Board of Canada* 21:1129–1154.

Hubbs, C. L., L. C. Hubbs, and R. E. Johnson. 1943. Hybridization in nature between species of catostomid fishes. *Contributions of the Laboratory of Vertebrate Biology, University of Michigan* 22:1–77.

Hubbs, C., R. A. Kuehne, and J. C. Ball. 1953. The fishes of the upper Guadelupe River, Texas. *Texas Journal of Science* 5:216–244.

Hudson, R. R. 1982. Estimating genetic variability with restriction endonucleases. *Genetics* 100:711–719.

Hunter, G. A. and E. M. Donaldson. 1983. Hormonal sex control and its application to fish culture. In *Fish Physiology*, vol. 9B, ed. W. S. Hoar, D. J. Randall, and E. M. Donaldson (Academic Press, New York), pp. 223–303.

Hunter, G. A., E. M. Donaldson, J. Stoss, and I. Baker. 1983. Production of monosex female groups of chinook salmon *(Oncorhynchus tshawytscha)* by the fertilization of normal ova with sperm from sex-reversed females. *Aquaculture* 33:355–364.

Hunter, R. and C. Markert. 1957. Histochemical demonstration of enzymes separated by zone electyrophoresis in starch gels. *Science* 125:1294.

Hurlbert, S. H., J. Zedler, and D. Fairbanks. 1972. Ecosystem alteration by mosquitofish *(Gambusia affinis)* predation. *Science* 175:639–641.

Hutchison, C. A., J. E. Newbold, S. S. Potter, and M. H. Edgell. 1974. Maternal inheritance of mammalian mitochondrial DNA. *Nature* 251:536–537.

Hynes, J. D., E. H. Brown, J. H. Helle, N. Ryman, and D. A. Webster. 1981. Guidelines for the culture of fish stocks for resource management. *Canadian Journal of Fisheries and Aquatic Sciences* 38:1867–1876.

Ihssen, P. 1976. Selective breeding and hybridization in fisheries management. *Journal of the Fisheries Research Board of Canada* 33:316–321.

Ihssen, P. E., H. E. Booke, J. M. Casselman, J. M. McGlade, N. R. Payne, and F. M. Utter. 1981. Stock identification: Materials and methods. *Canadian Journal of Fisheries and Aquatic Sciences* 38:1838–1855.

Ijiri, K. 1980. Gamma-ray irradiation of the sperm of the fish *Oryzias latipes* and induction of gynogenesis. *Journal of Radiation Research* 21:263–270.

Ijiri, K. and N. Egami. 1980. Hertwig effect caused by UV-irradiation of sperm of *Oryzias latipes* (Teleost) and its photoreactivation. *Mutation Research* 69:241–248.

Imhof, M., R. Leary, and H. Booke. 1980. Population of stock structure of lake whitefish, *Coregonus clupeaformis*, in northern Lake Michigan as assessed by isozyme electrophoresis. *Canadian Journal of Fisheries and Aquatic Sciences* 37:783–793.

Jacquard, A. 1975. Inbreeding: One word, several meanings. *Theoretical Population Biology* 7:338–363.

Jennrich, R. I. and M. L. Rahlston. 1979. Fitting nonlinear models to data. *Annual Reviews of Biophysics and Bioengineering* 8:195–238.

Jensen, G. L., W. L. Shelton, S.-L. Yong and L. O. Wilken. 1983. Sex reversal of gynogenetic grass carp by implantation of methyltestosterone. *Transactions of the American Fisheries Society* 112:79–85.

Johannsen, W. 1909. *Elemente der exakten Erblichkeitslehre*. Fischer, Jena, 516 pp.

Johansson, N. 1981. General problems in Atlantic salmon rearing in Sweden. In *Fish Gene Pools*, ed. N. Ryman, Ecological Bulletins (Stockholm) vol. 34, pp. 75–83.

Johnels, A. G. 1984. Masken som hotar laxen *(Gyrodactylus salaris)*, a parasite threatening the Atlantic salmon. *Svenskt Fiske* 9/84:42–44. In Swedish.

Johns, B. and G. L. G. Miklos. 1979. Functional aspects of satellite DNA and heterochromatin. *International Review of Cytology* 587:1–114.

Johnson, A. G., F. M. Utter, and H. O. Hodgins. 1975. Study of the feasibility of immunochemical methods for identification of pleuronectid eggs. *Journal du Conseil International pour l'Exploration de la Mer* 36:158–161.

Johnson, G. B. 1974. Enzyme polymorphism and metabolism. *Science* 184:28–37.

Johnson, L. 1980. The Arctic charr, *Salvelinus alpinus*. In *Charrs, Salmonid Fishes of the Genus* Salvelinus, ed. E. K. Balon (Dr. W. Junk Publishers, The Hague).

Johnson, M. J., D. C. Wallace, S. D. Ferris, M. C. Rattazzi, and L. L. Cavalli-Sforza. 1983. Radiation of human mitochondria DNA types analyzed by restriction endonuclease cleavage patterns. *Journal of Molecular Evolution* 19:255–271.

Johnson, R. A. and D. W. Wichern. 1982. *Applied Multivariate Statistical Methods*. Prentice Hall, Inc., Englewood Cliffs, N.J., 594 pp.

Jonsson, B., K. Hindar, and T. G. Northcote. 1984. Optimal age at sexual maturity of sympatric and experimentally allopatric cutthroat trout and Dolly Varden charr. *Oecologia* 61:319–325.

Jordan, D. S. 1885. A list of the fishes known from the Pacific coast of tropical America, from the Tropic of Cancer to Panama. *Proceedings of the U.S. National Museum* 8:361–394.

Jordan, D. S. 1908. The law of geminate species. *American Naturalist* 42:73–80.

Jordan, D. S. 1923. Classification of fishes including families and genera as far as known. *Stanford University Publications, Biological Sciences* 3:77–243.

Jordan, D. S. and B. W. Evermann. 1902. *American Food and Game Fishes*. Doubleday, Page and Co., New York, 573 pp.

Jorde, L. 1980. The genetic structure of subdivided human populations. In *Current Developments in Anthropological Genetics*, vol. 1, ed. J. H. Mielke and M. H. Crawford (Plenum Press, New York), pp. 135–208.

Kaplan, N. 1983. Statistical analysis of restriction enzyme map data and nucleotide sequence data. In *Statistical Analysis of DNA Sequence Data*, ed. B. S. Weir (Marcel Dekker, Inc., New York).

Kaplan, N. and C. H. Langley. 1979. A new estimate of sequence divergence of mitochondrial DNA using restriction endonuclease mapping. *Journal of Molecular Evolution* 13:295–304.

Kaplan, N. and K. Risko. 1981. An improved method for estimating sequence divergence of DNA using restriction endonuclease mappings. *Journal of Molecular Evolution* 17:156–162.

Karlin, S. 1982. Classification of selection-migration structures and conditions for protected polymorphism. *Evolutionary Biology* 14:61–204.

Karpov, L. K. and G. G. Novikov. 1980. Hemoglobin alloforms in cod, *Gadus morhua* (Gadiformes, Gadidae), their functional characteristics and occurrence in populations. *Journal of Ichthyology* 20:823–827.

Keigwin, L. D. 1978. Pliocene closing of the Isthmus of Panama based on biostratigraphy evidence from nearby Pacific Ocean and Caribbean Sea cores. *Geology* 6:630–634.

Keigwin, L. D. 1982. Isotopic paleooceanography of the Caribbean and East Pacific: Role of Panama uplift in Late Neocene time. *Science* 217:350–353.

Kendall, A. W., Jr. and R. J. Behnke. 1984. Salmonidae: Development and relationships. In *Ontogeny and Systematics of Fishes*, ed. H. E. Moser. American Society of Ichthyology and Herpetology, Special Publication No. 1.

Kerr, R. A. 1983. An early glacial two-step? *Science* 221:143–144.

Kim, Y. J., G. C. Gorman, T. Papenfuss, and A. K. Roychoudhury. 1976. Genetic relationships and genetic variation in the amphisbaenian genus *Bipes*. *Copeia* 1976:120–124.

Kimura, M. 1953. "Stepping stone" model of population. *Annual Report of the National Institute of Genetics, Japan* 3:62–63.

Kimura, M. 1968. Evolutionary rate at the molecular level. *Nature* 217:624–626.

Kimura, M. 1969. The rate of molecular evolution considered from the standpoint of population genetics. *Proceedings of the National Academy of Sciences U.S.A.* 63:1181–1188.

Kimura, M. and T. Ohta. 1971. *Theoretical Aspects of Population Genetics*. Princeton University Press, Princeton, New Jersey.

Kimura, M. and G. H. Weiss. 1964. The stepping stone model of population structure and the decrease of genetic correlation with distance. *Genetics* 49:561–576.

Kincaid, H. L. 1976a. Inbreeding in rainbow trout *(Salmo gairdneri)*. *Journal of the Fisheries Research Board of Canada* 33:2420–2426.

Kincaid, H. 1976b. Effects of inbreeding on rainbow trout populations. *Transactions of the American Fisheries Society* 105:273–280.

Kincaid, H. L. 1981. *Trout Strain Registry*. FWS/NFC-L/81-1, National Fisheries Center-Leetown, Kearneysville, West Va.

Kincaid, H. L. 1983. Inbreeding in fish populations used for aquaculture. *Aquaculture* 33:215–227.

King, J. and T. Ohta. 1975. Polyallelic mutational equilibria. *Genetics* 79:681–691.

King, J. L. 1973. The probability of electrophoretic identity of proteins as a function of amino acid divergence. *Journal of Molecular Evolution* 2:317–322.

King, J. L. and T. H. Jukes. 1969. Non-Darwinian evolution. *Science* 164:788–798.

King, M.-C. and A. C. Wilson. 1975. Evolution at two levels in humans and chimpanzees. *Science* 188:107–116.

Kirkpatrick, M. and R. K. Selander. 1979. Genetics of speciation in lake whitefishes in the Allegash Basin. *Evolution* 33:478–485.

Kirpichnikov, V. S. 1981. *Genetic Bases of Fish Selection*, translated by G. G. Gause (Springer-Verlag, Berlin, Heidelberg, New York), 410 pp.

Kitchell, J. A. and D. L. Clark. 1982. Late Cretaceous Paleocene paleogeography and paleocirculation: Evidence of north polar upwelling. *Paleography, Palaeoclimatology, Palaeoecology* 40:135–165.

Kligerman, A. D. and S. E. Bloom. 1977. Rapid chromosome preparations from solid tissues of fishes. *Journal of the Fisheries Research Board of Canada* 34:266–269.

Kobayashi, T., G. B. Milner, D. Teel, and F. M. Utter. 1984. Genetic basis for electrophoretic variation of adenosine deaminase in chinook salmon. *Transactions of the American Fisheries Society* 113:86–89.

Koehn, R. K. and W. F. Eanes. 1977. Subunit size and genetic variation of enzymes in natural populations of *Drosophila. Theoretical Population Biology* 11:330–341.

Koehn, R. K., A. J. Zera, and J. G. Hall. 1983. Enzyme polymorphism and natural selection. In *Evolution of Genes and Proteins,* ed. M. Nei and R. K. Koehn (Sinauer Associates, Sunderland, Mass.), pp. 115–136.

Kondo, K. 1980. The recovery of the Japanese sardine—the biological basis of stock-size fluctuations. *Rapports et Procès-verbaux des Réunions du Conseil International pour l'Exploration de la Mer* 177:332–354.

Konovalov, S. M. 1972. The structure of isolates of *Oncorhynchus nerka* in Azabachje Lake *Journal of Genetic Biology* (U.S.S.R.) 33:668–682. In Russian.

Konovalov, S. M. 1980. *Population Biology of Pacific Salmon.* Nauka, Leningrad, 237 pp. In Russian.

Konovalov, S. M., A. P. Shapira, and G. E. Leybovich. 1975. Exploitation of biological resources in connection with species spatial structure. *Marine Biology of the U.S.S.R.* 6:26–36. In Russian.

Korey, K. A. 1981. Species number, generation length, and molecular clock. *Evolution* 35:139–147.

Kornfield, I., K. Beland, J. Moring, and F. Kircheis. 1981a. Genetic similarity among endemic arctic char *(Salvelinus alpinus)* and implications for their management. *Canadian Journal of Fisheries and Aquatic Sciences* 38:32–39.

Kornfield, I., P. Gagnon, and B. Sidell. 1981b. Inheritance of allzymes in Atlantic herring *(Clupea harengus harengus). Canadian Journal of Genetics and Cytology* 32:715–720.

Kornfield, I., B. D. Sidell, and P. S. Gagnon. 1982a. Stock definition in Atlantic herring *(Clupea harengus harengus):* Genetic evidence for discrete fall and spring spawning populations. *Canadian Journal of Fisheries and Aquatic Sciences* 39:1610–1621.

Kornfield, I., D. Smith, P. S. Gagnon, and J. Taylor. 1982b. The cichlid fish of Cuatro Cinegas, Mexico: Direct evidence of conspecificity among distinct trophic morphs. *Evolution* 36:658–664.

Krasznai, Z., T. Marian, L. Burizs, and F. Ditroi. 1982. Production of sterile hybrid grass carp *(Ctenopharyngodon idella* Val. × *Aristichthys nobilis* Rich.) for weed control. In *Second International Symposium on Herbivorous Fish,* pp. 55–60.

Kreitman, M. 1983. Nucleotide polymorphism at the alcohol dehydrogenase locus of *Drosophila melanogaster. Nature* 304:412–417.

Kristiansson, C. and J. McIntyre. 1976. Genetic variation in chinook salmon *(Oncorhynchus tshawytscha)* from the Columbia River and three Oregon coastal rivers. *Transactions of the American Fisheries Society* 105:620–623.

Krueger, C. C., A. J. Gharrett, J. R. Dehring, and F. W. Allendorf. 1981. Genetic aspects of fisheries rehabilitation programs. *Canadian Journal of Fisheries and Aquatic Sciences* 38:1877–1881.

Lackey, R. T. and L. A. Nielson (eds). 1980. *Fisheries Management.* John Wiley and Sons, New York.

Lande, R. 1982. A quantitative genetic theory of life history evolution. *Ecology* 63:607–615.

Lande, R. and S. J. Arnold. 1983. The measurement of selection on correlated characters. *Evolution* 37:1210–1226.

Lane, S. 1984. The implementation and evaluation of a genetic mark in a hatchery stock of pink salmon *(Oncorhynchus gorbuscha)* in southeast Alaska. M.S. thesis, University of Alaska, Juneau, 107 pp.

Langley, C. H. and W. M. Fitch. 1974. An examination of the constancy of the rate of molecular evolution. *Journal of Molecular Evolution* 3:161–177.

Lansman, R. A., R. O. Shade, J. F. Shapira, and J. C. Avise. 1981. The use of restriction endonucleases to measure mitochondrial DNA sequence relatedness in natural populations. III. Techniques and potential applications. *Journal of Molecular Evolution* 17:214–226.

Lansman, R. A., J. F. Shapira, C. Aquadro, S. W. Daniel, and J. C. Avise. 1982. Mitochondrial DNA and evolution in *Peromyscus*: A preliminary report. In *Mitochondrial Genes,* ed. P. Slonimski, P. Borst, and G. Attardi (Cold Spring Harbor Lab., N.Y.).

Lansman, R. A., J. C. Avise, C. F. Aquadro, J. F. Shapira, and S. W. Daniel. 1983a. Extensive genetic variation in mitochondrial DNA's among geographic populations of the deer mouse, *Peromyscus maniculatus. Evolution* 37:1–16.

Lansman, R. A., J. C. Avise, and M. C. Huettel. 1983b. Critical experimental test of the possibility of "paternal leakage" of mitochondrial DNA. *Proceedings of the National Academy of Sciences U.S.A.* 80:1969–1971.

Larkin, P. A. 1963. Interspecific competition and exploitation. *Journal of the Fisheries Research Board of Canada* 20:647–678.

Larkin, P. A. 1977. An epitaph for the concept of maximum sustained yield. *Transactions of the American Fisheries Society* 106:1–11.

Larkin, P. A. 1978. Fisheries management—an essay for ecologists. *Annual Review of Ecology and Systematics* 9:57–74.

Larkin, P. A. 1981. A perspective on population genetics and salmon management. *Canadian Journal of Fisheries and Aquatic Sciences* 38:1469–1475.

Larson, A., D. B. Wake, and K. P. Yanev. 1984. Measuring gene flow among populations having high levels of genetic fragmentation. *Genetics* 106:293–308.

Latter, B. D. H. 1973a. The island model of population differentiation: A general solution. *Genetics* 73:147–157.

Latter, B. D. H. 1973b. Measures of genetic distance. In *Genetic Structure of Populations,* ed. N. E. Morton (University Press of Hawaii, Honolulu), pp. 27–37.

Latter, B. D. H. 1980. Genetic differences within and between populations of the major human subgroups. *American Naturalist* 116:220–237.

Leary, R. F. and F. W. Allendorf. 1983. *Genetic analysis of three hatchery stocks of North American Atlantic salmon,* Salmo salar. Genetics Report 83/1, Dept. of Zoology, University of Montana, 11 pp.

Leary, R. F., F. W. Allendorf, and K. L. Knudsen. 1983a. Consistently high meristic counts in natural hybrids between brook trout and bull trout. *Systematic Zoology* 32:369–376.

Leary, R. F., F. W. Allendorf, and K. L. Knudsen. 1983b. Developmental stability and enzyme heterozygosity in rainbow trout. *Nature* 301:71–72.

Leary, R. F., F. W. Allendorf, and K. L. Knudsen. 1984a. Major morphological effects of a regulatory gene, *Pgm1-t,* in rainbow trout. *Molecular Biology and Evolution* 1:183–194.

Leary, R. F., F. W. Allendorf, and K. L. Knudsen. 1984b. Superior developmental stability of enzyme heterozygotes in salmonid fishes. *American Naturalist* 124:540–551.

Leary, R. F., F. W. Allendorf, and K. L. Knudsen. 1985a. Inheritance of meristic variation and the evolution of developmental stability in rainbow trout. *Evolution* 39:308–314.

Leary, R. F., F. W. Allendorf, K. L. Knudsen, and G. H. Thorgaard. 1985b. Hetero-zygosity and developmental stability in gynogenetic diploid and triploid rainbow trout. *Heredity* 54:219–225.

Leary, R. F., F. W. Allendorf, and K. L. Knudsen. 1985c. Developmental instability as an indicator of the loss of genetic variation in hatchery trout. *Transactions of the American Fisheries Society* 114:230–235.

Leary, R. F., F. W. Allendorf, S. R. Phelps, and K. L. Knudsen. 1984. Introgression between westslope cutthroat and rainbow trout in the Clark Fork River drainage, Montana. *Proceedings of the Montana Academy of Sciences* 43:1-18.

Leary, R. F., F. W. Allendorf, S. R. Phelps, and K. L. Knudsen. In press. Genetic di-vergence among seven subspecies of cutthroat trout and the rainbow trout. *Transactions of the American Fisheries Society.*

Lee, R. 1912. *An investigation into the methods of growth determination in fishes.* Pub-lications de Circonstance Conseil Permanent International pour l'Exploration de la Mer 63, 35 pp.

Leider, S. A., M. W. Chilcote, and J. J. Loch. 1984. Spawning characteristics of sym-patric populations of steelhead trout *(Salmo gairdneri):* Evidence for partial re-productive isolation. *Canadian Journal of Fisheries and Aquatic Sciences* 41:1454–1462.

Leim, A. H. and W. B. Scott. 1966. Fishes of the Atlantic Coast of Canada. *Fisheries Research Board of Canada Bulletin* 155:1–485.

Leitritz, E. and R. C. Lewis. 1976. Trout and salmon culture (hatchery methods). *California Department of Fish and Game Fish Bulletin* No. 164.

Lemoine, H. L., Jr. and L. T. Smith. 1980. Polyploidy induced in brook trout by cold shock. *Transactions of the American Fisheries Society* 109:626–631.

Lessios, H. A. 1979. Use of Panamanian sea urchins to test the molecular clock. *Nature* 280:599–601.

Lessios, H. A. 1981. Divergence in allopatry: Molecular and morphological differentia-tion between sea urchins separated by the Isthmus of Panama. *Evolution* 35:618–634.

Levene, H. 1953. Genetic equilibrium when more than one ecological niche is available. *American Naturalist* 87:331–333.

Levine, L. and H. van Vunakis. 1967. Micro complement fixation. *Methods in En-zymology* 11:928–926.

Lewontin, R. C. 1964. The interaction of selection and linkage. I. General considera-tions; heterotic models. *Genetics* 49:49–67.

Lewontin, R. C. 1972. The apportionment of human diversity. *Evolutionary Biology* 6:381–398.

Lewontin, R. C. 1974. *The Genetic Basis of Evolutionary Change.* Columbia University Press, New York.

Lewontin, R. C. 1979. Fitness, survival and optimality. In *Analysis of Ecological Sys-tems,* ed. D. J. Horn, G. R. Stairs, and R. D. Mitchell (Ohio State University Press, Columbus), pp. 3–21.

Lewontin, R. C. 1984. Detecting population differences in quantitative characters as op-posed to gene frequencies. *American Naturalist* 123:115–124.

Lewontin, R. C. and J. Hubby. 1966. A molecular approach to the study of genic hetero-zygosity in natural populations. II. Amount of variation and degree of hetero-zygosity in natural populations of *Drosophila pseudoobscura. Genetics* 54:595–609.

Lewontin, R. C. and K. Kojima. 1960. The evolutionary dynamics of complex poly-morphisms. *Evolution* 14:450–472.

Lewontin, R. C. and J. Krakauer. 1973. Distribution of gene frequency as a test of the theory of the selective neutrality of polymorphisms. *Genetics* 74:175–195.

Li, W. H. 1976a. Effect of migration on genetic distance. *American Naturalist* 110:841–847.

Li, W. H. 1976b. Electrophoretic identity of proteins in a finite population and genetic distance between taxa. *Genetical Research* 28:119–127.

Li, W. H. 1986. Evolutionary change of restriction cleavage sites and phylogenetic inference. *Genetics* 113:187–213.

Li, C. C. and D. G. Horvitz. 1953. Some methods of estimating the inbreeding coefficient. *American Journal of Human Genetics* 5:107–117.

Li, W. H. and M. Nei. 1975. Drift variances of heterozygosity and genetic distance in transient states. *Genetical Research* 25:229–248.

deLigny, W. 1969. Serological and biochemical studies on fish populations. *Oceanography and Marine Biology: An Annual Review* 7:411–513.

deLigny, W. 1972. Blood groups and biochemical polymorphisms in fish. In *12th European Conference on Animal Blood Groups, Biochemistry, and Polymorphism*, ed. G. Kovacs and M. Papp (Dr. W. Junk, The Hague), pp. 55–65.

Lim, S. and G. Bailey. 1977. Gene duplication in salmonid fish: Evidence for duplicated but catalytically equivalent A(4) lactate dehydrogenases. *Biochemical Genetics* 15:707–721.

Lincoln, R. F. 1981a. The growth of female diploid and triploid plaice *(Pleuronectes platessa)* × flounder *(Platchthys flesus)* hybrids over one spawning season. *Aquaculture* 25:259–268.

Lincoln, R. F. 1981b. Sexual maturation in triploid male plaice *(Pleuronectes platessa)* and plaice × flounder *(Platichthys flesus)* hybrids. *Journal of Fish Biology* 19:415–426.

Lincoln, R. F. 1981c. Sexual maturation in female triploid plaice, *Pleuronectes platessa*, and plaice × flounder, *Platichthys flesus*, hybrids. *Journal of Fish Biology* 19:499–507.

Lincoln, R. F. and A. P. Scott. 1983. Production of all-female triploid rainbow trout. *Aquaculture* 30:375–380.

Lincoln, R. F. and A. P. Scott. 1984. Sexual maturation in triploid rainbow trout, *Salmo gairdneri* Richardson. *Journal of Fish Biology* 25:385–392.

Lindsey, C. C. 1981. Stocks are chameleons: Plasticity in gill rakers of coregonid fishes. *Canadian Journal of Fisheries and Aquatic Sciences* 38:1497–1506.

Loftus, K. H. 1976. Science for Canada's fisheries rehabilitation needs. *Journal of the Fisheries Research Board of Canada* 33:1822–1857.

Loftus, K. H. and H. A. Regier (eds.). 1972. *Proceedings of the Symposium on Salmonid Communities in Oligotrophic Lakes. Journal of the Fisheries Research Board of Canada* 29(6).

Lou, Y. D. and C. E. Purdom. 1984. Polyploidy induced by hydrostatic pressure in rainbow trout, *Salmo gairdneri* Richardson. *Journal of Fish Biology* 25:345–351.

Loudenslager, E. and R. Kitchin. 1979. Genetic similarity of two forms of cutthroat trout, *Salmo clarki*, in Wyoming. *Copeia* 4:673–678.

Loudenslager, E. J. and G. H. Thorgaard. 1979. Karyotypic and evolutionary relationships of Yellowstone *(Salmo clarki bouvieri)* and westslope *(S. c. lewisi)* cutthroat trout. *Journal of the Fisheries Research Board of Canada* 36:630–635.

Lowe, V. and A. Gardiner. 1974. A re-examination of the subspecies of red deer *(Cervus elaphus)* with particular reference to the stocks in Britain. *Journal of Zoology* (London) 174:185–201.

Lundqvist, J. 1965. The Quaternary of Sweden. In *The Quaternary*, ed. K. Rankama, vol. 1, pp. 139–198.

MacCrimmon, H. R. and B. L. Gots. 1979. World distribution of Atlantic salmon, *Salmo salar. Journal of the Fisheries Research Board of Canada* 36:422–457.

MacGregor, R. B. and H. R. MacCrimmon. 1977. Evidence of genetic and environmental influences on meristic variation in the rainbow trout, *Salmo gairdneri* Richardson. *Environmental Biology of Fishes* 2:25–33.

MacLean, C. J. and P. L. Workman. 1973a. Genetic studies on hybrid populations. I. Individual estimates of ancestry and their relation to quantitative traits. *Annals of Human Genetics* 36:341–351.

MacLean, C. J. and P. L. Workman. 1973b. Genetic studies on hybrid populations. II. Estimation of the distribution of ancestry. *Annals of Human Genetics* 36:459–465.

Mahalanobis, P. C. 1936. On the generalized distance in statistics. *Proceedings of the National Academy of Sciences, India* 2:49–55.

Makela, M. E. and R. H. Richardson. 1977. The detection of sympatric sibling species using genetic correlation analysis. I. Two loci. Two gamodemes. *Genetics* 86:665–678.

Malecot, G. 1948. *Les Mathématiques de l'Hérédité*. Masson, Paris.

Malecot, G. 1951. Some probabilistic schemes for the variability of natural populations. *Université de Lyon I. Département de Sciences de la Terre.* 13:37–60.

Malecot, G. 1959. Stochastic models in population genetics. *Publications de l'Institut de Statistique de l'Université de Paris* 8(F3):173–210.

Malfait, B. T. and M. G. Dinkelman. 1972. Circum-Caribbean tectonic and igneous activity and the evolution of the Caribbean plate. *Geological Society of America Bulletin* 83:251–272.

Malinovsky, A. A. and M. V. Mina. 1976. Review of Yu. P. Altukhov's book "Population genetics in fish." *Journal of Genetic Biology* (U.S.S.R.) 37:788–790. In Russian.

Mallet, A. L. and L. E. Haley. 1983. Effects of inbreeding on larval and spat performance in the American oyster. *Aquaculture* 33:229–235.

Mann, R. 1979. Exotic species in aquaculture: An overview of when, why and how. In *Exotic Species in Mariculture,* ed. R. Mann (MIT Press, Cambridge, Mass.), pp. 331–354.

Mantelman, I. I. 1980. First results obtained from usage of chemical mutagens for producing gynogenetic progeny in the Siberian whitefish, *Coregonus peled* Gm. In *Karyological Variability, Mutagenesis and Gynogenesis in Fishes* (Institute of Cytology, U.S.S.R. Academy of Sciences, Leningrad), pp. 82–85.

Margoliash, E. and E. L. Smith. 1965. Structural and functional aspects of cytochrome c in relation to evolution. In *Evolving Genes and Proteins,* ed. V. Bryson and H. J. Volgal (Academic Press, New York), pp. 221–242.

Margolis, L. 1963. Parasites as indicators of the geographical origin of sockeye salmon, *Oncorhynchus nerka* (Walbaum), occurring in the North Pacific Ocean and adjacent seas. *International North Pacific Fisheries Commission Bulletin* 11:101–156.

Marian, T. and Z. Krasznai. 1978. Kariological investigation of *Ctenopharyngodon idella* and *Hypophthalmichthys nobilis* and their cross-breeding. *Aquacultura Hungarica* 1:44–50.

Markert, C. and F. Moller. 1959. Multiple forms of enzymes: Tissue, ontogenetic and species specific patterns. *Proceedings of the National Academy of Sciences U.S.A.* 45:753–763.

Markoviac, V. V., R. G. Worton, and J. M. Berg. 1978. Evidence for the inheritance of silver stained nucleolar organizer regions. *Human Genetics* 41:181–187.

Marshall, J. T., Jr. and R. D. Sage. 1981. Taxonomy of the house mouse. *Symposia of the Zoological Society of London* 47:15–25.

Martin, F. D. and R. C. Richmond. 1973. An anlaysis of five enzyme gene loci in four etheostomid species (Percidae: Pisces) in an area of possible introgression. *Journal of Fish Biology* 5:511–517.

Maruyama, T. 1970. Effective number of alleles in a subdivided population. *Theoretical Population Biology* 1:273–306.

Maruyama, T. 1972. Rate of decrease of genetic variability in a two-dimensional continuous population of finite size. *Genetics* 70:639–651.

Mather, K. 1953. Genetic control of stability in development. *Heredity* 7:297–336.

Maxam, A. M. and W. Gilbert. 1977. A new method for sequencing DNA. *Proceedings of the National Academy of Sciences U.S.A.* 74:560–564.

Maxson, L. R. and R. D. Maxson. 1979. Comparative albumin and biochemical evolution in plethodontid salamanders. *Evolution* 33:1057–1062.

Maxson, L. R. and A. C. Wilson, 1974. Convergent morphological evolution detected by studying proteins of tree frogs in the *Hyla eximia* species group. *Science* 185:66–68.

Maxson, L. R. and A. C. Wilson. 1975. Albumin evolution and organismal evolution in tree frogs (Hylidae). *Systematic Zoology* 24:1–15.

Maxson, L. R., V. M. Sarich, and A. C. Wilson. 1975. Continental drift and the use of albumin as an evolutionary clock. *Nature* 255:397–400.

Maxson, L. R., R. Highton, and D. B. Wake. 1979. Albumin evolution and its phylogenetic implications in the plethodontid salamander genera *Plethodon* and *Ensatina*. *Copeia* 1979:502–508.

May, B. 1980. The salmonid genome: Evolutionary restructuring following a tetraploid event. Ph.D. dissertation, Pennsylvania State University, 199 pages.

May, B., J. Wright, and M. Stoneking. 1979. Joint segregation of biochemical loci in Salmonidae: Results from experiments with *Salvelinus* and review of the literature on other species. *Journal of the Fisheries Research Board of Canada* 36:1114–1128.

May, R. M. 1977. Thresholds and breakpoints in ecosystems with a multiplicity of stable states. *Nature* (London) 269:471–477.

Mayhew, D. A. 1983. A new hybrid cross, *Notropis atherinoides* × *Notropis volucellus* (Pisces: Cyprinidae), from the lower Monongahela River, western Pennsylvania. *Copeia* 1983:1077–1082.

Maynard Smith, J. 1970. Population size, polymorphism, and the rate of non-Darwinian evolution. *American Naturalist* 104:231–236.

Mayr, E. 1963. *Animal Species and Evolution.* Harvard University Press, Cambridge, Mass.

Mayr. E. 1969. *Principles of Systematic Zoology.* McGraw-Hill, New York.

Mayr, E. 1970. *Populations, Species and Evolution.* The Belknap Press of Harvard University Press, Cambridge, Mass., 453 pp.

McCleod, M. J., D. L. Wynes, and S. I. Guttman. 1980. Lack of biochemical evidence for hybridization between two species of darters. *Comparative Biochemistry and Physiology B* 67:323–325.

McCommas, S. A. 1982. Biochemical genetics of the sea anemone *Bunodosoma cavernata* and the zoogeography of the Gulf of Mexico. *Marine Biology* 68:169–173.

McCommas, S. A. 1983. A taxonomic approach to evaluation of the charge state model using twelve species of sea anemone. *Genetics* 103:741–752.

McCracken, F. D. 1958. On the biology and fishery of the Canadian Atlantic halibut, *Hippoglossus hippoglossus* L. *Journal of the Fisheries Research Board of Canada* 15:1269–1311.

McDonald, J. 1983. The molecular basis of adaptation: A critical review of relevant ideas and observations. *Annual Review of Ecology and Systematics* 14:77–102.

McHugh, J. L. 1980. Coastal fisheries. In *Fisheries Management*, ed. R. T. Lackey and L. A. Nielsen (John Wiley and Sons, New York).

Mendel, G. 1866. Versuche über Pflanzenhybriden. *Verhandlungen naturforschischer Vereinigkeit Brunn* 4:3–47.

Menzel, B. 1978. Three hybrid combinations of minnows (Cyprinidae) involving members of the common shiner species complex (genus *Notropis,* subgenus *Luxilus).* *American Midland Naturalist* 99:249–256.

Menzies, R. J. 1968. Transport of marine life between oceans through the Panama Canal. *Nature* 220:802–803.

Merriman, D. 1982. The history of Georges Bank. In *Georges Bank: Past, Present and Future of a Marine Environment,* ed. G. C. McLeod and J. H. Prescott (Westview Press, Boulder, Colo.), pp. 11–30.

Messieh, S. N. 1972. Use of otoliths in identifying herring stocks in the southern Gulf of St. Lawrence and adjacent waters. *Journal of the Fisheries Research Board of Canada* 29:1113–1118.

Messinger, H. B. and H. T. Bilton. 1974. Factor analysis in discriminating the racial origin of sockeye salmon *(Oncorhynchus nerka). Journal of the Fisheries Research Board of Canada* 31:1–10.

Michod, R. E. 1978. Evolution of life histories in response to age-specific mortality factors. Ph.D. dissertation, University of Georgia.

Millar, R. B. 1985. *Convergence results for the EM algorithm in the application of stock identification.* Report prepared for Washington State Department of Fisheries, Olympia, Wash.

Millar, R. B. In press. Maximum likelihood estimation of mixed fishery composition. *Canadian Journal of Fisheries and Aquatic Sciences.*

Millenbach, C. 1973. Genetic selection of steelhead trout for management purposes. *International Atlantic Salmon Journal* 4:253–357.

Miller, M., P. Patillo, G. B. Milner, and D. Teel. 1983. *Analysis of chinook stock composition in the May 1982 troll fishery off the Washington coast: An application of the genetic stock identification method.* Washington State Department of Fisheries Technical Report No. 74, 27 pp.

Miller, R. B. 1957. Have the genetic patterns of fishes been altered by introductions or by selective fishing? *Journal of the Fisheries Research Board of Canada* 14:797–806.

Miller, R. G. 1974. The jackknife—a review. *Biometrika* 61:1–15.

Milner, G. B., D. J. Teel, F. M. Utter, and C. L. Burley. 1981. *Columbia River stock identification study: Validation of genetic method.* Annual report of research (FY 80) NOAA, Northwest and Alaska Fisheries Center, Seattle, Wash.

Milner, G. B., D. J. Teel, F. M. Utter, and G. A. Winans. 1985. A genetic method of stock identification in mixed populations of Pacific salmon, *Oncorhynchus* spp. *Marine Fisheries Review* 47:1–8.

Mina, M. V. 1978. On the population structure of fish species. The evaluation of some hypotheses. *Zhurnal Obshchei Biologii* [Journal of Genetic Biology U.S.S.R.] 39:453–460. In Russian.

Misra, R. K. 1971. Statistical methods for analyzing hybrid indices for meristic traits. *Biometrische Zeitschrift* 13:329–334.

Misra, R. K. 1972. Analysis of a meristic trait for hybridization in sunfishes (genus *Lepomis.) Freshwater Biology* 2:321–324.

Misra, R. K., D. W. Bateson, and M. H. A. Keenleyside. 1970. Statistical methods for analysis of hybrid indices, with an example from fish populations. *Journal of Statistical Research* 4:50–70.

Mitton, J. B. 1977. Genetic differentiation of races of man as judged by single-locus and multilocus analysis. *American Naturalist* 111:203–212.

Miyata, T., H. Hayashida, R. Kikuno, M. Hasegawa, M. Kobayashi, and K. Koike.

1982. Molecular clock of silent substitution: At least six-fold preponderance of silent changes in mitochondrial genes over those in nuclear genes. *Journal of Molecular Evolution* 19:28–35.

Moav, R. and G. W. Wohlfarth. 1963. *Breeding schemes for the improvement of edible fish.* Progress Report, 1962, Fish Breeding Association of Israel, 44 pages.

Moav, R., T. Brody, G. W. Wohlfarth, and G. Hulata. 1976. Applications of electrophoretic genetic markers to fish breeding. I. Advantages and methods. *Aquaculture* 9:217–228.

Moav, R., T. Brody, and G. Hulata. 1978. Genetic improvement of wild fish populations. *Science* 201:1090–1094.

Monaco, P. J., E. M. Rasch, and J. S. Balsano. 1984. Apomictic reproduction in the Amazon molly, *Poecilia formosa,* and its triploid hybrids. In *Evolutionary Genetics of Fishes,* ed. B. J. Turner (Plenum Press, New York and London), pp. 311–328.

Montoya, J., G. L. Gaines, and G. Attardi. 1983. The pattern of transcription of the human mitochondrial rRNA genes reveals two overlapping transcription units. *Cell* 34:151–159.

Moore, W. S. 1977. An evaluation of narrow hybrid zones in vertebrates. *Quarterly Review of Biology* 52:263–277.

Moore, W. S. 1984. Evolutionary ecology of unisexual fishes. In *Evolutionary Genetics of Fishes,* ed. B. J. Turner (Plenum Press, New York and London), pp. 329–398.

Morizot, D. C. and M. J. Siciliano. 1982. Protein polymorphisms, segregation in genetic crosses and genetic distances among fishes of the genus *Xiphophorus* (Poeciliidae). *Genetics* 102:539–556.

Morizot, D. C. and M. J. Siciliano. 1984. Gene mapping in fishes and other vertebrates. In *Evolutionary Genetics of Fishes,* ed. B. J. Turner (Plenum Press, New York), pp. 173–234.

Mork, J. and T. Haug. 1983. Genetic variation in halibut *(Hippoglossus hippoglossus* L.) from Norwegian waters. *Hereditas* 98:167–174.

Mork, J. and G. Sundnes. 1985. Hemoglobin polymorphisms in Atlantic cod (*Gadus morhua*): Allele frequency variation between year classes in a Norwegian fjord stock. *Helgoländer Meeresuntersuchungen* 39:55–62.

Mork, J., C. Reuterwall, N. Ryman, and G. Ståhl. 1982. Genetic variation in Atlantic cod *(Gadus morhua* L.): A quantitative estimate from a coastal population. *Hereditas* 96:55–61.

Mork, J., R. Giskeödegard, and G. Sundnes. 1983. Haemoglobin polymorphism in *Gadus morhua:* Genotypic differences in maturing age and within-season gonad maturation. *Helgolaender Meeresuntersuchungen* 36:313–322.

Mork, J., P. Salemdal, and G. Sundnes. 1983. Identification of marine fish eggs: A biochemical genetics approach. *Canadian Journal of Fisheries and Aquatic Sciences* 40:361–369.

Mork, J., R. Giskeödegård, and G. Sundnes. 1984a. The hemoglobin polymorphism in Atlantic cod *(Gadus morhua* L.): Genotypic differences in somatic growth and in maturing age in natural populations. *Flodevigen Rapportser* 1:721–732.

Mork, J., R. Giskeödegård, and G. Sundnes. 1984b. Population genetic studies in cod *(Gadus morhua* L.) by means of the hemoglobin polymorphism: Observations in a Norwegian coastal population. *Fiskeridirektoratets Skrifter Serie Havundersokelser* 17:449–471.

Mourant, A. 1954. *The Distribution of the Human Blood Groups.* Charles C. Thomas, Springfield, Ill.

Moyle, P. B. 1969. Comparative behavior of young brook trout of domestic and wild origin. *Progressive Fish-Culturist* 31:51–56.

Muller, W. P., C. H. Thiebaud, L. Ricard, and M. Fischberg. 1978. The induction of triploidy by pressure in *Xenopus laevis*. *Revue Suisse Zoologie* 85:20–26.

Mulligan, T. J., L. Lapi, R. Kieser, S. B. Yamada, and D. L. Duewer. 1983. Salmon stock identification based on elemental composition of vertebrae. *Canadian Journal of Fisheries and Aquatic Sciences* 40:215–229.

Murphy, B. R., L. A. Nielsen, and B. J. Turner. 1983. Use of genetic tags to evaluate stocking success for reservoir walleyes. *Transactions of the American Fisheries Society* 112:457–463.

Murray, J. D. and R. M. Kitchin. 1976. Chromosomal variation and heterochromatin polymorphisms in *Peromyscus maniculatus*. *Experientia* 32:307–309.

Nagy, A. and V. Csanyi. 1982. Changes of genetic parameters in successive gynogenetic generations and some calculations for carp gynogenesis. *Theoretical and Applied Genetics* 63:105–110.

Nagy, A. and V. Csanyi. 1984. A new breeding system using gynogenesis and sex-reversal for fast inbreeding in carp. *Theoretical and Applied Genetics* 67:485–490.

Nagy, A., K. Rajki, L. Horvath, and V. Csanyi. 1978. Investigation on carp *Cyprinus carpio* L. gynogenesis. *Journal of Fish Biology* 13:215–224.

Nagy, A., M. Bercsenyi, and V. Csanyi. 1981. Sex reversal in carp *(Cyprinus carpio)* by oral administration of methyltestosterone. *Canadian Journal of Fisheries and Aquatic Sciences* 38:725–728.

Nagylaki, T. 1975. Conditions for the existence of clines. *Genetics* 80:595–615.

Nagylaki, T. 1983. The robustness of neutral models of geographical variation. *Theoretical Population Biology* 24:268–294.

Nathans, D. and H. O. Smith. 1975. Restriction endonucleases in the analysis and restructuring of DNA molecules. *Annual Review of Biochemistry* 44:273–293.

Neave, F. 1954. Principles affecting the size of pink salmon and chum salmon populations in British Columbia. *Journal of the Fisheries Research Board of Canada* 9:450–491.

Neff, N. A. and G. R. Smith. 1979. Multivariate analysis of hybrid fishes. *Systematic Zoology* 28:176–196.

Nei, M. 1969. Gene duplication and nucleotide substitution in evolution. *Nature* 221:40–42.

Nei, M. 1972 Genetic distance between populations. *American Naturalist* 106:283–292.

Nei, M. 1973a. Analysis of gene diversity in subdivided populations. *Proceedings of the National Academy of Sciences U.S.A.* 70:3321–3323.

Nei, M. 1973b. The theory and estimation of genetic distance. In *Genetic Structure of Populations*, ed. N. E. Morton (University Press of Hawaii, Honolulu), pp. 45–54.

Nei, M. 1975. *Molecular Population Genetics and Evolution*. American Elsevier, New York.

Nei, M. 1976. Mathematical models of speciation and genetic distance. In *Population Genetics and Ecology*, ed. S. Karlin and E. Nevo (Academic Press, New York), pp. 723–765.

Nei, M. 1977. F-statistics and analysis of gene diversity in subdivided populations. *Annals of Human Genetics* 41:225–233.

Nei, M. 1978a. Estimation of average heterozygosity and genetic distance from a small number of individuals. *Genetics* 89:583–590.

Nei, M. 1978b. The theory of genetic distance and evolution of human races. *Japanese Journal of Human Genetics* 23:341–369.

Nei, M. 1983. Genetic polymorphism and the role of mutation in evolution. In *Evolution of Genes and Proteins*, ed. M. Nei and R. K. Koehn (Sinauer Associates, Sunderland, Mass.), pp. 165–190.

Nei, M. In press. *Molecular Evolutionary Genetics.* Columbia University Press, New York.

Nei, M. and R. Chakraborty. 1973. Genetic distance and electrophoretic identity of proteins between taxa. *Journal of Molecular Evolution* 2:323–328.

Nei, M. and R. K. Chesser. 1983. Estimation of fixation indices and gene diversities. *Annals of Human Genetics* 47:253–259.

Nei, M. and M. W. Feldman. 1972. Identity of genes by descent within and between populations under mutation and migration pressures. *Theoretical Population Biology* 3:460–465.

Nei, M. and W.-H. Li. 1973. Linkage disequilibrium in subdivided populations. *Genetics* 75:213–219.

Nei, M. and W.-H. Li. 1979. Mathematical model for studying genetic variation in terms of restriction endonucleases. *Proceedings of the National Academy of Sciences U.S.A.* 76:5269–5273.

Nei, M. and A. K. Roychoudhury. 1974a. Sampling variances of heterozygosity and genetic distance. *Genetics* 76:379–390.

Nei, M. and A. K. Roychoudhury. 1974b. Genic variation within and between the three major races of man, Caucasoids, Negroids, and Mongoloids. *American Journal of Human Genetics* 26:421–443.

Nei, M. and F. Tajima. 1981a. DNA polymorphism detectable by restriction endonucleases. *Genetics* 97:145–163.

Nei, M. and F. Tajima. 1981b. Genetic drift and estimation of effective population size. *Genetics* 98:625–640.

Nei, M. and F. Tajima. 1983. Maximum likelihood estimation of the number of nucleotide substitutions from restriction sites data. *Genetics* 105:207–217.

Nei, M. and Y. Tateno. 1975. Interlocus variation of genetic distance and the neutral mutation theory. *Proceedings of the National Academy of Sciences U.S.A.* 72:2758–2760.

Nei, M., T. Maruyama, and R. Chakraborty. 1975. The bottleneck effect and genetic variability in populations. *Evolution* 29:1–10.

Nei, M., R. Chakraborty, and P. A. Fuerst. 1976a. Infinite allele model with varying mutation rate. *Proceedings of the National Academy of Sciences U.S.A.* 73:4164–4168.

Nei, M., P. A. Fuerst, and R. Chakraborty. 1976b. Testing the neutral mutation hypothesis by distribution of single locus heterozygosity. *Nature* 262:491–493.

Nei, M., P. A. Fuerst, and R. Chakraborty. 1978. Subunit molecular weight and genetic variability of proteins in natural populations. *Proceedings of the National Academy of Sciences U.S.A.* 75:3359–3362.

Nei, M., F. Tajima, and Y. Tateno. 1983. Accuracy of estimated phylogenetic trees from molecular data. II. Gene frequency data. *Journal of Molecular Evolution* 19:153–170.

Nelson, J. S. 1966. Hybridization between two cyprinid fishes, *Hybopsis plumbea* and *Rhinichthys cataractae*, in Alberta. *Canadian Journal of Zoology* 44:963–968.

Nelson, J. S. 1973. Occurrence of hybrids between longnose sucker *(Catostomus catastomus)* and white sucker *(C. commersoni)* in upper Kananaskis Reservoir, Alberta. *Journal of the Fisheries Research Board of Canada* 30:557–560.

Netboy, A. 1974. *The Salmon, Their Fight for Survival.* Houghton-Mifflin, Boston, 613 pp.

Nevo, E., Y. J. Kim, C. R. Shaw, and C. S. Thaeler, Jr. 1974. Genetic variation, selection and speciation in *Thomomys talpoides* pocket gophers. *Evolution* 28:1–23.

Niethammer, J. and F. Krapp. 1978. *Handbuch der Rodentia I.* Saugetiere Europas, vol. 1. Akademische Verlagsgesellschaft, Wiesbaden.

Nikolskii, G. V. 1969. *Theory of Fish Population Dynamics as the Biological Back-ground for Rational Exploitation and Management of Fishery Resources.* Translated by J. E. S. Bradley. Oliver and Boyd, Edinburgh.

Nordeng, H. 1983. Solution to the "char problem" based on Arctic char *(Salvelinus alpinus)* in Norway. *Canadian Journal of Fisheries and Aquatic Sciences* 40:1372–1387.

Nozawa, K., T. Shotake, Y. Ohkura, and Y. Tanabe. 1977. Genetic variations within and between species of Asian macaques. *Japanese Journal of Genetics* 52:15–30.

Nozawa, K., T. Shotake, Y. Kawamoto, and Y. Tanabe. 1982. Electrophoretically estimated genetic distance and divergence time between chimpanzee and man. *Primates* 23:432–443.

Odell, P. L. and J. P. Basu. 1976. Concerning several methods for estimating crop averages using remote sensing data. *Communications in Statistics A* 5:1091–1114.

Ohno, S. 1970. *Evolution by Gene Duplication.* Springer-Verlag, Berlin and New York.

Ohno, S., C. Stenius, E. Faisst, and M. T. Zenzes. 1965. Post-zygotic chromosomal rearrangements in rainbow trout *(Salmo irideus* Gibbons). *Cytogenetics* 4:117–129.

Ohta, T. 1982. Linkage disequilibrium due to random genetic drift in finite subdivided populations. *Proceedings of the National Academy of Sciences U.S.A.* 79:1940–1944.

Ohta, T. and M. Kimura. 1973. A model of mutation appropriate to estimate the number of electrophoretically detectable alleles in a finite population. *Genetical Research* 22:201–204.

Ojima, Y. and S. Makino. 1978. Triploidy induced by cold shock in fertilized eggs of the carp. A preliminary study. *Proceedings of the Japan Academy* 54:359–362.

Ojima, Y. and H. Ueda. 1982. A karyotypical study of the conger eel *(Conger myriaster)* in *in vitro* cells, with special regard to the identification of the sex chromosome. *Proceedings of the Japan Academy* 58:56–59.

Okazaki, T. 1978. Genetic differences of two chum salmon *(Oncorhynchus keta)* populations returning to the Tokachi River. *Far Seas Fisheries Research Bulletin* (Japan) 16:121–128.

Okazaki, T. 1982. Genetic study on population structure in chum salmon *(Onchorhynchus keta). Far Seas Fisheries Research Bulletin* (Japan) 19:25–116.

Onozato, H. 1982. The "Hertwig effect" and gynogenesis in chum salmon *(Oncorhynchus keta)* eggs fertilized with ^{60}Co-ray irradiated milt. *Bulletin of the Japanese Society of Scientific Fisheries* [Nihon Suisan Gakkai-shi] 48:1237–1244.

Patillo, P. and S. L. Varvio-Aho. 1980. On the estimation of population size from allele frequency changes. *Genetics* 95:1055–1057.

Parsons, J. E. and G. H. Thorgaard. 1984. Induced androgenesis in rainbow trout. *Journal of Experimental Zoology* 231:407–412.

Parsons, J. E. and G. H. Thorgaard. 1985. Production of androgenetic diploid rainbow trout. *Journal of Heredity* 76:177–181.

Paulik, G. J., A. S. Hourston, and P. A. Larkin. 1967. Exploitation of multiple stocks by a common fishery. *Journal of the Fisheries Research Board of Canada.* 24:2527–2537.

Payne, R. H., A. R. Child, and A. Forrest. 1971. Geographical variation in the Atlantic salmon. *Nature* 231:250–252.

Pella, J. J. and T. L. Robertson. 1979. Assessment of composition of stock mixtures. *U.S. Department of Commerce Fisheries Bulletin* 77(2):387–398.

Pelzman, R. J. 1980. Impact of Florida largemouth bass, *Micropterus salmoides floridanus,* introductions at selected northern California waters, with a discussion of the use of meristics for detecting introgression and for classifying individual fish of intergraded populations. *California Fish and Game* 66:133–162.

Penny, D. 1982. Towards a basis for classification: The incompleteness of distance measures, incompatibility analysis and phenetic classification. *Journal of Theoretical Biology* 96:129–142.

Penney, D. F. and E. G. Zimmerman. 1976. Genic divergence and local population differentiation by random drift in the pocket gopher genus *Geomys*. *Evolution* 30:473–483.

Peterman, R. M. 1977. A simple mechanism that causes collapsing stability regions in exploited salmonid populations. *Journal of the Fisheries Research Board of Canada* 34:1130–1142.

Phelps, S. R. and F. W. Allendorf. 1984. Genetic identity of pallid and shovelnose sturgeon *(Scaphirynchus albus* and *S. platorynchus). Copeia* 1983:696–700.

Philipp, D. P., W. F. Childers, and G. S. Whitt. 1983. A biochemical genetic evaluation of the northern and Florida subspecies of largemouth bass. *Transactions of the American Fisheries Society* 112:1–20.

Phillips, R. B. 1983. Chromosomal location of nucleolar organizer regions (NORs) in salmonids. *Genetics* 104:56–57.

Phillips, R. B. and K. D. Zajicek. 1982. Q band chromosomal banding polymorphisms in lake trout *(Salvelinus namaycush). Genetics* 101:227–234.

Phillips, R. B., K. D. Zajicek, and F. M. Utter. 1985. Q band chromosome polymorphisms in chinook salmon *(Oncorhynchus tshawytscha). Copeia* 1985:273–278.

Pirchner, F. 1969. *Population Genetics in Animal Breeding.* W. H. Freeman and Company, San Francisco.

Place, A. R. and D. A. Powers, 1979. Genetic variation and relative catalytic efficiencies: Lactate dehydrogenase B allozymes of *Fundulus heteroclitus. Proceedings of the National Academy of Sciences U.S.A.* 76:2354–2358.

Pollak, E. 1980. Effective population numbers and mean times to extinction in dioecious populations with overlapping generations. *Mathematical Biosciences* 52:1–25.

Potter, S. S., J. E. Newbold, C. A. Hutchison III, and M. H. Edgell. 1975. Specific cleavage analysis of mammalian mitochondrial DNA. *Proceedings of the National Academy of Sciences U.S.A.* 72:4496–4500.

Powell, J. R. 1983. Interspecific cytoplasmic gene flow in the absence of nuclear gene flow: Evidence from *Drosophila. Proceedings of the National Academy of Sciences U.S.A.* 80:492–495.

Prager, E. M. and A. C. Wilson. 1971. The dependence of immunological cross-reactivity upon sequence resemblance among lysozymes. *Journal of Biological Chemistry* 246:5978–5989.

Prager, E. M., A. H. Brush, R. A. Nolan, M. Nakanishi, and A. C. Wilson. 1974. Slow evolution of transferrin and albumin in birds according to micro-complement fixation analysis. *Journal of Molecular Evolution* 3:243–262.

Price, T. and P. Boag. 1984. Genetic changes in the morphological differentiation of Darwin's ground finches. In *Population Biology and Evolution* (Springer-Verlag Berlin, Heidelberg).

Prout, T. 1973. Appendix to the paper by J. B. Mitton and R. K. Koehn. *Genetics* 73:493–496.

Purdom, C. E. 1969. Radiation-induced gynogenesis and androgenesis in fish. *Heredity* 24:431–444.

Purdom, C. E. 1972. Induced polyploidy in plaice *(Pleuronectes platessa)* and its hybrid with the flounder *(Platichtys flesus). Heredity* 29:11–24.

Purdom, C. E. 1976. Genetic techniques in flatfish culture. *Journal of the Fisheries Research Board of Canada* 33:1088–1093.

Purdom, C. E. 1979. Genetics of growth and reproduction in teleosts. *Symposia of the Zoological Society of London* 44:207–217.

Purdom, C. E. 1983. Genetic engineering by the manipulation of chromosomes. *Aquaculture* 33:287–300.

Radinsky, L. 1978. Do albumin clocks run on time? *Science* 200:1182–1183.

Ralls, K. and J. Ballou. 1983. Extinction: Lessons from zoos. In *Genetics and Conservation: A Reference for Managing Wild Animal and Plant Populations*, ed. C. B. Schonewald-Cox, S. M. Chambers, B. MacBryde, and L. Thomas (Benjamin/Cummins, Menlo Park, Calif.), pp. 164–184.

Ramshaw, J. A. M., J. A. Coyne, and R. C. Lewontin. 1979. The sensitivity of gel electrophoresis as a detector of genetic variation. *Genetics* 93:1019–1037.

Rao, C. R., 1982a. Diversity: Its measurement, decomposition, apportionment and analysis. *Sankhya* A 44:1–22.

Rao, C. R., 1982b. Diversity and dissimilarity coefficients: A unified approach. *Theoretical Population Biology* 21:24–43.

Rao, S. S. 1984. *Optimization. Theory and Applications*, 2nd ed. John Wiley and Sons, New York, 747 pp.

Rasmuson, M. 1968. *Populationsgenetiska synpunkter på laxodlingsverksamheten i Sverige*. Swedish Salmon Research Institute Report 3/1968, 18 pp.

Reed, T. E. 1969. Caucasian genes in American Negroes. *Science* 165:762–768.

Reed, T. E. 1973. Number of gene loci required for accurate estimation of ancestral population proportions in individual human hybrids. *Nature* 244:575–576.

Refstie, T. 1983. Induction of diploid gynogenesis in Atlantic salmon and rainbow trout using irradiated sperm and heat shock. *Canadian Journal of Zoology* 61:2411–2416.

Reftie, T., V. Vassvik, and T. Gjedrem. 1977. Induction of polyploidy in salmonids by cytochalasin B. *Aquaculture* 10:65–74.

Reftie, T., J. Stoss, and E. M. Donaldson. 1982. Production of all female coho salmon *(Oncorhynchus kisutch)* by diploid gynogenesis using irradiated sperm and cold shock. *Aquaculture* 29:67–82.

Regier, H. A. 1973. Sequence of exploitation of stocks in multispecies fisheries in the Laurentian Great Lakes. *Journal of the Fisheries Research Board of Canada* 30:1992–1999.

Regier, H. A. and K. H. Loftus. 1972. Effects of fisheries exploitation in salmonid communities in oligotrophic lakes. *Journal of the Fisheries Research Board of Canada* 29:959–968.

Regier, H. A. and F. D. McCracken. 1975. Science for Canada's shelf-seas fisheries. *Journal of the Fisheries Research Board of Canada* 32:1887–1932.

Reisenbichler, R. R. 1981. Columbia River salmonid broodstock management: Annual progress report. Unpublished, National Fishery Research Center, Seattle, Wash.

Reisenbichler, R. R. and J. D. McIntyre. 1977. Genetic differences in growth and survival of juvenile hatchery and wild steelhead trout, *Salmo gairdneri. Journal of the Fisheries Research Board of Canada* 34:123–128.

Remington, C. L. 1968. Suture-zones of hybrid interaction between recently joined biotas. In *Evolutionary Biology*, ed. T. Dobzhansky, M. K. Hecht, and W. C. Steere (Appleton-Century-Crofts, New York), vol. 2, pp. 321–428.

Repenning, C. A. 1983. New evidence for the age of the Gubik Formation, Alaskan North Slope. *Quarternary Research* (New York) 19:356–372.

Reynolds, J., B. S. Weir, and C. C. Cockerham. 1983. Estimation of the coancestry coefficient: Basis for a short-term genetic distance. *Genetics* 105:767–779.

Ricker, W. E. 1958. Maximum sustained yields from fluctuating environments and mixed stocks. *Journal of the Fisheries Research Board of Canada* 15:991–1006.

Ricker, W. E. 1969. Effects of size-selective mortality and sampling bias in estimates of growth, mortality, production and yield. *Journal of the Fisheries Research Board of Canada* 26:479–541.

Ricker, W. E. 1972. Hereditary and environmental factors affecting certain salmonid populations. In *The Stock Concept of Pacific Salmon*, H. R. MacMillan Lectures in Fisheries, University of British Columbia, Vancouver, pp. 19–160.

Ricker, W. E. 1973. Two mechanisms that make it impossible to maintain peak-period yields from stocks of Pacific salmon and other fishes. *Journal of the Fisheries Research Board of Canada* 30:1275–1286.

Ricker, W. E. 1976. Review of the rate of growth and mortality of Pacific salmon in salt water, and noncatch mortality caused by fishing. *Journal of the Fisheries Research Board of Canada* 33:1483–1524.

Ricker, W. E. 1980a. *Changes in the age and size of chum salmon* (Oncorhynchus keta). Canadian Technical Report of Fisheries and Aquatic Sciences 930, 99 pp.

Ricker, W. E. 1980b. *Causes of the decrease in age and size of chinook salmon* (Oncorhynchus tshawytscha). Canadian Technical Report of Fisheries and Aquatic Sciences 944, 25 pp.

Ricker, W. E. 1981. Changes in the average size and average age of Pacific salmon. *Canadian Journal of Fisheries and Aquatic Sciences* 38:1636–1656.

Ricker, W. E. 1982. *Size and age of British Columbia sockeye salmon* (Oncorhynchus nerka) *in relation to environmental factors and the fishery*. Canadian Technical Report of Fisheries and Aquatic Sciences 115, 117 pp.

Ricker, W. E. and W. P. Wickett. 1980. *Causes of the decrease in size of coho salmon* (Oncorhynchus kisutch). Canadian Technical Report of Fisheries and Aquatic Sciences 917, 63 pp.

Ricker, W. E., H. T. Bilton, and K. V. Aro. 1978. *Causes of decrease in size of pink salmon* (Oncorhynchus gorbuscha). Fisheries and Marine Service (Canada) Technical Report 820, 93 pp.

Ridgway, G. J. 1957. The use of immunological techniques in racial studies. In *Contributions to the Study of Subpopulations of Fishes*, ed. J. C. Marr, U. S. Department of Interior Special Scientific Report-Fisheries 208, pp. 39–43.

Ridgway, G. J. and G. W. Klontz. 1960. *Blood types in Pacific salmon*. U.S. Department of Interior Special Scientific Report-Fisheries 324, 9 pp.

Ridgway, G. J., G. W. Klontz, and C. Matsumoto. 1962. Intraspecific differences in serum antigens of red salmon demonstrated by immunochemical methods. *International North Pacific Fisheries Bulletin* 8:1–13.

Rieger, R., A. Michaelis, and M. Green. 1976. *Glossary of Genetics and Cytogenetics*. Springer-Verlag, Berlin.

Roberts, D. F. 1955. The dynamics of racial intermixture in the American Negro: Some anthropological considerations. *American Journal of Human Genetics* 7:361–367.

Roberts, R. J. 1982. Restriction and modification enzymes and their recognition sequences. *Nucleic Acids Research* 10:117–144.

Robertson, F. 1972. Value and limitations of research in protein polymorphism. In *Proceedings of the 12th European Conference on Animal Blood Groups and Biochemical Polymorphism* (Dr. W. Junk, The Hague), pp. 41–54.

Robertson, F. W. 1955. Selection response and the properties of genetic variation. *Cold Spring Harbor Symposia on Quantitative Biology* 20:166–177.

Robertson, O. H. 1961. Prolongation of the life span of kokanee salmon *(Oncorhynchus nerka kennerlyi)* by castration before beginning of gonad development. *Proceedings of the National Academy of Sciences U.S.A.* 47:609–621.

Robinson, D. F. and L. R. Foulds. 1981. Comparison of phylogenetic trees. *Mathematical Bioscience* 53:131–147.

Rocchi, A., A. De Capoa, and F. Gigliani. 1971. Double satellites: Auto-radiographic study of a chromosomal marker observed in two generations. *Humangenetik* 14:6–12.

Roe, B. A., D. P. Ma, R. K. Wilson, and J. F. H. Wong. 1985. The complete nucleotide sequence of the *Xenopus laevis* mitochondrial genome. *Journal of Biological Chemistry* 260:9759–9779.

Roff, D. A. 1983. An allocation model of growth and reproduction in fish. *Canadian Journal of Fisheries and Aquatic Sciences* 40:1395–1404.

Rogers, A. R. and H. C. Harpending. 1983. Population structure and quantitative characters. *Genetics* 105:985–1002.

Rogers, J. S. 1972. Measures of genetic similarity and genetic distance. In *Studies in Genetics VII* (University of Texas Publication 7213, Austin), pp. 145–153.

Rohrer, R. L. 1982. *Evaluation of Henry's Lake management program*. Federal Aid to Fish and Wildlife Restoration Project F-73-R-4 Job Performance Report, Idaho Department of Fish and Game.

Romashov, D. D. and V. N. Belyaeva. 1964. Cytology of radiation gynogenesis and androgenesis in the loach *(Misgurnus fossilis* L.) *Doklady Akademii Nauk SSSR* 157(4):964–967.

Romashov, D. D. and V. N. Belyaeva. 1965. Analysis of diploidization induced by low temperature during radiation gynogenesis in loach. *Tsitologiya* 7(5):607–615.

Ros, T. 1981. Salmonids in the Lake Vänern area. In *Fish Gene Pools,* ed. N. Ryman. Ecological Bulletins (Stockholm), vol. 34, pp. 21–32.

Rose, M. R. 1982. Antagonistic pleiotropy, dominance, and genetic variation. *Heredity* 48:63–78.

Rosenblatt, R. H. 1967. The zoogeography relationships of the marine shore fishes of tropical America. *Proceedings of the International Conference on Tropical Oceanography* 5:579–592.

Rosenfield, A. and F. G. Kern. 1979. Molluscan imports and the potential for introduction of disease organisms. In *Exotic Species in Mariculture* (MIT Press, Cambridge, Mass.), pp. 165–191.

Rosenthal, H., D. F. Alderdice, and F. P. J. Velsen. 1978. *Cross-fertilization experiments using Pacific herring eggs and cryopreserved Baltic herring sperm.* Canada Fisheries and Marine Service Technical Report 844, 9 pp.

Ross, M. R. and T. M. Cavender. 1981. Morphological analysis of four experimental intergeneric cyprinid hybrid crosses. *Copeia* 1981:377–387.

Roughgarden, J. 1979. *Theory of Population Genetics and Evolutionary Ecology: An Introduction.* MacMillan Publ. Co., New York, 634 pp.

Rubinoff, R. W. and I. Rubinoff. 1968. Interoceanic colonization of a marine goby through the Panama Canal. *Nature* 217:476–478.

Rubinoff, R. W. and I. Rubinoff. 1971. Geographic and reproductive isolation in Atlantic and Pacific populations of Panamanian *Bathygobius. Evolution* 25:88–97.

Ruiz, I. R. G., M. Soma, and W. Becak. 1981. Nucleolar organizer regions and constitutive heterochromatin in polyploid species of the genus *Odontophrynus* (Amphibia, Anura). *Cytogenetics and Cell Genetics* 29:84–98.

Ryder, R. A., S. R. Kerr, W. W. Taylor, and P. A. Larkin. 1981. Community consequences of fish stock diversity. *Canadian Journal of Fisheries and Aquatic Sciences* 38:1856–1866.

Ryman, N. 1970. A genetic analysis of recapture frequencies of released young of salmon (*Salmo salar* L.). *Hereditas* 65:159–160.

Ryman, N. 1972. An attempt to estimate the magnitude of additive genetic variation of body size in the guppy-fish, *Lebistes reticulatus. Hereditas* 71:237–244.

Ryman, N. (ed.). 1981a. *Fish Gene Pools.* Ecological Bulletins (Stockholm), vol. 34.

Ryman, N. 1981b. Conservation of genetic resources: Experiences from the brown trout *(Salmo trutta).* In *Fish Gene Pools,* ed. N. Ryman. Ecological Bulletins (Stockholm), vol. 34, pp. 61–74.

Ryman, N. 1981c. Recommendations. In *Fish Gene Pools,* ed. N. Ryman. Ecological Bulletins (Stockholm), vol. 34, pp. 107–108.

Ryman, N. 1983. Patterns of distribution of biochemical genetic variation in salmonids: Differences between species. *Aquaculture* 33:1–21.

Ryman, N. and G. Ståhl. 1980. Genetic changes in hatchery stocks of brown trout *(Salmo trutta). Canadian Journal of Fisheries and Aquatic Sciences* 37:82–87.

Ryman, N. and G. Ståhl. 1981. Genetic perspectives of the identification and conservation of Scandinavian stocks of fish. *Canadian Journal of Fisheries and Aquatic Sciences* 38:1562–1575.

Ryman, N., F. W. Allendorf, and G. Ståhl. 1979. Reproductive isolation with little genetic divergence in sympatric populations of brown trout *(Salmo trutta). Genetics* 92:247–262.

Ryman, N., R. Baccus, C. Reuterwall, and M. H. Smith. 1981. Effective population size, generation interval, and potential loss of genetic variability in game species under different hunting regimes. *Oikos* 36:257–266.

Ryman, N., R. Chakraborty, and M. Nei. 1983. Differences in the relative distribution of human gene diversity between electrophoretic and red and white cell antigen loci. *Human Heredity* 33:93–102.

Ryman, N., U. Lagercrantz, L. Andersson, R. Chakraborty, and R. Rosenberg. 1984. Lack of correspondence between genetic and morphologic variability patterns in Atlantic herring *(Clupea harengus). Heredity* 53:687–704.

Sage, R. D. 1981. Wild mice. In *The Mouse in Biomedical Research,* ed. H.L. Foster, J. D. Small, and J. G. Fox (Academic Press, New York), vol. 1, pp. 39–90.

Sage, R. D. and R. K. Selander. 1975. Trophic radiation through polymorphism in cichlid fishes. *Proceedings of the National Academy of Sciences U.S.A.* 72:4669–4673.

Saito, T. 1976. Geologic significance of coiling direction in planktonic Foraminifera, *Pulleniatina. Geology* 4:305–309.

Saldanha, P. H. 1957. Gene flow from white into Negro populations in Brazil. *American Journal of Human Genetics* 9:299–309.

Salmenkova, E. A., V. T. Omel'chenko, T. V. Malinina, K. I. Afanasyev, and Yu.P. Altukhov. 1981. Population-genetic differences between successive generations of pink salmon spawning in rivers of the Asian coasts of the North Pacific. In *Genetics and Reproduction of Marine Organisms* (Vladivostok, USSR), pp. 95–104. In Russian.

Sanders, B. and J. Wright. 1962. Immunogenetic studies in two trout species of the genus *Salmo. Annals of the New York Academy of Sciences* 97:116–130.

Sanghvi, L. D. 1953. Comparison of genetical and morphological methods for a study of biological differences. *American Journal of Physical Anthropology* 11:385–404.

Sarich, V. M. 1969. Pinniped origins and the rate of evolution of carnivore albumins. *Systematic Zoology* 18:286–295.

Sarich, V. M. 1977. Rates, sample sizes, and the neutrality hypothesis for electrophoresis in evolutionary studies. *Nature* 265:24–28.

Sarich, V. M. and A. C. Wilson. 1966. Quantitative immunochemistry and the evolution of primate albumins: Micro-complement fixation. *Science* 154:1563–1566.

Sarich, V. M. and A. C. Wilson. 1967. Immunological time scale for hominid evolution. *Science* 158:1200–1203.

Saunders, R. L. 1966. Some biological aspects of Greenland salmon fishery. *Atlantic Salmon Journal* 1966:17–23.

Saunders, R. L. 1981. Atlantic salmon *(Salmo salar)* stocks and management implications in the Canadian Atlantic Provinces and New England, USA. *Canadian Journal of Fisheries and Aquatic Sciences* 38:1612–1625.

Saville, A. (ed.). 1980. *The Assessment and Management of Pelagic Fish Stocks.* Rapports et Procès-Verbaux des Réunions Conseil International pour l'Exploration de la Mer 177.

Schaffer, H. E. 1983. Determination of DNA fragment size from gel electrophoresis mobility. In *Statistical Analysis of DNA Sequence Data,* ed. B. S. Weir (Marcel Dekker, Inc., New York).

Schaffer, W. M. and P. F. Elson. 1975. The adaptive significance of variations in life history among local populations of Atlantic salmon in North America. *Ecology* 56:577–590.

Scheerer, P. D. and G. H. Thorgaard. 1983. Increased survival in salmonid hybrids by induced triploidy. *Canadian Journal of Fisheries and Aquatic Sciences* 40:2040–2044.

Schmidt, J. 1917. Racial investigations. I. *Zoarces viviparus* L. and local races of the same. *Comptes rendus des Travaux du Laboratoire Carlsberg* 14:1–14.

Schmidtke, J. and I. Kandt. 1981. Single-copy DNA relationship between diploid and tetraploid fish species. *Chromosoma* (Berlin) 83:191–197.

Schnell, G. and R. Selander. 1981. Environmental and morphological correlates of genetic variation in mammals. In *Mammalian Population Genetics,* ed. M. H. Smith and J. Joule (University of Georgia Press, Athens).

Schonewald-Cox, C. M. and J. W. Bayless. 1983. Questions posed by managers. In *Genetics and Conservation: A Reference for Managing Wild Animal and Plant Populations,* ed. C. M. Schonewald-Cox, S. M. Chambers, B. MacBryde, and W. L. Thomas (Benjamin/Cummings, Menlo Park, Calif.), pp. 485–499.

Schonewald-Cox, C. M., S. M. Chambers, B. MacBryde, and W. L. Thomas (eds.). 1983. *Genetics and Conservation: A Reference for Managing Wild Animal and Plant Populations.* Benjamin/Cummings, Menlo Park, Calif.

Schroder, S. 1982. The influence of intrasexual competition on the distribution of chum salmon in an experimental stream. In *Salmon and Trout Migratory Behavior Symposium,* ed. E. L. Brannon and E. O. Salo (University of Washington College of Fisheries, Seattle), pp. 175–285.

Schultz, R. J. 1977. Evolution and ecology of unisexual fishes. In *Evolutionary Biology,* vol. 10, ed. M. K. Hecht, W. C. Steere, and B. Wallace (Plenum Press, New York), pp. 277–331.

Schwartz, F. J. 1972. *World Literature to Fish Hybrids, With an Analysis by Family, Species, and Hybrid.* Publications of the Gulf Coast Research Laboratory Museum, No. 3, 328 pp.

Schwartz, F. J. 1981. *World Literature to Fish Hybrids, With an Analysis by Family, Species, and Hybrid: Supplement 1.* NOAA Technical Report NMFS SSRF-750, U. S. Dept. of Commerce, 507 pp.

Schweigert, J. F., F. J. Ward, and J. W. Clayton. 1977. Effects of fry and fingerling introductions on walleye *(Stizostedion vitreum vitreum)* production in West Blue Lake, Manitoba. *Journal of the Fisheries Research Board of Canada* 34:2142–2150.

Sederoff, R. R. 1984. Structural variation in mitochondrial DNA. *Advances in Genetics* 22:1–108.

Seeb, L. 1986. Biochemical systematics and evolution of the Scorpaenid genus *Sebastes.* Ph.D. dissertation, University of Washington. 176 pp.

Selgeby, J. H. 1982. Decline of lake herring *(Coregonus artedii)* in Lake Superior: An analysis of the Wisconsin herring fishery, 1936–78. *Canadian Journal of Fisheries and Aquatic Sciences* 39:554–563.

Setzer, P. Y. 1970. An analysis of a natural hybrid swarm by means of chromosome morphology. *Transactions of the American Fisheries Society* 99:139–146.

Shah, D. M. and C. H. Langley. 1979. Inter- and intraspecific variation in restriction maps of *Drosophila* mitochondrial DNAs. *Nature* 281:696–699.

Shaklee, J. B. 1983. The utilization of isozymes as gene markers in fisheries management and conservation. In *Isozymes: Current Topics in Biological and Medical Research,* vol. 11 (Alan R. Liss, New York), pp. 213–247.

Shaklee, J. and C. Tamaru. 1981. Biochemical and morphological evolution of Hawaiian bonefishes *(Albula). Systematic Zoology* 30:125–146.

Shaklee, J. B., C. S. Tamura, and R. S. Waples. 1982. Speciation and evolution of marine fishes studied by the electrophoretic analysis of proteins. *Pacific Science* 36:141–157.

Shaw, C. 1964. The use of genetic variation in the analysis of isozyme structure. In *Subunit Structure of Proteins, Biochemical and Genetic Aspects.* Brookhaven Symposia of Biology No. 17 (Brookhaven National Lab., New York), pp. 117–130.

Shireman, J. V., R. W. Rottmann, and F. J. Aldridge. 1983. Consumption and growth of hybrid grass carp fed four vegetation diets and trout chow in circular tanks. *Journal of Fish Biology* 22:685–693.

Shmalhausen, I. I. 1938. *Organism as a Unit in Individual and Historical Development.* Nauka, Moscow-Leningrad, 144 pp. In Russian.

Sibley, C. G. and J. E. Ahlquist. 1984. The phylogeny of the hominoid primates as indicated by DNA-DNA hybridization. *Journal of Molecular Evolution* 20:2–15.

Siciliano, M. and C. Shaw. 1976. Separation and visualization of enzymes of gels. In *Zone Electrophoresis, Chromatographic and Electrophoretic Techniques,* vol. 22, ed. I. Smith (Heinemann, London), pp. 185–209.

Siciliano, M. J., D. A. Wright, S. L. George, and C. R. Shaw. 1973. Inter- and intra-specific genetic distances among teleosts. In *17ième Congrès International de Zoologie, thème no. 5,* pp. 1–24.

Simon, R. C. and R. E. Noble. 1968. Hybridization in *Oncorhynchus* (Salmonidae). I. Viability and inheritance in artificial crosses of chum and pink salmon. *Transactions of the American Fisheries Society* 97:109–118.

Simpson, G. G. 1961. *Principles of Animal Taxonomy.* Columbia University Press, New York.

Singh, R. S., R. C. Lewontin, and A. A. Felton. 1976. Genetic heterogeneity within electrophoretic "alleles" of xanthine dehydrogenase in *Drosophila pseudoobscura. Genetics* 84:609–629.

Sinnock, P. and C. F. Sing. 1972. Analysis of multilocus genetic systems in Tecumseh, Michigan. II. Consideration of the correlation between nonalleles in gametes. *American Journal of Human Genetics* 24:393–415.

Skibinski, D. O. F. and R. D. Ward. 1982. Correlations between heterozygosity and evolutionary rate of proteins. *Nature* 298:490–492.

Slatkin, M. 1973. Gene flow and selection in a cline. *Genetics* 75:733–756.

Slatkin, M. 1981. Estimating levels of gene flow in natural populations. *Genetics* 99:323–335.

Slatkin, M. 1985a. Rare alleles as indicators of gene flow. *Evolution* 39:53–65.

Slatkin, M. 1985b. Gene flow in natural populations. *Annual Review of Ecology and Systematics* 16:393–430.

Slatkin, M. and T. Maruyama. 1975. The influence of gene flow on genetic distance. *American Naturalist* 109:597–601.

Slobodkin, L. B. 1968. How to be a predator. *American Zoologist* 8:43–51.

Smith, C. A. B. 1969. Local fluctuations in gene frequencies. *Annals of Human Genetics* 32:961–965.

Smith, D. G. and R. G. Coss. 1984. Calibrating the molecular clock: Estimates of ground squirrel divergence made using fossil and geological time markers. *Molecular Biology and Evolution* 1:249–259.

Smith, G. R. 1973. Analysis of several hybrid cyprinid fishes from western North America. *Copeia* 1973:395–410.

Smith, P. J., B. A. Wood, and P. G. Benson. 1979. Electrophoretic meristic separation of blue maomao and sweep. *New Zealand Journal of Marine Freshwater Research* 13:549–551.

Smith, R. H. and R. C. von Borstel. 1972. Genetic control of insect populations. *Science* 178:1164–1174.

Smith, S. H. 1968. Species succession and fishery exploitation in the Great Lakes. *Journal of the Fisheries Research Board of Canada* 25:667–693.

Smith, S. H. 1972. Factors of ecological succession in oligotrophic fish communities of the Laurentian Great Lakes. *Journal of the Fisheries Research Board of Canada* 29:717–730.

Smithies, O. 1955. Zone electrophoresis in starch gels: Group variations in the serum proteins of normal human adults. *Biochemical Journal* 61:629–641.

Smouse, P. E. and J. V. Neel. 1977. Multivariate analysis of gametic disequilibrium in the Yanomama. *Genetics* 85:733–752.

Smouse, P. E. and R. H. Ward. 1978. A comparison of the genetic infrastructure of the Ye'cuana and the Yanomama: A likelihood analysis of genotypic variation among populations. *Genetics* 88:611–631.

Smouse, P. E., J. Neel, and W. Liu. 1983. Multiple-locus departures from panmictic equilibrium within and between village gene pools of Amerindian tribes at different stages of agglomeration. *Genetics* 104:133–153.

Smouse, P. E., R. S. Spielman, and M. H. Park. 1982. Multiple locus allocation of individuals to groups as a function of the genetic variation within and differences among human populations. *American Naturalist* 119:445–463.

Sneath, P. H. A. and R. R. Sokal. 1973. *Numerical Taxonomy.* W. H. Freeman, San Francisco 573 pp.

Sokal, R. R. and F. J. Rohlf. 1981. *Biometry,* 2nd ed. W. H. Freeman, San Francisco.

Sokolov, V. E. 1981. The biosphere reserve concept in the USSR. *AMBIO* 10:97–101.

Sola, L., S. Cataudella, and E. Capanna. 1981. New developments in vertebrate cytotaxonomy. III. Karyology of bony fishes: A review. *Genetica* 54:285–328.

Solignac, M., M. Monnerot, and J.-C. Mounolou. 1983. Mitochondrial DNA heteroplasmy in *Drosophila mauritiana. Proceedings of the National Academy of Sciences U.S.A.* 80:6942–6946.

Soloman, D. J. and A. R. Child. 1978. Identification of juvenile natural hybrids between Atlantic salmon *(Salmo salar* L.) and trout *(Salmo trutta* L.). *Journal of Fish Biology* 12:499–501.

Soulé, M. 1976. Allozyme variation: Its determinants in space and time. In *Molecular Evolution,* ed. F. Ayala (Sinauer Associates, Sunderland, Mass.), pp. 60–77.

Soulé, M. E. 1980. Thresholds for survival: Maintaining fitness and evolutionary potential. In *Conservation Biology: An Evolutionary-Ecological Perspective,* ed. M. E. Soulé and B. A. Wilcox (Sinauer Associates, Sunderland, Mass.), pp. 151–170.

Soulé, M. E. 1982. Allomeric variation. 1. The theory and some consequences. *American Naturalist* 120:751–764.

Soulé, M. E. and J. Cuzin-Roudy. 1982. Allomeric variation. 2. Developmental instability of extreme phenotypes. *American Naturalist* 120:765–786.

Soulé, M. E. and B. A. Wilcox (eds.). 1980. *Conservation Biology: An Evolutionary-Ecological Perspective.* Sinauer Associates, Sunderland, Mass.

Southern, E. M. 1975. Detection of specific DNA sequences among DNA fragments separated by gel electrophoresis. *Journal of Molecular Biology* 98:503–517.

Spangler, G. R., N. R. Payne, J. E. Thorpe, J. M. Byrne, H. A. Regier, and W. J.

Christie. 1977. Responses of percids to exploitation. *Journal of the Fisheries Research Board of Canada* 34:1983–1988.

Spangler, G. R., A. H. Berst, and J. F. Koonce. 1981. Perspectives and policy recommendations on the relevance of the stock concept to fishery management. *Canadian Journal of Fisheries and Aquatic Sciences* 38:1908–1914.

Sparks, B. W. and R. G. West. 1972. *The Ice Age in Britain.* Methuen, London, 302 pp.

Spolsky, C. and T. Uzzell. 1984. Natural interspecies transfer of mitochondrial DNA in amphibians. *Proceedings of the National Academy of Sciences U.S.A.* 81:5802–5805.

Stabell, O. 1984. Homing and olfaction in salmonids: A critical review with special references to the Atlantic salmon. *Biological Reviews and Biological Proceedings of the Cambridge Society* 59:333–388.

Ståhl, G. 1981. Genetic differentiation among natural populations of Atlantic salmon *(Salmo salar)* in northern Sweden. In *Fish Gene Pools,* ed. N. Ryman. Ecological Bulletins (Stockholm). vol. 34, pp. 95–105.

Ståhl, G. 1983. Differences in the amount and distribution of genetic variation between natural populations and hatchery stocks of Atlantic salmon. *Aquaculture* 33:23–32.

Ståhl, G., E. J. Loudenslager, R. L. Saunders, and E. J. Schofield. 1983. *Electrophoretic study on Atlantic salmon populations from the Miramichi River (New Brunswick) system, Canada.* International Council for the Exploration of the Sea, C. B. M 20.

Stanley, J. G. 1976. Production of hybrid, androgenetic, and gynogenetic grass carp. *Transactions of the American Fisheries Society* 105:10–16.

Stanley, J. G. and J. B. Jones. 1976. Morphology of androgenetic and gynogenetic grass carp, *Ctenopharyngodon idella* (Valenciennes). *Journal of Fish Biology* 9:523–528.

Stanley, J. G. and K. E. Sneed. 1974. Artificial gynogenesis and its application in genetics and selective breeding of fishes. In *The Early Life History of Fish,* ed. J. H. S. Blaxter (Springer-Verlag, Berlin and New York), pp. 527–536.

Starobogatov, Ya.I. 1975. Review of Yu.P. Altukhov's book "Population Genetics of Fish." *Journal of Genetic Biology* (U.S.S.R.) 36:626–628. In Russian.

Stauffer, J. R., Jr., C. H. Hocutt, and R. F. Denoncourt. 1979. Status and distribution of the hybrid *Nocomis micropogon* × *Rhinichthys cataractae,* with a discussion of hybridization as a viable mode of vertebrate speciation. *American Midland Naturalist* 101:355–365.

Stearns, S. C. 1976. Life history tactics: A review of the ideas. *Quarterly Review of Biology* 51:3–47.

Stearns, S. C. 1980. A new view of life-history evolution. *Oikos* 35:266–281.

Steinberg, A. G., H. K. Bleibtreu, T. W. Kurczynski, A. O. Martin, and E. M. Kurczynski. 1967. Genetic studies on an inbred human isolate. In *Proceedings of the 3rd International Congress on Human Genetics,* ed. J. F. Crow and J. V. Neel (Johns Hopkins Press, Baltimore, Md.), pp. 267–289.

Stevenson, M. M. and T. M. Buchanon. 1973. An analysis of hybridization between the cyprinodont fishes *Cyprinodon variegatus* and *C. elegans. Copeia* 1973:682–692.

Stoneking, M., B. May, and J. Wright. 1981. Loss of duplicate gene expression in salmonids: Evidence for a null allele polymorphism at the duplicate aspartate aminotransferase loci in brook trout *(Salvelinus fontinalis). Biochemical Genetics* 19:1063–1077.

Stormont, C., R. Owen, and M. Irwin. 1951. The B and C systems of bovine blood groups. *Genetics* 36:134–161.

Streisinger, G., C. Walker, N. Dower, D. Knauber, and F. Singer. 1981. Production of clones of homozygous diploid zebra fish *(Brachydanio rerio). Nature* (London) 291:293–296.

Sullivan, C. R. 1984. Striped bass and the Chesapeake. *Fisheries: A Bulletin of the American Fisheries Society* 9(4), editorial page.

Svärdson, G. 1945. *Chromosome studies on Salmonidae.* Reports and Short Papers, Institute of Freshwater Research Drottningholm 23, 151 pp.

Svärdson, G. 1979. *Speciation of Scandinavian Coregonus.* Reports and Short Papers, Institute of Freshwater Research Drottningholm 57, 95 pp.

Sved, J. A. 1968. The stability of linked systems of loci with a small population size. *Genetics* 59:543–563.

Svetovidov, A. N. 1948. *Gadiformes.* English translation., U. S. Department of Commerce, Springfield, Virginia, OTS 63–10071.

Svetovidov, A. N. 1952. *Clupeidae.* English translation, U.S. Department of Commerce, Springfield, Virginia, OTS 61–11435.

Swarup, H. 1959. Production of triploidy in *Gasterosteus aculeatus* L. *Genetics* 56:129–142.

Tajima, F. 1983. Evolutionary relationship of DNA sequences in finite populations. *Genetics* 105:437–460.

Takahata, N. 1983. Gene identity and genetic differentiation of populations in the finite island model. *Genetics* 28:114–138.

Takahata, N. and M. Nei. 1984. F_{ST} and G_{ST} statistics in the finite island model. *Genetics* 107:501–504.

Takahata, N. and M. Slatkin. 1984. Mitochondrial gene flow. *Proceedings of the National Academy of Sciences U.S.A.* 81:1764–1767.

Taniguchi, N., K. Sumantadinata, and S. Iyama. 1983. Genetic change in the first and second generations of hatchery stock of black seabream. *Aquaculture* 35:309–320.

Tateno, Y., M. Nei, and F. Tajima. 1982. Accuracy of estimated phylogenetic trees from molecular data. I. Distantly related species. *Journal of Molecular Evolution* 18:387–404.

Templeton, A. R. 1980. Modes of speciation and inferences based on genetic distances. *Evolution* 34:719–729.

Templeton, A. R. 1983a. Convergent evolution and nonparametric inferences from restriction data and DNA sequences. In *Statistical Analysis of DNA Sequence Data,* ed. B. S. Weir (Marcel Dekker, New York), pp. 151–179.

Templeton, A. R. 1983b. Phylogenetic inference from restriction endonuclease cleavage site maps with particular reference to the evolution of humans and apes. *Evolution* 37:221–244.

Templeton, A. R. and B. Reed. 1983. The elimination of inbreeding depression in a captive herd of Speke's gazelle. In *Genetics and Conservation,* ed. C. M. Schonewald-Cox, S. M. Chambers, B. MacBryde, and W. L. Thomas (Benjamin/Cummings, Menlo Park, Calif.) pp. 241–261.

Therman, E. 1980. *Human Chromosomes: Structure, Behavior, Effects.* Springer-Verlag, New York.

Thomas, W. K., R. E., Withler, and A. T. Beckenbach 1986. Mitochondrial DNA analysis of Pacific salmonid evolution. *Canadian Journal of Zoology* 64.1058–1064.

Thompson, D. and A. P. Scott. 1984. An analysis of recombination data in gynogenetic diploid rainbow trout. *Heredity* 53:441–452.

Thompson, D., C. E. Purdom, and B. W. Jones. 1981. Genetic analysis of spontaneous gynogenetic diploids in the plaice *Pleuronectes platessa. Heredity* 47:269–274.

Thompson, E. A. 1973. The Icelandic admixture problem. *Annals of Human Genetics* 37:69–80.

Thompson, G. 1981. A review of theoretical aspects of HLA and disease association. *Theoretical Population Biology* 20:168–208.

Thorgaard, G. H. 1976. Robertsonian polymorphism and constitutive heterochromatin distribution in chromosomes of the rainbow trout *(Salmo gairdneri)*. *Cytogenetics and Cell Genetics* 17:174–184.

Thorgaard, G. H. 1983a. Chromosome set manipulation and sex control in fish. In *Fish Physiology*, vol. 9B, ed. W. S. Hoar, D. J. Randall, and E. M. Donaldson (Academic Press, New York), pp. 405–534..

Thorgaard, G. H. 1983b. Chromosomal differences among rainbow trout populations. *Copeia* 1983:650–662.

Thorgaard, G. H. and G. A. E. Gall. 1979. Adult triploids in a rainbow trout family. *Genetics* 93:961–973.

Thorgaard, G. H., M. E. Jazwin, and A. R. Stier. 1981. Polyploidy induced by heat shock in rainbow trout. *Transactions of the American Fisheries Society* 110:546–550.

Thorgaard, G. H., P. S. Rabinovitch, M. W. Shen, G. A. E. Gall, J. Propp, and F. M. Utter. 1982. Triploid rainbow trout identified by flow cytometry. *Aquaculture* 29:305–309.

Thorgaard, G. H., F. W. Allendorf, and K. L. Knudsen. 1983. Gene-centromere mapping in the rainbow trout: High interference over long map distances. *Genetics* 103:771–783.

Thorgaard, G. H., P. D. Scheerer, and J. E. Parsons. 1985. Residual paternal inheritance in gynogenetic rainbow trout: Implications for gene transfer. *Theoretical and Applied Genetics* 71:119–121.

Thorpe, J. E. and K. A. Mitchell. 1981. Stocks of Atlantic salmon *(Salmo salar)* in Britain and Ireland: Discreteness and current management. *Canadian Journal of Fisheries and Aquatic Sciences* 38:1576–1590.

Thorpe, J. E., R. I. G. Morgan, C. Talbot, and M. S. Miles. 1983. Inheritance of developmental rates in Atlantic salmon, *Salmo salar* L. *Aquaculture* 33:119–128.

Thorpe, J. E. and J. F. Koonce (with D. Borgeson, B. Henderson, A. Lamsa, P. S. Maitland, M. A. Ross, R. C. Simon, and C. Walters). 1981. Assessing and managing man's impact on fish genetic resources. *Canadian Journal of Fisheries and Aquatic Sciences* 38:1899–1907.

Thorpe, J. P. 1979. Enzyme variation and taxonomy: The estimation of sampling errors in measurements of interspecific genetic similarity. *Journal of the Linnaean Society* 11:369–386.

Thorpe, J. P. 1982. The molecular clock hypothesis: Biological evolution, genetic differentiation and systematics. *Annual Review of Ecology and Systematics* 13:139–168.

Timofeev-Resovsky, N. V., N. N. Vorontsov, and A. V. Yablokov. 1977. *A Brief Survey of the Evolutionary Theory*. Biologiia Moria Nauka [Marine Biology USSR Publication], Moscow, 301 pp. In Russian.

Todd, I. and P. Larkin. 1971. Gillnet selectivity on sockeye *(Oncorhynchus nerka)* and pink salmon *(O. gorbuscha)* of the Skeena River system, British Columbia. *Journal of the Fisheries Research Board of Canada* 28:821–842.

Todd, T. N., G. R. Smith, and L. E. Cable. 1981. Environmental and genetic contributions to morphological differentiation in ciscoes (Coregoninae) of the Great Lakes. *Canadian Journal of Fisheries and Aquatic Sciences* 38:59–67.

Tompkins, R. 1978. Triploid and gynogenetic diploid *Xenopus laevis*. *Journal of Experimental Zoology* 203:251–256.

Tsoi, R. M. 1969. Action of nitrosomethylurea and dimethylsulfate on the sperm cells of rainbow trout and the pcled. *Doklady Akademii Nauk SSSR* 189(1):411–414. In Russian.

Tsuyuki, H. and J. Westrheim. 1970. Analysis of the *Sebastes aleutianus–S. melanostomus* complex and description of a new scorpaenid species, *Sebastes caenaematicus,* in the northeast Pacific Ocean. *Journal of the Fisheries Research Board of Canada* 27:2233–2254.

Tsuyuki, H., E. Roberts, and E. A. Best. 1969. Serum transferrin systems and the hemoglobins of the Pacific halibut *(Hippoglossus stenolepsis). Journal of the Fisheries Research Board of Canada* 26:2351–2362.

Turner, B. J. 1974. Genetic divergence of Death Valley pupfish species: Biochemical versus morphological evidence. *Evolution* 28:281–294.

Turner, B. J. 1982. The evolutionary genetics of a unisexual fish, *Poecilia formosa.* In *Mechanisms of Speciation,* ed. C. Barigozzi (A. R. Liss, New York), pp. 265–305.

Turner, B. J. and D. J. Grosse. 1980. Trophic differentiation in *Illyodon,* a genus of stream-dwelling goodeid fishes: Speciation versus ecological polymorphism. *Evolution* 34:259–270.

Turner, H. J. 1949. *Report on investigations of methods of improving the shellfish resources of Massachusetts.* Department of Conservation, Division of Marine Fisheries Communications, Massachusetts, 22 pp.

Turner, J. L. 1977. Changes in the size structure of cichlid populations of Lake Malawi resulting from bottom trawling. *Journal of the Fisheries Research Board of Canada* 34:232–238.

Ueda, T., Y. Ojima, R. Sato, and Y. Fukuda. 1984. Triploid hybrids between female rainbow trout and male brook trout. *Bulletin of the Japanese Society of Scientific Fisheries [Nihon Suisan Gakkai-Shi]* 50:1331–1336.

United Nations. 1980. *Genetic resources.* United Nations Environment Program Report 5.

Upholt, W. B. 1977. Estimation of DNA sequence divergence from comparisons of restriction endonuclease digests. *Nucleic Acids Research* 4:1257–1265.

Upholt, W. B. and I. B. Dawid. 1977. Mapping of mitochondrial DNA of individual sheep and goats: Rapid evolution in the D-loop region. *Cell* 11:571–583.

Utter, F. M. 1972. Phosphoglucomutase and esterase polymorphism in Pacific herring in Washington waters. In *The Stock Concept in Pacific Salmon,* ed. R. C. Simon and P. A. Larkin (H. R. MacMillan Lectures in Fisheries, University of British Columbia, Vancouver, B. C.), pp. 191–197.

Utter, F. M. 1981. Biological criteria for definition of species and distinct intraspecific populations of anadromous salmonids under the U.S. Endangered Species Act of 1973. *Canadian Journal of Fisheries and Aquatic Sciences* 38:1626–1635.

Utter, F. M. and H. Hodgins. 1972. Biochemical genetic variants at six loci in four stocks of rainbow trout. *Transactions of the American Fisheries Society* 101(3):494–502.

Utter, F. M., F. W. Allendorf, and H. O. Hodgins. 1973. Genetic variability and relationships in Pacific salmon and related trout based on protein variation. *Systematic Zoology* 22:257–270.

Utter, F. M., H. O. Hodgins, and F. W. Allendorf. 1974. Biochemical genetic studies of fishes: Potentialities and limitations. In *Biochemical and Biophysical Perspectives in Marine Biology,* ed. D. C. Malins and J. R. Sargent (Academic Press, New York), vol. 1, pp. 213–238.

Utter, F. M., F. W. Allendorf, and B. May. 1976. The use of protein variation in the management of salmonid populations. *Transactions of the 41st North American Wildlife and Natural Resources Conference* 41:373–384.

Utter, F. M., F. W. Allendorf, and B. May. 1979. Genetic basis of creatine kinase isozymes in skeletal muscle of salmonid fishes. *Biochemical Genetics* 17:1079–1091.

Utter, F. M., D. Campton, S. Grant, G. B. Milner, J. Seeb, and L. Wishard. 1980. Population structures of indigenous salmonid species of the Pacific Northwest. In

Salmonid Ecosystems of the North Pacific, ed. W. J. McNeil and D. C. Himsworth (Oregon State University Press, Corvallis), pp. 285–304.

Utter, F. M., O. W. Johnson, G. H. Thorgaard, and P. S. Rabinovitch. 1983. Measurement and potential applications of induced triploidy in Pacific salmon. *Aquaculture* 35:125–135.

Utter, F. M., P. Aebersold, J. Helle, and G. Winans. 1984. Genetic characterization of populations in the southeastern range of sockeye salmon. In *Proceedings of the Olympic Wild Fish Conference,* ed. J. Walton and D. Houston, pp. 17–32.

Uwa, H. 1965. Gynogenetic haploid embryos of the medaka *(Oryzias latipes).* *Embryologia* 9:40–48.

Valenti, R. J. 1975. Induced polyploidy in *Tilapia aurea* (Steindachner) by means of temperature shock treatment. *Journal of Fish Biology* 7:519–528.

Valentine, D. W., and M. Soulé. 1973. Effects of p.p'-DDT on developmental stability of pectoral fin rays in the grunion, *Leuresthes tenius.* *U.S. National Marine Fisheries Service Fishery Bulletin* 71:921–926.

Valentine, D. W., M. E. Soulé, and P. Samollow. 1973. Asymmetry analysis in fishes: A possible statistical indicator of environmental stress. *U.S. National Marine Fisheries Service Fishery Bulletin* 71:357–370.

Van Valen, L. 1962. A study of fluctuating asymmetry. *Evolution* 16:125–142.

Vasetskii, S. G. 1967. Changes in the ploidy of sturgeon larvae induced by heat treatment of eggs at different stages of development. *Doklady Akademii Nauk SSSR* 172(5):1234–1237.

Vasil'ev, V. P., A. P. Makeeva, and I. N. Ryabov. 1975. On the triploidy of remote hybrids of carp *(Cyprinus carpio* L.) with other representatives of Cyprinidae. *Genetika* 11(8):49–56.

Vasil'ev, V. P., V. N. Ivanov, and L. K. Polykarpova. 1980. Frequencies of chromosomal morphs in various size groups of black sea bass *(Spicara flexulosa* Raf., Centracanthidae). In *Karyological Variability, Mutagenesis and Gynogenesis in Fishes* (Institute of Cytology, USSR Academy of Sciences, Leningrad), pp. 43–49.

Vawter, A. T., R. Rosenblatt, and G. C. Gorman. 1980. Genetic divergence among fishes of the Eastern Pacific and the Caribbean: Support for the molecular clock. *Evolution* 34:705–711.

Vincent, R. E. 1960. Some influences of domestication upon three stocks of brook trout *(Salvelinus fontinalis* Mitchell). *Transactions of the American Fisheries Society* 889:35–52.

Vithayasai, C. 1973. Exact critical values of the Hardy-Weinberg test statistic for two alleles. *Communications in Statistics* 1:229–242.

Vogt, P. R., P. T. Taylor, L. C. Kovacs, and G. L. Johnson. 1979. Detailed aeromagnetic investigation of the Arctic Basin. *Journal of Geophysical Research* 84:1071–1089.

Vrijenhoek, R. C. 1984. The evolution of clonal diversity in *Poeciliopsis.* In *Evolutionary Genetics of Fishes,* ed. B. J. Turner (Plenum Press, New York and London), pp. 399–429.

Vrijenhoek, R. C. and S. Lerman. 1982. Heterozygosity and developmental stability under sexual and asexual breeding systems. *Evolution* 36:768–776.

Vrijenhoek, R. C., M. E. Douglas, and G. K. Meffe. 1985. Conservation genetics of endangered fish populations in Arizona. *Science* 229:400–402.

Vuorinen, J. 1982. Little genetic variation in the Finnish Lake salmon, *Salmo salar sebago* (Girard). *Hereditas* 97:189–192.

Vuorinen, J. 1984. Reduction of genetic variability in a hatchery stock of brown trout, *Salmo trutta. Journal of Fish Biology* 24:339–348.

Wahlund, S. 1928. The combination of populations and the appearance of correlation examined from the standpoint of the study of heredity. *Hereditas* 11:65–106. In German.

Wallace, D. G., M.-C. King, and A. C. Wilson. 1973. Albumin differences among ranid frogs: Taxonomic and phylogenetic implications. *Systematic Zoology* 22:1–13.

Wallace, D. G., L. R. Maxson, and A. C. Wilson. 1971. Albumin evolution in frogs: A test of the evolutionary clock hypothesis. *Proceedings of the National Academy of Sciences U.S.A.* 68:3127–3129.

Walters, C. J. and R. Hilborn. 1976. Adaptive control of fishing systems. *Journal of the Fisheries Research Board of Canada* 33(1):145–159.

Walters, C. J. and R. Hilborn. 1978. Ecological optimization and adaptive management. *Annual Review of Ecology and Systematics* 9:157–188.

Ward, O. G. 1977. Dimorphic nucleolar organizer regions in the frog *Rana blairi*. *Canadian Journal of Genetics and Cytology* 19:51–57.

Ward, R. D. 1977. Relationship between enzyme heterozygosity and quaternary structure. *Biochemical Genetics* 15:123–135.

Ward, R. D. 1978. Subunit size of enzymes and genetic heterozygosity in vertebrates. *Biochemical Genetics* 16:799–810.

Ward, R. D. and R. A. Galleguillos. 1976. Protein variation in the plaice, dab and flounder, and their genetic relationships. In *Marine Organisms: Genetics, Ecology and Evolution,* ed. B. Battaglia and J. A. Beardmore (Plenum Press, New York), pp. 71–93.

Ward, R. H., and C. F. Sing. 1970. A consideration of the power of the chi-square test to detect inbreeding effects in natural populations. *American Naturalist* 104:355–365.

Watson, J. and F. Crick. 1953. Genetic implications of the structure of deoxyribonucleic acid. *Nature* 171:737–738.

Wattendorf, R. J. and R. S. Anderson. In press. *Hydrilla* consumption by triploid grass carp in aquaria. In *Proceedings of the Annual Conference of the Southeast Association of Fish and Wildlife Agencies.*

Webster, T. 1975. An electrophoretic comparison of the Hispaniolan lizards *Anolis cybotes* and *A. marcanoi*. *Breviora Museum of Comparative Zoology* 431:1–8.

Weir, B. S. 1979. Inferences about linkage disequilibrium. *Biometrics* 35:235–254.

Weir, B. S. and C. C. Cockerham. 1979. Estimation of linkage disequilibrium in randomly mating populations. *Heredity* 42:105–111.

Weir, B. S. and C. C. Cockerham. 1984. Estimating F-statistics for the analysis of population structure. *Evolution* 38:1358–1370.

Westrheim, J. and H. Tsuyuki. 1967. *Sebastodes reedi,* a new scorpaenid fish in the northeast Pacific Ocean. *Journal of the Fisheries Research Board of Canada* 24:1945–1954.

White, M. J. D. 1973. *Animal Cytology and Evolution*. University Press Cambridge, London.

Whitmore, D. H. 1983. Introgressive hybridization of smallmouth bass *(Micropterus dolomieui)* and Guadelupe bass *(M. treculi)*. *Copeia* 1983:672–679.

Whitmore, D. H. and W. Butler. 1982. Interspecific hybridization of smallmouth and Guadelupe bass *(Micropterus):* Evidence based on biochemical genetic and morphological analyses. *Southwestern Naturalist* 27:99–106.

Wiley, E. O. 1981. *Phylogenetics. The Theory and Practice of Phylogenetic Systematics*. John Wiley and Sons, New York.

Williams, G., R. Koehn, and J. Mitton. 1973. Genetic differentiation without isolation in the American eel, *Anquilla rostrata*. *Evolution* 27:192–204.

Wilson, A. C., S. S. Carlson, and T. J. White. 1977. Biochemical evolution. *Annual Review of Biochemistry* 46: 573–639.

Wilson, A. C., R. L. Cann, S. M. Carr, M. George, Jr., U. B. Gyllensten, K. M. Helm-Bychowski, R. G. Higuchi, S. R. Palumbi, E. M. Prager, R. D. Sage, and M. Stoneking. 1985. Mitochondrial DNA and two perspectives on evolutionary genetics. *Biological Journal of the Linnaean Society* 26:375–400.

Wilson, G. M., W. K. Thomas, and A. T. Beckenbach. 1985. Intra- and interspecific mitochondrial DNA sequence divergence in *Salmo:* rainbow, steelhead, and cutthroat trout. *Canadian Journal of Zoology* 63:2088–2094.

Wilson, G. M., W. K. Thomas, and A. T. Beckenbach. 1986. Mitochondrial DNA analysis of Pacific Northwest populations of *Oncorhynchus tshawytscha*. *Canadian Journal of Fisheries and Aquatic Sciences*, in press.

Winans, G. 1980. Geographic variation in the milkfish *Chanos chanos*. I. Biochemical evidence. *Evolution* 34:558–574.

Wishard, L. N., J. E. Seeb, F. M. Utter, and D. Stefan. 1984. A genetic investigation of suspected redband trout populations. *Copeia* 1984:120–132.

Withler, F. C. 1982. *Transplanting Pacific salmon*. Canadian Technical Report of Fisheries and Aquatic Sciences 1079, 27 pp.

Wohlfarth, G., R. Moav, and G. Hulata. 1975. Genetic differences between the Chinese and European races of the common carp. 2. Multicharacter variation—a response to the diverse methods of fish cultivation in Europe and China. *Heredity* 34:341–350.

Wolfe, P. 1959. The simplex method for quadratic programming. *Econometrica* 27 (3):382–398.

Wolters, W. R., G. S. Libey, and C. L. Chrisman. 1981. Induction of triploidy in channel catfish. *Transactions of the American Fisheries Society* 110:310–312.

Wolters, W. R., G. S. Libey, and C. L. Chrisman. 1982. Effect of triploidy on growth and gonadal development of channel catfish. *Transactions of the American Fisheries Society* 111:102–105.

Woodruff, D. S. 1973. Natural hybridization and hybrid zones. *Systematic Zoology* 22:213–218.

Workman, P. L., and J. D. Niswander. 1970. Population studies on Southwestern Indian tribes. II. Local genetic differentiation in the Papago. *American Journal of Human Genetics* 22:24–49.

Wright, J., K. Johnson, A. Hollister, and B. May. 1983. Meiotic models to explain classical linkage, pseudolinkage and chromosome pairing in tetraploid derivative salmonids. In *Isozymes: Current Topics in Biological and Medical Research*, vol. 10, ed. M. Rattazzi, J. Scandalios, and G. Whitt (Alan R. Liss, New York), pp. 239–260.

Wright, S. 1921. Systems of mating. *Genetics* 6:111–178.

Wright, S. 1931. Evolution in Mendelian populations. *Genetics* 16:97–159.

Wright, S. 1932. The roles of mutation, inbreeding, crossbreeding, and selection in evolution. In *Proceedings of the 6th International Congress of Genetics*, vol. 1, pp. 356–366.

Wright, S., 1943. Isolation by distance. *Genetics* 28:114–138.

Wright, S. 1951. The genetical structure of populations. *Annals of Eugenics* (London) 15:323–354.

Wright, S. 1965. The interpretation of population structure by F-statistics with special regard to systems of mating. *Evolution* 9:395–420.

Wright, S. 1969. *The Theory of Gene Frequencies*. Evolution and the Genetics of Populations, vol. 2. Chicago University Press, Chicago, 511 pp.

Wright, S. 1978. *Variability Within and Among Natural Populations*. Evolution and the Genetics of Populations, vol. 4. University of Chicago Press, Chicago-London.

Wyles, J. and G. C. Gorman. 1980. The albumin immunological and Nei electrophoretic distance correlation: A calibration for the saurian gene in *Anolis* (Iguanidae). *Copeia* 1980:66–71.

Wyles, J. S., J. G. Kunkel, and A. C. Wilson. 1983. Birds, behavior, and anatomical evolution. *Proceedings of the National Academy of Sciences U.S.A.* 80:4394–4397.

Yamazaki, F., H. Onozato, and K. Arai. 1981. The chopping method for obtaining permanent chromosome preparation from embryos of teleost fishes. *Bulletin of the Japanese Society of Scientific Fisheries [Nihon Suisan Gakkai-Shi]* 47:963.

Yang, S. Y. and J. L. Patton. 1981. Genic variability and differentiation in the Galapagos finches. *The Auk* 98:230–242.

Yonekawa, H., K. Moriwaki, O. Gotoh, J.-I. Hayashi, J. Watanabe, N. Miyashita, M. L. Petras, and Y. Tagashira. 1981. Evolutionary relationships among five subspecies of *Mus musculus* based on restriction enzyme cleavage patterns of mitochondrial DNA. *Genetics* 98:801–816.

Zain, B. S. and R. J. Roberts. 1977. A new specific endonuclease from *Xanthomonas badrii*. *Journal of Molecular Biology* 115:249–255.

Zaleski, M. B., S. Dubiski, E. G. Niles, and R. K. Cunningham. 1983. *Immunogenetics.* Pitman Publishing Company, Marshfield, Mass.

Zaret, T. M. and R. T. Paine. 1973. Species introduction in a tropical lake. *Science* 182:449–455.

Zimmerman, E. G. and M. E. Nejtek. 1977. Genetics and speciation of three semi-species of Neotoma. *Journal of Mammalogy* 58:391–402.

Zouros, E. 1979. Mutation rates, population sizes, and amounts of electrophoretic variation of enzyme loci in natural populations. *Genetics* 92:623–646.

Zouros, E., S. M. Singh, and H. E. Miles. 1980. Growth rate in oysters: An overdominant phenotype and its possible explanation. *Evolution* 34:856–867.

Zuckerkandl, E. and L. Pauling. 1962. Molecular disease, evolution and genetic heterozygosity. In *Horizons in Biochemistry,* ed. M. Kasha and B. Pullman (Academic Press, New York), pp. 189–225.

SPECIES INDEX

www.ingramcontent.com/pod-product-compliance
Lightning Source LLC
Chambersburg PA
CBHW060746220326
41598CB00022B/2339